SYMPOSIA OF THE
SOCIETY FOR EXPERIMENTAL BIOLOGY

NUMBER XXIII

Other Publications of The Company of Biologists

JOURNAL OF EXPERIMENTAL BIOLOGY
JOURNAL OF CELL SCIENCE
JOURNAL OF EMBRYOLOGY AND EXPERIMENTAL MORPHOLOGY

SYMPOSIA

I	NUCLEIC ACID
II	GROWTH, DIFFERENTIATION AND MORPHOGENESIS
III	SELECTIVE TOXICITY AND ANTIBIOTICS
IV	PHYSIOLOGICAL MECHANISMS IN ANIMAL BEHAVIOUR
V	FIXATION OF CARBON DIOXIDE
VI	STRUCTURAL ASPECTS OF CELL PHYSIOLOGY *
VII	EVOLUTION
VIII	ACTIVE TRANSPORT AND SECRETION
IX	FIBROUS PROTEINS AND THEIR BIOLOGICAL SIGNIFICANCE
X	MITOCHONDRIA AND OTHER CYTOPLASMIC INCLUSIONS
XI	BIOLOGICAL ACTION OF GROWTH SUBSTANCES
XII	THE BIOLOGICAL REPLICATION OF MACROMOLECULES
XIII	UTILIZATION OF NITROGEN AND ITS COMPOUNDS BY PLANTS
XIV	MODELS AND ANALOGUES IN BIOLOGY
XV	MECHANISMS IN BIOLOGICAL COMPETITION
XVI	BIOLOGICAL RECEPTOR MECHANISMS
XVII	CELL DIFFERENTIATION
XVIII	HOMEOSTASIS AND FEEDBACK MECHANISMS
XIX	THE STATE AND MOVEMENT OF WATER IN LIVING ORGANISMS
XX	NERVOUS AND HORMONAL MECHANISMS OF INTEGRATION
XXI	ASPECTS OF THE BIOLOGY OF AGEING
XXII	ASPECTS OF CELL MOTILITY

The Journal of Experimental Botany
is published by the Oxford University Press
for the Society for Experimental Biology

SYMPOSIA OF THE
SOCIETY FOR EXPERIMENTAL BIOLOGY

NUMBER XXIII

DORMANCY AND SURVIVAL

*Published for The Company of Biologists
on behalf of the Society for Experimental Biology*

ACADEMIC PRESS, INC., PUBLISHERS
NEW YORK, NEW YORK
1969

Published by the Syndics of the Cambridge University Press
Bentley House, 200 Euston Road, London N.W. 1

The Symposia of the Society for Experimental Biology are
published in the U.S.A. by:
ACADEMIC PRESS INC.
111 Fifth Avenue, New York, New York 10003

© The Society for Experimental Biology 1969

Printed in Great Britain
at the University Printing House, Cambridge
(Brooke Crutchley, University Printer)

CONTENTS

Preface *page* vii
   by H. W. WOOLHOUSE

Dormancy as an adaptive strategy 1
   by R. LEVINS

Developmental changes accompanying the breaking of the dormant state in bacteria 11
   by W. STEINBERG, J. IDRISS, S. RODENBERG *and* H. O. HALVORSON

The biochemistry of amoebic encystment 51
   by R. J. NEFF *and* R. H. NEFF

Dormancy and survival in nematodes 83
   by C. ELLENBY

The dormancy and germination of fungus spores 99
   by A. S. SUSSMAN

Survival of algae under adverse conditions 123
   by G. E. FOGG

Seed germination and the capacity for protein synthesis 143
   by A. MARCUS

Seed dormancy and oxidation processes 161
   by E. H. ROBERTS

Light-controlled germination of seeds 193
   by M. BLACK

The problem of dormancy in potato tubers and related structures 219
   by L. RAPPAPORT *and* N. WOLF

The control of bud dormancy in seed plants 241
   by P. F. WAREING

Diapause and seasonal synchronization in the adult Colorado beetle (*Leptinotarsa decemlineata* Say) 263
   by J. DE WILDE

Photoperiodism and the endocrine aspects of insect diapause *page* 285
 by C. M. WILLIAMS

Diapause and photoperiodism in the parasitic wasp *Nasonia vitripennis*, with special reference to the nature of the photoperiodic clock 301
 by D. S. SAUNDERS

The survival of insects at low temperatures 331
 by R. W. SALT

Some mechanisms of mammalian tolerance to low body temperatures 351
 by R. K. ANDJUS

Growth and survival of plants at extremes of temperature—a unified concept 395
 by J. LEVITT

The physiology of dormancy and survival of plants in desert environments 449
 by D. KOLLER

The dormancy and survival of plants in the humid tropics 471
 by K. A. LONGMAN

Hyperresponsiveness in hibernation 489
 by C. P. LYMAN *and* R. C. O'BRIEN

Some interrelations of reproduction and hibernation in mammals 511
 by W. A. WIMSATT

Principles and further problems in the study of dormancy and survival 551
 by L. IRVING

Author Index 565

Subject Index 579

PREFACE

The twenty-third symposium of the Society for Experimental Biology was held in Norwich at the School of Biological Sciences, University of East Anglia, from 2 to 6 September 1968, at the invitation of the University. The meeting included papers dealing with mechanisms of survival of extreme conditions in micro-organisms and in many different groups of higher organisms. Where a dormant state is involved in the survival mechanisms there arise interesting questions of how it is evolved. The introductory paper by Levins points the way towards the use of strategic analysis in the understanding of these processes. Whilst it is clear that for largely technical reasons survival problems in micro-organisms are now being tackled at a more biochemical level than is the case with higher organisms, it seems that considerable stimulus to new experimental work in all groups can be gained from a comparative reading of the work with a wide range of organisms. Moreover, it is clear that even in bacteria and protozoa we have as yet no really comprehensive picture of the chemistry of the dormancy-locking mechanism or of its induction and release. Several papers in this volume show how temperature and drought tolerance are beginning to be studied in terms of molecular structure. It is interesting to note, however, that progress in this direction may require much more basic chemical work concerning changes in the tertiary structure and aggregation of macromolecules, and in particular of proteins, under these conditions.

I am indebted to colleagues and contributors for advice with the planning of the meeting, particularly acknowledging the help of Professor P. J. Syrett, the Symposium Secretary, Dr P. Croghan who acted as local secretary, and my wife, Leonie, for the checking of references. Finally, it is a pleasure to record the help and co-operation of the Cambridge University Press in the preparation of the volume.

H. W. WOOLHOUSE
Editor of the twenty-third Symposium of the
Society for Experimental Biology

DORMANCY AS AN ADAPTIVE STRATEGY

By R. LEVINS

The University of Chicago Committee on Mathematical Biology,
939 East 57th Street, Chicago, Illinois 60637

The rate of increase of a population depends on the parameter r, the intrinsic rate of increase. For example, in the simplest case of density-independent continuous growth of a population of size x:

$$\frac{dx}{dt} = rx. \tag{1}$$

But for the populations with discrete generations r is given by

$$r = \frac{\log R}{T}, \tag{2}$$

where R is the average number of offspring and T the generation time. Cole (1954) and Lewontin (1965) have pointed out some of the consequences of equation (2). Of special importance is the observation that r is more sensitive to changes in T than changes in R. Selection would be expected to act most strongly on the rate of development, minimizing T. Thus, while it is relatively easy to carry out selection procedures in the laboratory for increasing total egg production in *Drosophila*, selection to shorten the generation-time is much more difficult. Presumably the populations have already approached their limits in this respect.

Despite the very strong selective value attached to short generation-time, reproduction is in fact delayed beyond the physiologically necessary minimum in many groups of organisms. For example, Panamanian lizards retain their eggs at the beginning of the dry season (Q. D. Sexton, personal communication), many insects enter a regular winter or dry-season diapause, some plants produce seed that may be dormant for long periods of time, soil bacteria are probably dormant most of the time, and winter hibernation has arisen in four orders of mammals. Indeed, all metazoan development is a delay of reproduction compared to forms which begin meiosis immediately after fertilization.

Suppose that we modify some biological parameter, B, which affects both the number of offspring and the generation time. Then r will be:

$$\frac{\partial r}{\partial B} = \frac{1}{RT}\frac{\partial R}{\partial B} - \frac{1}{T^2}\frac{\partial T}{\partial B}. \tag{3}$$

A change which increases R and T will be advantageous only if

$$\frac{1}{R}\frac{\partial R}{\partial B} > \frac{1}{T}\frac{\partial T}{\partial B}. \tag{4}$$

Therefore delayed reproduction is most likely to be advantageous when R is small, when R is greatly increased compared to the change in T, and when T is already large.

In order to understand the adaptive significance of dormancy, it must be placed within the broader context of strategic analysis. Unlike the physiological viewpoint, which is concerned with the mechanisms of dormancy and the breaking of dormancy, of heat resistance and cold hardiness, strategic analysis is concerned with their adaptive significance. It is based on four principles.

A. *The principle of allocation*

A strategy is an allocation of some resource among several alternatives. In some cases the total resource is fixed, as when a given number of seed are divided among germination classes or a fixed amount of lactic de-

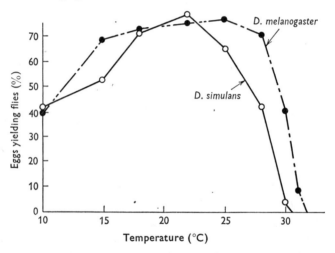

Fig. 1. The temperature/emergence relation for two species of *Drosophila*. Clearly *D. melanogaster* has a broader temperature niche than *D. simulans*. (After Tantaway & Mullah, 1961, fig. 3.1.)

hydrogenase may be divided among several isoenzymic forms. In other cases the resource to be allocated is some fitness parameter which varies with the environment in such a way that it cannot be maximal simultaneously in all environments. Figure 1 shows the viability of *Drosophila* eggs as a function of temperature. The broken curve is broad and relatively flat. It corresponds to *D. melanogaster*, a broadly adapted species. The solid curve, for *D. simulans*, is narrower but with a higher peak. It is a narrower-niched, or specialized, species. It seems to be a general rule that the breadth of the fitness curve must be purchased at the cost of height, see, for example, Sacher (1966). More qualitatively, Levitt in this symposium has pointed out that the presence of a large number of SH bonds in the proteins of plants is associated with rapid growth but increases the sensitivity to heat and cold. Thus we cannot simultaneously maximize growth-rate and resistance.

For purposes of mathematical analysis, these qualitative observations are idealized to give formal constraints on fitness parameters. Thus we may assume that the area under the curves of Fig. 1 is fixed but that the shape may be altered freely subject to this constraint. Or, as in Levins (1968), we may fix the shape of the curve and move the location of the peak along the environmental axis. Different models have been considered, and more or less uniform results obtained.

B. *The principle of optimization*

Given that it is not possible to maximize fitness simultaneously in all environments, the optimum strategy allocates resources in such a way as to maximize some combination of fitness in different environments and giving proper attention to the relative frequencies of these environments.

The appropriate combination of fitnesses depends on the 'grain' of the environment for the species in question. In a fine-grained environment the individual is exposed to the various environmental factors in small doses; that is, to short-term fluctuations or wanderings through many patches of an environmental mosaic. In the limiting case each individual experiences the same range of environments in the same frequencies and there is no uncertainty. The appropriate fitness measure here is

$$W = \Sigma p_i W_i, \tag{5}$$

where p_i is the frequency of environment i, and W_i is fitness in that environment.

In a coarse-grained environment each individual spends its whole life, or a long period at least, in a single alternative, either because the environment varies from generation to generation or because the spatial patches of the environment mosaic are large. In the case of dormancy the environment is coarse-grained since a whole generation either includes or does not include winter or a dry season. The appropriate fitness measure here is

$$W = \Pi W_i^{p_i}, \tag{6}$$

the product of the fitnesses in each environment raised to the power of the frequencies of each. A more convenient form of this equation is

$$\log W = \Sigma p_i \log W_i. \tag{7}$$

An optimum strategy then is that allocation of resource, subject to some constraint, which maximizes the appropriate fitness measure. We cannot claim that natural selection will always produce an optimum strategy. However, we do assert the weaker proposition: that populations will differ in nature in the same direction that their strategic optima differ.

Analysis of fitness in a heterogeneous environment has identified several types of optimal strategies:

(1) If the alternative environments are not too different compared to the tolerance conferred by a given biological state—that is to say, by a particular physiological condition or phenotype—for non-optimal environments, the optimum strategy is a single type which is best suited to some intermediate environment. This optimum varies continuously with the probabilities of occurrence of the environments.

(2) If the alternative environments are very different compared to the tolerance of the single type, and if the environment acts in a fine-grained way, the optimum is a single type specialized to the more common environment. There is no change in the optimum as the probabilities of the environments change until some threshold is reached, and then the optimum shifts abruptly to the opposite specialization.

(3) If the alternative environments are very different compared to the tolerance of a single type, and if the environment is coarse-grained, the optimum strategy is polymorphism, a mixture of types specialized each to one of the possible environments. The proportions of morphs will change in a continuous way with changes in the p_i.

C. *Strategy as a norm of reaction*

The p_i of equations (5) and (7) represents the probabilities of the organism finding itself in environment i. These probabilities are modified by the present environment. For instance, the conditional probability of frost within a few weeks given a day-length of 12 hr. is greater than the same conditional probability with a day-length of 14 hr. Therefore the optimum strategy is an allocation of resource as a function of the environment at the time the strategy is determined. Figure 2 shows three norms of reaction, the curves 1, 2 and 3 representing the optimal norms of reaction corresponding to the optimal strategies (1), (2) and (3) defined in the preceding section. If the environments are not too different compared to the range of tolerance of the individual phenotype, the optimum norm of reaction is one in which phenotype varies as a continuous function of the probabilities of occurrence of particular environmental conditions. This is what Schmalhausen (1949) calls 'dependent development', Fig. 2 curve 1. If the environments are very different compared to the tolerance of the individual, the optimum norm of reaction depends on the grain of the environment. In a fine-grained environment it will be as in curve 2 of Fig. 2, in which phenotype changes discontinuously at a threshold value of the environment (Schmalhausen's autonomous regulative development). Finally, in a coarse-grained environment the optimum is a mixed strategy, with the proportions of the different phenotypes varying as a function of the environment, Fig. 2 curve 3.

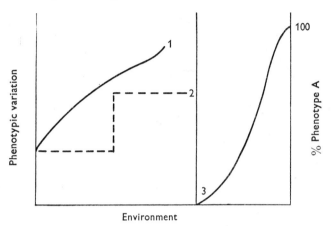

Fig. 2. Norms of reaction. Curve 1, Dependent development; Curve 2. Autonomous regulative development; Curve 3, mixed strategy.

The norm of reaction is subject to natural selection, so that the phenotype/environment curve will be modified in the course of evolution. At each value of environmental variable there will be a certain genetically determined phenotypic variation. Much of the selection which takes place is of the stabilizing type, reducing the variation about the norm for each environment. But the total selection will be greater in the environments which occur frequently than in the rare environments. And the norm of reaction in those laboratory environments which never occur in nature will not be selected for directly but will evolve as an indirect result of the modification of the norm in other environments.

There are two major consequences of the pattern of selection. First, the norm of reaction will be most stabilized for the normal environments of a population and will show greatest variation about the norm in those environments which only appear rarely or not at all. Secondly, the norm of reaction itself will reflect the ecological situation for the common environments, while in unusual or artificial environments it will indicate more about the physiological mechanisms at work but will lack ecological significance.

D. *The environment as information and uncertainty*

In most multivoltine insects, several generations develop in the course of the summer. The last generation, usually in response to the shorter day-length, enters diapause or lays diapausing eggs. It might seem therefore that a long day is a physiological necessity, and in its absence development is halted. However, in the silk-worm moth *Bombyx mora* the situation is reversed. A long day evokes diapause whereas a short day results in direct development. This insect has only two generations per year. The first generation emerges in early spring when the days are still short, and the short day induces the mechanism that results in another generation. The second generation develops under conditions of maximum day-length in mid summer, preventing the production of a third generation. In all cases the response is not to light as an environmental condition to be coped with, but as information, as a prediction of the conditions which will obtain when the response is being evaluated; that is, when there is differential mortality or reproduction.

Koller's paper in this volume shows other examples of environmental tokens being used to predict the environment. Thus in some desert plants the amount of light reaching the seed is indicative of depth in the soil rather than season.

Once an environmental stimulus penetrates an organism, its effects may spread out along many pathways, some of which may have opposite effects. Selection can act to enhance some of these pathways and inhibit others, so that almost any stimulus may be linked to any response system. Therefore we assert the hypothesis that *there is no necessary relation between the physical form of a stimulus and the response it evokes or the environmental factor to which this response is an adaptation.* Dormancy may arise as an adaptation to extreme conditions or merely to extreme uncertainty in the environment. Further, we have to distinguish among recurrent extreme conditions which appear at a regular stage in the life-cycle of all generations and those which occur irregularly, and these latter are further divided into predictable and unpredictable environments.

1. *Uncertain environments*

All organisms face a diversity of environments and may respond to this diversity by spreading fitness over the various phases of environment by specializing to a restricted range of environments, or by avoiding some of the environments. Consider first a model in which the average number of progeny in environment s, $W(s)$, is subject to the restriction

$$\int W(s)\, ds = C. \tag{8}$$

In an environmental pattern in which environment S occurs with frequency $P(S)$, the appropriate measure of fitness is

$$W = \int \log [W(s)] P(s) ds. \tag{9}$$

The strategic problem here is the maximization of W subject to the restriction (8), and the well-known solution is $W(s) = CP(s)$.

Thus fitness is spread over the environments in the proportion of their frequency. At this optimal strategy the fitness measure is

$$W = \int \log [CP(s)] P(s) ds, \tag{10}$$

which is

$$W = \log C + \int \log [P(s)] P(s) ds. \tag{11}$$

But the second term is the negative of a familar measure of the uncertainty of a distribution so that

$$W = \log C - U \tag{12}$$

and the maximum fitness attainable is the maximum fitness in a constant environment minus the uncertainty of the environment.

The situation is changed when we allow dormancy. Suppose for convenience that there are N equally frequent environments. Then each has a frequency $1/N$ and the maximum fitness is $\log C - \log N$. The population can only increase if $\log C$ exceeds N.

Now suppose that the organism is dormant, and will only develop when the appropriate environments are present. Let it accept any K of the N environments. For these environments, each of equal frequency, the fitness is $\log C - \log K$. However, there is a waiting period before an appropriate environment is found. The fitness is therefore

$$W = \frac{\log C - \log K}{T}, \tag{13}$$

where T is the waiting time. The average waiting time is N/K, so that a first approximation is

$$W = \log C - \log K \, (K/N). \tag{14}$$

This can now be maximized as a function of K:

$$\frac{\partial W}{\partial K} = \frac{\log C - \log K}{N} - \frac{1}{N}, \tag{15}$$

and setting $\partial W/\partial K = 0$ we obtain the optimum

$$\hat{K} = C/e, \tag{16}$$

where e is the base of the natural logarithms. From this it follows that the optimum strategy involves accepting the fraction $C/(eN)$ of the environments and remaining dormant during $(eN-C)/N$ of the time.

The final fitness will be $W = C/(eN)$.

From this it follows that organisms with a high reproductive potential C will spread their fitness over more environments and be dormant less of the time.

2. Diapause

In the previous model there was no difficulty in identifying the favourable environment when it occurred. In the case of diapause, however, the appropriate response must be made before the onset of the unfavourable conditions. Therefore the strategic problem is one of predicting whether or not there is yet time for another generation before winter hits. The prediction is based on some signals, such as day-length, which are correlated with the season. Therefore the model specifies: let p = the probability of satisfactory conditions persisting long enough to complete another generation, W_S = average number of offspring of individuals which develop directly into adults under satisfactory conditions, W_E = average number of offspring under extreme conditions, W_D = viability of diapausing forms (taken to be the same under both sets of conditions), q = proportion of the population developing directly. Then the rate of increase under good conditions is $\log[W_D + q(W_S - W_D)]$, and under extreme conditions is $\log[W_D - q(W_D - W_E)]$. Hence the average fitness is

$$r = p \log[W_D + q(W_S - W_D)] + (1-p) \log. \tag{17}$$

To find the optimum q, we differentiate (17) and solve, getting

$$\hat{q} = W_D \left(\frac{p}{W_D - W_E} - \frac{1-p}{W_S - W_D} \right), \tag{18}$$

provided this falls between 0 and 1. Hence

$$\hat{q} = \begin{cases} 0 & p \geq (W_D - W_E)/(W_S - W_D) \\ W_D \left(\dfrac{p}{W_D - W_E} - \dfrac{1-p}{W_S - W_D} \right) & \text{for } \dfrac{W_D - W_E}{W_S - W_D} \leq p \leq \dfrac{W_S}{W_D}\left(\dfrac{W_D - W_E}{W_E - W_S}\right) \\ 1 & p > \dfrac{W_S}{W_D} \dfrac{W_D - W_E}{W_S - W_D} \end{cases} \tag{19}$$

Thus for $W_S = 1$, $W_E = 0{\cdot}1$, $W_D - 0{\cdot}9$. The optimum q is 0 when p is less than about 0·89 and reaches 1 at about $p = 0{\cdot}97$. For a larger W_S the transition from no diapause to total diapause is less abrupt. Thus for $W_S = 10$ and the other parameters unchanged, $\hat{q} = 0$ up to $p = 0{\cdot}08$, and rises to 1 at $p = 0{\cdot}89$.

Cohen (1967) studied the adaptive significance of delayed germination in the seeds of desert plants. In his model the viability Y of seeds which have germinated and the viability V of seed in the ground are treated as independent random variables. He lets G, the proportion of seed germinating every year, be the parameter to optimize and finds that in order for G to be less than 1 (that is, in order for some dormancy to be optimal) we must have

$$E(V) > \frac{1}{E(1/Y)}. \qquad (20)$$

The harmonic mean of Y depends mostly on its smallest values. Therefore the variance enters in: the more variable Y the smaller the harmonic mean and the greater the likelihood that dormancy will be advantageous. Once again, delayed reproduction appears as a response to environmental uncertainty.

DISCUSSION

The approach presented here is amenable to several kinds of testing.

First, there are qualitative predictions which can be made based on the information provided by various environmental signals. For example, under conditions of the North American Midwest, temperature would be a poor indicator of season. The topography is relatively uniform, and the onset of frost varies little within a county. On the other hand, in the northeast, topography results in a range of several weeks in the onset of frost, and temperature may indicate topography much better than season. Therefore we could predict a greater role for temperature in the New England area than in the Midwest in the determination of diapause.

Since light is the major determinant of diapause because of its high information content, we could use artificial lights to extend the day-length and delay diapause as a method of control. However, once day-length ceases to be a good predictor of winter the response would be expected to change, and the local population might become obligate diapausers with one generation per year.

Secondly, quantitative estimates of the parameters of equation (16) could be made and the optimum strategy calculated numerically in order to compare to the actual diapause pattern.

Finally, the hypothesis that there is no necessary relation between the physical form of a signal and the response it evokes suggests many laboratory experiments in which we select for slower development (pupal dormancy or pseudo-diapause) using different signals. The same signal can be used with opposite information. Thus after a short heat-signal we could select only the last flies to emerge and after a cold signal only the first. Or we could reverse this. If this line of investigation is successful we could give more complex signals so determining the collating and resolution limits of the insect genotype.

In those insects which have an obligate diapause, the hormonal and other conditions during diapause constitute the environment within which other processes occur. After these, other processes have evolved in the diapause environment for a long enough period, diapause will become a developmental necessity and its subsequent evolution may be controlled by other factors. There is no evidence for this in hibernating mammals, but winter hibernation seems to be a relatively recent acquisition since the appearance of the glacial climate. In plants and insects where dormancy may have been adaptive in relation to drought or food shortage before cold became an issue it is more likely to be seen.

REFERENCES

COHEN, D. (1967). *J. theor. Biol.* **1**, 14.
COLE, L. (1954). *Rev. Biol.* **29**, 103.
LEVINS, R. (1968). *Evolution in Changing Environments.* Princeton University Press.
LEWONTIN, R. C. (1965). *In The Genetics of Colonizing Species,* ed. Baker and Stebbins. Head Press.
SACHER, G. (1966). The complementarity between development and ageing. *N.Y. Acad. Sci. Conf. on Interdisciplinary Perspectives of Time.*
SCHMALHAUSEN, I. I. (1949). *Factors of Evolution.* Philadelphia: Blakiston.
TANTAWAY, D. C. & MULLAH, G. S. (1961). *Evolution, Lancaster, Pa.* **15**, 1.

DEVELOPMENTAL CHANGES ACCOMPANYING THE BREAKING OF THE DORMANT STATE IN BACTERIA*

By W. STEINBERG,† J. IDRISS, S. RODENBERG‡
AND H. O. HALVORSON§

Department of Bacteriology and Laboratory of Molecular Biology,
University of Wisconsin, Madison, Wisconsin 53706

INTRODUCTION

The suppression of protein synthesis during cryptobiosis is a well-known phenomenon. The nature of the block in protein synthesis in some dormant systems is at the level of transcription (Kobayashi et al. 1965; Hirikoshi, Ohtaka & Ikeda, 1965; Henney & Storck, 1964), and in others at the level of translation (Terman & Gross, 1965; Marcus & Feeley, 1966; Bell, Humphreys, Slayter & Hall, 1965). Within a short period after dormancy is broken, a rapid increase is observed in the protein-synthesizing capacity of these systems.

The emergence of a metabolically active bacterial cell from a state (the spore) where protein synthesis is blocked provides a unique system where one can examine those features that have been built into a dormant structure which ensure that the transition to active growth will occur in an orderly fashion. The profound differences in the structural organization and activity of bacterial spores and vegetative cells (Sussman & Halvorson, 1966) makes it evident that a number of metabolic changes must occur in order to accomplish the transformation from dormancy. In spores of the genus *Bacillus* the transition to active growth occurs in three distinct phases: (1) activation, a period in which those mechanisms responsible for initiating germination are potentiated; (2) germination, a stage characterized by degradative reactions which break the dormant state; (3) outgrowth, a stage which is dependent on the commencement of biosynthetic activity and which covers the period of development after germination until the first cell division. A simplified diagram illustrating the sequence of these phases has been constructed in order to facilitate a comparison with the

---

* This work was supported in part by research grants from the National Institutes of Health (GM-12332, AI-01459 and GM-686).
† Present address: Centre de Génétique Moléculaire, C.N.R.S., 91 Gif-sur-Yvette, France.
‡ Present address: Department of Biology, University of Pennsylvania, Philadelphia.
§ National Institutes of Health Research Career Professor.

terminology used for other 'analogous' biological systems: conidia and seeds (Fig. 1).

Activation represents the initial event in the breaking of dormancy and may be accomplished by several methods. The 'heat-shock' or heat-activation procedure greatly reduces the nutritional requirements for

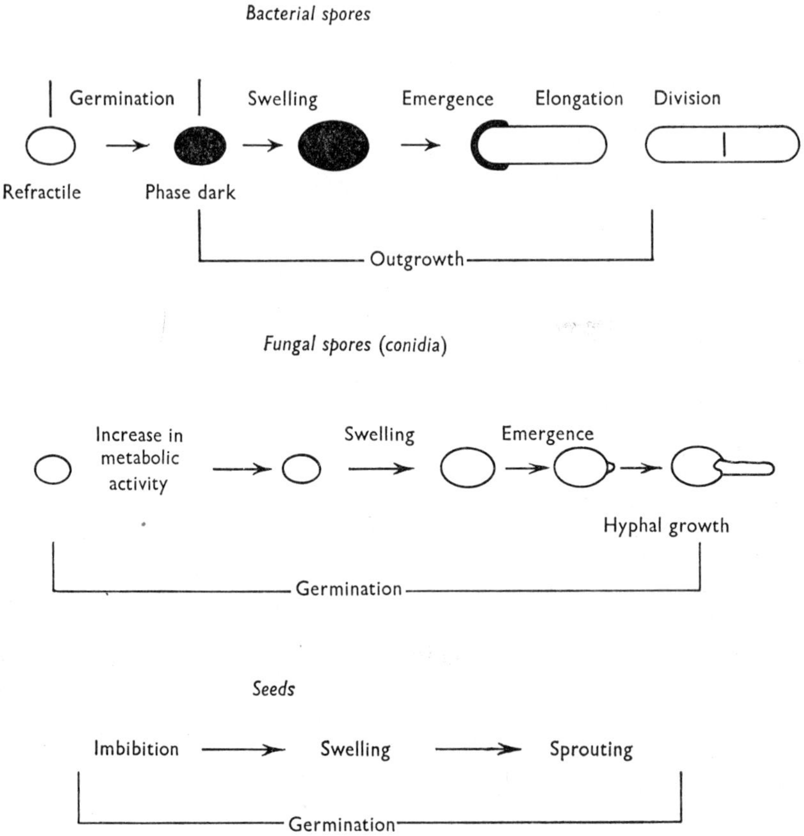

Fig. 1. Comparison of bacterial spore germination and outgrowth with analogous biological systems: conidia and seeds.

germination. Both the temperature and duration of heat-treatment directly affect the rate and extent of germination (Curran & Evans, 1945; Desrosier & Heiligman, 1956). During heat-activation a number of the properties of spores are altered. Gerhardt & Black (1961) observed an increase in permeability. This is accompanied by a release of amino acids and a non-dialysable peptide (Powell & Strange, 1953) and by a partial loss of calcium dipicolinate (Harrell & Mantini, 1957). In addition to these phenomena, Church & Halvorson (1957) and Krishna-Murty (1957) observed that the glucose-oxidizing enzymes normally dormant in intact spores of *Bacillus cereus* could be activated in the absence of any detectable germination. This 'unmasking' of enzyme systems is one of the most recognizable events associated with heat-activation. Since this occurs in the absence of detectable protein synthesis, it is clear that these metabolic systems are pre-existent in the dormant state (Steinberg, Halvorson, Keynan & Weinberg, 1965).

When heat-activated spores have been placed in an appropriate environment the return to vegetative growth occurs rapidly and synchronously. Although many enzyme systems have been shown to exist in the germinating spore, their relationship to germination is still unknown (Halvorson, Vary & Steinberg, 1966). The existence of a lag period between the addition of the germinating agent and the first manifestation of germination, however, indicates that this stage is the outcome of a multiple-step metabolic chain (Halmann & Keynan, 1962). Recently, a statistical analysis of the initial events in spore germination has led to the proposal that spores contain an allosteric enzyme that controls the rate-limiting step in germination (Woese, Vary & Halvorson, 1968).

The changes occurring during and after germination can be grouped into either of two major processes: (1) the breakdown of spore structure; (2) the synthesis of vegetative cell material. After a lag period following the addition of germinating agents, the spore sequentially loses its heat resistance, dipicolinic acid, impermeability to dyes, Ca^{2+}, refractility and optical density to visible light (Levinson & Hyatt, 1966). Under favourable conditions, germination is accompanied by swelling and dissolution of the spore cortex. As dissolution of the cortex proceeds, the spore coat is ruptured and the spore-core membrane becomes the cell wall of the emerging vegetative cell, which then elongates and divides (Kawata, Inoue & Takagi, 1963).

That germination and outgrowth are entirely distinct and separate processes has been demonstrated by differences in their respective metabolic activities and nutritional requirements. Demain & Newkirk (1960), using auxotrophic mutants of *Bacillus subtilis*, were able to dissociate spore germination from outgrowth. They showed that the imposition of a nutritional requirement via mutation did not affect the ability of spores to germinate, but completely inhibited outgrowth. An examination of the nutritional requirements for germination and outgrowth led Levinson & Hyatt (1963) to a similar conclusion. They noted that utilization of a particular nutrient for germination did not necessarily imply that the substance could support outgrowth. A further indication of the differences between these two events is shown by the fact that spores will germinate normally in the presence of compounds which inhibit protein and nucleic acid synthesis in vegetative cells (Keynan & Halvorson, 1965). Stuy (1956) found that antibiotics such as aureomycin, terramycin and chloramphenicol, which inhibited vegetative growth, had no effect on germination. Treadwell, Jann & Salle (1958) had similar results with subtilin, streptomycin and penicillin. Also, normal germination occurred in the presence of 2,4-dinitrophenol, cyanide and iodoacetate; agents which inhibited outgrowth. These results demonstrate that (*a*) the system responsible for germination is built into the spore and is not synthesized *de novo* and (*b*) that protein and nucleic acid synthesis, though not required for germination, are essential for outgrowth.

ONSET OF PROTEIN AND RIBONUCLEIC ACID (RNA) SYNTHESIS FOLLOWING GERMINATION

Given optimal germination stimulants (e.g. L-alanine and adenosine) the rate of germination of spores of *Bacillus cereus* T is unaffected by nutrients and by the concentration of spores over the range of 30–3000 μg./ml. (J. C. Vary, unpublished results). To explore the relationship between germination and macromolecular synthesis, spores were germinated in CDGS medium (Nakata, 1964) supplemented with the 20 amino acids and the four bases of RNA. The incorporation of saturating concentrations of phenylalanine and uracil into protein and RNA, respectively, was examined as a function of time after the addition of the germination stimulants. The results, shown in Fig. 2, are plotted as counts/min. uncorrected for zero time absorption to illustrate the lag in RNA and protein synthesis. Germination commences after 2 min. and is essentially complete by 6 min. RNA synthesis is detectable after 4 min. Also shown in Fig. 2 are the kinetics of uracil incorporation predicted on the basis that the rate of incorporation is constant per cell and proportional to the number of germinated spores present. As can be seen, the predicted kinetics parallel the observed incorporation, preceding it by approximately 25 sec., which closely approaches the microgermination time in this system (Vary & Halvorson, 1965). To measure this time more accurately the refractility changes in a single spore were followed as a function of time during germination. The typical germination time for a single spore is about 36 sec. (Rodenberg *et al.* 1968). Thus, within 25 sec. after a spore enters microgermination a linear rate of transcription commences.

The initial rate of RNA synthesis following germination is high, and unless the medium is supplemented with added bases the rate rapidly declines during the first hour. Thus the endogenous reserves and the rates

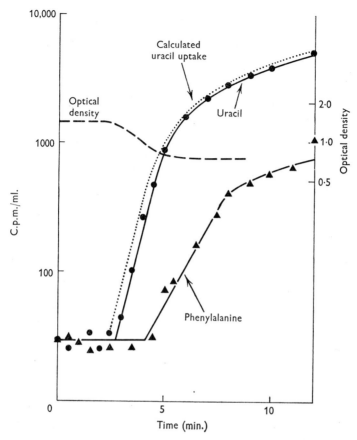

Fig. 2. Kinetics of RNA and protein synthesis immediately following germination. Heat-activated spores were germinated at 30° C. at 500 μg./ml. in CDGS+glucose medium supplemented with amino acids and bases and containing either [³H]uracil (5 μc/46·2 μg./ml.) or [¹⁴C]-L-phenylalanine (0·1 μc/23·8 μg./ml.). At intervals 1 ml. samples were removed and the radioactivity incorporated into hot or cold trichloroacetic acid (TCA)-precipitable materials was determined. Germination was measured continuously by following the decrease in optical density (- - -) at 625 mμ in a Gilford Recording Spectrophotometer with the cuvette compartment maintained at 30° C. The theoretical kinetics of uracil uptake were calculated by assuming that $dRNA/dt = k$ (germinated spores), where $k = 600$ counts/min. incorporated/ml. and the number of germinated spores was calculated from the optical density curve (Vary & Halvorson, 1965).

of syntheses of nucleotides and amino acids are insufficient to maintain maximal rates of RNA and protein synthesis. In a medium which is not completely supplemented with amino acids and bases there is no direct relationship between the rate of total RNA synthesis and the rate of protein synthesis (Rodenberg et al. 1968).

Dependence of protein synthesis on RNA transcription during outgrowth: absence of stable, functional messenger RNA (mRNA) in dormant spores

During outgrowth of bacterial spores the synthesis of all classes of RNA (transfer, ribosomal and messenger) have been observed (Doi & Igarashi,

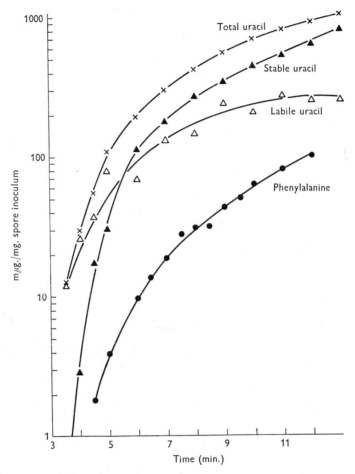

Fig. 3. Kinetics of RNA and protein synthesis during early outgrowth. Heat-activated spores were germinated at 30° C. at 500 μg./ml. in CDGS+glucose medium supplemented with amino acids and containing [³H]uracil (2 μc/20 μg./ml.) or [¹⁴C]phenylalanine (0·2 μc/20 μg./ml.). For uracil incorporation, two 1 ml. samples were removed at intervals. One was added immediately to cold TCA to measure total uracil incorporation. The second was added to a tube containing 10 μg. actinomycin D. The tube was shaken for 20 min. in the dark to permit degradation of labile RNA, and the stable cold TCA-precipitable activity was then determined. Labile uracil was calculated from the difference between total and stable RNA. For phenylalanine incorporation, 1 ml. samples were removed at intervals and the radioactivity incorporated into hot TCA-insoluble material determined. The mμg. incorporated per mg. spores was calculated after a correction for absorption was applied.

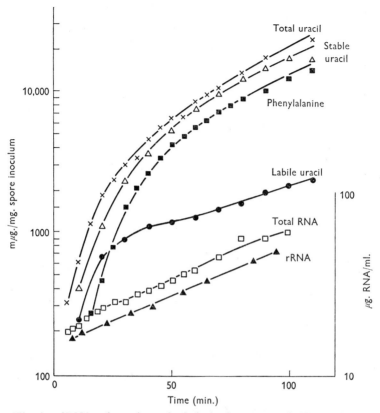

Fig. 4. Kinetics of RNA and protein synthesis during late outgrowth. Heat-activated spores (0·25 mg./ml.) were germinated at 30° C. in CDGS+glucose plus amino acids (Steinberg & Halvorson, 1968a) containing [³H]uracil (2 μc/2 μg./ml.) or [¹⁴C]-L-phenylalanine (1 μc/20 μg./ml.). In this medium cell division occurred at 75–85 min. For the details of uracil and phenylalanine incorporation see legend to Fig. 3. For total RNA determinations, 1 ml. aliquots were removed at intervals to tubes containing 3 ml. cold 10 % TCA and allowed to stand in ice for 30 min. The precipitates were collected by centrifugation, washed once with cold TCA and the pellets assayed for RNA by the orcinol reaction. To determine the formation of ribosomal RNA, 200 ml. aliquots were removed, the cells collected by filtration, washed by centrifugation with cold growth medium and the pellets frozen. The extraction and separation of the RNA classes by chromatography was performed on methylated albumin–kieselguhr columns according to Donnellan, Nags & Levinson (1965).

1964a; Donnellan, Nags & Levinson, 1965; Armstrong & Sueoka, 1968; Rodenberg et al. 1968). In spores of B. cereus T, initially over 95 % of the total radioactive uracil that is taken up by the germinated spore is incorporated into labile RNA (Fig. 3). Ribosomal RNA synthesis begins almost immediately after germination and the proportion of total stable RNA species rises continuously during outgrowth (Fig. 4). Similar observations on the synthesis of ribosomal RNA have been made by Fitz-James (1955a), Woese (1961), Balassa (1963) and Donnellan et al. (1965). The synthesis of soluble RNA (4+5 S) also commences early as judged from the incorporation of radioactive uracil into these components during the first 8 min. after the initiation of germination (H. O. Halvorson & R. Epstein, unpublished results).

To what extent is protein synthesis in outgrowth dependent on the transcription of new messenger RNA? If a stable mRNA produced late in sporulation was incorporated into the spore and carried over through the dormant state, then one might expect that it would be available for protein synthesis during outgrowth. A convincing test for the absence of stable functional mRNA in spores is apparent from studies with the antibiotic actinomycin D. This compound completely inhibits deoxyribonucleic acid (DNA)-dependent RNA synthesis in vegetative cells of *Bacillus* (Levinthal, Keynan & Higa, 1962). In the presence of actinomycin D germination occurs normally without any detectable RNA or protein synthesis. If the protein-synthesizing system during outgrowth were dependent on new mRNA synthesis, then one would expect that the half-life of amino acid incorporation and mRNA, in the presence of actinomycin D, would be of the same order (a few minutes) as that observed in vegetative cells. When heat-activated spores were germinated in the presence of [^{14}C]-L-leucine, addition of 20 μg./ml. of actinomycin D with the germinating agents completely blocked leucine incorporation, supporting the view that dormant spores are devoid of stable functional mRNA (Halvorson *et al.* 1966). When actinomycin D was added 10 or 19 min. after the onset of germination, leucine incorporation continued in each case for 3 min. and then ceased. The apparent half-life of protein-synthesizing ability (1·5 min.) is in agreement with the half-life of pulse-labelled RNA (2 min.) synthesized during outgrowth (Kobayashi *et al.* 1965).

It therefore appears that the unstable part of the protein-synthesizing machinery, mRNA, is synthesized after germination and that the new protein synthesized during outgrowth is dependent upon new mRNA transcription. The simplest mechanism predicts that initial protein synthesis would take place on ribosomes pre-existing in dormant spores and that this would be under the direction of newly formed mRNA. On this basis one would expect that amino acid incorporation should start at approximately the same time as RNA synthesis. However, as seen above (Fig. 2), there was a short delay in the start of protein synthesis and phenylalanine incorporation lagged behind RNA synthesis. From these considerations it appeared possible that other factors in addition to mRNA synthesis may be involved in the start of protein synthesis, and we decided to examine whether or not a functional protein synthesizing system was present in dormant spores.

Development of protein synthesizing capacity during outgrowth

Several lines of evidence, including the classes of RNA present in spores (Doi & Igarashi, 1964b) and the inhibition of RNA, protein and enzyme synthesis by actinomycin D indicated that spores of *Bacillus cereus* were devoid of stable functional mRNA and that protein synthesis was dependent upon new transcription (Halvorson *et al.* 1966). If the absence of protein synthesis in dormant spores was due solely to the absence of mRNA, then one would predict that cell-free amino acid-incorporating systems derived from dormant spores should be devoid of basal incorporating activity and that they should recover full capacity when supplemented with exogenous mRNA. To follow the development of the protein-synthesizing system, and to determine whether other blocks in the protein-synthesizing machinery existed in addition to the apparent absence of mRNA, Kobayashi & Halvorson (1968) prepared ribosomal and supernatant fractions from sand-ground spores of *B. cereus* T and of cells at several stages of outgrowth. The [^{14}C]phenylalanine-incorporating activity was determined in the presence or absence of poly U. The first column of Table 1 shows the basal activity of these preparations, i.e. phenylalanine incorporation in the absence of poly U. The incorporation in dormant and heat-activated spores was less than 10 $\mu\mu$moles/mg. ribosomal protein. This approximates the level of detection of radioactivity in the zero time control, and we have considered for all practical purposes that it is zero. Ten minutes after the

Table 1. *Development of protein-synthesizing activity during outgrowth**

	Phenylalanine incorporated ($\mu\mu$moles/mg. ribosomal protein)	
Extracts	−poly U	+poly U
Dormant spores	9·9†	23
Heat-activated spores	6·4†	66
Germinated spores		
5 min.		4262
10 min.	50	3550
20 min.		5210
40 min.	38	4438
Vegetative cells	88	9485

* Reaction mixture contained, in μmoles/ml.: 45 Tris, pH 7·8; 10 magnesium acetate, 40 NH$_4$Cl, 30 KCl, 0·2 GTP, 0·8 ATP, 4 PEP, 13·6 μg. PEP kinase, 40 mercaptoethanol, 1 spermidine, 0·04 amino acid mixture, 0·02 [^{14}C]-L-phenylalanine (10 μc/μmole), 600 μg. *B. cereus* sRNA; and, where indicated, 100 μg. poly U. Total volume was 0·5 ml. A quantity of 55 μg. of the supernatant fraction (S-105-S) and 23 μg. of ribosomal (W-Rib-S) protein per 0·5 ml. of the reaction mixture was used. Samples were incubated at 36° C. for 60 min., and the hot trichloro-acetic acid-insoluble radioactivity was determined.

† These values are a limit of detection, and, therefore, for all practical purposes, are zero.

initiation of germination the system has about 60% of the efficiency of that from vegetative cells. The inactivity in dormant spores cannot be due to a limitation in the amount of mRNA, since the addition of saturating levels of poly U to cell-free systems prepared from germinated spores or vegetative cells increased phenylalanine incorporation nearly 100-fold, while systems from dormant and heat-activated spores responded very poorly. It should be mentioned here that the requirements and characteristics of amino acid-incorporating systems from germinated spores and vegetative cells are essentially the same (Kobayashi & Halvorson, 1968).

During the transition from the dormant to germinated state an increase in amino acid-incorporating activity is readily observed (Table 1). Within 5 min. of the initiation of germination the system responds dramatically to the addition of poly U. The addition of chloramphenicol and/or actinomycin D did not inhibit the appearance of incorporating activity, thereby indicating that increased protein-synthesizing capacity following germination did not involve *de novo* synthesis of mRNA.

Stability of ribosomes from dormant spores

The above results show that when extracts of *B. cereus* T are prepared from dormant and germinated spores by the same procedure of sand grinding, a large increase in ribosome incorporating activity accompanied germination. Since this same extraction procedure yielded highly active ribosomes from vegetative cells, it was concluded that spores contained defective ribosomes. Recent findings in *B. cereus* T of J. Idriss & H. O. Halvorson (unpublished results) provide a further insight into this problem. They observed that in contrast to the behaviour of vegetative cells, the activity of ribosomes from dormant spores might vary according to the method used for their preparation. Low activity preparations were obtained by sand grinding or treatment with a Mini Mill. However, when spores were disrupted by high pressure in the frozen state by means of the Eaton Press, a low yield of ribosomes (5%) was obtained and these were as active as those from germinated spores. These may have been extracted from a subclass of immature or partially germinated spores. Subsequent freezing and thawing lowers the activity of ribosomes from dormant spores but not from vegetative cells, suggesting that they are inherently more unstable.

A similar analysis has recently been reported in *Bacillus megaterium* by Kornberg, Spudich, Nelson & Deutschar (1968). These workers prepared extracts from spores treated with urea-mercaptoethanol by treatment with lysozyme. The resulting ribosome preparation was fully active in the presence of high concentrations of added mRNA. From these and other findings they suggest that the low *in vitro* activity of ribosomes from spores may be due to the presence of high levels of ribonuclease.

The results from *B. megaterium* do not clearly define the nature of ribosomes in dormant spores since many of the characteristics of germination accompany lysozyme treatment. If germination of *B. megaterium* spores precedes lysis by lysozyme, active ribosomes may well be expected. When dormant spores of *B. megaterium* were physically disrupted by various methods the ribosome preparations had low activity (like those reported above for *B. cereus*) even in the presence of high concentrations of poly U (Idriss & Halvorson, unpublished).

On the hypothesis that ribonuclease is the cause of inactive ribosome preparations from dormant spores of *B. cereus* T, it is difficult to understand why, when the same extraction conditions are employed, germination yields fully active ribosomes in spite of the fact that the ribonuclease level actually increases (E. Dickinson, unpublished results). Apparently other factors are involved in the inherent instability of ribosomes from dormant spores. To understand this phenomenon, further chemical and physical characterization of ribosomes from dormant and germinated spores are required.

CONTROL OF PROTEIN SYNTHESIS DURING OUTGROWTH

After the initiation of germination the time required before the first dividing cell is observed is influenced by the type of carbon source and the supply of phosphate (Rodenberg *et al.* 1968; MacKechnie & Hanson, 1968). These two variables are also influential in determining the direction in which the emerging cell will develop. At some point a decision must be made by the emerging cell whether to enter vegetative growth and divide, or whether to initiate those processes leading to microsporogenesis (Vinter & Slepecky, 1965). From this observation it is clear that the germinated spore is subject to some form of regulation prior to the first cell division. How does this developmental control manifest itself?

During outgrowth of *Bacillus subtilis* spores the chromosome replicates in an ordered sequential manner (Yoshikawa, O'Sullivan & Sueoka, 1964), and, as we have shown, protein synthesis at this time appears to be transcriptionally controlled (Kobayashi *et al.* 1965; Rodenberg *et al.* 1968). The apparent dependence of outgrowth on transcriptional products permits an interesting examination of the question as to whether the developmental changes observed are the result of an ordered transcription of the genome, and if this is governed by DNA replication. One approach to this problem has been to determine whether the classes of proteins synthesized during two different intervals of outgrowth are the same (random transcription and translation) or different (ordered transcription and translation). Because the mRNA fraction is unstable during this period and dormant spores are devoid of a stable functional messenger fraction, the degree to which classes of proteins synthesized during outgrowths are ordered serves as an index of the order existing in the synthesis of new mRNA. (It should be pointed out that ordered transcription does not necessarily imply sequential transcription of the genome.) Several findings

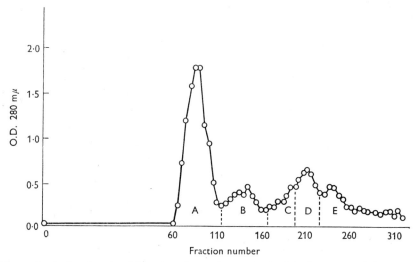

Fig. 5. Sephadex G-200 gel filtration of proteins synthesized during two periods in outgrowth. Activated spores were germinated (30° C.) in two flasks of CDGS medium (550 ml./flask, each containing 0·5 mg. spores/ml.). After 30 min. of incubation a 50 ml. aliquot was removed from the first flask and incubated for 5 min. with 50 μc. [^{14}C]-L-phenylalanine. The radioactive cells were poured over ice ($-20°$ C.), washed twice by centrifugation, recombined with the remaining washed, unlabelled cells and lyophilized. The second culture was treated similarly at 120 min. with 1 mc. of [^{3}H]-L-phenylalanine. The two lyophilized preparations were combined, suspended in Tris-maleate (TM) buffer and disrupted in the Mini Mill. The broken cell suspension was then centrifuged to remove cell debris, the pellet re-extracted twice with TM buffer and the combined supernatants centrifuged at 100,000g for 1 hr. Ribosomal pellets were discarded (they contained 5–10 % of the total ^{14}C or ^{3}H radioactivity) and the supernatant dialysed against [^{12}C]-L-phenylalanine and concentrated by lyophilization (23·2 mg. protein, 71,000 counts/min. ^{14}C and 140,000 counts/min. ^{3}H in 5 ml.). This preparation of soluble proteins was then placed on a Sephadex G-200 column and eluted with TM buffer. The optical density (O.D.) at 280 mμ was measured for each 2 ml. fraction and the major peaks (A–E) were pooled for further fractionation.

reveal that the developmental changes observed in outgrowth do involve an ordering of the appearance of gene products (Kobayashi *et al.* 1965; Torriani & Levinthal, 1967).

In more recent experiments spores were germinated and labelled at an early period in outgrowth (30 min.) with [^{14}C]-L-phenylalanine and at a later period (120 min.) with [^{3}H]-L-phenylalanine. The cultures were combined and the extracted soluble proteins separated according to molecular weight on a column of Sephadex G-200 (Fig. 5). To increase resolution, each of the major peaks from the Sephadex column were subsequently fractionated on diethylaminoethyl (DEAE)-cellulose. Figure 6 represents the DEAE elution profile of proteins appearing on the Sephadex G-200 void volume (peak A). It is observed that the newly synthesized

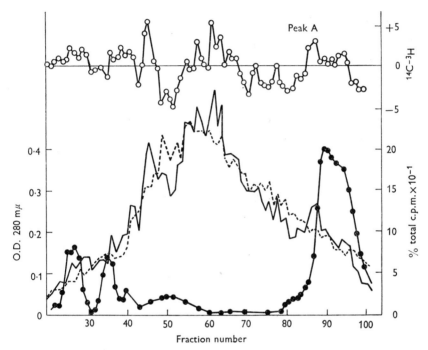

Fig. 6. Chromatography of Sephadex G-200 peak A on DEAE cellulose. Peak A from the Sephadex G-200 column (see Fig. 5) was loaded on a DEAE cellulose column and eluted with a linear 0–0·75 M-NaCl TM buffer gradient (100 ml. in each reservoir). Fractions of 2 ml. each were collected and measured for their optical density at 280 mμ. Bovine serum albumin (100 μg.) was added to each fraction and the TCA precipitable materials were collected and radioactivity determined. The ^{14}C and ^{3}H radioactivities are plotted as per cent of total counts and as per cent ^{14}C-^{3}H. (—, ^{14}C; – –, ^{3}H.)

proteins (^{3}H and ^{14}C counts) do not elute with the major protein peaks. In addition, the radioactivity profiles of ^{3}H and ^{14}C differ in a significant number of ways, indicating that different patterns of proteins were synthesized during the two intervals.

In a second experiment spores were germinated and incubated in CDGS medium for 180 min. At 15 min. intervals thereafter, 30 ml. aliquots were removed and pulse labelled for 3 min. with [^{14}C]-L-leucine. Protein extracts were prepared, dialysed, and fractionated by disk electrophoresis on polyacrylamide gels. At 180 min. the cells were not yet free of their spore coats. By 255 min. the cells had elongated, septum formation was observed but very few cells had completed their first division. To test the initial rates of labelling of isolated protein classes, three of the prominant protein bands were extracted and monitored for radioactivity. As can be seen in Fig. 7, the rate of leucine incorporation into each band varies up to fivefold over the interval examined. The kinetics of leucine incorpor-

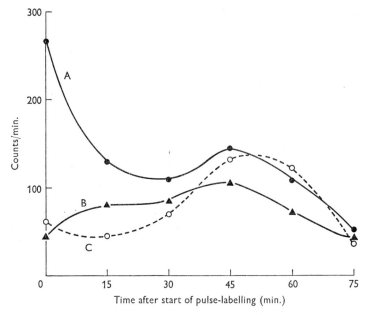

Fig. 7. Kinetics of leucine incorporation into selected proteins during late outgrowth. Activated spores (125 mg.) were germinated in 250 ml. of CDGS medium containing 0·8 mg. L-leucine/ml. After 180 min. (zero time) and at 15 min. intervals thereafter, 30 ml. samples were removed and labelled for 3 min. with [^{14}C]-L-leucine (0·33 μc./ml.). Protein extracts were prepared as described (Hoyem et al. 1968). 100 μl. of each of the samples (7000–13,700 counts/min.; 12 mg. protein/ml.) were subjected to gel electrophoresis. Three distinct bands (a, R_F 0·95; b, R_F 0·65; c, R_F 0·39) were selected for analysis. These bands were cut out with a razor blade, extracted with hydrogen peroxide and the radioactivities determined. The radioactivities were normalized to an initial input of 10,000 counts/min. per gel.

ation into each band follows a separate course. Since only approximately 30 major protein bands are detected on the polyacrylamide gels, it is likely that each is composed of a number of protein species. In spite of this heterogeneity, however, differences can be detected, demonstrating that the rates of amino acid incorporation into selected protein classes differs as a function of time in outgrowth. Increases in radioactivity indicate periods of preferential synthesis, and decreases indicate periods in which incorporation into a particular band decreases in proportion to the over-all rate of protein synthesis. From these results it appears that after germination the genome is gradually available for transcription and translation. During the early stages of outgrowth only selected classes of proteins are synthesized, later the pattern becomes more complex (Torriani & Levinthal, 1967; Hoyem, Rodenberg, Douthit & Halvorson, 1968).

Are there regulatory controls on enzyme synthesis during outgrowth?

The previous results showed that during outgrowth different classes of proteins were preferentially synthesized at different intervals. An examination of enzyme-timing during this period was therefore undertaken to

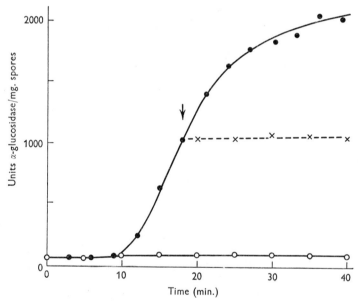

Fig. 8. Effect of actinomycin D and chloramphenicol on the induced synthesis of α-glucosidase. Heat-activated spores were germinated at 500 μg./ml. in CDGS+histidine and maltose. The medium was supplemented with 0·5 mg. of adenosine per ml. Actinomycin D (10 μg./ml.) or chloramphenicol (100 μg./ml.) was added at the points indicated and 5 ml. samples were removed and prepared for enzyme assays according to the procedure described in the text. ●, Control, no addition; ○, actinomycin D at zero time; ×, chloramphenicol at 18 min.

learn whether the apparent controls on transcription and translation could be altered by conditions of induction or repression and to clarify the level at which this control was exercised. When heat-activated spores were germinated in a minimal medium containing maltose, induced α-glucosidase synthesis began at 10–12 min.; 20 min. later the rate of enzyme synthesis declined appreciably (Fig. 8). Actinomycin D added at zero time completely inhibited the induction of this enzyme. Thus dormant spores do not contain a stable mRNA for α-glucosidase synthesis and increased enzyme activity is not a result of an 'unmasking' of some dormant enzyme system (Steinberg et al. 1965). That enzyme induction during this period is dependent on protein synthesis and is not a result of activation is shown by the fact that chloramphenicol addition immediately blocks α-glucosidase formation. The time required for enzyme synthesis represents only a small fraction of the 220 min. which ensue before the first signs of cell division can be observed.

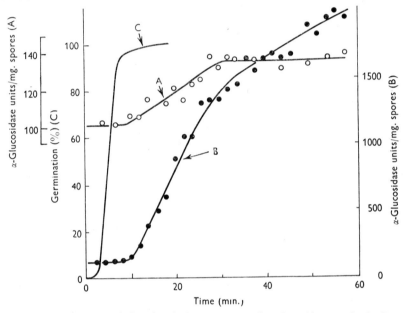

Fig. 9. Timing of maltose-induced and glucose-repressed α-glucosidase synthesis. Spores were heat-activated and germinated at 500 μg./ml. in the following media: (A) CDGS-Glu; (B) CDGS+histidine and maltose (0·4% unpurified maltose). Both media were supplemented with 0·5 mg. adenosine/ml. and 100 μg. of L-alanine/ml. At the indicated intervals during outgrowth, 5 ml. samples were removed and prepared for enzyme assays.

Under fully repressed conditions (i.e. in the presence of glucose) the period of α-glucosidase synthesis occurs over the same interval as that observed for the induced enzyme, even though the induced level of α-glucosidase is 15- to 20-fold greater (Fig. 9). Thus conditions of induction or repression do not alter the periodicity of enzyme synthesis. The timing of histidine-induced and maltose-repressed histidase synthesis has also been examined (Steinberg & Halvorson, 1968a). The first signs of histidase synthesis are observed at 34–36 min. and enzyme synthesis is continuous until 60 min. Again, as for α-glucosidase, conditions of induction or repression do not alter the timing of enzyme synthesis.

An examination of the syntheses of three different enzymes—α-glucosidase, histidase and L-alanine dehydrogenase—reveals that enzyme synthesis is ordered during outgrowth (Fig. 10). Each enzyme is synthesized at a specific time and synthesis continues for only a brief period. Induced histidase appeared approximately 25 min. after the initiation of induced α-glucosidase synthesis, and occurred at a time when the rate of L-alanine dehydrogenase synthesis had fallen to zero (the nature of the stepwise

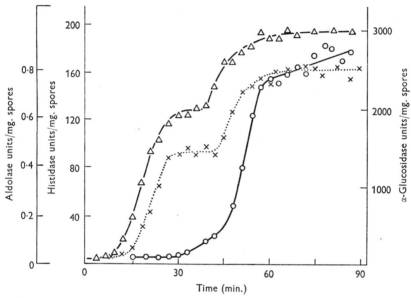

Fig. 10. Ordering of α-glucosidase, L-alanine dehydrogenase, and histidase synthesis during outgrowth. Spores were heat-activated and germinated at 500 μg./ml. in either CDGS+histidine (○) or CDGS+histidine and maltose. Both media were supplemented with 0·5 mg. adenosine/ml. and 100 μg. L-alanine/ml. At the indicated intervals during outgrowth 5 ml. samples were removed and prepared for enzyme assays. △, Induced α-glucosidase; ○, induced histidase. L-Alanine dehydrogenase activity (×) was assayed on samples from CDGS+histidine and maltose.

synthesis is not understood). Clearly, if genes were continually available for transcription and regulation during outgrowth, then enzyme periodicity should have been abolished; all induced enzymes would have appeared at the same time and enzyme synthesis would have been continuous. The delay in the appearance of histidase is not a result of a permeability phenomenon since radioactive histidine is incorporated into protein as early as 8 min. after the initiation of germination, and the addition of the inducer 16 min. after germination does not delay the appearance of the enzyme (Steinberg & Halvorson, 1968a).

Availability of the genome for transcription and translation during outgrowth

If the ordered enzyme synthesis one observes is a result of a mechanism which selectively controls gene expression at the level of mRNA synthesis, then enzyme inducers would only function during the period in which the particular cistron was being transcribed into the unstable message. To verify this hypothesis it is necessary to know the time of mRNA synthesis

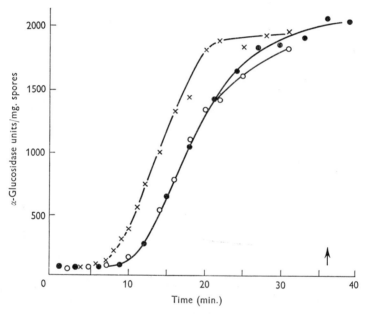

Fig. 11. Timing of induced α-glucosidase mRNA synthesis. Heat-activated spores were germinated at 500 μg./ml. in CDGS+histidine and maltose. The medium was supplemented with 100 μg. L-alanine and 0·5 mg. adenosine/ml. At the indicated intervals 5 ml. samples were removed and added to: ●, 4 ml. of cold phosphate buffer (0·067 M, pH 6·8) containing chloramphenicol (225 μg./ml.): ○, puromycin dihydrochloride (200 μg./ml.); and (×) actinomycin D (10 μg./ml.). The puromycin- and actinomycin D-treated samples were incubated to 37 min., at which time enzyme synthesis was arrested by the addition of 3 ml. of cold phosphate buffer (0·067 M, pH 6·8) containing chloramphenicol (250 μg./ml.). The cells were collected by centrifugation and prepared for enzyme assays. ↑ Indicates time of addition of chloramphicol-containing buffer.

for both induced α-glucosidase and histidase, and to show that the RNA templates for these enzymes had short half-lives.

Figure 11 represents an experiment designed to measure the timing and stability of the message for induced α-glucosidase. Two control experiments were employed in which inhibitors of protein synthesis were used. In the first, samples were removed at intervals in outgrowth and added to cold buffer containing chloramphenicol. For the second control, the samples were added to puromycin at intervals and incubation in the inducing medium was continued until 37 min. The experiment consisted of adding actinomycin D at 1, 2 or 3 min. intervals and allowing incubation to proceed until 37 min. At this time both the puromycin control and the actino-

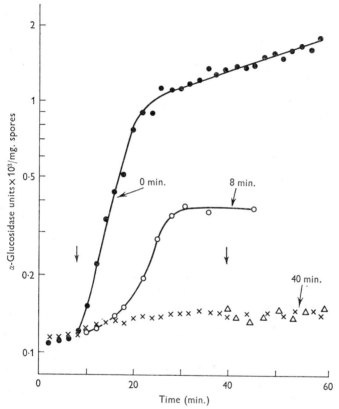

Fig. 12. α-Glucosidase inducibility during outgrowth. Spores were heat-activated and germinated at 500 μg./ml. in either CDGS-Glu or in CDGS medium containing 0·4 % unpurified maltose in place of glucose. Both media were supplemented with 0·5 mg. adenosine/ml. At 8 and 40 min., 60 ml. samples were removed from the CDGS-Glu medium; the cells were washed once by centrifugation at room temperature with CDGS salts, and were resuspended in CDGS+maltose (0·4 %, unpurified). At intervals, 5 ml. samples were removed and cells were collected and prepared for enzyme assays. ×, Repressed α-glucosidase; ●, maltose added at zero-time; ○, maltose added at 8 min.; △, maltose added at 40 min.

mycin D experiments were stopped by the addition of cold buffer containing chloramphenicol.

If there were stable RNA templates for α-glucosidase present at any time after the addition of actinomycin D, then at these time periods the level of enzyme activity should have been the same as that observed at 37 min. This is clearly not the case. Actinomycin D stopped α-glucosidase synthesis within 3–4 min. and this was identical to the stability of pulse-labelled RNA made during this period (Kobayashi et al. 1965). α-Glucosidase mRNA transcription began 5–6 min. after the initiation of germination and stopped 15–16 min. later. The appearance of enzyme activity occurred at 10 min. and synthesis continued until 30 min. A similar experiment conducted to determine the time of synthesis of induced histi-

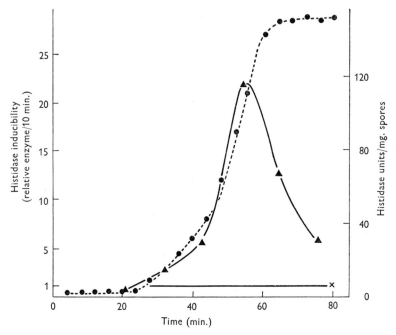

Fig. 13. Histidase inducibility during outgrowth. Spores were heat-activated and germinated at 250 μg./ml. in CDGS+glutamate supplemented with 0·5 mg. adenosine/ml. At intervals during outgrowth, 10 ml. samples were added to flasks containing L-histidine (3·23 × 10⁻² M). After a 10 min. incubation period enzyme synthesis was arrested by the addition of 8 ml. of cold phosphate buffer (0·067 M, pH 6·8) containing chloramphenicol (225 μg./ml.). The cells were collected and prepared for enzyme assays. ×, Control, histidase activity in CDGS+glutamate; ●, induced histidase synthesis during outgrowth; ▲, relative histidase inducibility plotted as a function of the time at which the 10 min. induction was terminated.

dase mRNA and measure its stability showed that actinomycin D shuts off induced histidase synthesis within 1–2 min. These experiments further demonstrate that the delay between the appearance of induced α-glucosidase and induced histidase is not a result of a delay in the translation of a previously synthesized message, nor is it due to delayed assembly of histidase subunits (Hartwell & Magasanik, 1963).

After we determined the time of synthesis of the mRNA's for α-glucosidase and histidase and showed that these messages were unstable, it was possible to test whether specific regions of the genome were continually available for induction. α-Glucosidase inducibility was tested by adding maltose at different intervals in outgrowth (Fig. 12). When the inducer was added at zero time, the level of enzyme synthesized was 15- to 20-fold greater than that produced under repressed conditions. On the other hand, when cells were removed from the repressing medium part-way through the period of gene expression and transferred to the inducing medium the level of enzyme synthesized was reduced to 40% of the control. Addition of the inducer after the period of gene expression (e.g. at 40 min.) had no significant effect on the level of enzyme produced.

To test histidase inducibility during outgrowth, samples were removed from the germination medium (containing glutamate as the main carbon source) at various times and incubated for 10 min. in the presence of histidine. Enzyme synthesis was arrested by the addition of cold phosphate buffer containing chloramphenicol. Figure 13 shows that during outgrowth the capacity for histidase induction reached a peak at the same time that histidase mRNA was being transcribed. It is obvious from these experiments that the inducer can only function for a limited time period, for if it were able to overcome transcriptional controls, then for both experiments we would have observed the *same* capacity for enzyme synthesis regardless of the time at which it was added. In contrast, when the inducers for α-glucosidase and histidase were added to an *asynchronous* vegetative culture, the synthesis of both enzymes began at essentially the same time (Steinberg & Halvorson, 1968a). Thus the formation of one enzyme is not dependent on the previous synthesis of another nor is there a difference in the capacity of all or some of the cells in the population to synthesize α-glucosidase and histidase. The induction of these two enzymes therefore represents two independent events.

Relationship between ordered enzyme synthesis and DNA replication

During outgrowth of *B. subtilis* spores the chromosome replicates in an ordered sequential manner (Yoshikawa *et al.* 1964). The simplest model which could explain the ordered appearance of enzymes during outgrowth is that transcription for these enzymes follows, or is governed by, chromosomal (DNA) replication. Since the initial phase of outgrowth is directly dependent on new mRNA transcription, this hypothesis can be tested directly.

DNA synthesis during outgrowth

When an estimation of DNA synthesis was made during outgrowth of spores of B. cereus T by employing a colorimetric assay, it was found that significant DNA synthesis did not occur until a period just prior to the onset of cell division (Steinberg & Halvorson, 1968b). In contrast, radioactive thymidine incorporation was observed at any time. This incorporation could be inhibited by mitomycin C, a potent inhibitor of DNA syn-

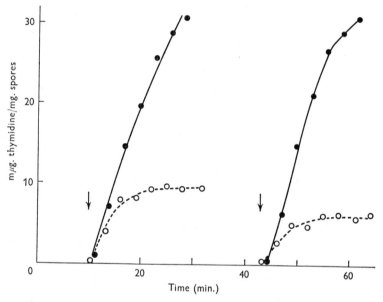

Fig. 14. Effect of mitomycin C on DNA synthesis during outgrowth. Heat-activated spores were germinated at 500 μg./ml. in CDGS+histidine containing 0·5 mg. adenosine/ml. and 100 μg. L-alanine/ml. as germinating agents. At the points indicated (10 and 43 min.), samples were withdrawn from the main growth flask and added to flasks containing either [³H]thymidine, 1·43 μc/2·9 μg./ml. (●), or [³H]thymidine plus mitomycin C, 10 μg./ml. (○). Samples (0·6 ml.) were removed at the indicated intervals and the radioactivity incorporated into cold trichloroacetic acid precipitates was measured.

thesis (Fig. 14). To further characterize the nature of the material into which tritiated thymidine was incorporated, heat-activated spores were germinated in the presence of [³H]thymidine and incorporation was stopped at the end of 30 min. The DNA isolated from the germinated spores was analysed by equilibrium sedimentation in CsCl. The results showed that all radioactivity coincided with the optical density (at 260 mμ) profile. This material was alkali stable, deoxyribonuclease-sensitive and ribonuclease-resistant, thereby confirming that [³H]thymidine incorporation was a result of DNA synthesis (Steinberg & Halvorson, 1968b).

Since the amount of DNA synthesized within this period represented

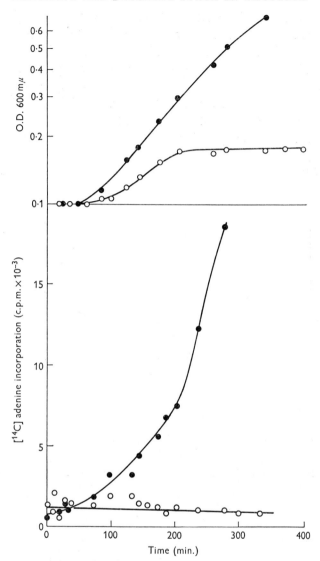

Fig. 15. Test of the thy^- mutant for 'leakiness'. Vegetative cells of the thy^- strain were grown in CDGS medium plus thymidine (200 μg./ml.) at 30° C. The cells were harvested while in logarithmic growth by filtration, washed with CDGS medium and resuspended in CDGS containing saturating levels of [^{14}C]adenine (1·3 μc/48 μg./ml.). Samples (1·0 ml.) were removed at the indicated intervals during outgrowth and the radioactivity incorporated into alkali-stable material in the absence (○—○) and presence (●—●) of thymidine (200 μg./ml.) was measured as previously described (Steinberg & Halvorson, 1968b). The background was not subtracted. Optical density (O.D.) at 600 mμ was determined with a Bausch and Lomb Spectronic 20 colorimeter (data from C. Anderson, Physiology Class, Woods Hole, 1967).

only a few per cent of the total DNA present in the spore (Fitz-James, 1955b), it was imporatnt to know whether or not this represented DNA replication or DNA synthesis caused by extensive repair (turnover) processes (Pettijohn & Hanawalt, 1964). To resolve this question, and at the same time establish a system with which the DNA replication model for ordered enzyme synthesis could be tested, a thymidine-requiring auxotroph of *B. cereus* T was isolated. To check the possibility that the mutant might be 'leaky', either by having some alternate pathway for thymidylate

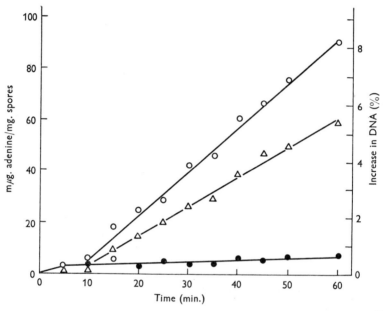

Fig. 16. DNA synthesis during outgrowth of wild-type and thymine mutant spores. Heat-activated spores of the thymine-requiring auxotroph and wild-type strains were germinated at 500 μg./ml. in CDGS+glucose supplemented with L-alanine (270 μg./ml.), inosine (15 μg./ml.), and saturating levels of [^{14}C]adenine (0·63 μc/34·5 μg./ml). Samples (0·6 ml.) were removed at the indicated intervals during outgrowth. The radioactivity in-corporated into alkali-stable material in the absence and presence of thymidine (200 μg./ml.) was measured. The results are plotted as millimicrograms of adenine incorporated per milli-gram of spores and as the percentage increase in total DNA. ○, Thymine-mutant spores plus thymidine; △, wild-type spores, no thymidine; ●, thymine-mutant spores, no thymidine.

synthesis, an endogenous supply of thymine, or some system of continuous DNA breakdown and re-utilization of end-products (Hewitt, Suit & Billen, 1967; Wilson, Farmer & Rothman, 1966), the following experiment was conducted with vegetative cells. A log. phase culture of the *thy⁻* strain grown in a minimal medium supplemented with thymidine was washed free of its auxotrophic requirement and divided into two flasks, only one of which contained thymidine. [^{14}C]Adenine was added to both cultures and the radioactivity incorporated into alkali-stable material (DNA) was measured (Fig. 15). The results show that the mutant is not leaky—there is no DNA synthesized in the absence of exogenous thymidine. The extent of DNA synthesis during outgrowth of spores of the mutant and wild-type strains was also examined (Fig. 16). In the absence of added thymidine, the rate of DNA synthesis in the mutant was only 4% of the control value (thy^- spores plus exogenous thymidine), and only 6% of the wild-type rate (no thymidine added). In the presence of thymidine, the total amount of DNA synthesized after 60 min. of outgrowth represented only 8% of that initially present in the spore. Although these are minimal estimates we believe that they approximate net DNA synthesis since saturating levels of isotope were employed and significant amounts of DNA synthesis (10%) would have been seen in the chemical assays for DNA.

Is DNA synthesis during early outgrowth repair or replication?

Since outgrowing spores of the thy^- strain synthesized DNA at 4% of the control rate, it was important to test whether the incorporation of DNA precursors during outgrowth represented repair (turnover) synthesis or replication. If a culture of spores were labelled during outgrowth with [^{3}H]bromodeoxyuridine ([^{3}H]BUdR) and the DNA isolated, sheared and subjected to equilibrium sedimentation in CsCl, then one could expect the following patterns: DNA synthesis resulting from repair would have all the radioactivity associated with the normal density (light–light DNA), while DNA synthesis resulting from replication would have all the radioactivity associated with the hybrid density (light–heavy) DNA (Pettijohn & Hanawalt, 1964). The results of such an experiment are shown in Fig. 17. Spores of the thymidine requiring auxotroph were germinated and incubated for 60 min. in the presence of [^{3}H]BUdR. The DNA was isolated and centrifuged to equilibrium in a CsCl density gradient. Essentially all of the radioactivity is associated with the hybrid region, thus during outgrowth the low rate of DNA synthesis one observes represents primarily replication and not repair synthesis.

Ordered transcription and DNA replication

Outgrowth of spores of *B. cereus* T is characterized by an ordered synthesis of both basal and induced enzymes which is transcriptionally controlled. Does the slow rate of DNA replication during outgrowth serve as the mechanism for ordering the RNA transcription necessary for enzyme synthesis? In an experiment where mutant spores were germinated in the

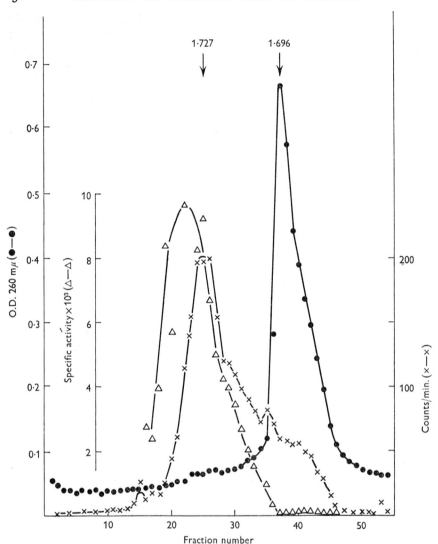

Fig. 17. CsCl density-gradient centrifugation of [³H]bromodeoxyuridine ([³H]BUdR) DNA synthesized during outgrowth. Thymine-mutant spores (500 mg.) were heat-activated and germinated in 500 ml. of CDGS+glutamate containing adenosine (0·5 mg./ml.), L-alanine (100 μg./ml.), and saturating concentrations of [³H]BUdR (0·95 μc./15 μg./ml.). Growth was arrested after 60 min. by immersing the culture flask in an ice–salt bath. Cells were collected by centrifugation and DNA was isolated as described. Approximately 100 μg. of the [³H]BUdR-labelled DNA was added to 3 ml. of CsCl (starting density of 1·70 g./c.c.). The density gradient was formed as described. Fractions of 5 drops each were collected and the radioactivity and optical density (O.D.) (260 mμ) were measured for each tube. The density gradient was determined by measuring the refractive index of several undiluted fractions with a refractometer (Steinberg & Halvorson, 1968b).

presence and absence of exogenous thymidine, it was found that the absence of DNA replication had no effect on RNA synthesis as estimated by [³H]-uracil incorporation (Steinberg & Halvorson, 1965b). An examination of the timing of induced enzyme synthesis during outgrowth of spores of the thymidine mutant showed that both induced α-glucosidase and histidase have the same periodicity under conditions of thymidine starvation (Fig. 18).

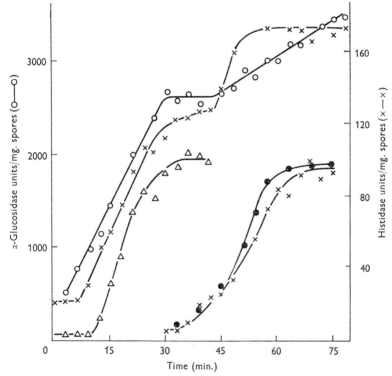

Fig. 18. Effect of thymine addition on the timing of induced enzyme synthesis during outgrowth of spores of the thymine auxotroph. Spores of the wild-type and thymine mutant were heat-activated and germinated at 500 μg./ml. in CDGS+histidine and maltose containing L-alanine (100 μg./ml.) and adenosine (0·5 mg./ml.). At the indicated intervals during outgrowth, 5 ml. samples were removed and assayed for α-glucosidase and histidase. Thymine-requiring mutant: ×, Induced α-glucosidase and histidase, thymine (200 μg./ml.) added at zero-time; ●, induced α-glucosidase and histidase, no thymine added. Wild type: △, induced α-glucosidase, ○, no thymine added.

The timing of α-glucosidase synthesis was the same as that observed for wild-type spores even though thy^- spores initially have a sixfold higher enzyme level. It is clear, then, that ordered transcription accompanying outgrowth is not dependent upon DNA replication.

An attempt was made to ascertain how ordering could be sustained in the absence of DNA replication. The DNA inhibitors, mitomycin C, which cross-links complementary DNA strands (Szybalski & Iyer, 1964) and

5-fluoro-2′-deoxyuridine (FUdR), a competitive inhibitor of thymidylate synthetase (Hartmann & Heidelberger, 1961) were employed. When mitomycin C was added at 29 min. or at intervals thereafter, induced histidase synthesis was terminated within a period of 4–5 min. (Fig. 19). The time required for enzyme inhibition is the same as that previously found for inhibition of DNA synthesis (Fig. 14). On the other hand, the initial rates of RNA and protein synthesis were not inhibited (Steinberg, 1967).

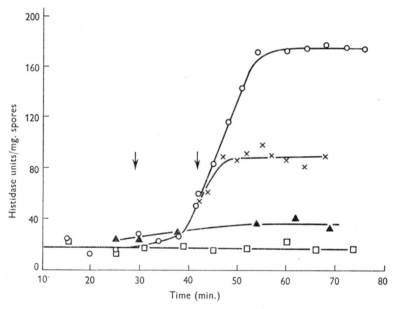

Fig. 19. Effect of the time of addition of mitomycin C (MC) on histidase induction. Heat-activated spores were germinated at 0·5 mg./ml. in CDGS+histidine or CDGS+histidine and maltose. Both media contained adenosine (0·5 mg./ml.) and L-alanine (100 μg./ml.). At the points indicated (29 and 41 min.) a portion of the control culture was added to a flask containing mitomycin C (10 μg./ml.), and, at intervals thereafter, 5 ml. samples were removed and assayed for histidase. ○, Induced histidase, control; □, maltose-repressed histidase, mitomycin C added at 29 min.; ▲, induced histidase, mitomycin C added at 29 min.; ×, induced histidase, mitomycin C added at 41 min.

The effect of mitomycin C on induced histidase appeared to be in conflict with the conclusions reached with the thymidine auxotroph, i.e. DNA replication was not necessary for enzyme timing. The use of FUdR, however, provided an experimental system which would mimic the conditions employed with the thy^- strain; inhibition of DNA synthesis at the enzymic level. Under conditions where the rate of DNA synthesis in the presence of FUdR was only 20% of the control value it was found that histidase could be induced to the control level (Fig. 20). Apparently, the effect of mitomycin C on enzyme synthesis was the result of some 'physical modification' in the DNA (presumably cross-linking), while inhibition of DNA synthesis by interfering with an enzymic mechanism had no effect on enzyme timing.

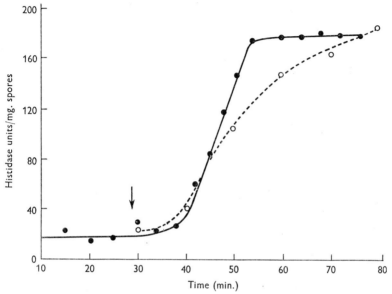

Fig. 20. Effect of 5-fluoro-2′-deoxyuridine (FUdR) on histidase induction during outgrowth. Heat-activated spores were germinated at 0·5 mg./ml. in CDGS+histidine containing adenosine (0·5 mg./ml.) and L-alanine (100 μg./ml.). Uridine (100 μg./ml.) was added to the culture at 28 min. after the initiation of germination (↓). At 29 min. part of this control culture was added to a flask containing FUdR (10 μg./ml.) (○). Samples (5 ml.) were removed from both flasks at the indicated intervals and assayed for histidase. (●), Induced histidase, control.

DISCUSSION

The system that is built into the dormant spore must allow it to undergo an ordered and rapid metabolic switch when placed in a favourable environment. It is interesting therefore that, in addition to the absence of functional mRNA, ribosomes in dormant spores may be inherently unstable. The stabilization of these ribosomes during germination and before macromolecular synthesis begins suggests that an activation process may be involved. Similar increases in the metabolic activities of dormant spores after activation and germination have been described (Steinberg et al. 1965). This additive limitation on development could be readily understood if it represented a block imposed late in sporulation, during which limitations in protein synthesis are known to occur. How the dormant state is established may well dictate, in part, the controls over the re-establishment of vegetative growth.

From quantitative considerations it appears that the labile class of RNA synthesized during outgrowth (this class of RNA includes mRNA and in the absence of protein synthesis, which occurs at the very early stages, unstable ribosomal precursors) regulates the rate of protein synthesis. At present it is not known what proportion of the labile RNA is represented by mRNA, but it appears likely that the rate-limiting step in protein synthesis during outgrowth is the synthesis of mRNA and not some process associated with the maturation of the ribosomal system. This is supported by the fact that the rate of protein synthesis at any time during outgrowth is directly proportional to the size of the labile RNA fraction, while there is no such relationship to either total or newly synthesized ribosomal RNA, and consequently, ribosomes (Rodenberg et al. 1968).

Transcriptional controls during outgrowth

The ordered enzyme synthesis observed during outgrowth can be explained by either of two distinct hypotheses: (i) transcriptional processes are ordered and each gene is transcribed and translated at a specific period (Halvorson et al. 1964); (ii) a gene is always available for transcription—enzymes can be induced or derepressed at all times. Enzyme periodicity is a result of a cyclic variation in repressor or co-repressor levels produced in response to enzyme function (Kuempel, Masters & Pardee, 1965; Pardee, 1965). It should not be overlooked, however, that other physiological controls operating at a secondary level could act to modify the initial products of transcription and translation. Such controls could influence the assembly, activation or conformation of the translated polypeptide, or even the translation process itself (Grobstein, 1963; Griffin & Cox; 1966; Cline & Bock, 1966; Berberich, Kovach and Goldberger, 1967). These latter factors apparently are not operating in the present system since for the simplest form of translational control (random transcription of mRNA with ordered translation) the mRNA would be expected to be stable.

The timing of enzyme synthesis during outgrowth appears to be transcriptionally controlled, and during this phase the genome is periodically and not continuously available for induction. This conclusion is based on the following observations. (*a*) The timing of enzyme synthesis during outgrowth was not altered by conditions of induction or repression. (*b*)The level of enzyme synthesized in response to an inducer was a function of the time of inducer addition. If the inducer was added part-way through the period of gene expression, the level of enzyme induction was diminished. Addition of inducer at a time following the period of gene expression had no significant effect. (*c*) If enzymes could be induced at any time during outgrowth, then in the presence of several inducers, enzyme order would be abolished and all enzymes would show the same periodicity. This was not observed. (*d*) The order of enzymes synthesized during outgrowth was the same as that found during the first few synchronous divisions in vegetative growth (Kobayashi *et al.* 1965; Steinberg *et al.* 1965), thereby indicating that the 'metabolic' shift to vegetative growth was not involved in the ordering process.

It is generally recognized that the genetic determinants in DNA are arranged in a linear manner. It appears obvious therefore that some type of polarized transcription of the genetic material would aid in establishing an ordered appearance of gene products. At least for one genetic region, the tryptophan operon in *Escherichia coli*, DNA–RNA hybridization has directly shown that the synthesis of its mRNA is sequential (Imamoto, 1968). More indirectly, it has also been demonstrated that the mRNA for induced enzymes of the *lac* operon of *E. coli* is transcribed in the same sequence as the gene order found within this operon (Alpers & Tomkins, 1966). In *Bacillus subtilis* it is not yet clear whether RNA synthesis is ordered, however, during the early period following germination in both *B. cereus* T (R. Epstein & H. O. Halvorson, unpublished) and *B. subtilis* (Donnellan, Nags & Levinson, 1965; Armstrong & Sueoka, 1968) 4, 16 and 23 S RNA are all synthesized. The clustering of the cistrons for these RNA species close to the replicating origin of the chromosome of *B. subtilis* (Oishi & Sueoka, 1965; Dubnau, Smith & Marmur, 1965; Smith, Dubnau, Morell & Marmur, 1968) probably explains their coordinate synthesis.

In contrast to the apparent discontinuous synthesis of the mRNA's for individual enzymes, ribosomal RNA (rRNA) appears to be synthesized continuously during outgrowth (Donnellan, Nags & Levinson, 1965; Rodenberg *et al.* 1968). This suggests several possibilities for the control of rRNA synthesis: (i) there are numerous cistrons coding for ribosomal RNA; (ii) the number of rRNA cistrons are limited and these are randomly located on the chromosome; (iii) the number of rRNA cistrons are limited and transcription is slow; (iv) there are a limited number of rRNA cistrons but they are continuously available for transcription. This last possibility seems the most reasonable for the following reasons. In *B. subtilis* the number of 16 and 23 S RNA cistrons has been estimated at 9–10 for each species. Furthermore, these cistrons are not scattered, but appear to be confined to only two regions of the chromosome (Smith, Dubnau, Morell & Marmer, 1968). Finally, the dormant spore contains approximately 20,000 copies of rRNA and this more than doubles before the first division (Fig. 4). Since the number of ribosomal cistrons in *B. cereus* is only about 2·5 times that for *B. subtilis* (Halvorson, unpublished results), one must conclude that this region is always open to transcription (see below).

Since the idea of a region of DNA continuously available to transcription conflicts with the conclusions drawn from the enzyme induction experiments, it might be suggested that transcription of the rRNA cistrons is regulated in a different manner. For example, there might be something special about the region of DNA at which they are located (e.g. these cistrons are near the chromosome origin) or the DNA is physically altered. One such possibility is that a 'transcriptional loop' is formed where RNA polymerase can be 'locked in' and transcribe rRNA continuously. On the other hand, RNA polymerase may have a high affinity for the rRNA cistrons leading to repeated rounds of transcription. Periodic rounds of transcription are known to occur during derepression of the tryptophan operon in *E. coli* (Imamoto, 1968; Baker & Yanofsky, 1968), and under appropriate conditions purified *E. coli* RNA polymerase is apparently able to select which regions (initiating points) of native double-helical DNA are to be copied (Maitra & Hurwitz, 1965; Geiduschek, Snyder, Colvill & Sarnat, 1966).

The polymerase for synthesizing all three classes of RNA (ribosomal, transfer and messenger) pre-exists in dormant spores. Two findings support this view. First, spores germinated in the presence of chloramphenicol can synthesize RNA. Secondly, a DNA-dependent RNA polymerase has been demonstrated in the extracts of spores of *B. subtilis* (Kerjean & Szulmajster, 1966). This enzyme has a lower molecular weight and differs in its heat stability from the enzyme of the vegetative cell, but the specific activity of all the molecular forms observed is the same (Kerjean, Marchetti & Szulmajster, 1967). During outgrowth the kinetics of RNA synthesis are precisely those predicted on the basis that once a spore is germinated, the rate of total RNA synthesis (as measured by uracil incorporation) is linear with respect to time. What does this suggest concerning spore RNA polymerase and the regulation of transcription? A linear rate of transcription could mean the following.

(i) RNA polymerase is limiting in the spore. The simplest theory would allow for only one enzyme molecule. This is unlikely on the grounds that one would not expect a cell to base its survival on the presence of one molecule of such an important enzyme and the fact that thousands of copies of RNA are synthesized during outgrowth.

(ii) Spores contain large numbers of polymerase molecules but only one functional initiating site for a polycistronic rRNA. The remaining polymerase molecules do not have a 'substrate' site or are prevented from attaching by the presence of an inhibitor or limitation in an attachment factor (crude extracts of dormant spores are apparently inhibitory for RNA polymerase (Kerjean *et al.* 1967)).

(iii) If both initiator sites and polymerase were present in multiple copies then all enzyme molecules would either initiate transcription at the same time or transcription would be controlled so that new DNA regions would be transcribed only upon the completion of others. If additional sites for transcription became available during outgrowth, this would manifest itself as a change in the rate of RNA synthesis (this may be true in the case of rRNA).

On the basis of what has been said previously about the control of rRNA transcription, and the following, it would seem that the last proposal probably best represents the situation. Let us assume that 20,000 copies of rRNA are made before the first division and that the 25 cistrons for rRNA in *Bacillus cereus* (16 + 23 S) are located at one site. We can then make these predictions. (*a*) One RNA polymerase molecule would have to transcribe this region at a minimal rate of 8 times/min. in order to produce 20,000 copies by the first division. Even though this rate is relatively 600 times faster than that of *in vivo* mRNA transcription by *E. coli* polymerase (Imamoto, 1968; Baker & Yanofsky, 1968), this would still mean that during the entire period of outgrowth one enzyme molecule would be concerned solely with rRNA transcription. (The rate of 8 copies/min./RNA polymerase molecule seems extremely high in comparison with mRNA transcription. However, it would only be 3–6 times greater than the rate of DNA replication during vegetative growth with only 1 or 2 replication points.) (*b*) Alternatively it might be argued that a molecule of RNA polymerase has an efficiency several times that of DNA polymerase. This would account for the production of all the rRNA, but this region represents only 0·4% of the *Bacillus* chromosome (Oishi & Sueoka, 1965; Smith *et al.* 1968; Halvorson, unpublished). (*c*) During logarithmic growth of *E. coli* the transcription time for the mRNA of the tryptophan operon, having a molecular weight 20 times smaller than the *Bacillus cereus* rRNA region, is 5–10 min. (Imamoto, 1968; Baker & Yanofsky, 1968). If the rate of RNA transcription were the same for rRNA during the period of outgrowth, then, by the first division, one polymerase molecule would produce only 10 copies of rRNA. This is 2000 times less than what is needed!

The above arguments support the hypothesis that there is a large or multiple number of RNA polymerase molecules in the spore and that during outgrowth there probably are multiple sites of transcription. Experimental support (though weak) for this proposal can be drawn from the results with mitomycin C. This antibiotic blocked DNA synthesis and the induction of histidase within 5 min. (Fig. 14, 19), yet RNA and protein synthesis continued in its presence. It seems logical that this synthesis should be attributed to activity at a locus other than that for histidase. Presumably this is either a region which is not appreciably cross-linked or one where mitomycin C-induced cross-links can be repaired (Mahler, 1966). A second alternative implied by these results is that transcription is re-initiated, either at the origin or at some locus where cross-links are not extensive enough to completely inhibit RNA polymerase. If this explanation is correct it would imply that several RNA templates could be transcribed at the same time from different regions of the chromosome. The observation of an ordered synthesis of enzymes would then mean that we are selectively looking at those RNA templates that were transcribed at different intervals. That reinitiation or new rounds of transcription could occur during outgrowth is implied from the work of Torriani & Levinthal (1967). They showed that 'groups' of proteins are ordered during outgrowth and that the initial period after germination was limited to the synthesis of only a few proteins. When the period of outgrowth was prolonged, however, as it would be in a minimal medium, then the same 'groups' of proteins were resynthesized.

'Rounds' of mRNA transcription have recently been demonstrated by Imamoto (1968) and it appears that the DNA molecule may indeed contain multiple sites for the initiation of transcription (Bremer, Konrad & Bruener, 1966). Bremer et al. (1966) have estimated that there are approximately 180 active RNA polymerase units bound per T_4 phage DNA unit. Since the saturation of DNA sites by the polymerase was much less than that predicted by complete filling of the DNA it was suggested that the binding reaction was specific.

Ordered transcription and the 'state' of spore DNA

It was concluded from the evidence presented that the initiation and maintenance of ordered transcription during outgrowth was not dependent upon the replication of the genome. However, the effect of cross-linking DNA with mitomycin C indicated that transcription requires an unmodified, native DNA structure. Since enzyme inducibility during outgrowth was limited to specific time intervals, it seems likely that a mechanism which either alters the capacity of DNA to serve as a template for RNA transcription or which limits RNA transcription itself may be involved in the ordering process. Could this regulation of transcription simply be a function of the physical state of the DNA? If the chromosome contained within the bacterial spore is in the completed form (Yoshikawa et al. 1964) two simple models can be proposed by which ordered transcription can be controlled. First, the state of the DNA within the spore may change in some polarized manner during outgrowth. This could serve as a 'signal' for initiating transcription. The alteration may be polarized in one direction or several initiation points can be present. A second possibility is that the order of transcription is dictated by some temporal (clock) mechanism. How might these controls be achieved?

On the basis of photochemical evidence Stafford & Donnellan (1968) have suggested that the DNA in dormant spores, germinating spores and vegetative cells may exist in different conformations; a transition from one form to another occurring during germination. In comparison with vegetative cells the intracellular water concentration of at least part of the structure of dormant spores approaches that of an 'anhydrous' system (Sussman & Halvorson, 1966). During germination the spore absorbs a quantity of water approximately 1·5 times greater than its dry weight (M. Kirschner, unpublished), yielding a cell which is about 70 % water by weight. What effect this has on spore DNA is not known, but it is not unreasonable to suggest that hydration *in vivo* may alter DNA conformation in much the same manner as that observed for *in vitro* studies. This in turn might serve as both a 'signal' to initiate transcription and as a mechanism controlling the availability of transcription sites. Similarly, the spore DNA might be in an unusual 'tight' or 'compact' configuration. Subsequent physical 'unfolding' after the initiation of germination would then limit both the rate of replication and the capacity of the DNA to serve as a template for RNA synthesis.

Ordered enzyme synthesis may be a consequence of the emergence of the vegetative genome from a state where structural modifications or complexing with inhibitors makes it accessible for transcription in an ordered manner. That this condition alone governs transcription appears to be unlikely on several grounds. First, the same periodicity in enzyme synthesis has been observed during the first few cell divisions following outgrowth (Kobayashi *et al.* 1965; Steinberg *et al.* 1965). Secondly, during outgrowth of *B. subtilis* the genes replicate in the same sequential order as that observed for vegetative cells (Oishi, Yoshikawa & Sueoka, 1964).

Finally, any one of the above mechanisms might cause a temporary unwinding of the two DNA strands in a small region, allowing one of the strands to be used as a template for the polymerization of RNA (Chamberlin & Berg, 1964). It is known that both single and double-stranded DNA's serve as equally efficient primers for the spore DNA-dependent RNA polymerase (Kerjean, Marchetti & Szulmajster, 1967).

It should be pointed out that very little is known about the state of DNA within the spore and what happens to it during germination, so that at best these proposals are highly speculative.

DNA synthesis during outgrowth

The development of the vegetative state has been shown to involve the sequential appearance of RNA, protein and DNA synthesis. Spores of *Bacillus cereus* T are apparently devoid of stable mRNA and protein synthesis after germination is completely dependent upon the transcription of mRNA. At the same time the absence of DNA synthesis during outgrowth of spores of *B. subtilis* and the low level of DNA synthesis observed for *B. cereus* over the major part of outgrowth does not have any counterpart in the vegetative growth cycle. In *B. subtilis*, DNA synthesis is not observed until 120–150 min. after the initiation of germination (Woese & Forro, 1960; Wake, 1967) even though DNA polymerase and enzymes involved in the synthesis of deoxyribonucleotides are present in the dormant spore (Falaschi, Spudich & Kornberg, 1965). In addition to these enzymes it is likely that DNA repair enzymes are also present (Yoshikawa, 1965); however, repair synthesis may be below the threshold level necessary for detection (Wake, 1967). Since enzymes required for DNA synthesis are initially present in the spore, it is of interest that during early outgrowth of *B. subtilis* no replication occurs, and in *B. cereus* T there is only a slow rate of DNA replication until a period just prior to the first cell division. The possibility exists that spores of *B. cereus* contain an incomplete chromosome and that DNA synthesis observed during outgrowth may represent terminal steps in chromosome replication. At present, those factors limiting DNA synthesis during outgrowth are unknown, but it is not unlikely that since we are dealing with a system of ordered transcription that some protein essential for DNA replication is synthesized in outgrowth.

SUMMARY

The bacterial spore is a highly differentiated structure and represents one of the most dormant biological systems known. In spores of *Bacillus cereus* strain T, RNA synthesis commences approximately 3 min. after the initiation of germination, followed about 1 min. later by protein synthesis. The inhibition of RNA, protein and enzyme synthesis by actinomycin D throughout the period of outgrowth, and the finding that protein synthesis is proportional to the amount of labile RNA, indicate that spores are devoid of a stable, functional mRNA and that protein synthesis during outgrowth is dependent upon new transcription. Cell-free amino acid incorporating systems prepared from dormant spores lack basal amino acid incorporating activity and possess only limited capacity when supplemented with exogenous mRNA. The low activity was attributed to unstable ribosomes in the dormant spore. Ribosomes are repaired during heat-activation and germination in the absence of macromolecular synthesis.

The apparent dependence of the developmental changes in outgrowth on RNA transcription permits an examination of the question as to whether the proteins synthesized during this period result from an ordered transcription and translation of the genome or from an ordered use of randomly produced transcriptional products. Electrophoretic and chromatographic separation of soluble proteins labelled with radioactive precursors indicate that different classes of proteins are made at different stages in outgrowth. During this period the pattern of enzyme synthesis is also varied with respect to time. The syntheses of α-glucosidase, L-alanine dehydrogenase and histidase are ordered events; each begins at a specific time and synthesis continues for only a brief period. The timing of induced α-glucosidase and histidase has been examined in order to determine if specific regions of the genome are continuously available for transcription and regulation after germination. Enzyme synthesis can be induced during outgrowth, but the period of induction is restricted to the same time interval as that found for uninduced and catabolically repressed cultures. Since mRNA formed during outgrowth had a half-life of only a few minutes, it is concluded that ordered enzyme synthesis is a result of ordered transcription and translation of the genome.

Experiments were carried out to determine if DNA replication served as the mechanism by which ordered transcription was controlled. During outgrowth, significant synthesis of DNA does not occur until just prior to the onset of cell division. However, incorporation of radioactive thymidine into DNA is observed within 5–10 min. after the initiation of germination. This incorporation appears to be a result of DNA replication and not repair synthesis. In the early stages of outgrowth, ordered enzyme synthesis is not dependent upon the replication of the genome, since thymidine addition to outgrowing spores of a thymidine auxotroph does not alter the periodicity of induced α-glucosidase and histidase. It appears likely that some mechanism which alters the capacity of DNA to serve as a template for RNA transcription may be involved in the ordering process.

REFERENCES

ALPERS, D. H. & TOMKINS, G. M. (1966). *J. biol. Chem.* **241**, 4434.
ARMSTRONG, R. L. & SUEOKA, N. (1968). *Proc. natn. Acad. Sci. U.S.A.* **59**, 153.
BAKER, R. F. & YANOFSKY, C. (1968). *Proc. natn. Acad. Sci. U.S.A.* **60**, 313.
BALASSA, G. (1963). *Biochim. biophys. Acta* **72**, 497.
BELL, E., HUMPHREYS, T., SLAYTER, H. S. & HALL, C. E. (1965). *Science, N.Y.* **148**, 1739.

BERBERICH, M. A., KOVACH, J. S. & GOLDBERGER, R. F. (1967). *Proc. natn. Acad. Sci. U.S.A.* **57**, 1857.
BREMER, H., KONRAD, M. & BRUENER, R. (1966). *J. molec. Biol.* **16**, 104.
CHAMBERLIN, M. & BERG, P. (1964). *J. molec. Biol.* **8**, 297.
CHURCH, B. D. & HALVORSON, H. (1957). *J. Bact.* **73**, 470.
CLINE, A. L. & BOCK, R. M. (1966). *Cold Spring Harb. Symp. quant. Biol.* **31**, 321.
CURRAN, H. R. & EVANS, F. R. (1945). *J. Bact.* **49**, 335.
DEMAIN, A. L. & NEWKIRK, J. F. (1960). *J. Bact.* **79**, 783.
DESROSIER, N. W. & HEILIGMAN, F. (1956). *Fd Res.* **21**, 54.
DOI, R. H. & IGARASHI, R. T. (1964*a*). *Proc. natn. Acad. Sci. U.S.A.* **52**, 755.
DOI, R. H. & IGARASHI, R. T. (1964*b*). *J. Bact.* **87**, 323.
DONNELLAN, J. E., Jr., NAGS, E. H. & LEVINSON, H. S. (1965). In *Spores*, vol. III. Ed. L. L. Campbell and H. O. Halvorson. Ann Arbor: American Society for Microbiology.
DUBNAU, D., SMITH, I. & MARMUR, J. (1965). *Proc. natn. Acad. Sci. U.S.A.* **54**, 724.
FALASCHI, A., SPUDICH, J. & KORNBERG, A. (1965). In *Spores*, vol. III. Ed. L. L. Campbell and H. O. Halvorson. Ann Arbor: American Society for Microbiology.
FITZ-JAMES, P. C. (1955*a*). *Can. J. Microbiol.* **1**, 525.
FITZ-JAMES, P. C. (1955*b*). *Can. J. Microbiol.* **1**, 502.
GEIDUSCHEK, E. P., SNYDER, L., COLVILL, A. J. E. & SARNAT, M. (1966). *J. molec. Biol.* **19**, 541.
GERHARDT, P. & BLACK, S. H. (1961). *J. Bact.* **82**, 750.
GRIFFIN, M. J. & COX, R. P. (1966). *Proc. natn. Acad. Sci. U.S.A.* **56**, 946.
GROBSTEIN, C. (1963). In *Cytodifferentiation and Macromolecular Synthesis*, ed. M. Locke. New York: Academic Press.
HALMANN, M. & KEYNAN, A. (1962). *J. Bact.* **84**, 1187.
HALVORSON, H. O., GORMAN, J., TAURO, P., EPSTEIN, R. & LABERGE, M. (1964). *Fedn Proc. Fedn Am. Socs. exp. Biol.* **23**, 1002.
HALVORSON, H. O., VARY, J. C. & STEINBERG, W. (1966). *A. Rev. Microbiol.* **20**, 169.
HARRELL, W. K. & MANTINI, E. (1957). *Can. J. Microbiol.* **3**, 735.
HARTMANN, K. V. & HEIDELBERGER, C. (1961). *J. biol. Chem.* **236**, 3006.
HARTWELL, L. H. & MAGASANIK, B. (1963). *J. molec. Biol.* **7**, 401.
HENNEY, H. R. & STORCK, R. (1964). *Proc. natn. Acad. Sci. U.S.A.* **51**, 1050.
HEWITT, R., SUIT, J. C. & BILLEN, D. (1967). *J. Bact.* **93**, 86.
HORIKOSHI, K., OHTAKA, Y. & IKEDA, Y. (1965). *J. agric. Chem. Soc. Japan* **29**, 724.
HOYEM, T., RODENBERG, S., DOUTHIT, H. A. & HALVORSON, H. O. (1968). *Archs Biochem. Biophys.* **125**, 964.
IMAMOTO, F. (1968). *Proc. natn. Acad. Sci. U.S.A.* **60**, 305.
KAWATA, T., INOUE, T. & TAKAGI, A. (1963). *Jap. J. Microbiol.* **7**, 23.
KERJEAN, P., MARCHETTI, J. & SZULMAJSTER, J. (1967). *Bull. Soc. Chim. biol.* **49**, 1139.
KERJEAN, P. & SZULMAJSTER, J. (1966). *C. r. hebd. Séanc. Acad. Sci., Paris* **262**, 312.
KEYNAN, A. & HALVORSON, H. (1965). In *Spores*, vol. III. Ed. L. L. Campbell and H. O. Halvorson. Ann Arbor: American Society for Microbiology.
KOBAYASHI, Y. & HALVORSON, H. O. (1968). *Archs Biochem. Biophys.* **123**, 622.
KOBAYASHI, Y., STEINBERG, W., HIGA, A., HALVORSON, H. O. & LEVINTHAL, C. (1965). In *Spores*, vol. III. Ed. L. L. Campbell and H. O. Halvorson. Ann Arbor: American Society for Microbiology.

Kornberg, A., Spudich, J. A., Nelson, D. L. & Deutschar, M. P. (1968). *A. Rev. Biochem.* **37**, 51.
Krishna-Murty, G. G. (1957). In *Spores*, ed. H. O. Halvorson. Washington: American Institute for Biological Science.
Kuempel, P. L., Masters, M. & Pardee, A. B. (1965). *Biochem. biophys. Res. Commun.* **18**, 858.
Levinson, H. S. & Hyatt, M. T. (1963). *Ann. N.Y. Acad. Sci.* **102**, 773.
Levinson, H. S. & Hyatt, M. T. (1966). *J. Bact.* **91**, 1811.
Levinthal, C., Keynan, A. & Higa, A. (1962). *Proc. natn. Acad. Sci. U.S.A.* **48**, 1631.
MacKechnie, I. & Hanson, R. S. (1968). *J. Bact.* **95**, 355.
Mahler, I. (1966). *Biochem. biophys. Res. Commun.* **25**, 73.
Maitra, U. & Hurwitz, J. (1965). *Proc. natn. Acad. Sci. U.S.A.* **54**, 815.
Marcus, A. & Feeley, J. (1966). *Proc. natn. Acad. Sci. U.S.A.* **56**, 1770.
Nakata, H. M. (1964). *J. Bact.* **88**, 1522.
Oishi, M. & Sueoka, N. (1965). *Proc. natn. Acad. Sci. U.S.A.* **54**, 483.
Oishi, M., Yoshikawa, H. & Sueoka, N. (1964). *Nature, Lond.* **204**, 1069.
Pardee, A. B. (1965). In *Control of Energy Metabolism*, ed. C. Britton, R. W. Estabrook and J. R. Williamson. New York: Academic Press.
Pettijohn, D. & Hanawalt, P. (1964). *J. molec. Biol.* **9**, 395.
Powell, J. F. & Strange, R. E. (1953). *Biochem. J.* **54**, 205.
Rodenberg, S., Steinberg, W., Piper, R. J., Nickerson, K., Vary, J., Epstein, R. & Halvorson, H. O. (1968). *J. Bact.* **96**, 492.
Smith, I., Dubnau, D., Morell, P. & Marmur, J. (1968). *J. molec. Biol.* **33**, 123.
Stafford, R. S. & Donnellan, J. E. Jr. (1968). *Proc. natn. Acad. Sci. U.S.A.* **59**, 822.
Steinberg, W. (1967). Ph.D. thesis, University of Wisconsin.
Steinberg, W. & Halvorson, H. O. (1968*a*). *J. Bact.* **95**, 469.
Steinberg, W. & Halvorson, H. O. (1968*b*). *J. Bact.* **95**, 479.
Steinberg, W., Halvorson, H. O., Keynan, A. & Weinberg, E. (1965). *Nature, Lond.* **208**, 710.
Stuy, J. (1956). *Antonie van Leeuwenhoek* **22**, 337.
Sussman, A. S. & Halvorson, H. O. (1966). *Spores, Their Dormancy and Germination.* New York: Harper and Row.
Szybalski, W. & Iyer, V. N. (1964). *Fedn Proc. Fedn Am. Socs. exp. Biol.* **23**, 946.
Terman, S. A. & Gross, P. R. (1965). *Biochem. biophys. Res. Commun.* **21**, 595.
Torriani, A. & Levinthal, C. (1967). *J. Bact.* **94**, 176.
Treadwell, P. E., Jann, G. J. & Salle, A. J. (1958). *J. Bact.* **76**, 549.
Vary, J. C. & Halvorson, H. O. (1965). *J. Bact.* **89**, 1340.
Vinter, V. & Slepecky, R. (1965). *J. Bact.* **90**, 803.
Wake, R. G. (1967). *J. molec. Biol.* **25**, 217.
Wilson, M. C., Farmer, J. L. & Rothman, F. (1966). *J. Bact.* **92**, 186.
Woese, C. R. (1961). *J. Bact.* **82**, 695.
Woese, C. R. & Forro, J. R. (1960). *J. Bact.* **80**, 811.
Woese, C. R., Vary, J. C. & Halvorson, H. O. (1968). *Proc. natn. Acad. Sci. U.S.A.* **59**, 869.
Yoshikawa, H. (1965). *Proc. natn. Acad. Sci. U.S.A.* **53**, 1476.
Yoshikawa, H., O'Sullivan, A. & Sueoka, N. (1964). *Proc. natn. Acad. Sci. U.S.A.* **52**, 973.

THE BIOCHEMISTRY OF AMOEBIC ENCYSTMENT

By R. J. NEFF and R. H. NEFF

Department of Molecular Biology, Vanderbilt University,
Nashville, Tennessee 37203

INTRODUCTION

The biological process to be considered in this paper is the encystment of a soil amoeba, *Acanthamoeba* sp. Encystment is the process whereby an active vegetative cell is transformed into a dormant cell encased in a wall of visible microscopic thickness.

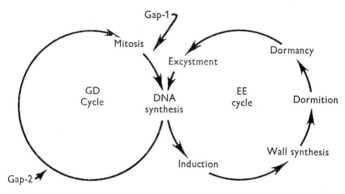

Fig. 1. The growth–division cycle and the encystment–excystment cycle in *Acanthamoeba* sp.

The cyst is truly dormant as evidenced by the fact that respiration of mature cysts a week or more old is below the limits of measurement by Warburg respirometry. This is in contrast to vigorous respiration of equal numbers of vegetative cells (Neff, Neff & Taylor, 1958). Other life functions which we normally associate with living cells, such as growth, division, and movement of cellular organelles, are also absent. The mature cyst wall is tough and it completely resists usual homogenization techniques. Mature cysts survive years of drying-out in shelf cultures and also survive vacuum desiccation.

Encystment is the only specialized function this amoeba can perform. The relationship between the growth–division cycle (GDC) and the encystment–excystment cycle (EEC) as we interpret it is illustrated in Fig. 1.

Encystment bears a striking resemblance to certain aspects of embryonic differentiation. Two such properties common to differentiating cells are seen from a perusal of the encystment part of the diagram in Fig. 1 and are that cells leave the GDC and do not divide and that there is a distinct period of induction closely followed by a period of sequential structural change resulting in a new cell type. Two additional properties associated with known differentiating systems as well as encystment are: new types of macromolecules are present in the transformed cell—the macromolecules of the cyst wall in this case; and the process is blocked by inhibitors of RNA and protein synthesis. Encystment has been considered to be a single cell differentiation by other workers as well (Trager, 1963; Willmer, 1963; Sussman, 1965; Wright, 1966; Bullough, 1967).

Encystment has been demonstrated to be under nuclear control (Neff & Neff, unpublished). Single cells were cut in half freehand or with the aid of a micromanipulator. Cutting was performed on cells in droplets of a saline which supported 50–60% encystment of mass cultures. The sister halves, one with a nucleus and one without, were observed for several days. Approximately 60% of the nucleated halves encysted while *none* of the enucleated halves formed cyst walls. We can therefore conclude that encystment is a single cell differentiation which is under nuclear control and whose end-product is a dormant cell. The mature cyst with its tough wall and its dormant cell inside is an ideal system for surviving long periods of time, perhaps even in space.

The objectives in writing this paper were twofold. The first was to pull together data on encystment in order to determine what is presently known and to provide a basis for deciding what might be investigated profitably in the future. The second was to present recent data on the induction phase of encystment and from it propose a hypothesis as to the mechanism of induction. In what follows we shall attempt to examine the conditions favouring encystment, the structural changes and the timing in the synchronous encystment, the chemistry of the mature cyst, the sequential biochemical events during wall synthesis, and the mechanism of the induction process.

ENVIRONMENTAL CONDITIONS FAVOURING ENCYSTMENT

It has long been known that small free-living soil amoebae encyst when confronted with 'adverse' environmental factors such as starvation, desiccation, temperature elevation, accumulation of metabolic wastes and overcrowding. We found that ageing cells spontaneously encysted in either stationary or aerated cultures.

Our attack upon the differentiation problem was facilitated by the accidental observation that growing and dividing amoebae promptly encyst if they are transferred from growth medium (GM) to an inorganic salt solution (EM). Indeed, it was found that a *single* isolated cell can encyst in EM, provided that it is protected from drying.

These two observations lead to three generalizations about encystment. First, encystment of soil amoebae is induced by the *depletion* or *absence* of some factor from the medium. The process behaves as a system under negative feedback control, i.e. the absence of an inhibiting factor 'turns on' the encystment process. Secondly, the induction and completion of encystment is endogenous; it occurs in the total absence of externally supplied energy sources or organic metabolites. The entire cyst wall can be synthesized from pre-existing constituents of the normal vegetative cell. Thirdly, encystment is a self-contained, single-cell differentiation. An encysting cell requires no stimulus or contribution from any other cell to initiate and complete its encystment.

We have developed a technique for the synchronous induction and differentiation of a mass culture of *Acanthamoeba* sp. (Neff, Ray, Benton & Wilborn, 1964). The environmental conditions favouring a highly synchronized differentiation in a large percentage of the population are discussed at length in that publication. The most favourable conditions are summarized in Table 1. We have made no improvement on our basic procedure; it is used whenever large numbers of encysting cells of known age are required.

Table 1. *Requirements and optimal conditions for synchronized encystment in* Acanthamoeba *sp.*

Factor	Optimum response observed when:
1. Cells	The GT is 8 hr. or less, and cells are harvested at stationary phase after growth on full medium
2. Encystment medium (EM)	The EM is a buffered, isotonic salt solution, 0·1 M in either KCl or NaCl; $Mg^{2+}:Ca^{2+}$ = 20:1 when Mg^{2+} is 0·01 M
3. pH	The pH is 7·0 during induction and 8·9 during wall synthesis
4. Temperature	The temperature is 29–30° C.
5. Oxygen	The EM is saturated with air. Oxygen is an absolute requirement. The actual rate of aeration depends on the geometry and the volume of the container

STRUCTURAL CHANGES AND TIMING IN SYNCHRONIZED ENCYSTMENT

Our current concept of the *Acanthamoeba* EEC is represented in Fig. 2. Here we have opened out the cycle and have shown the succeeding stages as segments of a time sequence reading from left to right. We have, whenever possible, indicated the duration of each stage. We have made no attempt to list all events, processes or syntheses occurring in each stage, but have listed only the marker events by which each stage can be defined.

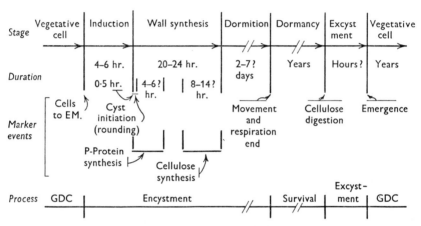

Fig. 2. The encystment–excystment cycle in *Acanthamoeba* sp. GDC = growth division cycle. → = time moves from left to right.

Much of the proposed scheme is conjectural; the dormition stage and dormant stage are almost completely uncharacterized, in spite of their inherent interest as examples of 'turned-off' cells. Our knowledge of excystment is qualitative only, based upon casual observations. Although it is widely assumed that excystment is a reversed encystment, there is no evidence that this is true.

During the past decade we have focused our attention upon the first two stages of the EEC, primarily because of our interest in studying a cellular control mechanism which can determine whether a given cell shall grow and divide or encyst and remain dormant.

We have used our technique for inducing a mass culture of growing cells to encyst in synchrony to study the morphological and biochemical changes occurring in the differentiating cells. The morphological changes typical of the first 24–36 hr. have been earlier described (Neff *et al.* 1964). In the past 3 years our increased understanding of the differentiation has compelled us to revise our terminology, as shown in Fig. 2. We propose that the entire encystment process (i.e. from vegetative cell to dormant cell) be divided into three stages. The first 4–6 hr. after cells are transferred from GM to EM (or inhibitor is added) is now the induction stage; it was formerly known as pre-encystment. The second period constitutes the wall-synthesis stage. It lasts from 20 to 24 hr. and is subdivided into two phases. The first phase is characterized by secretion of the outer cyst wall. After a short pause a second phase begins when the inner layer of the cyst wall is added, and sculpturing is completed. The third stage, the *dormition stage*, is of unknown duration but is believed to occupy between 2 and 7 days. Finally, we now restrict the term *cyst initiation* to the 30 min. period when vegetative cells lose their motility and assume a permanently spherical form.

The intracellular structural changes accompanying encystment are summarized in Table 2. It is seen that the cellular mitochondria decrease in size and number during encystment and that the density of their matrix material increases. Lamellate intracristal bodies form. Vickerman (1962) has suggested that the decreased mitochondrial volume and surface may be related to the decreasing respiration during encystment and the negligible respiration of the mature cysts. This is in agreement with Klein & Neff's report (1960) that the respiration of isolated *Acanthamoeba* mitochondria is reduced when their volume is decreased osmotically.

Also noteworthy is the report by Bowers & Korn (1967) of an increase in the activity of a Golgi system in the encysting cell. They describe the accumulation of dense material in the cisternae, which is pinched off into vesicles, carried to the surface and extruded to the outside. We cannot determine *when* the Golgi becomes active or to what layer of wall the dense material is contributed, because the authors did not report the age of the cells in which the Golgi activity was found.

Bowers & Korn (1967) also describe the appearance of autolysosomes 'early' in encystment. This provides a morphological confirmation at the EM level of our earlier conclusions (based upon biochemical measurements) that cellular organelles and storage materials are degraded early in cyst induction to provide raw materials and energy supplies for the differentiating cells. The possible significance of the observation is that lysosomal activation may be the first step in induction; that is, a gene activation, similar to the case reported by Weissmann, Troll, Brittenger & Hirschhorn (1967), who demonstrated that lysosomal breakdown follows phytohamagglutinin activation of lymphocytes *in vitro*. A vigorous nuclear RNA synthesis follows. Autolysosomal debris is incorporated into the cyst wall.

We are unable to confirm Volkonsky's report (1931) that chromatin

Table 2. Changes in intracellular structures of Acanthamoeba sp. during encystment

Structure or property	Vegetative stage	Induction stage	Wall-synthesis stage	Mature cyst	Ref.
		Microscopy of living cells: phase-contrast			
Behaviour	Active; motile, feed, pinocytose, etc.	No change	Round up; form clumps	Spherical cysts; clumps separate	1
Cytoplasmic churning	Vigorous	No change	No change	Gradually diminishes and ceases	1
Nucleus	Usually mononucleate; may be polynucleate	No change	No change	(Optics unfavourable?)	1
		Microscopy of living cells: interference			
Nucleolus	Large; centrally located in nucleus	Mass and volume constant, or increase slightly	Mass and volume decrease gradually	May be absent	2
		Cytochemistry: standard microscopy of stained cells			
DNA stain (Feulgen)	Diffuse chromatin in peripheral area of interphase nucleus	No change	Chromatin migrates to cyst wall and is added to it	—	3
DNA stain (Feulgen)	Diffuse chromatin in peripheral area of interphase nucleus	No change	No change	In majority of cysts no change. In 4–5 % of cysts; nuclear reorganizational forms, many small nuclei	1
Cytoplasmic basophilia	Intense	Decreases gradually	Decreases gradually	—	1

		Electron microscopy		
Cytoplasm (general)	—	Decrease in size; food residues discarded; 'active dehydration' by formation and extrusion of 'water' vacuoles	—	4
		Decrease in size; total cytoplasmic density increases		5
Mitochondria	0·5–1·5 μ long by 0·5–1·0 μ thick; dense matrix, cristae; may (rarely) see intracristal bodies	—	Decrease in size; intracristal bodies increase in size	4
	Tubular cristae embedded in a dense matrix	A lamellated intracristal body forms; it may be expelled into cytoplasm, or retained in mitochondrion	—	5
Endoplasmic reticulum	Flattened vesicles and tubuli, 200 Å apart; agranular	No change	No change	5
Golgi	—	Amount of Golgi increases; a dense material is secreted into vesicles which migrate to cell surface and contribute dense material to the wall	—	5
Autolysosomes	—	Form 'early' in encystment; autolysed cell organelle debris may be trapped in the cyst wall	—	5
Nucleus	—	Decreases in size	Refractile bodies appear on nuclear membrane	4

1, Neff (unpub.); 2, Park (1965); 3, Volkonsky (1931); 4, Vickerman (1962); 5, Bowers & Korn (1967).

dispersed during encystment migrated to the cell periphery and was added to the cyst wall. We found no change in cell nuclei during the induction and wall synthesis stages, but did observe a nuclear rearrangement in a small portion (i.e. no more than 4–5%) of the population during dormition. In these cells, chromatin is redistributed into many (up to 30) tiny nuclei, each about 1–2 μ in diameter and containing a nucleolus. The significance of this nuclear fragmentation is not known. No sexual process has been observed in *Acanthamoeba*; we do not believe them to be reproductive cells. The process resembles merozooite formation in *Plasmodium*.

Table 3 summarizes changes occurring at the cell surface during encystment. It is gratifying to observe that the cytochemical analyses of the cyst walls and the recent electron-microscope studies are consistent with our previous conclusions on the nature of the cyst wall, based upon phase-contrast microscopy and biochemical analyses.

For example, the outer cyst wall can first be stained with alcian blue when the cell rounds up at cyst initiation; similarly, the ninhydrin–Schiff reaction for protein also becomes strong in the same outer layer just after cyst initiation. Although the Alcian blue stain is most frequently used to localize acid mucopolysaccharides, the high phosphorus content of the outer wall (see below) suggests that in *Acanthamoeba* the alcian blue stain is due to a reaction with phosphate or phosphonate linkages in a phosphoprotein. We conclude that the encysting cell rounds up *because* the first outer wall substance is secreted. This substance could account for the large negative surface charge of cysts (R. N. Band, personal communication). Both the zinc-chloro-iodide stain for cellulose and the periodic acid–Schiff (PAS) test for carbohydrate become positive only when secretion of the inner cyst wall begins. These observations certainly indicate the beginning of cellulose secretion. It should be noted that neither of the carbohydrate stains is consistent or permanent unless it is preceded by a lipid extraction. This fact suggests that the bound lipids reported in the chemical analyses may be complexed in some way to either cellulose or phosphoprotein layers.

Finally, the cyst wall structure as described by the electron microscopists is entirely consistent with earlier conclusions about the structure of the cyst wall. The significance or function of the 'food debris' of Vickerman or the 'autolysosomal debris' of Bowers and Korn in wall structure remain to be investigated.

Table 3. *Changes in the cell surface and formation of extracellular layers during encystment of Acanthamoeba sp.*

Structure or property	Vegetative stage	Induction stage	Wall-synthesis stage		Mature cyst	Ref.
			Outer layer	Inner layer		
			Phase-contrast microscopy			
Appearance	No extracellular surface coat, slime layer, or test; pseudopodia	Same; cells round up at end of induction stage	Thin extracellular layer secreted	Thick extracellular layer secreted inside outer layer	Double cyst wall ($1-1.5\mu$), thick with typical corrugated surface; pores or mamillae	1, 2
Stickiness	None	None	Sticky; cells clump	Sticky; cells clump	None; clumps disperse	1
			Electrophoresis			
Charge	Small negative charge	—	—	—	Large negative charge	3
		Cytochemistry: reaction to each of these staining procedures				
Alcian blue*	—	—	+	+	Outer layer, + Inner layer, —	4
Ninhydrin–Schiff†, **	—	—	+	+	Outer layer, + Inner layer, —	4
Periodic acid–Schiff‡, **	—	—	—	—	Outer layer, — Inner layer, +	4
Zinc-chloro-iodide§, **	—	—	—	—	Outer layer, — Inner layer, +	4
Molybdate‖	—	—	+	+	Outer layer, + Inner layer, —	4
			Electron microscopy			
Fine structure	Unit membrane: no extracellular material	—	.	.	Two wall layers: each consists of multilayered sheets or fibres, about 100 Å, thick, arranged parallel to cell surface: space between layers may contain 'debris' or 'water vacuoles'; layers join at pores	65,

References: 1, Neff (unpublished observations); 2, Volkonsky (1931); 3, Band (1963 and personal communication); 4, Neff and Benton (unpublished observations); 5, Bowers & Korn (1967); 6, Vickerman (1962).
* Method of Lison (1954). † Method of Yasuma & Ichikawa (1953). ‡ Hotchkiss (1948); McManus (1948). § Jensen (1962). ‖ Thomson (1960).
** After lipid extraction; stains inconsistent if lipids are present.

BIOCHEMICAL STUDIES

The chemical composition of the mature cyst wall

A mature cyst contains two end-products of differentiation: a dormant cell and a cyst wall. We believe the wall to be the first product completed, while the dormant cell is the final one. In our work on differentiation in *Acanthamoeba* we have analysed the cyst wall in the belief that if we can define the end-product we will know which materials and/or processes to seek in earlier stages. For the cyst wall this approach has proved fruitful. The dormant cell has received less attention.

Analyses of the mature cyst wall were made possible by Tomlinson's (1962) development of a successful method for the isolation of cyst walls from mature cysts in high yield and purity. Analyses have been performed on walls prepared by this method. The composition of isolated mature cyst walls is summarized in Table 4. Cellulose was the first major constituent to be identified (Tomlinson & Jones, 1962). They reported that one-third of the isolated dry wall was insoluble in hot alkali; the residual material was identified as cellulose on the basis of its solubility, hydrolysis products, infrared spectrum of its acetylated derivative, resistance to alpha and beta amylases and by the absence of nitrogen and sulphur. Linkage studies were not performed; they should be completed before this fraction is unequivocally identified as cellulose. These authors reported that the cyst wall accounted for about 37% of the dry weight of the mature cyst.

Table 4. *The chemical composition of isolated mature cyst walls of* Acanthamoeba *sp.*

Constituent	Dry weight (%)	Ref.
Cellulose	33·3	Tomlinson & Jones (1962)
Total carbohydrate*	35·0	
Total lipid†	4·5–6·5	
Unbound lipid	1·5–2·5	
Bound lipid	3–4	
Protein‡	33·0	Neff, Benton & Neff (1964)
Nitrogen§	6·25	
Ash	7·8	
Phosphorus||	2·1	
Total constituents identified	80–82·5	
Acid-insoluble residual protein	54–58	Neff, Neff, Benton & Hale (unpublished)

* Anthrone reaction, glucose standard (Seifter, Seymour, Novic & Muntwyler, 1950).
† Extracted in hot ethanol and chloroform–methanol. Unbound lipid extracted before cellulose removal; bound lipid extracted after cellulose removal.
‡ Lowry, Rosebrough, Farr & Randall (1951); albumin standard.
§ Micro-Kjeldahl procedure (Pregl & Grant, 1951).
\|\| Method of Fiske & Subbarrow (1925).

Finally, they identified glycine, phenylalanine, and tyrosine in hydrolysed whole walls.

Neff, Benton & Neff (1964) attempted a complete quantitative analysis of isolated cyst walls (Table 4). Their figure for total carbohydrate agrees with Tomlinson's measurement for cellulose. Protein accounted for 33% of the total wall material when determined by the method of Lowry et al. (1951) or for 40% when calculated from nitrogen determinations by the micro-Kjeldahl procedure. The phosphorus content is high; it can account for most of the ash if it is present as phosphate. Approximately 80% of the total wall components were accounted for.

More recently, Neff, Neff, Benton & Hale (unpublished) have found that cellulose can be selectively extracted from isolated cyst walls by treating them with concentrated sulphuric acid at 0° C. (Jayme & Lang, 1963; Saeman, Moore & Millett, 1963). Cellulose is solubilized, leaving an acid-insoluble residue behind. After it is washed in water, dried, and its bound lipids extracted, this residual fraction amounts to 54% of the original mass. The acid residue contains all of the nitrogen and phosphorus present in the total wall. The N:P ratio remains unchanged. When this acid-insoluble residue is hydrolysed for 24 hr. in 6 N-HCl under vacuum (method of Moore & Stein, 1963), the hydrolysate yields 17 amino acids. These have been identified by paper chromatography, and confirmed using an amino acid analyser. The quantitative results are shown in Table 5. In addition to these identified amino acids, amino acid-containing spots which also react with the molybdate reagent for phosphate were found; these spots have

Table 5. *Amino acids of the acid insoluble residue of the cyst wall of* Acanthamoeba *sp.*

Amino acid	μ moles/ 7 mg. sample	Ratio relative to Methionine
Lysine	3·41	15·5
Histidine	1·14	5·2
Ammonia	5·93	—
Arginine	1·45	6·6
Aspartic Acid	3·64	16·5
Threonine	5·84	26·5
Serine	2·01	9·1
Glutamic Acid	4·16	18·9
Proline	2·66	12·1
Glycine	2·68	12·1
Alanine	1·77	8·0
1/2 Cystine	2·49	11·3
Valine	1·36	6·2
Methionine	0·22	1·0
Isoleucine	1·00	4·5
Leucine	1·44	6·5
Tyrosine	0·77	3·5
Phenylalanine	1·00	4·5
Galactosamine	2·36	10·7

been rechromatographed but not yet identified. Finally, several unidentified u.v.-absorbing and fluorescing materials were found on the chromatograms.

These analyses show that the acid-insoluble component of the cyst wall is probably a phosphorus-containing protein which is unusual in its high threonine and phosphorus content and in the stability of the linkage(s) between phosphorus and amino acid. Most phosphoric acid esters are split rapidly under milder hydrolysis conditions than the ones here employed. The stability of the linkage in the wall phospho-protein suggests that a stable linkage is involved—perhaps a phosphonate. Either a phosphate or a phosphonate linkage could account for the alcian blue staining reaction of the outer wall, and for the strong negative charge of the mature cyst. It should be noted that none of the common wall components of other organisms, such as teichoic acid, hemicellulose, pectin or lignin, were found in cyst walls.

THE CHEMICAL COMPOSITION OF THE MATURE CYST CONTENTS

All of the major organelles found in vegetative cells are present in the cyst cell. It appears from the EM studies that the ground cytoplasm and mitochondrial matrix are more concentrated in the cyst than in vegetative cells. Similar observations on a variety of encysting systems have been reviewed by Van Wagtendonk (1955). Klein (1959) has studied the water content of vegetative cells and cysts. The water content of wet-packed cells was 84·4% for vegetative cells and 90·3% for mature cysts. These data are in conflict with microscopists' observations that the cyst is 'concentrated' or 'condensed'. However, the increased water content of the cysts may be more apparent than real because it included water occluded between the cysts, in addition to the water of encysted cell and wall. Since the increase recorded is slight, it may be largely due to occluded water. However, the interesting possibility should be considered that even if the dormant cell is 'condensed' the cyst wall—especially the inner layers—may be highly hydrated. Since about one-third of the mass of a dry cyst can be attibuted to its wall, a thick water-saturated cyst wall could easily account for the slightly increased water content of the cysts.

Klein (1959) has also compared the concentration of monovalent cations in vegetative cells and mature cysts. He reports that the concentration of Na^+ in mature cysts is only 30% of that in vegetative cells; for K^+ the concentration decreases to 10% of the vegetative cell. The ratio of $K^+:Na^+$ shifts from 1·77 in vegetative cells to 0·58 in mature cysts.

Biochemical changes during synchronous encystment

Three general types of changes have been followed during synchronous encystment; these are quantitative changes in the major cell components, changes in synthetic rates of major components, and variations in enzymic activities.

Changes in major cell components

The results of a series of experiments on the synchronized encystment system are summarized in Figs. 3–5. Measurements were made on aliquots of the same batch of encysting cells.

Figure 3 represents a differential count during the synchronous encystment. The percentage of vegetative cells, young cysts (rounded forms), mature cysts, and total cysts at successive time intervals are plotted. Total cell numbers remained constant throughout the experiment. The vertical lines at 7 and 16 hr. mark the time when young cysts, or cysts with detectable wall thickenings, reached a level of 10%. These times lag by more than 2 hr. behind the times at which cell rounding and wall thickening are first detected in individual cells of the population.

In Fig. 4 changes in major constituents during the first 24 hr. are normalized as the percentage of the maximum value. The maximum value is the initial value (i.e. the value at the time of placing cells in EM) for all except cellulose, which reaches its maximal value between the 24th and 36th hr. Excepting cellulose, *all* of the constituents decrease during encystment. The rate of decrease for proteins, lipid, DNA and RNA is greatest for a few hours beginning immediately after cyst initiation. We believe this period of rapid decline corresponds to the period of synthesis of the phosphorus-containing wall protein, and possibly to the time of extrusion of autolysosomal debris. The rate of decline of the constituents decreases shortly before or at the time of initiation of cellulose synthesis.

The loss of DNA deserves special comment. DNA was extracted from cells by the method of Schneider (1945) and assayed by the diphenylamine method (Dische, 1962b), reading at two wavelengths. The same pattern has been obtained by M. Loggins (thesis in preparation), using the method of Ogur & Rosen (1950) for the determination of DNA and RNA. The decrease of DNA thus appears to be real. There are several possible explanations of this result. Cells in aerated cultures show varying amounts of polynucleation. Cultures close to the stationary phase may have as many as 40% of their cells with two or more nuclei. The loss in DNA might be accounted for by the breakdown of superfluous nuclei. A second more likely possibility is that a mass destruction of mitochondria accounts

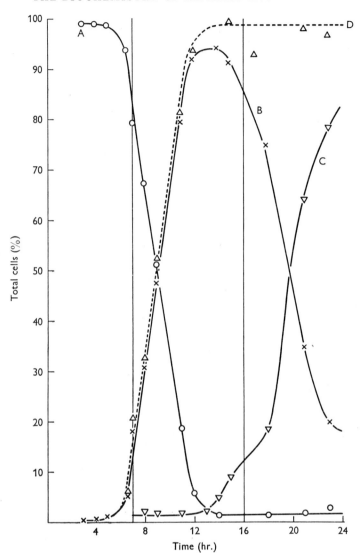

Fig. 3. A synchronized encystment of an *Acanthamoeba* culture in EM. The experiment begins when the cells are transferred from GM to EM. Total cell numbers remain constant. Symbols: A, vegetative cells; B, young cysts; C, mature cysts; D, total cysts.

for the DNA loss. Klein & Neff (1960) have found that mitochondria may account for 25% or more of the vegetative cell volume and some of these are destroyed during encystment. The amount of DNA per mitochondrion is not yet known, so that it cannot yet be decided whether the number of mitochondria lost can account for the DNA decrease. Thirdly, DNA loss may result from a destruction of redundant genes or of partially replicated

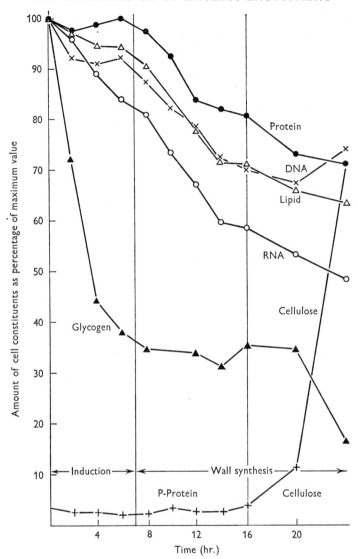

Fig. 4. Changes in the content of six cellular constituents during the synchronized encystment of *Acanthamoeba* in EM. Protein was assayed by the phenol method of Lowry, Rosebrough, Farr & Randall (1951). DNA and RNA were extracted by the Schneider (1945) procedure: DNA was assayed by the diphenylamine reaction (Dische, 1962b). RNA was assayed by the orcinol reaction (Dische, 1962a). Lipid was extracted and measured by the method of Enteman (1957). Glycogen and cellulose were assayed by the anthrone method (Seifter, Seymour, Novic & Muntwyler, 1950).

chromosomes. The requirement of *blocked* DNA synthesis for encystment, to be described in the following section, indicates that further study of this decrease in DNA may be rewarding.

The pattern of glycogen decrease is different from that of other components. Over 60% of the glycogen is lost during the induction stage. Glycogen is constant during outer-wall synthesis, and decreases again during cellulose synthesis when the glucose equivalents of glycogen that disappear are just balanced by the glucose equivalents of cellulose that are

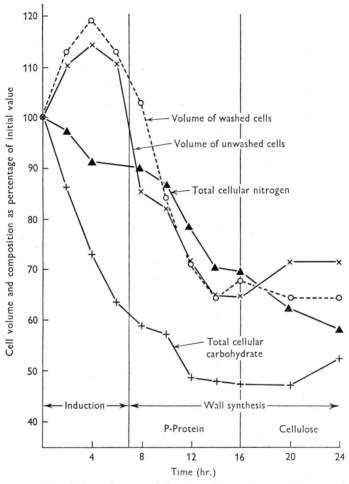

Fig. 5. Changes in cellular volume, and in two cell constituents during synchronized encystment of *Acanthamoeba* in EM. Nitrogen was determined by the micro-Kjeldahl procedure as outlined in Pregl & Grant (1951). Total carbohydrate was assayed by the anthrone method (Seifter *et al.* 1950).

formed, indicating that glycogen is the source of the glucose units of cellulose. The onset of the synthesis of cellulose coincides almost exactly with the visible beginning of wall thickening, and the cytochemical detection of cellulose in the inner wall layer. The reason for the large initial drop in glycogen remains a mystery. Glucose, as measured by the glucostat method, decreases progressively. There seems to be no causal relationship between the glycogen and cellulose changes and the glucose level.

If we compare the slopes of the curve showing the protein decline (Fig. 4) and the total cell nitrogen decline (Fig. 5) it can be seen that protein decreases at a slower rate than total nitrogen. In general, we have found that the protein loss accounts for about half the total nitrogen lost at any one time.

Approximately 80% of the nitrogenous materials lost from the cell were found in the supernatant encystment medium (Neff, Neff & Benton, 1964). The total amount recovered increased progressively. After 2 days it amounted to 40–50% of the original cell nitrogen. Of the supernatant nitrogen, 10% was in the form of peptides or proteins, 40% as ammonia, and the remaining 50% was amino acids and unidentified materials. Seven amino acids were identified in the supernatant. The discarding of amino acids is surprising in view of the concurrent synthesis of the protein-containing layer of the cyst wall; they are presumably surplus amino acids not used in wall protein.

The cell volume increases during induction (Fig. 5). Since cells remain in osmotic equilibrium, this indicates an increase in the total number of osmotically active molecules per cell. The increase could be due to a net inflow of monovalent inorganic ions from the medium, or to the rapid accumulation of small molecular weight degradation products. The total number of solute particles must decrease during the outer wall synthesis, since the entire compensating loss of excess volume occurs during this period. As soon as cellulose synthesis begins, the volume becomes constant or increases slightly before it stabilizes. A volume change of the cyst cell would hardly be detectable in any case, since the cellulose wall is unlikely to be osmotically extensible.

In general, it appears that the cyst-induction and wall-synthesis stages are primarily characterized by the massive breakdown of major vegetative-cell constituents. Cellulose synthesis proceeds under these otherwise erosive conditions. The outer-wall synthesis must occur as well, but was not assayed in these studies.

Changes in synthetic rates of major cell components

RNA and protein synthesis during the periods of cyst induction and wall synthesis have been investigated. Park (1965) supplied pulses of radioactively labelled uracil to synchronously encysting amoebae during the first 24 hr. of encystment. He found uracil was incorporated into RNA throughout the entire period tested. Loggins (unpublished) also found that labelled uridine and labelled cytidine were incorporated into cellular RNA for the first 24 hr. of encystment. She found that precursor incorporation into

RNA could be partially blocked by actinomycin D (AD). The inhibitor reduced the percentage of incorporation by about the same amount throughout the test period. It should be noted that incorporation was not completely blocked by AD even at a high inhibitor concentration of 100 μg./ml. The high concentration of AD required to affect RNA synthesis in *Acanthamoeba* is undoubtedly similar to the case reported by Mittermayer, Braun & Rusch (1965, 1966). These authors found that in *Physarum* it was necessary to raise the AD concentration to 150 μg./ml. in order to block 92% of the *in vivo* incorporation of uridine into RNA, whereas a much lower concentration, namely 0·1 μg./ml., blocked 50% of the RNA synthesis by isolated nuclei supplied with labelled ribonucleotide triphosphates. It is probable that in *Acanthamoeba*, as in *Physarum*, an uncommonly high concentration of AD is needed to block a substantial portion of the RNA synthesis *in vivo*, because of poor penetration of the inhibitor into the cell. We conclude that RNA synthesis occurs in encysting *Acanthamoeba* cells during the induction and wall-synthesis stages, that the *de novo* RNA synthesis is AD-sensitive, and therefore, in all likelihood, directed by DNA. It should be emphasized that the new RNA is formed during a period when the total cellular RNA is decreasing to about half its initial value.

Park (1965) also studied the uptake and incorporation of labelled leucine and valine into cellular protein in encysting amoebae. In pulse-labelling experiments he found a pattern of amino acid uptake and incorporation into protein qualitatively similar to the uptake and incorporation of labelled uracil into RNA. In a continuous labelling experiment he found a net increase of label incorporation into protein continuing to the tenth hour, a time when the encysting cells had completed cyst initiation and were well into wall synthesis. The label incorporated into protein decreased during the remainder of the experiment, indicating a secondary breakdown of the protein synthesized during the first 10 hr. Park also demonstrated in an radioautography experiment that the incorporation of labelled amino acids into cellular protein could be reduced by puromycin. Thus, it has been demonstrated that protein synthesis occurs during the first 24 hr. of encystment and that the newly synthesized protein may be subject to immediate breakdown.

The above data indicate that two special syntheses—that of new RNA and new protein—can occur in encysting *Acanthamoeba* in the face of a net decrease in total amounts of both these constituents, and in the face of a massive destruction of cell organelles and storage materials. It seems probable that these new syntheses represent a sequence of syntheses of DNA-directed mRNA's, and their related enzymes, in a pattern similar to that found in the sporulation of bacteria by Halvorson (1965).

Changes in enzyme activity

The changes in activity of several enzymes during the induction and wall-synthesis stages have been followed and are summarized in Table 6. In the first class are enzymes whose activities decrease progressively and follow curves similar to those described for RNA, lipid and protein (Fig. 4). In the second class are enzymes that increase in activity at some time during these two stages. Of these, isocitratase, UDPG pyrophosphorylase, and NAD-malic dehydrogenase decrease initially and then rise to peak activities at the times shown and then decrease again. The enzyme 'cellulose synthetase', studied by Tomlinson (1962), showed little or no activity until the 14th or 15th hr. of encystment. It then increased in activity parallel to the increased number of maturing cysts in the population. It would appear that the timing of the onset of cellulose synthesis, and the appearance of the terminal enzyme that synthesizes it, coincide.

Two mitochondrial enzymes of the electron transport chain—NADH-cytochrome-c-oxidoreductase and succinate-cytochrome-c-oxidoreductase—have been assayed by Hryniewiecka (1967) in vegetative cells and young cysts harvested from spent media after natural encystment. The activities per unit of mitochondrial protein of the $NADH_2$ and succinate enzymes were reduced to one-thirtieth and one-half respectively in the cyst. The endogenous oxygen reduction by the cyst mitochondria was reduced to

Table 6. *Changes in enzyme activities of* Acanthamoeba *sp. during encystment*

Enzymes and enzyme systems which continuously decline in activity during encystment	Ref.	Enzymes and enzyme systems which increase in activity or are modulated through a peak activity during encystment	Time of peak activity (hr.)	Ref.
Respiration*	1	Isocitratase	10	2
Glucose-1-phosphatase†	1	UDPG-pyrophosphorylase	15	2
Esterase‡	1	UDPG-cellulose trans-	—‖	2
Acid phosphatase§	1	glucosylase (= cellulose		
Malate synthetase	2	synthetase)		
NADP isocitric dehydrogenase	3	NAD-malic dehydrogenase	'1st wall cysts = 7?'	3
NADH-cytochrome-c-oxido-reductase	4			
Succinate-cytochrome-c-oxidoreductase	4			

References: 1, Neff & Ray (unpublished observations); 2, Tomlinson (1962); 3, Meyers & Jensen (1967); 4, Hryniewiecka (1967).
 * Method of Umbreit, Burris & Stauffer (1947).
 † Method of Swanson (1955). ‡ Method of Huggins & Lapides (1947).
 § Method of Gutman & Gutman (1940).
 ‖ A peak was not observed with this enzyme. It appeared at the 15th hr. and increased continuously to the 24th hr., when the measurements were ended.

zero, suggesting that not only do the enzymes per mitochondrion decrease, but that possibly their decrease is preceded by elimination of endogenous substrate.

We may summarize this section as follows. During at least the first two stages of encystment there is a gross decrease in all major cell components. In the face of this massive erosion of cellular constituents there is evidence for synthesis of RNA, of protein, and polysaccharide. Several enzymes decrease in activity in parallel with the major cell constituents, others show modulation in activity, presumably related to wall synthesis. In the case of the cellulose synthesizing enzyme, there is little doubt of the correlation of the increase in activity with the onset of cellulose synthesis.

These data, along with the knowledge that new RNA and protein are synthesized, show that *sequential* enzyme induction or depression is possible; indeed it seems to be the operating mechanism in this differentiating cell. By applying suitable concentrations of specific RNA and protein-synthesis inhibitors at intervals throughout encystment, it should now be possible to decide whether the changes in enzyme activities represent new syntheses or a change in activity of pre-existing enzyme molecules.

THE MECHANISM OF INDUCTION OF CYST FORMATION

In normal cultures cells begin to encyst 2 or 3 days after they enter the stationary phase of culture growth. The stationary phase is therefore the gateway to encystment. James (1966) has demonstrated that stationary-phase cells undergo unbalanced growth in that, on a per nucleus basis, the cell volume of stationary-phase cells became 5·45 times, the cell protein 5·32 times, and the cell RNA 2·7 times as great as that of exponential phase cells. The DNA content of stationary and exponential phase cells are, however, identical (Neff, unpublished). Unbalanced growth is not, however, an absolute prerequisite for encystment. In the synchronized encystment system, cells do not increase in mass, and their induction requires only 5–6 hr. The simultaneous depletion of all nutrients appears to speed the cell through induction.

Our interest in and our understanding of induction has been increased by the observation that certain inhibitors added to actively growing cultures in GM induce cells to encyst. We have identified several classes of inhibitors with regard to their ability to induce encystment. In decreasing order of effectiveness, they are: inhibitors of DNA replication; inhibitors of deoxynucleotide synthesis; and inhibitors of protein and RNA synthesis. Polyamines such as histone or poly-L-lysine induce no encystment.

The effects of mitomycin C on cyst induction

Mitomycin C is a typical inhibitor of DNA replication. The growth and encystment of *Acanthamoeba* as a function of mitomycin C concentration are shown in Fig. 6. Growth is completely inhibited at 20 μg./ml., or higher. We refer to the concentration which just halts population growth as the critical concentration. Encystment is already 40% at a level of 5 μg./ml., and has reached 99% by 20 μg./ml. Encystment remains near 100% through all concentrations up to 250 μg./ml., but decreases to values of 45% and 9% at 500 and 1000 μg./ml., respectively. Unbalanced growth of

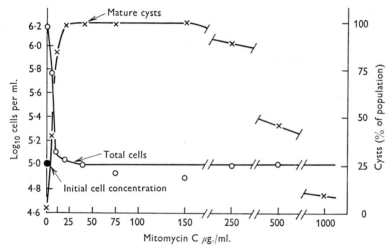

Fig. 6. The growth and encystment of *Acanthamoeba* in GM (at 72 hr.) as a function of mitomycin C concentration.

the encysting cells in the 5–250 μg./ml. range is extreme, amounting to at least 8 times the volume of the vegetative cells. Even at the two highest concentrations there is little cell destruction. Over a wide concentration range the degree encystment is independent of concentration.

The effects of the mitomycins on biological systems and the mechanism of action have been recently reviewed by Szybalski & Iyer (1967). In bacteria, mitomycin C blocks cell division. DNA synthesis is inhibited immediately but RNA and protein synthesis continue and cell enlargement occurs, resulting in unbalanced growth. In some bacteria, apparently those with a repair mechanism, DNA breakdown may subsequently result. In lysogenic bacteria, mitomycin may induce the lytic phase. The molecular mechanism inducing these effects is, *in vivo*, the cross-linking of complementary strands of DNA.

It is apparent from the data above that division blockage, unbalanced growth and encystment result from the application of mitomycin to growing cultures. We have found by radioautography that mitomycin blocks incorporation of [^{3}H]thymidine into nuclear DNA in the concentration range where encystment is induced. We conclude that unbalanced growth and encystment in *Acanthamoeba* is a result of blocked DNA synthesis. The basic molecular mechanism inducing unbalanced growth and encystment is very likely a cross-linking of complementary strands of DNA in *Acanthamoeba* as it is in bacteria. We suspect that there must be an additional or secondary effect, however, because *all* members of the amoeba population—i.e. cells in *all* stages of the GDC—are induced to differentiate. We propose that cross-linking of complementary DNA strands results in a conformational change in the chromosomes, so that encystment genes are exposed for sequential transcription. The encystment sequence is the same whether it is induced by ageing, by starvation, or by inhibitors. If the sequence is independent of the inducing agent, it follows that all agents affect the same initial cellular mechanism.

The effect of 5-fluoro-2'-deoxyuridine (FUdR) on cyst initiation

The inhibitor FUdR has been chosen as typical of the second group of inhibitor compounds—those which block deoxyribonucleotide synthesis. Growth and encystment of *Acanthamoeba* in GM as a function of FUdR concentration is shown in Fig. 7. The critical concentration is about 1 μM. Above this concentration all divisions are blocked after 40% of the population has divided once. Encystment is maximal close to the critical concentration. By 3 μM encystment is about 50% and it remains at this level at least to 1 mM—the highest concentration tested. Cells that encyst show unbalanced growth. Cells that divide once never divide again and never encyst. Instead they grow smaller—to one-tenth or less of their normal exponential phase volume—and eventually lyse. We refer to this behaviour as negative growth. Over a very wide concentration range the percentage of the population dividing and the percentage encysting are independent of inhibitor concentration.

The effect of FUdR on a variety of biological systems and the mechanism of its action have been reviewed by Heidelberger (1965). FUdR also blocks DNA synthesis and induces unbalanced growth in bacteria (Cohen *et al.* 1958) and in tissue culture cells (Lindner *et al.* 1963). In tissue culture cells the G-2 cells grow, divide and proceed through G-1 to the beginning of S and block in S or late G-1 in the absence of exogenous thymidine. Cells in S at the time of FUdR addition are blocked in S.

Thymidine 'reverses' the FUdR block. Thymidine is phosphorylated to thymidylate by thymidine kinase, and then, by two additional kinases, to thymidine triphosphate. The thymidine triphosphate is then incorporated into DNA by DNA polymerase. It is obvious from these considerations that thymidine does not 'reverse' FUdR; it *by-passes* the inhibitor instead.

The mechanism of FUdR action is by blocking the enzyme thymidylate synthetase. FUdR is phosphorylated to FUdR-P (the active form) by thymidine kinase. The combination of inhibitor to enzyme is very tight, and can be completely reversed only by a very high concentration of the normal substrate, deoxyuridine monophosphate. For most practical purposes the inhibition by FUdR is not reversible.

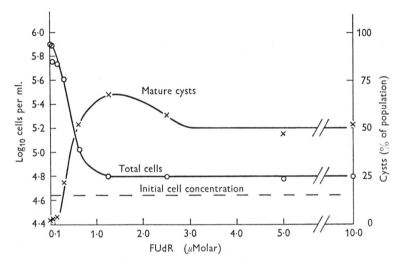

Fig. 7. The growth and encystment of *Acanthamoeba* in GM (at 96 hr.) as a function of FUdR concentration.

It appears that in the amoeba encystment system FUdR blocks cells in S and these cells undergo unbalanced growth and then encyst. G-2 cells divide, block in G-1 and undergo negative growth. This interpretation seems clear but there are complications.

We found that the response of the cell to FUdR depended upon the past history of that cell. The data presented in Fig. 7 were taken from an experiment using stationary-phase cells placed in fresh growth-medium when the inhibitor was added. With stationary phase cells we routinely found that, above the critical concentration, 30–50% of the population divided and 20–50% encysted.

With exponential-phase cells (cells which had been in exponential growth for at least four generations without being allowed to increase over 10^5 cells/ml.) we found that, above the critical concentration, 60–80% of the population divided but that only 0–5% encysted. For both stationary- and exponential-phase cultures the cells that encyst undergo unbalanced growth, and those that divide go into severe negative growth. The unsettling thing about these results is that the percentage of cells which can be induced to encyst differs so greatly, and is dependent upon the past history of the inoculum. If the hypothesis that S-phase cells are induced to encyst by FUdR is correct, then the extent of the S period must be different in stationary- and exponential-phase cells. This interpretation is obvious, and is true *provided* FUdR is specific for thymidylate synthetase and that it behaves in *Acanthamoeba* as it does in other cells. Let us examine our available data to determine if FUdR meets these criteria in our system.

First, FUdR *blocks* DNA synthesis in the concentration range which induces encystment. This has been shown by preincubating cells in FUdR for varying periods of time before adding [^{3}H]thymidine. Radioautographs of the treated cells show that the percentage of labelled nuclei is directly proportional to the time of preincubation of the cells in FUdR. It is clear that FUdR blocks DNA synthesis and traps cells in S and/or G-1.

Secondly, FUdR is *stable* in *Acanthamoeba*. There are several indications that this is true, but the most convicing one is that uridine does not potentiate the effect of FUdR. As summarized by Heidelberger (1965) in liver cells, and in some bacteria, FUdR is degraded by nucleoside phosphorylase to fluorouracil and deoxyribose. The phosphorylase is inhibited by uridine, and therefore maximal inhibition by FUdR is obtained when uridine is present. No potentiating effect of uridine on the FUdR inhibition of growth or the induction of encystment of *Acanthamoeba* has been observed at any of a number of combinations of FUdR and uridine concentrations.

Thirdly, FUdR is *specific* for the blocking of thymidylate synthesis by the amoeba's thymidylate synthetase. Thymidine was the only nucleoside of nine naturally occurring nucleosides tested that restored growth of a FUdR-inhibited culture (or had any effect) at a concentration 100-fold greater than that of the FUdR. Iododeoxyuridine (IUdR), used at a 100-fold excess, permitted about 1 division.

Fourthly, the high *sensitivity* of the thymidine synthetase to FUdR was demonstrated by S. Block (unpublished) in our laboratory, who showed that deoxyuridine reversed the FUdR-inhibition of growth and the induction of encystment only when added at a 20,000-fold molar excess of FUdR. Deoxyuridine monophosphate is the normal substrate for thymidylate synthetase.

Finally, low concentrations of thymidine *specifically prevent the induction of encystment* of stationary-phase cells by equimolar concentrations of

FUdR without permitting culture growth. The concentration of thymidine (about 10^{-5} M) is too low to support growth, but it permits cells that would have otherwise encysted to divide, and enter a negative phase of growth. This experiment also indicates that there is little or no available thymidine in GM.

Thus it is concluded that in *Acanthamoeba*, FUdR is *stable* and *blocks* DNA synthesis; that thymidylate synthetase is *sensitive* to FUdR; that FUdR is a *specific inhibitor* of the enzyme and that thymidine *prevents* encystment. These facts strengthen the argument that FUdR-induced encystment occurs only during the S period of the GDC. Additional support would be provided for this hypothesis if evidence could be cited that S is short in exponential phase cells and long in stationary-phase cells.

We have evidence that S is very short in exponential-phase cells from radioautographic studies with [^{3}H]thymidine. We find routinely that 2–4% of the population is in S. Using the Sisken (1964) method, we have found that the G-2 period is 75–80% of the GDC, and that S occurs during the first 20% of the cycle. This result correlates well with our FUdR data, where 60–80% of the population divides (G-2 cells) and 0–5% encyst (S cells).

FUdR studies with isolated single exponential-phase cells of known age in the GDC suggests that cells enter S when they are between 2 and 20% of the GDC in age. For this experiment a dividing cell was transferred aseptically to a drop of fresh medium under sterile mineral oil in a polystyrene dish. At intervals in the following GDC the medium around a sister pair was exchanged for fresh GM(controls) or for fresh GM containing 10^{-4} M FUdR. The time of division, if any, and the fate of each pair was determined by continuing microscopic observations. The result of this experiment is shown in Fig. 8. For FUdR-treated cells, no cell less than 2% of the GDC divided, but all cells older than 20% divided. In this experiment, no FUdR-treated cells encysted.

The extent of S and its position in the GDC of stationary-phase cells has not yet been measured radioautographically. However, single cell experiments like those above but using stationary-phase cells after they begin to divide in fresh GM have been begun in our laboratory by Frazier, whose preliminary results show that many of the cells in the *central* portion of the GDC encyst when GM is exchanged for GM containing FUdR. The suggestion is clear that in stationary-phase cells the S period is shifted toward a central position and is extended in length.

The results from an experiment using the amino acid analogue canavanine are presented as typical of the results obtained with inhibitors of protein synthesis. Growth and encystment in GM as a function of canava-

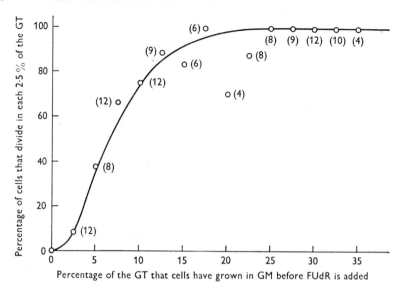

Fig. 8. The relation of a cell's age in the GDC to the FUdR inhibition of division in isolated sister pairs of *Acanthamoeba* sp. No cell that divides once in FUdR divides a second time; daughter cells undergo negative growth and lyse. All cells which have grown in GM for 22 % of the GDC or longer divide once. No cell in this experiment encysted. Conditions: FUdR concentration was 100 μM, in GM. Control cells, for determining average generation time, were 31 sister pairs. Experimental cells, treated with FUdR, were 60 pairs. Figures in parentheses indicate the number of cells tested in a given 2·5 % of the average GT (8·5 hr.).

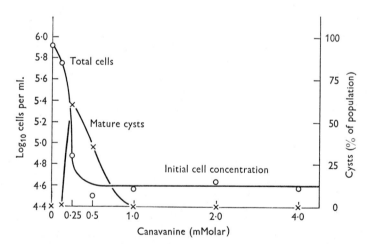

Fig. 9. The growth and encystment of *Acanthamoeba* in GM (at 108 hr.) as a function of canavanine concentration.

nine concentration are shown in Fig. 9. The critical concentration is 0·5 mM. Encystment is maximal at 0·25 mM; it is restricted to a limited concentration range close to the critical concentration. Cells that encyst undergo unbalanced growth. Above the critical concentration all cells fail to divide and undergo negative growth.

Canavanine is a structural analogue of arginine; its specific mechanism of action as an inhibitor of protein synthesis is not clear. We suggest that inhibitors of protein synthesis induce *Acanthamoeba* to encyst by preventing the synthesis of the so called 'initiator' proteins, or by producing abnormal or malfunctioning initiator proteins. Initiator proteins are proteins which must be synthesized by a growing cell before DNA replication can proceed. They have been demonstrated in bacteria (Maaløe & Hanawalt, 1961; Maaløe & Kjeldgaard, 1966) and in eucaryotes by Cummins & Rusch (1966). Rasmussen & Zeuthen (1966) have also demonstrated that a 'division protein' must be synthesized before division can occur in *Tetrahymena*. There is no evidence that such a protein fraction occurs in *Acanthamoeba*, but we believe a mechanism similar to this can best explain our data.

DISCUSSION

The data presented relate mainly to the first two stages of encystment, induction and wall synthesis. Little information is available on the last two stages—dormition and dormancy. In *Acanthamoeba* the dormant cell appears to be the opposite of a vegetative cell in all life functions. How these life functions are temporarily terminated during dormition and how cell cytoplasm and organelles survive intact during years of dormancy remain to be discovered.

Wall formation involves the sequential synthesis of two new classes of macromolecules. The first wall layer formed is chiefly protein, with a high content of tightly bound phosphorus. Its appearance at the outer surface of the cell constitutes the first outward sign that differentiation has begun; it is, in fact, the first differentiative product. The second product of differentiation that can be identified is an alkali-insoluble polysaccharide with the properties of cellulose. It is localized in the inner layer of the wall and its completion signals the end of wall synthesis. Cellulose imparts the toughness, rigidity and resistance to abrasion characteristic of the mature cyst.

In addition to the wall constituents RNA and protein are synthesized throughout the induction and wall synthesis stages, although the total amounts of both decrease. These special syntheses may continue into dormition. It is assumed that the RNA and protein syntheses are causally related to the formation of enzymes that synthesize the wall macromolecules. Several sequential increases in enzyme activities have been observed. In no case has an increase in enzyme activity been proven to be due to *de novo* synthesis. However, this seems likely because the differentiation is sensitive to AD as shown by Neff & Neff (unpublished). Since AD acts primarily by blocking DNA-directed RNA synthesis (Reich, 1966) it seems likely that new enzyme synthesis is involved. However, a demonstration of a specific enzyme synthesis at successive times during encystment will be required to prove this point.

Encystment is endogenous in that the massive breakdown of major cell constituents during induction and wall synthesis (and on into dormition?) can serve as the source of building materials for the wall and for energy-rich metabolites. Details of the breakdown, salvage and resynthesis pathways remain unknown.

Perhaps our greatest surprise in this study was that blocking of DNA synthesis induces encystment. In hindsight, we can now see that the 'normal' process of encystment involves an inhibition of DNA synthesis relative to RNA, protein and glycogen synthesis, so that a stationary-phase cell undergoes unbalanced growth. This unbalanced growth, when caused by inhibitors of DNA synthesis, also results in encystment. Unbalanced growth due to blocked DNA synthesis was first noted by Cohen & Barner (1954) for a thymine-requiring mutant of *Escherichia coli*. Identical growth effects on *E. coli* were later obtained with FUdR inhibition by Cohen *et al.* (1958). FUdR-induced unbalanced growth has been noted by Lindner (1963) in tissue-culture cells. Szybalski & Iyer (1967) have reviewed unbalanced growth patterns in bacteria treated with mitomycin.

A possible key to understanding natural induction and why DNA synthesis is blocked here comes from the work of James (1966), who studied the entrance of *Acanthamoeba* sp. into stationary phase and reported that RNA was synthesized proportionally less than protein. The proportionally smaller increase in RNA suggests a retrenchment in nucleotide synthetic pathways. It is a well-known fact (Scarano & Augusti-Tocco, 1967; Schmidt, 1966) that deoxynucleotides are synthesized from ribonucleotides; the deoxynucleotide level therefore depends upon an adequate pool of ribonucleotides. Therefore, if the ribonucleotide pool becomes depleted, the deoxyribonucleotides are also depleted. If deoxynucleotides are the first limiting factor in a growing culture, the progress of any cell through S will be slowed relative to its progress through G-2, M or G-1, and the proportion of the population in S will increase. In fact, if all exogenous precursors are removed (as they are when a population is transferred from GM to EM) one could visualize a whole stationary phase population stalled in S. Perhaps the 4–6 hr. required for induction in the synchronized encystment system is the time required for all cells to 'realize' that they are out of deoxynucleotides. The breakdown of DNA during induction and early wall synthesis may represent a salvage mechanism where a starving cell can provide enough nucleotides endogenously to permit cells to enter and stall in S in the synchronized culture. The relationship of the GDC to FUdR-induced encystment has already been considered.

The mitomycin data suggest that induction involves a conformational change in DNA which permits the DNA to be transcribed in a sequential fashion. We can only guess as to what the molecular mechanism might be, but we would expect it to be a change so that a specific mRNA could be formed. For example, the mechanism of action could be similar to the one described by Allfrey *et al.* (1966), who demonstrated that proteolytic digestion of isolated nuclei leads to a greater dispersal of chromatin and to a several-fold increase in RNA synthesis. Hydrolytic enzymes may be released early in induction, since lysosomes are active during induction and wall-synthesis stages (Bowers & Korn, 1967). Or, as Beermann (1966) has shown with dipteran salivary-gland chromosomes, as soon as the condensed bands have reached the diffuse or puffed state induced by a hormone, massive RNA synthesis occurs.

The mechanism of sequential reading of genes is no more known in *Acanthamoeba* than it is in other differentiating systems. However, the process follows the same sequence regardless of the mechanism of induction—whether by natural depletion of GM, by starvation in salt solution, or by mitomycin or FUdR inhibition in GM. Thus, our system is an excellent one for the continuing study of mechanisms of sequential gene induction in eucaryotic cells.

Much of the authors' work reported in this paper was supported by PHS-NIH research grant no. 06731. The remainder was performed during the tenure of a PHS-NIH Special Fellowship, no. 1-F3-GM-28567. The experiments with inhibitors were done at the Carlsberg Foundations' Biological Institute in Copenhagen, Denmark, and will be reported in detail in their publication. Figures 6–9 are from this report submitted to the Comptes Rendus des Travaux du Laboratoire Carlsberg. The helpful suggestions and criticisms of Dr Erik Zeuthen are gratefully acknowledged.

The authors also wish to acknowledge the skilful technical assistance of Mr Will F. Benton and Mrs Darlene Hale, and to thank Mr Stephen Block and Mr William Frazier for permission to cite their experiments.

REFERENCES

ALLFREY, V. G., POGO, B. G. T., POGO, A. O., KLEINSMITH, L. J. & MIRSKY, A. E. (1966). In *Histones, Their Role in the Transfer of Genetic Information*, ed. A. V. S. de Reuck and J. Knight. London: Ciba Foundation.
BAND, R. N. (1963). *J. Protozool.* **10**, 14.
BEERMANN, W. (1966). In *Cell Differentiation and Morphogenesis*, p. 24. Ed. W. Beermann et al. Amsterdam: North-Holland Publishing Co.
BOWERS, B. & KORN, E. D. (1967). *J. biophys. biochem. Cytol.* **35**, 15A.
BULLOUGH, W. S. (1967). *The Evolution of Differentiation*. New York: Academic Press.
COHEN, S. S. & BARNER, H. D. (1954). *Proc. natn. Acad. Sci. U.S.A.* **40**, 885.
COHEN, S. S., FLAKS, J. G., BARNER, H. D., LOEB, M. R. & LICHTENSTEIN, J. (1958). *Proc. natn. Acad. Sci. U.S.A.* **44**, 1004.
CUMMINS, J. E. & RUSCH, H. P. (1966). *J. biophys. biochem. Cytol.* **31**, 577.
DISCHE, Z. (1962a). In *Methods in Carbohydrate Chemistry*, ed. R. L. Whistler and M. L. Wolfrom. New York: Academic Press.
DISCHE, Z. (1962b). In *Methods in Carbohydrate Chemistry*, ed. R. L. Whistler and M. L. Wolfrom. New York: Academic Press.
ENTEMAN, C. (1957). *Meth. Enzym.* **3**, 310.
FISKE, C. H. & SUBBARROW, Y. (1925). *J. biol. Chem.* **66**, 375.
GUTMAN, E. B. & GUTMAN, A. B. (1940). *J. biol. Chem.* **136**, 201.
HALVORSON, H. O. (1965). *Symp. Soc. gen. Microbiol.* **15**, 343.
HEIDELBERGER, C. (1965). In *Progress in Nucleic Acid Research and Molecular Biology*, ed. J. N. Davidson and W. E. Cohn. New York: Academic Press.
HOTCHKISS, R. D. (1948). *Archs Biochem.* **16**, 131.
HRYNIEWIECKA, L. (1967). *Bull. Soc. Amis Sci. Lett. Poznań* **8**, 125.
HUGGINS, C. & LAPIDES, J. (1947). *J. biol. Chem.* **170**, 467.
JAMES, T. E. (1966). M.A. Thesis, The Ohio State University, Columbus, Ohio, U.S.A.
JAYME, G. & LANG, F. (1963). In *Methods in Carbohydrate Chemistry*, ed. R. L. Whistler. New York: Academic Press.
JENSEN, W. A. (1962). *Botanical Histochemistry*, p. 202. San Francisco: W. H. Freeman and Co.
KLEIN, R. L. (1959). *J. cell. comp. Physiol.* **53**, 241.
KLEIN, R. L. & NEFF, R. J. (1960). *Expl Cell Res.* **19**, 133.
LINDNER, A., KUTHAM, T., SANKARANARAYANAN, K., RUCKER, K. & ARRANDONDO, J. (1963). *Expl Cell Res.* (Suppl.) **9**, 485.
LISON, L. (1954). *Stain Technol.* **29**, 131.
LOWRY, O. H., ROSEBROUGH, N. J., FARR, A. L. & RANDALL, R. J. (1951). *J. biol. Chem.* **193**, 265.
MAALØE, O. & HANAWALT, P. C. (1961). *J. molec. Biol.* **3**, 144.
MAALØE, O. & KJELDGAARD, N. O. (1966). *Control of Macromolecular Synthesis*. New York: W. A. Benjamin, Inc.
MCMANUS, J. F. A. (1948). *Stain Technol.* **23**, 99.
MEYERS, D. & JENSEN, T. (1967). *J. Protozool.* **14**, 11 (Suppl.).
MITTERMAYER, C., BRAUN, R. & RUSCH, H. P. (1965). *Expl Cell Res.* **38**, 33.
MITTERMAYER, C., BRAUN, R. & RUSCH, H. P. (1966). *Biochem. biophys. Acta* **114**, 536.
MOORE, S. & STEIN, W. H. (1963). *Meth. Enzym.* **6**, 819.
NEFF, R. J., BENTON, W. F. & NEFF, R. H. (1964). *J. biophys. biochem. Cytol.* **23**, 66A.

Neff, R. H., Neff, R. J. & Benton, W. F. (1964). *J. Protozool.* **11**, 20 (Suppl.).
Neff, R. J., Neff, R. H. & Taylor, R. E. (1958). *Physiol. Zoöl.* **31**, 73.
Neff, R. J., Ray, S. A., Benton, W. F. & Wilborn, M. (1964). In *Methods in Cell Physiology*, ed. D. M. Prescott. New York: Academic Press.
Ogur, M. & Rosen, G. (1950). *Archs Biochem.* **25**, 262.
Park, H. Z. (1965). M.A. thesis, Vanderbilt University, Nashville, Tennessee, U.S.A.
Pregl, F. & Grant, J. (1951). *Quantitative Organic Microanalysis*, p. 97. Philadelphia: The Blakiston Co.
Rasmussen, L. & Zeuthen, E. (1966). *C. r. Trav. Lab. Carlsberg* **35**, 85.
Reich, E. (1966). *Symp. Soc. gen. Microbiol.* **16**, 266.
Saeman, J. F., Moore, W. E. & Millett, M. A. (1963). In *Methods in Carbohydrate Chemistry*, vol. III, 54. Ed. R. L. Whistler. New York: Academic Press.
Scarano, E. & Augusti-Tocco, G. (1967). In *Comprehensive Biochemistry*, vol. **28**, 55. Ed. M. Florkin and E. H. Stotz. New York: Elsevier Publ. Co.
Schmidt, R. R. (1966). In *Cell Synchrony, Studies in Biosynthetic Regulation*, p. 189. Ed. I. L. Cameron and G. M. Padilla. New York: Academic Press.
Schneider, W. C. (1945). *J. biol. Chem.* **161**, 293.
Seifter, S., Seymour, S., Novic, B. & Muntwyler, E. (1950). *Archs Biochem.* **25**, 191.
Sisken, J. E. (1964). In *Methods in Cell Physiology*, ed. D. M. Prescott.
Sussman, M. (1965). *A. Rev. Microbiol.* **19**, 59.
Swanson, M. A. (1955). *Meth. Enzym.* **2**, 540.
Szybalski, W. & Iyer, V. N. (1967). In *Antibiotics, Mechanism of Action*, ed. D. Gottlieb and P. D. Shaw. New York: Springer Verlag.
Thomson, R. Y. (1960). In *Chromatographic and Electrophoretic Techniques*, ed. I. Smith. New York: Interscience Publishers.
Tomlinson, G. (1962). Ph.D. thesis, Vanderbilt University, Nashville, Tennessee, U.S.A.
Tomlinson, G. (1967). *J. Protozool.* **14**, 114.
Tomlinson, G. & Jones, E. (1962). *Biochim. biophys. Acta* **63**, 194.
Trager, W. (1963). *J. Protozool.* **10**, 1.
Umbreit, W. W., Burris, R. H. & Stauffer, J. F. (1947). *Manometric Techniques*. Minneapolis: Burgess Publ. Co.
Van Wagtendonk, W. J. (1955). *Protozoa*. New York: Academic Press.
Vickerman, K. (1962). *Expl Cell Res.* **26**, 497.
Volkonsky, M. (1931). *Archs Zool. exp. gén.* **72**, 317.
Weissmann, G., Troll, W., Brittinger, G. & Hirschhorn, R. (1967). *J. biophys. biochem. Cytol.* **35**, 140A.
Willmer, E. N. (1963). *Symp. Soc. expl Biol.* **17**, 215.
Wright, B. E. (1966). *Science, N.Y.* **153**, 830.
Yasuma, A. & Ichikawa, T. (1953). *J. Lab. clin. Med.* **41**, 296.

DORMANCY AND SURVIVAL IN NEMATODES

By C. ELLENBY

Department of Zoology, The University, Newcastle upon Tyne

INTRODUCTION

Nematodes are aquatic organisms, completely dependent on being in at least a film of water for their active life. As they are universally distributed, there are some situations in which they are particularly vulnerable. The group includes some of the most important of plant and animal parasites, and the evolution of a viable parasitic life-history has frequently involved the development of the ability to survive in a dormant state at stages along the life-cycle, in the transition from one host to another, or from a free-living to a parasitic mode of life. In some free-living forms, some dormancy may be involved in the transfer from one place to another, or as a response to a sudden change in the environment. Frequently, this will involve survival in a dry state, but sometimes the organism may be dormant but fully hydrated. These two states of dormancy clearly involve different problems, although it is possible that some factors may be common to them both. Although information is available about the ability to survive, very little is known about the mechanisms involved in either case.

As far as survival in the hydrated state is concerned, attention has been directed mainly to the stimulation required to break the dormancy. This subject has been very adequately reviewed recently, particularly in relation to the infective stage of parasitic nematodes, by Rogers & Sommerville (1963), leading workers in this field. They view the infective stage as representing a break in the continuous development and growth of an organism; the break is overcome by some change in the environment, sometimes in the form of a stimulus from the host. But how the organisms survive in the dormant state is almost a complete mystery.

THE SURVIVAL OF DRY NEMATODES

For those nematodes which survive dry, information available is largely concerned with their abilities rather than with mechanisms. The use of interference microscopy to determine the water content of single nematode worms (Ellenby, 1968 a, b) has enabled a relatively considerable advance to be made: but this is largely because, in the almost complete absence of any clue at all, the first clue can seem to have very great importance.

Very few species of nematodes are able to survive desiccation. This is not surprising when it is remembered that the vast majority of species live in water, if only water in the form of a film. It is also not surprising that those which possess outstanding ability to survive desiccation in cryptobiosis (Keilin, 1959) are forms which parasitize aerial parts of plants; even closely related forms which live in plant roots have very limited powers of this sort. A species of *Tylenchulus* survived for 39 years dry on a herbarium specimen (Steiner & Albin, 1946); the cockle nematode of wheat, *Anguina tritici*, survived 28 years and the stem eelworm, *Ditylenchus dipsaci*, for 23 years in dry-plant material (Fielding, 1951). Although there is no information as to the state of dryness of the organisms enclosed in dry-plant material for all those years, it is probable that they would have been very dry indeed.

There are suggestions from work with other kinds of organisms that the presence of certain substances in the tissues may be concerned with their recovery from dehydration. For example, Clegg (1967) has suggested that glycerol may be concerned in some way with the survival of the brine shrimp, *Artemia salina*. There is some evidence that constitution in this sense may also play a part in nematodes. This will be referred to later. But in some species of nematodes it is clear that whether individuals of the same species survive desiccation well or poorly will depend on physical factors affecting the drying process. This is particularly well shown by the potato-root eelworm, *Heterodera rostochiensis*.

Heterodera rostochiensis

The infective larva of the potato-root eelworm is the second-stage larva, the first moult having occurred in the egg. It attacks the roots of the potato plant and feeds on its tissues, leading a more or less sedentary life inside the root. During growth the larva moults three more times before reaching maturity. The mature male leaves the roots, but the female swells up and ruptures the superficial root layers to project into the soil; with her head still embedded in the root tissue, she continues to feed. There, the male fertilizes her. After fertilization the female worm gradually becomes transformed into the 'cyst' which is such an important feature of the *Heterodera* life-history. Her internal organs, except her reproductives, degenerate, and the spherical body now becomes filled with eggs. Finally, as the female dies, her cuticle becomes tanned by a polyphenol oxidase (Ellenby, 1946a) into a hard resistant cyst wall.

I have gone into some detail about cyst formation because of its importance. The cyst, which is around 0·5 mm. in diameter, may contain approximately 500 eggs, each soon containing a second-stage larva; the highest number of eggs recorded is 969, for a cyst 0·67 mm. in diameter (Ellenby, 1946b). Some of the larvae will hatch and emerge from the cyst in the absence of the host plant, though even in this case there is some evidence that the emergence is not 'spontaneous', but due to a stimulus possibly from micro-organisms (Ellenby, 1963; Ellenby & Smith, 1967). Most emergence takes place in response to a stimulus from the host plant, the

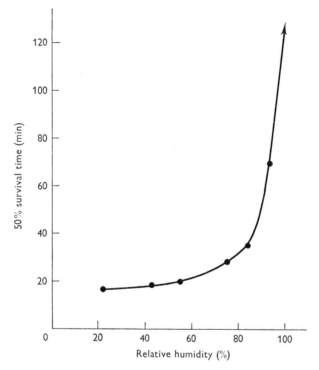

Fig. 1. The recovery of second-stage larvae of *Heterodera rostochiensis* was determined after exposure to different relative humidities for varying times. Tests were repeated until the time of exposure which 50 % of the larvae would survive was established for each relative humidity. (Redrawn from Ellenby, 1968b.)

roots of which give off an exudate containing the so-called hatching factor, the nature of which is still unknown. Even in the presence of the host plant, not all the larvae are stimulated to hatch in a single year. There is no doubt that it is due to the cyst that the eggs are able to survive for so long in the soil. With the related species, *H. glycines*, parasitic on soya, Endo (1962) has shown that, at low humidity, cysts in soil and free cysts will remain viable for months and weeks, while free eggs and free larvae survive only for days and minutes, respectively. *H. rostochiensis* is far better: in the absence of the host plant, it has been shown that cysts will still contain viable eggs for at least 8 years (Franklin, 1937).

Here then we have an organism which is able to remain in a state of dormancy while fully hydrated: it is also able to survive when dry. It is this aspect I wish mainly to discuss.

There are interesting differences between the different species of *Heterodera* parasitizing different hosts, in the ability of the contents of their cysts to survive dry storage. The potato-root eelworm is among the best in this

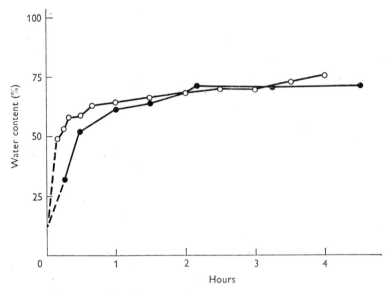

Fig. 2. Rate of water uptake of dry second-stage larvae of *Heterodera rostochiensis* still in the egg-shell (●), and free (○). ● is based on determinations on a series of larvae freed from the egg-shell after varying periods of soaking: ○ is based on the water uptake of single free larvae followed continuously. Values for zero time from estimations on other specimens in liquid paraffin. (Redrawn from Ellenby, 1968*b*.)

respect, and I have found that 5 years of dry storage produces no detectable difference in the emergence of larvae from the cysts (Ellenby, 1955). However, the free larvae will survive only brief exposures to a dry atmosphere (Kämpfe, 1959; Ellenby, 1968*b*). Figure 1 shows the survival of larvae which have hatched from the egg and emerged from the cyst. They have been exposed to different relative humidities (R.H.) for varying times until, for each R.H., the exposure time which 50% of the larvae would survive (S 50) was established. Clearly, survival, except at the highest values for R.H., is very poor indeed: even at 75% R.H. the S 50 value is only 28 min. This is trivial compared with the 5-year survival of the unhatched larvae inside the dry stored cyst.

Some clue to this profound difference is provided by a study of the rate of water movement into and out of the free larva and the unhatched larva still coiled inside the egg-shell. Figure 2 shows the rate at which water is taken up by the dry free larva and by the dry unhatched larva. The former graph is derived from single larvae studied continuously as water was taken up from an aqueous solution of bovine ox plasma; the latter graph is derived from larvae liberated from eggs soaked for different times. Clearly,

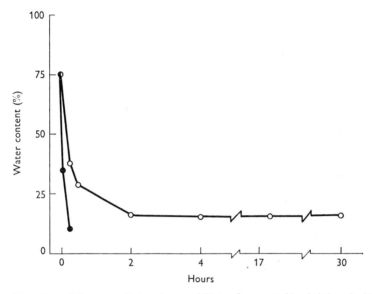

Fig. 3. Water loss of free second-stage larvae of *Heterodera rostochiensis* (●) and of larvae still inside the egg-shell (○). (Redrawn from Ellenby, 1968b.)

the free and unhatched larvae take up water at about the same rate; under these conditions the egg-shell is freely permeable to water. However, an entirely different picture is given if the experiment is carried out the other way—if the rate of water loss is measured instead of the rate of water uptake. The unhatched larva loses water more slowly than the free larva; Fig. 3 shows that after 30 hr. at 50% R.H. it still contains more water than the free larva contains after 15 min. exposure.

The egg-shell itself is freely permeable to water for, as Fig. 2 shows, both the free larva and the unhatched larva still inside the egg-shell take up water with equal rapidity. Nor are we dealing, with the egg-shell, with a membrane which is differentially permeable in two directions, as envisaged by some for insects (see Beament, 1965, for discussion). Figure 3 shows that the unhatched larva at first loses water almost as rapidly as the free larva. The slow rate at which the unhatched larva dries subsequently must therefore be largely due to a change in the permeability of the drying egg-shell.

There is no doubt that the exposed unhatched larva ultimately becomes as dry as the exposed free larva. Estimation of the water content of the dry larva still coiled inside the egg-shell is difficult, as I have as yet found it impossible to free completely dry unhatched larvae from their egg-shell. On two occasions, however, enough of a larva for an estimation has projected from the shell of an egg ruptured in a non-aqueous medium; both of these larvae were almost completely dry, and certainly as dry as exposed free larvae. The difference in their survival, therefore, is due, not to their ultimate water content, but to the speed at which it is attained. Devices to slow the rate of drying have been found in other species.

Ditylenchus dipsaci

D. dipsaci, the stem eelworm, exists as a number of strains specialized to various hosts. The narcissus strain, a particularly important horticultural pest, possesses outstanding ability to survive in the dry state. As already mentioned, specimens have revived after being stored in dry-plant material for 23 years (Fielding, 1951). At 50% R.H. the isolated fourth-stage or infective larva will survive for at least a month (Wallace, 1962). 'Eelwormwool' consists of massive aggregations of, largely, this larval stage found on dry bulbs of the narcissus. The viability of the organisms in the wool declines with time, although, stored at 2–4° C., they remain viable for 7 years (Bosher, 1960).

The wool aggregations are artificial in the sense that they are only formed when the masses of larvae leave infected bulbs when they are set out to dry: under these conditions they fortuitously become aggregated in masses. There is no doubt, however, of the survival value of the occurrence. Isolated single fourth-stage larvae will survive 3 days exposure in a desiccator (Ellenby, 1968b); and survival of the worms in small fragments of wool is about the same (Webster, 1964). However, in larger fragments the survival is much better, and almost 40% of the worms in a mass survived no less than 5 weeks in a desiccator (Ellenby, 1968b).

Examination of fragments taken from the periphery and centre of eelworm-wool masses showed that, as one would perhaps expect, death occurred first in the periphery. For example, after 1 week in a desiccator all worms in the centre of a mass revived, but only 30% of the worms from its outside. After 2 weeks practically all the worms from central fragments revived, but none from the periphery. There is little doubt that the differences in the survival of peripheral and central animals is related to desiccation, but it is difficult to establish this. Measurement is difficult, as I have not been able to separate the individual worms in a mass in a non-aqueous medium; but estimates from portions of animals protruding from fragments are in keeping with the obvious view that the outside of the mass dries first.

Almost certainly, as the periphery dries the rate of drying of the central regions is slowed. The mass which was kept in a desiccator for 5 weeks contained about 10,000 worms; I have estimated that if the mass were a sphere, and all the 40% survivors in the central zone, they would be surrounded by a coat of dead, dry worms about ten individuals deep. Their death would help those more fortunately situated to survive (Ellenby, 1968b).

In the mass, then, the same sort of situation develops as in the individual egg of *Heterodera rostochiensis*. In *Heterodera* the rate of drying of the unhatched larva is slowed as the egg-shell dries; similarly, in *Ditylenchus* the rate of drying of the worms in the centre of the mass is slowed by the drying of their superficial fellows. But the individual fourth-stage *Ditylenchus dipsaci* larva, unlike the free *H. rostochiensis* larva, possesses remarkable powers of survival. Here, too, there are indications that the same sort of mechanism may be involved.

The cuticle of the dry *Ditylenchus* larva is freely permeable to water: as Fig. 4 shows, larvae immersed in a solution of bovine ox plasma take up water at about the same rate as the dry *H. rostochiensis* larvae shown in Fig. 2. An important difference is revealed, however, if the rate at which the two species lose water is compared.

Second-stage larvae of *H. rostochiensis* are considerably smaller than those of fourth-stage larvae of *D. dipsaci*, so that direct comparison of the two is difficult, because of their different surface–volume relations; larvae of the free-living form *Panagrellus redivivus*, of approximately the same size as fourth-stage *D. dipsaci*, were therefore included in the comparison. Animals of each sort were dried at 53% R.H. for periods of 5–10 min. before their water content was estimated. Results presented in Fig. 5 show that there are marked differences between the species. *Panagrellus* dries most rapidly and *Ditylenchus* least rapidly. However, from our present point of view the most interesting feature of the results is that the rate of drying of *Ditylenchus* appears to slow down more than that of *Heterodera* or *Panagrellus*, so that the difference between *Ditylenchus* and *Heterodera* is more pronounced after 10 min. than after 5 min. In fact, statistical analysis of the results shows that the differences between these two latter species is not significant after 5 min., but that it is highly significant after 10 min. ($P < 0.01$).

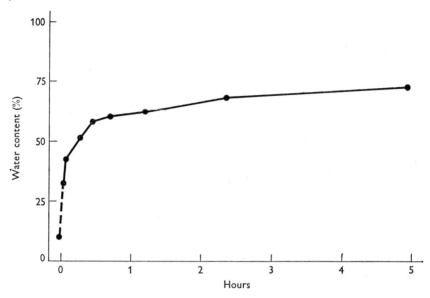

Fig. 4. Rate of water uptake of dry fourth-stage larvae of *Ditylenchus dipsaci*. Based on the water uptake of single larvae, followed continuously. Value for zero time from estimations on other specimens in liquid paraffin. (Redrawn from Ellenby, 1968b.)

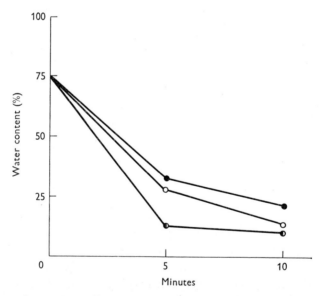

Fig. 5. Rate of water loss of larvae of *Heterodera rostochiensis* (○), *Ditylenchus dipsaci* (●), and *Panagrellus redivivus* (◐). *P. redivivus* larvae, of the same size as *D. dipsaci*, dry most rapidly. For the first 5 min. *Haemonchus* and *Ditylenchus* lose water at the same rate, but the rate of drying of the latter species then slows relatively. (Redrawn from Ellenby, 1968b.)

Almost certainly the fall in the rate at which water is lost from the unhatched *Heterodera* larvae was due to a change in the permeability of the egg-shell as it dried. Is there any evidence that some similar mechanism is operating in the isolated *Ditylenchus* larva? I have suggested that there might be (Ellenby, 1968b): the evidence is presented in Plate 1 (facing p. 96) for a series of isolated air-dried larvae of *Ditylenchus* further dried by transfer to a desiccator for from 1 to 3 days. The interference microscope shows that in an air-dried specimen, A, displacement at the periphery of the specimen is far greater than at its centre. In the specimen B, dried for 1 day in the desiccator, the displacement at the centre is somewhat greater than in A, and therefore the disparity between periphery and centre is less. The disparity is further reduced in C, dried for 2 days in a desiccator; in D, dried for 3 days, the centre has more or less overtaken the periphery.

Isolated larvae of *Ditylenchus* will only survive 3–4 days in a desiccator, like the larvae in small fragments of eelworm-wool (Webster, 1964). Death therefore appears to occur when the centre parts of the animal are as dry as the periphery; it is possible that structures more vital to survival than the cuticle may be more centrally located. If the cuticle dries first, as suggested by this series, it would indeed slow the rate of drying of the organism. But can one argue from this series to the fully hydrated larva losing water as it is exposed to the atmosphere, as in Fig. 5? Even if the same mechanism were operating under these conditions, the difference in refractive index between the periphery and the centre would clearly be slight at first, and only make a small difference to the angle of the displaced fringe. Moreover, as I have pointed out elsewhere (Ellenby, 1968c), the series in Plate 1 is not entirely satisfactory: these animals contain large quantities of highly refractile globules which, in a dry specimen, could produce the same peripheral–central disparities. Usually, these globules are large and readily avoided, but there remains the suspicion that they might be very finely dispersed and produce the same effect. It is largely for this reason that the work with the infective larval stage of *Haemonchus contortus* was undertaken.

Haemonchus contortus

The infective stage of *H. contortus* is the third-stage larva. It retains the second-stage larval cuticle. Exsheathment normally takes place in the rumen of the host (Sommerville, 1954) under the influence of a number of factors, chiefly un-ionized carbonic acid and dissolved gaseous carbon dioxide (Rogers, 1960).

The exsheathed larva can be obtained by incubating the ensheathed forms in CO_2-saturated 50% Ringer solution at 37° C. It is well known that the ensheathed larva survives desiccation better than the exsheathed form, as can be seen in Fig. 6. After 1 hr. exposure to 47% R.H. the survival of the exsheathed larvae is poor, and after 8 hr. there is no recovery; on the other hand, the ensheathed larvae recover after 30 hr. exposure and even after 4 weeks some worms recover. Compared with the infective larvae of *D. dipsaci*, this ability of the ensheathed larva is poor; but there is no doubt of its superiority over that of the exsheathed form.

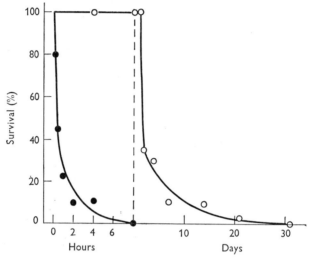

Fig. 6. Percentage survival of exsheathed (●) and ensheathed (○) larvae of *Haemonchus contortus* exposed to 47% R.H. at 18° C. Broken line indicates change in scale of exposure time (Redrawn from Ellenby, 1968c.)

Rogers & Sommerville (1960) showed that the sheath is permeable to water. There is little difference in the rate at which water enters dry ensheathed and exsheathed larvae, as shown in Fig. 7. Such difference as there is, is almost certainly due to the ensheathed larva contracting away from the sheath when it dries: water enters the ensheathed larva very rapidly indeed if it is examined before this contraction has taken place (Ellenby, 1968c). However, although the sheath is freely permeable to water, the rate of water loss of the ensheathed larva is far slower than that of the exsheathed form. Figure 8 shows that after 1 hr. at 47% R.H. the water content of the exsheathed larva is about 10%. The ensheathed form also dries rapidly at first and loses about half its water in the first 15 min. The rate of water loss then slows considerably, and even after 24 hr. the ensheathed larva still contains about 30% of water.

The contraction of the drying larva away from the sheath will undoubtedly help to slow the rate of drying. This contraction takes place, at 47% R.H.,

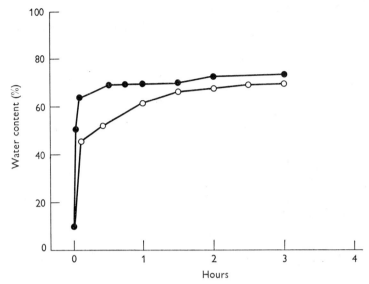

Fig. 7. Rate of water uptake of dry exsheathed (●) and ensheathed (○) larvae of *Haemonchus contortus*. Based on the water uptake of single larvae followed continuously. Values for zero time from other specimens in liquid paraffin. (Redrawn from Ellenby, 1968c.)

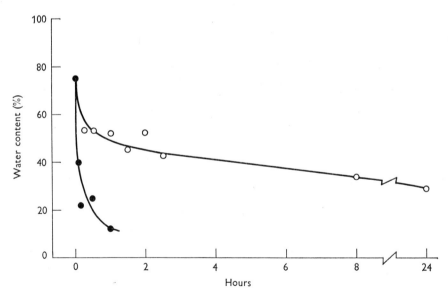

Fig. 8. Water loss of exsheathed (●) and ensheathed (○) larvae of *Haemonchus contortus*. (Redrawn from Ellenby, 1968c.)

after about 2 hr.: it is clear from Fig. 8, however, that the rate of drying slows before that. Undoubtedly this is due to a decrease in the permeability of the drying sheath.

Plate 2(a) is a photograph of an ensheathed larva dried for 15 min. at 47% R.H. The light liquid paraffin in which it is mounted has a refractive index of 1·478; if a specimen matched it in refractive index, it would have a water content of about 20%. In the Plate the first two interference fringes, reading from the left, are displaced to the right by the tail, showing that the second-stage larval tail is very dry: it has a water content of less than 10%. The next fringe is also displaced to the right at the periphery; that is, in the sheath; but it turns back to the left as it traverses the breadth of the animal. The sheath therefore has dried more rapidly than the larva it contains: in fact this larva had a water content of more than 50%. Clearly, even though the sheath is so thin, when it dries it becomes a very effective barrier to the loss of water from the larva.

DISCUSSION

The results with *Haemonchus* fully confirm the conclusions drawn from the observations with *Heterodera*: a thin membrane, such as the sheath of the one or the egg-shell of the other, freely permeable to water when wet, can become an effective barrier to water loss when dry. The ultimate water content, after many hours, of the dried unhatched larva of *Heterodera* is no different from that of the free larva dried for a matter of minutes, yet the former survives but the latter perishes. For this form at least there cannot be any differences in composition, such as the glycerol content which might be involved in the survival of *Artemia salina* (Clegg, 1967); the rate at which water is lost is clearly the decisive factor.

Within a confined space bounded by a relatively impermeable membrane, a very high R.H. will soon be established, even from contained objects of low water content. The unhatched *Heterodera* larva after 30 min. contains about 30% of water. Gels with the low water content of 20, 30 and 40%, sealed in glass containers at 24° C., soon established constant R.H.s of 80, 90 and 96%, respectively. At these levels of R.H., further water loss from the gels would be very slow indeed: and these are the sort of conditions which would be established inside the drying egg-shell by the partially-dried unhatched larva.

Although the second-stage cuticular sheath of the infective larva of *Haemonchus* is very closely applied to the cuticle of the third-stage larva, it may be argued that there is a discontinuity between them, and that the relationship is therefore different from that of the cuticle of the *Ditylenchus* larva to the tissues underlying it. It may be thought therefore that to argue from *Haemonchus* to *Ditylenchus* is unjustified. It cannot, of course, be justified completely; but it is possible to show, with another organism, that differences in the relative permeability of structures in continuity can lead to differences in rate of water loss.

The vinegar eelworm, *Turbatrix aceti*, bears living young. The closely related *Panagrellus redivivus* exhibits endotokia matricida; living young are produced, but they are not born, but escape by destroying the mother. Neither species is good at surviving desiccation, yet both are apparently spread on the feet of flies (Henneberg, 1900; Lees, 1953). There is little doubt that their 'viviparous' habit is of very great adaptive significance in this connection, for the larvae survive inside the body of the dried mother (Aubertot, 1923; Lees, 1953). As Plate 2(*b*) shows, the larvae do not dry as rapidly as the mother: although in continuity with the maternal tissues there is a sharp discontinuity in the direction of the interference fringes as they meet the contained larvae, indicating a sharp difference in the water content of the maternal and larval tissues. The specimen of *Turbatrix* shown was dried at 40% R.H. for 10 min.; the water content of the female was then about 12% and that of the larvae about 33%. The water content has therefore fallen throughout the mother-larval system but it has passed less rapidly through the cuticle of the contained larvae.

The decline in the rate of water-loss with time, as shown for *Panagrellus* in Fig. 5, may take place even without a change in the permeability of the outer covering; the remaining water may be less free to move. Indeed, this may itself lead to a drying out of the superficial layers. But the drying of the cyst of *Heterodera rostochiensis* under completely artificial experimental conditions shows that a change in the permeability of superficial layers can take place even when it encloses free water. The data for Fig. 9 were obtained from a cyst to which a water-filled capillary was attached; water was free to enter the cyst from the capillary as it evaporated from the cyst wall. The rate of water loss declined with time. The change in the appearance of the cyst wall during these experiments was striking; at first a moist structure, it was obvious that it became dry as the experiment proceeded. The rate of water loss from the cyst surface was greater than could be replaced from within, from the freely available water.

That the rate of drying is of importance to survival of desiccation has been known for some time. Lees (1953) showed that some specimens of *Panagrellus silusiae*, which normally survives drying very poorly, survived 2 months in a desiccator when embedded in a 2 cm.^{3} agar block. And the species which survive desiccation well all appear to possess devices to slow the rate of drying. The reason for the association, however, is obscure.

There is evidence that some water is required for the structural integrity of proteins (Klotz, 1958) and of amino acids (Falk, Hartman & Lord, 1963; Bather, Webb & Cunningham, 1965). Perhaps the rate at which it is removed may affect the structural integrity at the molecular or ultrastructural level: one is tempted to see some analogy with survival to freezing which is dependent on rate of cooling. According to Keilin (1959), Doyère, who showed that graded drying of Tardigrades improved the

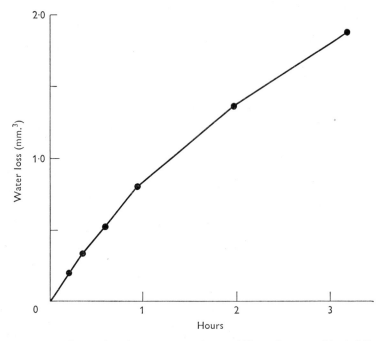

Fig. 9. The rate of water loss from an exposed cyst of *Heterodera rostochiensis* falls as its wall dries, although its inside is kept filled with water. (Redrawn from Ellenby, 1946.)

survival rate, already suggested in 1842 that 'organisms which revived had preserved the molecular composition of their tissues and their organic arrangements'.

Although rate of drying is of great importance it clearly is not the only factor, although, as already pointed out, it is difficult to see what other factors could be operating in *Heterodera rostochiensis*. I have found peripheral/central differences on drying in the cockle nematode, *Anguina tritici* (unpublished), similar to those I have reported for *Ditylenchus*. *Anguina* will survive for 7 days in a desiccator, while *Ditylenchus* survives for 3 days: I have not been able to detect any differences between the species in the rate at which they dry. It is, of course, possible that exceedingly subtle differences in rate of drying may be of importance. More-

PLATE I

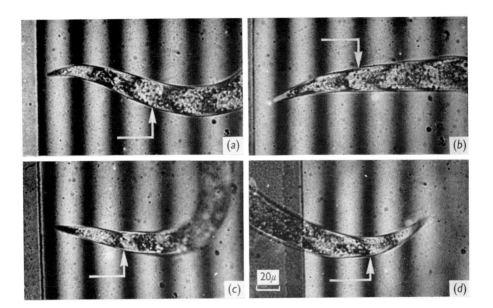

All specimens shown were mounted in liquid paraffin and photographed through a fringe-field eyepiece of an interference microscope. Greater refractive index is shown by displacement of the fringes to the right. A scale of 20 μ is common to all specimens. (a)–(d), tail regions of dry fourth-stage larvae of *Ditylenchus dipsaci*: (a) air-dried specimen; (b)–(d) air-dried specimens after subsequent 1, 2 and 3 days, respectively, in a desiccator. In (a) note that there is more displacement of the fringe at the periphery of the specimen than at its centre, indicating that the periphery is the drier. The difference between periphery and centre becomes less after each day in the desiccator. In each case, displacement at the periphery is indicated by ↑. (From Ellenby, 1968b.)

PLATE 2

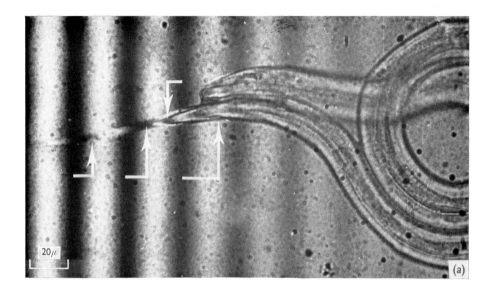

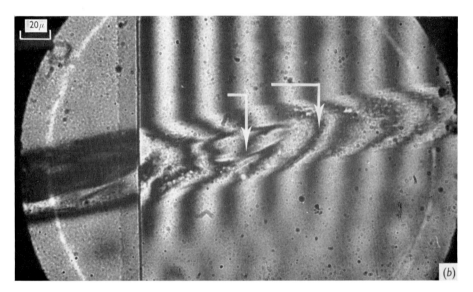

(a) Ensheathed larva of *Haemonchus contortus* dried for 15 min. at 47% R.H.; mounted in liquid paraffin. Greater refractive index (less water) is shown by displacement to the right. The second-stage cuticle has a higher refractive index at the tail (left and central lower arrows) and also where it ensheaths the third-stage larva (right lower arrow); the larva itself, however, has a lower refractive index than the medium (note displacement to left indicated by upper arrow), showing its higher water content. (Adapted from Ellenby, 1968c.)

(b) Dried female of *Turbatrix aceti* with larvae *in utero*. The higher water content of a larva is indicated by the smaller displacement of the fringe where it is present (left arrow). In liquid paraffin. (Adapted from Ellenby, 1968a.)

over, the rate at which the infective *Haemonchus* larva dries is far slower than either of these forms, yet its survival is trivial by comparison. Perhaps the differences between these forms may be due to differences in composition, such as the glycerol content which Clegg (1967) considers may be involved in *Artemia* survival. Bather *et al.* (1965) have shown that inositol will protect bacterial and tumour cells against desiccation and Warner (1962) has advanced a sterochemical explanation for the ability of this and similar compounds to act in the same way as water; indeed, it is considered that they produce a much more stable system with protein layers and other biological macromolecules. Comparison of the forms considered in this work, from this point of view, may be very fruitful.

REFERENCES

AUBERTOT, T. (1923). *C. r. hebd. Séanc. Acad. Sci., Paris* **176**, 1257.
BATHER, R., WEBB, S. J. & CUNNINGHAM, T. A. (1965). *Nature, Lond.* **207**, 30.
BEAMENT, J. W. L. (1965). *Symp. Soc. exp. Biol.* **19**, 273.
BOSHER, J. E. (1960). *Proc. helminth. Soc. Wash.* **27**, 127.
CLEGG, J. S. (1967). *Comp. Biochem. Physiol.* **20**, 801.
ELLENBY, C. (1946a). *Nature, Lond.* **157**, 302.
ELLENBY, C. (1946b). *Ann. appl. Biol.* **33**, 433.
ELLENBY, C. (1955). *Ann. appl. Biol.* **43**, 1.
ELLENBY, C. (1963). *Nature, Lond.* **198**, 1110.
ELLENBY, C. (1968a). *Experientia* **24**, 84.
ELLENBY, C. (1968b). *Proc. R. Soc.* B **169**, 203.
ELLENBY, C. (1968c). *J. exp. Biol.* **49**, 469.
ELLENBY, C. & SMITH, L. (1967). *Ann. appl. Biol.* **59**, 283.
ENDO, B. Y. (1962). *Phytopathology* **52**, 80.
FALK, M., HARTMAN, K. A. Jun. & LORD, R. C. (1963). *J. Am. chem. Soc.* **85**, 387.
FIELDING, M. J. (1951). *Proc. helminth. Soc. Wash.* **18**, 110.
FRANKLIN, M. T. (1937). *J. Helminth.* **15**, 69.
HENNEBERG, W. (1900). *Zentbl. Bakt. ParasitKde* (Abt. II), **6**, 180.
KÄMPFE, L. (1959). *Verh. zool.-bot. Ges. Wien* **25**, 378.
KEILIN, D. (1959). *Proc. R. Soc.* B **150**, 149.
KLOTZ, I. M. (1958). *Science, N.Y.* **128**, 815.
LEES, E. (1953). *J. Helminth.* **27**, 95.
ROGERS, W. P. (1960). *Proc. R. Soc.* B **152**, 367.
ROGERS, W. P. & SOMMERVILLE, R. I. (1960). *Parasitology* **50**, 329.
ROGERS, W. P. & SOMMERVILLE, R. I. (1963). *Advances in Parasitology.*
SOMMERVILLE, R. I. (1954). *Nature, Lond.* **174**, 751.
STEINER, G. & ALBIN, F. E. (1946). *J. Wash. Acad. Sci.* **36**, 97.
WALLACE, H. R. (1962). *Nematologica* **7**, 91.
WARNER, D. T. (1962). *Nature, Lond.* **196**, 1055.
WEBSTER, J. M. (1964). *Pl. Path.* **13**, 151.

THE DORMANCY AND GERMINATION OF FUNGUS SPORES

By A. S. SUSSMAN

The Department of Botany, University of Michigan,
Ann Arbor, Michigan

INTRODUCTION

As has been pointed out elsewhere (Sussman, 1968a), dormant stages of fungi, in contrast to those of bacteria, are formed in response to a wide variety of environmental stimuli. Thus, by far the most important such stimulus in bacteria is the nutritional status of the environment, the selective advantage of which is related to the types and sizes of the niches occupied by these organisms, and to their relatively rapid generation time (Stanier, 1953; Sussman, 1968a). However, sporulation in fungi may be induced by diverse physical and chemical factors and, frequently, by the interaction of these.

A similarly extensive range of treatments serve to break the dormancy of fungi, or to 'activate' them, including extremes of temperature and humidity, changes in the light regime and the application of both synthetic and naturally occurring chemicals (Sussman & Halvorson, 1966). Furthermore, a combination of treatments is often effective and may, in fact, be the method of choice in nature. Therefore, this paper will be devoted to a comparison between the mechanism of action of a physical means of activation (heat-shock) and a chemical one (furfural) and of their interaction in a single fungal system.

THE ORGANISM

We are concerned with the ascospore of *Neurospora*, which is the product of meiosis, whose pattern of segregation has provided so many insights into biochemical genetics. It is a large, black, football-shaped cell, about 25 μ long and 15 μ wide, with a complex series of walls (Plate 1). One of these walls is melanized and brittle and the outermost one can be removed without killing the cell (Lowry & Sussman, 1958) and probably consists of uronic acid residues in large part. This structural feature of *Neurospora* ascospores probably accounts for their notorious impermeability (Sussman, Holton & von Böventer-Heidenhain, 1958) which, in turn, may account for their great longevity (Sussman, 1965, 1968b). Dormant ascospores appear to lack endoplasmic reticulum and have a small number of very large mitochondria (Lowry & Sussman, 1968) whose endogenous content of cytochrome c is too low to support oxidations *in vitro* (Holton, 1960). It is of interest to note that a number of other dormant fungal spores lack endoplasmic reticulum and have few mitochondria (Sussman, 1966a) so there may be some generality in these observations. Finally, the feature that has made these cells of interest in studies of dormancy is their inability to develop unless certain treatments are administered; and once these treatments are applied germination proceeds rapidly and synchronously, even in distilled water, so all of the nutrients required for this process are contained endogenously.

METHODS OF ACTIVATION

Activation of *Neurospora* ascospores can be effected by a heat-shock or chemical means, but the former has been used almost exclusively by geneticists since the discovery by Shear & Dodge (1927) of this phenomenon. Heat-activation was studied in further detail by Goddard (1935, 1939) and co-workers, who found that temperatures between 50 and 60° C. for up to 30 min. were optimal. Activation by chemical means was first discovered by Mary Emerson (1948), who used furfural successfully in this way. Since then several furans (Table 1) and other heterocyclic five-membered ring compounds have been shown to be effective (Emerson,

Table 1. *Furans which are active in breaking the dormancy of* Neurospora *ascospores*

Substance	Formula*	Relative activity (%)†	Reference
Furfural	RCHO	100	Emerson, 1948
2-Furfuryl alcohol	RCH_2OH	77	Emerson, 1948
2-Methyl furan	RCH_3	76	
2-Furfuryl methyl ether	RCH_2OCH_3	9	
2-Furfural acrolein	RCHCHCHO	51	Sussman, 1953a
2-Furfuryl diacetate	RCH(COOCH$_3$)(COOCH$_3$)	113	
5-Methyl furfural	CH_3RCHO	66	Emerson, 1954
5-Hydroxymethyl furfural	$CH_2OHRCHO$	9	Emerson, 1954
Methyl 2-furoate	$RCOOCH_3$	36	
n-Propyl furoate	$RCOOCH_2CH_2CH_3$	10	
2-Furfuryl isobutyrate	$RCH_2OOC \cdot CH(CH_3)(CH_3)$	53	Sussman et al. 1959
2-Furfuryl-n-butyrate	$R \cdot CH_2OOC \cdot CH_2CH_2CH_3$	114	

* R = (furan ring)

† Calculated as a percentage, using furfural as 100%.

1954; Sussman, 1953a), from which the following generalizations can be derived: (1) only fully unsaturated heterocyclics are active; (2) the heteroatom may be either oxygen, nitrogen or sulphur; (3) conjugated derivatives like haemoglobin, haemin, furoin or furil are inactive. A number of other fungus spores are activated by furfural and, in addition, some bacterial spores are similarly affected (Mefferd & Campbell, 1951). In addition,

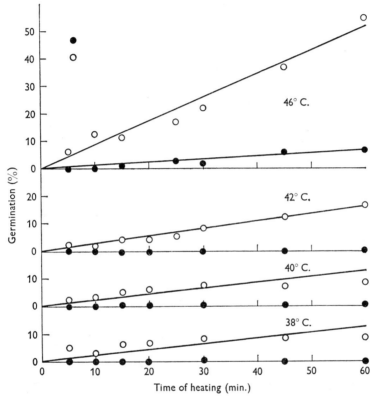

Fig. 1. The effect of exposure to temperatures from 38–46° C. upon the response of ascospores of *Neurospora tetrasperma* to furfural (1×10^{-3} M). Ascospores were used 28 months after harvest. ●, Distilled water; ○, 1×10^{-3} M furfural. (From Sussman, 1954a.)

several classes of aliphatic compounds will break dormancy, including aldehydes, ketones, esters and alcohols (Sussman, Lowry & Tyrrell, 1959). Whereas the heterocyclics activate in concentrations between 1×10^{-3} and 1×10^{-6} M, the aliphatics are required in molar concentrations, and higher, so the specificity of their effect is in question.

Although ascospores of *Neurospora* respond to heat-activation even after years of storage, their sensitivity to chemical activators is transitory. Thus, they respond maximally to furfural within a few weeks of being shed from the ascus and become refractory to this substance several months later (Emerson, 1954). But, as the data in Fig. 1 show, sensitivity to chemical activators is regained by these spores when they are exposed to temperatures between 45 and 50° C. for a short time, a treatment that does not activate by itself (Sussman, 1954a).

In view of the several activation treatments available to ascospores of *Neurospora*, which one is used in nature? Although I know of no studies that have approached this question directly, one of the first reports of the isolation of this organism mentions that it appears as a pink bloom on burned-over areas in south-east Asia (Went, 1901). It is likely that activation by heat, taken together with great resistance to high temperatures, may account for the widespread appearance of *Neurospora* under these circumstances. Moreover, furfural is one of the most ubiquitous of all organic compounds, being present in any habitat containing decaying plant materials, so that any compost heap or dung pile may supply both the heat and chemical ingredients for activation.

EFFECTS OF ACTIVATION

Once an effective activation treatment is applied a programmed series of developmental changes is triggered in ascospores. At the microscopic level no changes can be discerned until a germ tube appears about 2 hr. after activation. However, several distinct ultrastructural changes are induced (Lowry & Sussman, 1968), as can be seen by a comparison of Plate 1(*a*) and (*b*). The germinating spore contains numerous mitochondria of varying shapes (Plate 1(*b*)) which are unlike the large swollen ones found in the dormant spore (Plate 1(*a*)). Although vacuoles of the two types found in dormant ascospores are also present in germinating ones, the latter have much more endoplasmic reticulum and nuclei are more numerous as well. Moreover, a complex of concentric membranes that is continuous with the endoplasmic reticulum at one or more points appears in ascopores 3-6 hr. after activation. It is possible that these bodies generate endoplasmic reticulum but the data are too few to permit this conclusion to be made with assurance (Lowry & Sussman, 1968).

A series of physiological and biochemical changes also is triggered by the activation of *Neurospora* ascospores. For example, sensitivity to heat is markedly engendered by as little as 20-30 min. incubation at 27° C. after activation (Y. Lingappa & Sussman, 1959). Furthermore, it has been shown that heat-activated ascospores incubated anaerobically (Goddard, 1939) or at 4° C. (Sun & Sussman, 1960) are reversibly deactivated; that is, they revert to the dormant condition and can be

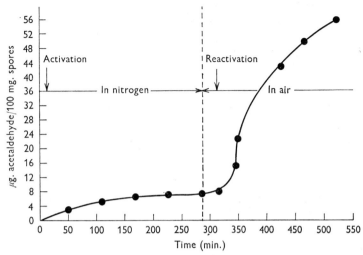

Fig. 2. Effect of anaerobiosis and subsequent reactivation and return to air upon acetaldehyde production by ascospores of *Neurospora tetrasperma*. (From Sussman, Distler & Krakow, 1956.)

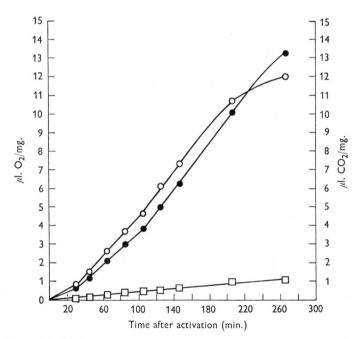

Fig. 3. Effect of furfural upon the respiration of freshly harvested ascospores of *Neurospora tetrasperma*. ●, O_2 uptake of spores treated with 5×10^{-4} M furfural; ○, CO_2 output of spores treated with 5×10^{-4} M furfural; □, O_2 uptake of inactive spores. (From Sussman, 1953b.)

reactivated. This effect is demonstrated in Fig. 2, wherein the release of acetaldehyde has been used to follow the progress of activation. However, under no circumstances has it been possible to deactivate chemically activated ascospores reversibly.

Another effect induced by heat- (Goddard, 1939) or furfural-activation (Sussman, 1953b) is a large increase in respiratory rate, which may exceed by 20-fold that of dormant ascospores (Fig. 3). In fact, such effects upon respiratory capacity is one of the most general results of the disruption of dormancy in microbial spores (Sussman & Halvorson, 1966). As the data in Table 2 disclose, the respiratory quotient (R.Q.) changes from 0·6 in dormant ascospores to about 1·0 in germinating ones (Goddard, 1939; Holton, 1958). The programming of some of these events is well illustrated in Table 2, which shows the stages at which the respiratory and physiological events described above occur. Accompanying these changes is the greatly enhanced accumulation of fermentation products like ethanol and acetaldehyde, which accumulate only after activation and until shortly before the germ tube protrudes. Indeed, only then does the full complement of Krebs-cycle acids appear (Sussman, Distler & Krakow, 1956).

Table 2. *Summary of the stages between activation and germination and the physiological markers which characterize them in ascospores of* Neurospora tetrasperma

Condition of spores	Min. after activation at 25° C.	Q_{O_2}	R.Q.*	Thermal resistance†	Deactivability at 4° C. or in N_2
Dormant	.	0·3	0·6	+	−
Activated					
Stage 1	0–20	0·5–4	1·2	+	+
Stage 2	30–60	4–10	1·0	−	+
Stage 3–n	60–150	15–30	1·9	−	−
Germinating	150	30	0·6	−	−

* Data taken from B. T. Lingappa & Sussman (1959), Goddard (1939) and Holton (1958).
† Data taken from Y. Lingappa & Sussman (1959).

The great self-sufficiency of these spores is underscored by the fuels that are the source of the products whose accumulation has been described above. Thus, dormant ascospores can survive for years by respiring their endogenous supply of lipids (B. T. Lingappa & Sussman, 1959). But as soon as an activation treatment is applied, the sizeable store of trehalose is used for the first time and is completely dissimilated within a few hours (Fig. 4). It should be noted that the spores continue to use lipids even after activation and that this dual substrate requirement appears to be characteristic of other fungus spores as well (Cochrane, Cochrane, Simon &

Spaeth, 1963). Therefore, an important aspect of dormancy in *Neurospora* ascospores is the nature of the restraint upon trehalose catabolism, a restraint that is not lifted until activation occurs. This question will be considered later.

Although dormant and activated ascospores of *Neurospora* can subsist entirely upon their endogenous supply of nutrients, except for water, they

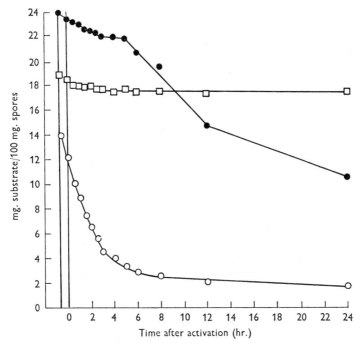

Fig. 4. Effect of heat-activation upon the amount of lipids and carbohydrates in ascospores of *Neurospora tetrasperma*. Ascopores began to germinate about 2·5 hr. after activation. □, Acid-soluble carbohydrates; ●, totallipids; ○, alcohol-soluble carbohydrates. (From B. T. Lingappa & Sussman, 1959.)

do utilize exogenous sugar if it is provided. Such studies reveal something of the metabolic capacity of spores at different stages, so the results of some of this work are outlined in Table 3. A much higher proportion of the entering label is produced as CO_2 in dormant cells than in activated or germinating ones. This difference is probably due to the greater ability of [^{14}C]glucose to supplant endogenous lipid as the substrate than other sugars, such as trehalose. Very few other differences are apparent in the over-all pattern of recovery of label except the very high proportion of sugar that goes into ethanol-insoluble materials, a result that suggests the rapid synthesis of wall material at this stage. But there are only a few qualitative differences in the synthetic capacities of dormant and activated ascospores, for label appeared in amino and organic acids, sugars, proteins and structural polysaccharides in all cases, albeit in different proportions in the several stages. An interesting exception to the qualitative similarity in the destination of exogenous substrate is a substance that may be a precursor of wall material that piles up only in dormant cells, suggesting that they lack the capacity to assemble this substance into finished walls (Budd, Sussman & Eilers, 1966).

Table 3. *Utilization of exogenously added [^{14}C]glucose by ascospores of* Neurospora tetrasperma*

	Stage of development		
	Dormant	Activated	Germinating
Label as CO_2 (%)	57–66	0·6–4·1 (0–2 hr.)	18–22 (3–5 hr.)
Inhibition of endogenous respiration (%)	56	5	41–70
Label in lipids (%)	4 (5 hr.)	2 (2·5 hr.)	2 (5 hr.)
Label in ethanol-soluble fraction (%)	27 (5 hr.)	28 (2·5 hr.)	7 (5 hr.)
Label in ethanol-insoluble fraction (%)	30 (5 hr.)	29 (2·5 hr.)	52 (5 hr.)

* Calculated from data of Budd, Sussman & Eilers (1966).

Nucleic acid metabolism in ascospores has been studied by Henney & Storck (1963a), who have reported that messenger RNA (mRNA) is not stored in dormant cells, although all stages that were studied appeared to have the same population of ribosomes and of total ribosomal and soluble RNA (Henney & Storck, 1963b). On the other hand, dormant ascospores did not contain polyribosomes, which appear first in the early stages of germination (Henney & Storck, 1964). Thus germination may be accompanied by mRNA synthesis and the activation of chromosomes.

MECHANISMS OF ACTIVATION

Thus a precisely delineated series of events transpires upon activation and these have been described at several organizational levels and suggest that deep-seated changes occur in response to this stimulus. What is the nature of the primary event from which the others flow, if, indeed, a single event is responsible for these multiple effects?

The following explanations have been advanced to explain dormancy and its release in micro-organisms: (1) permeability changes, (2) anhydrobiosis and its reversal, (3) metabolic changes. Of course it is an oversimplification to treat these as independent possibilities, because they may be interrelated. For instance, to the extent that the machinery of uptake is enzymic, permeability and metabolism are congruent. However, purely physical changes in membranes can be imagined that might have no direct relation to metabolism, so it is worth maintaining the dichotomy. What is known about these mechanisms in ascospores of *Neurospora*?

The fact that washed cells can germinate in water of conductivity grade (Sussman, 1954b) argues strongly that permeability changes other than to water are not important to the activation process. Moreover, detailed studies of the uptake of anions (Sussman, 1954b), cations (Sussman & Lowry, 1955) and neutral substances like glucose (Sussman et al. 1958) show that marked changes in permeability occur only at the time when the germ tube protrudes, which is about 2 hr. after activation.

As for water affecting dormancy, my colleague, Dr R. J. Lowry, and I have shown in unpublished work that ascospores remain dormant but viable when placed in concentrated glycerol even though the cell walls collapse because of the loss of turgor. Moreover, when ascospores which have been dehydrated in this manner are returned to distilled water they regain water very rapidly and, after activation, germinate. So, water can enter or leave ascospores readily without impairing their viability or affecting the dormant state. Therefore, unless there is a small anhydrous 'core' (Lewis, Snell & Burr, 1960) which is not in equilibrium with liquid water, an explanation of the dormant state must be sought in the metabolic machinery of these cells.

METABOLIC HYPOTHESES TO EXPLAIN ACTIVATION

The most striking metabolic change occurring in activated ascospores is their suddenly acquired ability to use trehalose. As a consequence of this change quantitative and qualitative shifts in the concentration of intermediates occur, shifts that might, in turn, lead to a series of other changes. Consequently, the hydrolysis of trehalose appears to be the key to the activation process, so we have investigated the means through which it is metabolized. Its hydrolysis is catalysed by an enzyme, trehalase, which has been crystallized and shown to be absolutely specific for trehalose (Hill & Sussman, 1963). Its activity is at a minimum shortly after the germination of conidia and reaches a peak after these spores are formed (Fig. 5). But the observation that is important for our purpose is the fact that there is considerable trehalase activity in extracts of dormant ascospores and, although there is an increase after activation, it is not immediate nor proportional to the respiratory increases observed.

With this background it is now possible to examine the mechanisms

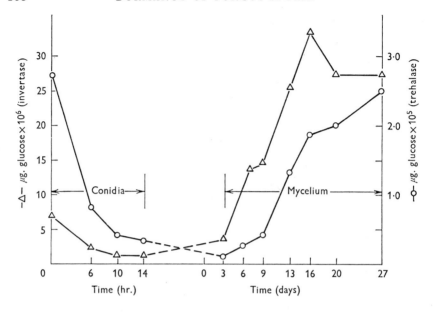
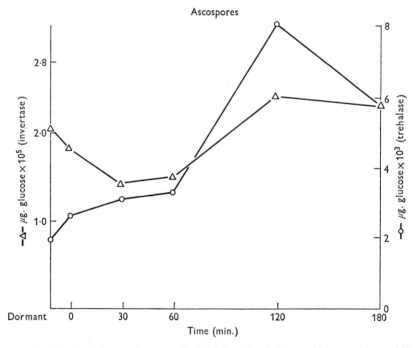

Fig. 5. Summary of the total recoverable activity of trehalase and invertase in conidia, mycelium and ascospores of *Neurospora*. (From Hill & Sussman, 1964.)

through which the metabolic changes which accompany activation might be accomplished. Among the possible means through which such changes could be engendered are the following:

(1) Inhibition of destructive enzyme(s) or removal of endogenous inhibitor.

(2) Synthesis of an enzyme (*a*) from precursor; (*b*) *de novo* synthesis: (i) unique enzyme, (ii) isozymes.

(3) Contiguity changes: (*a*) enzyme-substrate, (*b*) allosteric effectors.

Some of these alternatives have been studied with *Neurospora* ascospores but, as we shall see, a final selection from among them is impossible as yet.

The possibility that an endogenous inhibitor is removed has been tested by the addition of concentrated extracts of the spores to intact ones, but no effects have been observed. Moreover, the percentage germination is unaffected by changes in the concentration of spores over a wide range, so this alternative probably can be ruled out for *Neurospora* ascospores, although endogenous inhibitors have been described for a number of other fungus spores (Sussman, 1966*b*; Sussman & Halvorson, 1966).

On the other hand, changes in the activity of potentially destructive enzymes as affected by activation have not been studied. Thus, a heat-labile proteolytic enzyme might prevent the accumulation of another enzyme whose functioning is essential to the activation process (Swartz, Kaplan & Frech, 1956). The only data which bear on this possibility are those that show little change in trehalose activity upon activation (Fig. 5), thereby making improbable both aspects of alternative (1).

De novo synthesis of trehalase in direct response to activation seems unlikely for the same reason that part of alternative (1) was discarded; that is, the presence of activity even in dormant ascospores. However, the possibility that a precursor is transformed into active enzyme cannot be so easily eliminated, because the act of grinding spores, in order to prepare extracts for assay, may serve to accomplish changes in enzyme conformation that could transform an inactive protein into an active one. That heat can cause conformational changes in proteins is well known, and chemical treatments such as with surface-active agents (Kates, 1957) or solvents (Ryan & Mavrides, 1959) may activate enzymes by changing protein configurations at interfaces. So this possibility cannot be ruled out at this time for no good evidence exists on this subject for ascospores. An interesting aspect of the possibility of enzyme activation concerns whether isozymic forms exist. For example, it can be imagined that although trehalase activity exists in dormant ascospores it is not the *right* enzyme, for a different form of it must be made available to effect activation. Our discovery of active subunits of this enzyme from *Neurospora* mycelium (Plate 2) encourages this possibility, but more experiments need to be done with ascospores.

The final alternative is that somehow a substrate (trehalose) is sequestered from its hydrolase (trehalase) and activation brings them together. Perhaps heat-activation melts, or otherwise alters, a membrane such that it no longer can separate the enzyme and substrate. For a while we believed that this might be the case because of the existence of two separate pools of trehalose in ascospores. These were postulated on the basis of the data in Table 4, which reveal that only about 0·2% of the trehalose contained in dormant ascospores is extractable from these cells with hot 80% ethanol even though other substances, and some of large molecular weight, are removed by this means. In addition, the small amount of trehalose that can be extracted by this means is very highly labelled after the provision of [^{14}C]glucose, whereas the remainder is much less accessible to exogenous substrates. Accordingly, it appears as though dormant ascospores maintain two compartments of trehalose and that the larger one is inaccessible to solvents and exogenously added substrates. The presence of only one such pool in macroconidia of *Neurospora*, which do not require special treatment to germinate, made this observation all the more intriguing (Sussman, 1966a). However, recent experiments have revealed that the major compartment is not rendered more easily extractable by activation, so that its relevance to dormancy is questionable. If there is a separation of enzyme and substrate of the kind postulated, other methods will have to be devised to prove it, including, perhaps, *in vivo* studies using histochemical and radioautographic techniques.

Another type of compartmentation can be considered which is based upon the well-known effect of metabolites, other than substrates, upon

Table 4. *Proportion of trehalose extracted from, and label in, intact and ground spores of* Neurospora

	Total trehalose (%)	C.p.m./μmole
Conidia (*N. crassa*)		
Extract* from intact cells	99·9	.
Extract* from ground cells	0·1	.
Ascospores† (*N. tetrasperma*)		
Extract‡ from intact cells	0·2	119,000
Extract‡ from ground cells	99·8	56

* Hot water extract.   † Data of Budd, Sussman & Eilers (1966).
‡ Hot 80% ethanol extract.

enzyme activity. Such an 'allosteric' effect might be triggered by heat-activation if a previously separated metabolite and an enzyme to which it can attach were to be brought together. Inasmuch as allosteric effectors cause conformational changes this mechanism overlaps with mechanism 2, wherein such changes in a precursor are included. Although we have found that trehalase activity may be influenced by metabolites other than its substrate the effects have not yet been studied in relation to dormancy.

DIFFERENCES BETWEEN HEAT- AND CHEMICAL-ACTIVATION

The work described above mainly relates to cells activated by heat. What about those activated by chemical means? Although the over-all respiratory and structural effects are known to be similar, enough differences have appeared to merit a detailed comparison of the mechanisms of action of the two means of activation. Thus, it has already been pointed out that heat-activated ascospores may be reversibly deactivated when incubated anaerobically or at 4° C., whereas furfural-activated ones cannot be. Also noted above was the tendency of ascospores upon ageing to become unresponsive to chemical activators while retaining their heat-activability for years. Other differences exist and are outlined in Table 5, wherein it can be seen that there are differences in response to poisons and concentration of spores as well.

Table 5. *Differences between ascospores of* Neurospora *that have been heat- or chemically activated**

Effect	Activated by	
	Heat	Furfural
Poisoned by $NaHSO_3$	+	−
Poisoned by NaN_3	+	−
Reversible deactivation by CN^- or 4° C.	+	−
Influence of concentration of spores	−	+
Loss in activability with age	−	+

* Taken from Sussman (1961).

Respiratory differences also can be perceived when rates are measured by the oxygen electrode and infrared gas analyser (Figs. 6, 7). These data of my student, Mr Fred Eilers, indicate that the increase in the rate of CO_2 release attendant upon activation is more rapid in the case of heat-treated spores (Fig. 6) than in those to which furfural is added (Fig. 7). Furthermore, the maximum increase in the rate of O_2 uptake lags even further behind in furfural-activated ascospores.

In an attempt to explain the failure of aged ascospores to respond to furfural, respirometric studies were performed (Fig. 8). These data show that the rise in O_2 uptake associated with activation (Fig. 3) does not occur when old spores are used and that after 120 min., when the germ tube

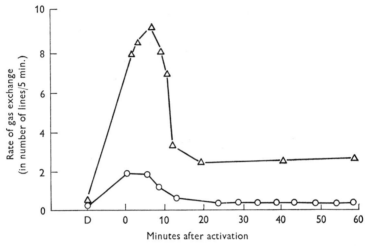

Fig. 6. Respiratory rates of heat-activated ascospores of *Neurospora tetrasperma*. Measurements made with infrared analyser (CO_2) and oxygen electrode. The following symbols were used: CO_2 (△); O_2 (○). (From Eilers, 1968.)

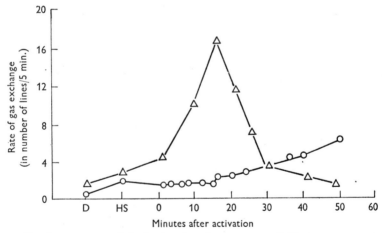

Fig. 7. Respiratory rates of furfural-activated ascospores of *Neurospora tetrasperma*. Measurements and symbols as in Fig. 6. (From Eilers, 1968.)

protrudes, a definite levelling-off in rate occurs. When excessive amounts of furfural (5×10^{-2} M) are added, as in the experiments summarized in Figs. 9 and 10, the O_2 uptake of aged spores increases faster than that of young spores after an initial lag, but falls again after about 120 min., but

DORMANCY OF FUNGUS SPORES

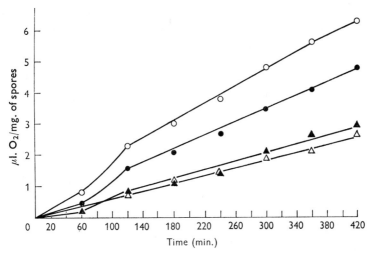

Fig. 8. Effect of furfural (1×10^{-3} M) upon the O_2 uptake of aged ascospores of *Neurospora tetrasperma*. ○, 1×10^{-2} M furfural; ●, 1×10^{-3} M furfural; ▲, 1×10^{-4} M furfural; △, H_2O control. (From Sussman, 1953 b.)

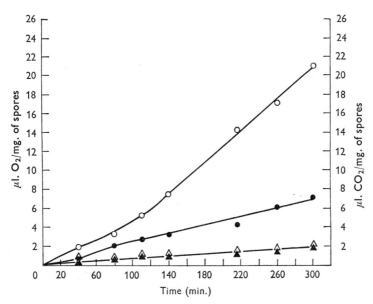

Fig. 9. Effect of high concentrations of furfural (5×10^{-2} M) upon the respiration of freshly harvested ascospores of *Neurospora tetrasperma*. ●, O_2 consumption of 'fresh' spores +furfural (5×10^{-2} M); ○, CO_2 output of 'fresh' spores+furfural (5×10^{-2} M); ▲, O_2 consumption of 'fresh' spores+H_2O; △, CO_2 output of 'fresh' spores+H_2O.

the uptake over-all is comparable. As for CO_2 release there is a lag soon after activation and a rapid increase between 60–90 min., after which there is a distinct inhibition. It is possible that the differences between aged and fresh spores may be due to slower uptake in the former at the start, but the diminished respiratory activity after 120 hr. argues for a more basic difference. These observations have not been pursued further but it is possible that by defining the lesion in aged spores the key step(s) in the activation of fresh ascospores will become more apparent.

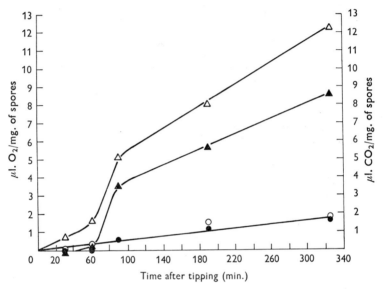

Fig. 10. Effect of high concentrations of furfural (5×10^{-2} M) upon the respiration of aged ascospores of *Neurospora tetrasperma*. ●, 'Old' spores+H_2O (oxygen uptake); ○, 'old' spores+H_2O (CO_2 output); ▲, 'old' spores+5×10^{-2} M furfural; △, 'old' spores +5×10^{-2} M furfural (CO_2 output). (From Sussman, 1953b.)

INTERMEDIARY METABOLISM AND ACTIVATION

The heat-induced increase in the production of CO_2, ethanol, acetaldehyde and pyruvate from a hitherto unavailable pool of trehalose suggest the de-repression of glycolysis, a pentose shunt, or both. Therefore, the concentration of intermediates in these pathways was studied in dormant ascospores and as a function of time after activation. These analyses were carried out by Mr Eilers, with the help of my colleague Dr Hiroshi Ikuma, utilizing the fluorimetric determination of changes in pyridine nucleotide concentrations in a linked series of enzymic reactions whereby both sensitivity and specificity are assured (Bergmeyer, 1963). An example of the data that can be obtained by these means is provided in Fig. 11, which reveals the percentage increase in ethanol, malate and pyridine nucleotides which result from furfural activation. Ascospores that were heat-sensitized as described before (Fig. 1) were used as the basis for comparison, thereby cancelling out the effect of this treatment. These results show that there is a rapid increase in malate and in reduced pyridine nucleotides followed by a slower increase in ethanol. Clearly, the accumulation of reducing

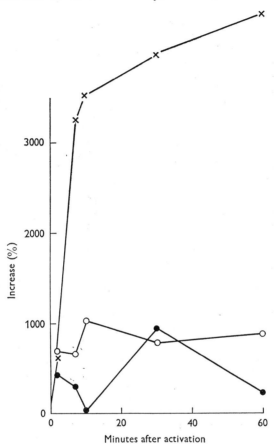

Fig. 11. Percentage increase in ethanol, malate and reduced pyridine nucleotides after furfural-activation of ascospores of *Neurospora tetrasperma*. The reference values were taken from heat-sensitized spores. ×, ethanol; ○, NADH+NADPH; ●, malate. (From Eilers, 1968.)

power in the form of NADPH and NADH appears to lead to the enhanced formation of ethanol—a relatively reduced product.

When data such as those in Fig. 11 for other metabolic intermediates are consolidated, using heat-activated ascospores, Fig. 12 is obtained. Such a 'phase-plot' uses different glyphs to designate the concentration of intermediates, which are arranged in the direction of a metabolic sequence, at various times after activation so that the dynamics of the changes can be illustrated. The concentrations of glycolytic intermediates are the only ones provided, because the levels of those in the pentose pathway are too low to be measured by our techniques, suggesting that glycolysis is the principal dissimilative sequence in ascospores of *Neurospora*. Of the intermediates tested, only glucose-6-phosphate (G-6-P), fructose-6-phosphate

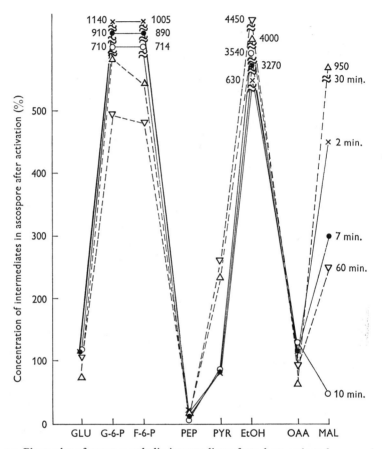

Fig. 12. Phase-plot of some metabolic intermediates from heat-activated spores during the first 60 min. after activation using the concentrations in dormant spores as the reference point (100%). The following symbols designate the times after activation: ×, 2 min.; ●, 7 min.; △, 30 min.; ▽, 60 min. GLU=glucose; PYR=pyruvate; EtOH=ethanol; OAA=oxaloactic acid; MAL=malate. (From Eilers, 1968.)

(F-6-P) and ethanol increased almost in proportion to the observed respiratory changes (Fig. 3) *and within 2 min. of measurements beginning*. Somewhat later, malate showed a smaller increase. The results obtained with furfural-activated ascospores are similar except that smaller and less rapid changes occur; moreover, phosphoenolpyruvate (PEP) accumulates more rapidly than ethanol, in contrast to heat-activated cells.

Data like those in Fig. 13 have been widely used for the detection of rate-limiting steps in metabolism through the application of the 'cross-over theorem' (Chance, Holmes, Higgins & Connelly, 1958). It holds that control is exerted in a metabolic pathway at a point where there is a sharp change in the concentration of metabolites arranged in a 'phase-plot'. The cross-over theorem can be applied when the levels of intermediates are measured during a transition between two steady states, provided the total concentrations of the intermediates are fixed and the only variation is in the 'substrate/product' ratio, that there are no branch points or loops in the pathway, and that a single pool exists for each intermediate. The latter

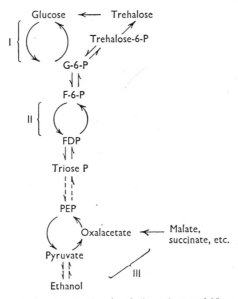

Fig. 13. Possible control points in the glycolytic pathway of *Neurospora* ascospores.

two conditions can only be guessed at in ascospores but the first condition is met approximately when the amount of substrate used is computed. Therefore not all the prerequisites for analysis by the cross-over theorem are known to hold, but since not all have been fulfilled in any study reported up to now (Scrutton & Utter, 1968), and the technique is one of the few available for such analyses, it will be used—but gingerly.

The observations recorded in Fig. 12 might be explained by the presence of a cross-over point between glucose and G-6-P (Fig. 13), probably mediated by hexokinase. Another point of control probably exists between F-6-P and PEP, at the level of phosphofructokinase, which may be inhibited or is present in limiting amounts. As ethanol accumulates it may cause the accumulation of a small amount of pyruvate, which leads to an increase in the amount of malate. A comparison between Figs. 6 and 12 shows that CO_2 follows the ethanol concentration profile, wherein the rate of production is fastest between 2 and 7 min., slowing down thereafter. These conclusions can be drawn from the data on furfural as well, which will be reported in more detail elsewhere (Eilers, 1968), except that pyruvate kinase may be another control point because PEP and pyruvate concentrations oscillate 180° out of phase. Another difference is that malate accumulates faster than ethanol after furfural-activation, suggesting the possibility that furfural affects malate dehydrogenase, condensing enzyme or others at this point in metabolism.

SITE OF ACTIVATION BY FURFURAL

Because the effects described in the previous section come after the initial event in activation—that is, the adoption of trehalose as a substrate—we must look further for the essential difference between the types of activators. Therefore the immediate site of action of furfural was sought.

Studies of the uptake of labelled furfural reveal that 0·0076 mg./100 mg. ascospores, or 25% of that added, is irreversibly bound, at 4° C. as well as room temperature. Moreover, the kinetics of uptake resemble those of an adsorption isotherm and repeated additions result in very little further uptake (Sussman, 1961), suggesting binding to the exterior of the cell. This possibility is supported by experiments in which ascospores to which labelled furfural had been added were disrupted and separated into wall and soluble components. The former were combusted in the presence of acid and shown to possess the majority of furfural that had been removed from solution (Table 6). Therefore, it seems likely that the irreversibly bound furfural is responsible for activation and that its site of action is the cell surface.

Table 6. *Furfural conversion and uptake in cell-free extracts of* Neurospora *ascospores**

Time of incubation (min.)	Wall fraction* (mg.)	Intermediate fraction† (mg.)	Supernatant fraction‡ (mg.)
2	0·163	0·032	0·00
30	0·166	0·025	0·00
60	0·157	0·030	0·00
120	0·093	0·025	0·00
180	0·000	0·029	0·00

* Material sedimenting at 1000*g* for 10 min.
† Material sedimenting between 1000*g* and 15,000*g* for 15 min.
‡ Supernatant fraction—material not sedimenting at 15,000*g* for 15 min.

PLATE I

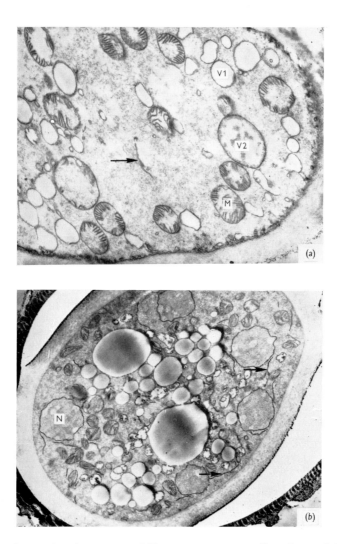

Electron micrographs of ascospores of *Neurospora tetrasperma* (from Lowry & Sussman, 1968). (*a*) Dormant ascospores showing two types of vacuoles, V_1 and V_2, and endoplasmic reticulum-like membrane at arrow. Note also the swollen mitochondria (*M*). × 12,000. (*b*) A cracked, germinating ascospore showing typical endoplasmic reticulum (arrows) and a nucleus (N). Vacuoles and mitochondria are also present: × 9,000.

(*Facing p.* 118)

PLATE 2

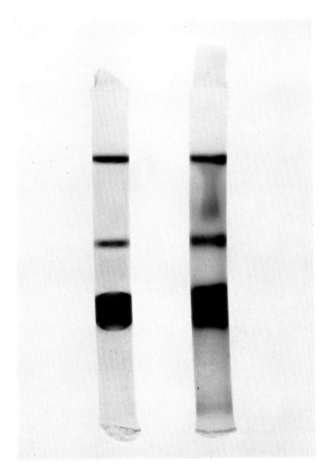

Acrylamide gel cylinders of trehalase from mycelium of *Neurospora crassa*. Extract was treated with 8 M urea. The native form of the enzyme is the fastest running band and subunits run behind it. Origin is at top of figure.

SUMMARY AND CONCLUSIONS

The diagram in Fig. 14 summarizes some of the data presented above. Thus, it is likely that both heat- and furfural-activation are accomplished by a two-step process, but different ones. When heat is applied a reaction that is reversible at 4° C. and in a nitrogen atmosphere is begun. A 'point-of-no-return' is reached if the conditions of incubation are adequate, upon which the cell loses its heat resistance and is ready to germinate. Inasmuch as reversibly deactivated ascospores may be reactivated by furfural the surface sites to which this compound are attached are unaffected by heat-activation, implying that the latter treatment affects internal sites, or different ones on the surface.

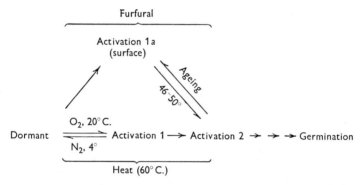

Fig. 14. Comparison of pathways through which heat- and furfural-activation are accomplished in *Neurospora* ascospores.

In contrast to heat-activation, that by furfural involves an irreversible first step which probably occurs at the cell surface. The nature of the reaction which occurs at the surface which causes the many changes described is unknown, but may be one like those discussed on page 109. It is possible that the lack of effect of azide and other inhibitors upon chemically activated cells (Table 5) is due to the change in the cell surface caused by this treatment. The irreversible first step is followed by a step that is blocked in aged spores, which can be overcome by heat-sensitization. The nature of this block is unknown but it appears to involve mainly the mechanisms that lead to enhanced O_2 uptake in activated cells.

The slowness of the respiratory responses of furfural-activated ascospores, as compared to heat-treated ones, may be due to an effect on the reactions beyond pyruvate (Fig. 13). Thus the formation of large amounts of ethanol (Fig. 12) after heat-activation is followed by a rapid increase in O_2 uptake (Fig. 6), probably due to the induction or stimulation of terminal oxidations via the cytochrome system (Sussman *et al.* 1956;

Holton, 1960). However, O_2 uptake lags much farther behind CO_2 evolution in furfural-activated ascospores (Fig. 7), so that this substance may inhibit a reaction that is required for the development of their full potential for O_2 uptake. One such possibility is the paucity of cytochrome c in mitochondria of dormant ascospores (Holton, 1960), whose structural peculiarities already have been discussed (Plate 1, a–b). But this is only one of the possibilities that must be studied along with the plethora of other questions which remain to be answered about these developmental events that occur in a single cell.

The experimental work reported herein has been supported by research grant GB 6811 from the National Science Foundation. In addition, the assistance of Miss Theresa Jepson, Mrs Anne Whitten and Dr S. A. Yu, along with that of my colleagues Drs R. J. Lowry and H. Ikuma, is gratefully acknowledged. In particular, I am grateful to Mr F. I. Eilers, who permitted me to use some of his data before they were published.

REFERENCES

BERGMEYER, H. U. (1963). *Methods of Enzymatic Analysis.* New York: Academic Press.
BUDD, K., SUSSMAN, A. S. & EILERS, F. I. (1966). *J. Bact.* **91**, 551.
CHANCE, B., HOLMES, W., HIGGINS, J. J. & CONNELLY, C. M. (1958). *Nature, Lond.* **182**, 1190.
COCHRANE, J. C., COCHRANE, V. W., SIMON, F. G. & SPAETH, J. (1963). *Phytopathology* **53**, 1155.
EILERS, F. I. (1968). *The Effect of Furfural upon Ascospores of Neurospora.* Thesis, University of Michigan, Ann Arbor.
EMERSON, M. R. (1948). *J. Bact.* **55**, 327.
EMERSON, M. R. (1954). *Pl. Physiol., Lancaster* **29**, 418.
GODDARD, D. R. (1935). *J. gen. Physiol.* **19**, 45.
GODDARD, D. R. (1939). *Cold Spring Harb. Symp. quant. Biol.* **7**, 362.
HENNEY, H. & STORCK, R. (1963a). *Science, N.Y.* **142**, 1675.
HENNEY, H. & STORCK, R. (1963b). *J. Bact.* **85**, 822.
HENNEY, H. & STORCK, R. (1964). *Proc. natn. Acad. Sci. U.S.A.* **51**, 1050.
HILL, E. P. & SUSSMAN, A. S. (1963). *Archs Biochem. Biophys.* **102**, 389.
HILL, E. P. & SUSSMAN, A. S. (1964). *J. Bact.* **88**, 1556.
HOLTON, R. (1958). Pyruvate metabolism and electron transport in *Neurospora tetrasperma.* Thesis, University of Michigan, Ann Arbor.
HOLTON, R. (1960). *Pl. Physiol., Lancaster* **35**, 757.
KATES, M. (1957). *Can. J. Biochem. Physiol.* **35**, 127.
LEWIS, J. C., SNELL, N. S. & BURR, H. K. (1960). *Science, N.Y.* **132**, 544.
LINGAPPA, B. T. & SUSSMAN, A. S. (1959). *Pl. Physiol., Lancaster* **34**, 466.
LINGAPPA, Y. & SUSSMAN, A. S. (1959). *Am. J. Bot.* **46**, 671.
LOWRY, R. J. & SUSSMAN, A. S. (1958). *Am. J. Bot.* **45**, 397.
LOWRY, R. J. & SUSSMAN, A. S. (1968). *J. gen. Microbiol.* **51**, 403.
MEFFERD, R. D. JUN. & CAMPBELL, L. L. (1951). *J. Bact.* **62**, 130.

RYAN, M. T. & MAVRIDES, C. A. (1959). *Science, N.Y.* **131**, 101.
SCRUTTON, M. C. & UTTER, M. F. (1968). *A. Rev. Biochem.* **37**, 249.
SHEAR, C. L. & DODGE, B. O. (1927). *J. agric. Res.* **34**, 1019.
STANIER, R. Y. (1953). In *Adaptation in Microorganisms*, p. 1. Cambridge University Press.
SUN, C. Y. & SUSSMAN, A. S. (1960). *Am. J. Bot.* **47**, 589.
SUSSMAN, A. S. (1953a). *J. gen. Microbiol.* **8**, 211.
SUSSMAN, A. S. (1953b). *Am. J. Bot.* **40**, 401.
SUSSMAN, A. S. (1954a). *Mycologia* **46**, 143.
SUSSMAN, A. S. (1954b). *J. gen. Physiol.* **38**, 59.
SUSSMAN, A. S. (1961). In *Spores*. Minneapolis: Burgess.
SUSSMAN, A. S. (1965). In *Ecology of Soil-borne Plant Pathogens—Prelude to Biological Control*, p. 99. Berkeley: National Academy of Sciences of the U.S.A.
SUSSMAN, A. S. (1966a). In *The Fungus Spore*, p. 235. London: Butterworth.
SUSSMAN, A. S. (1966b). In *The Fungi*. New York: Academic Press.
SUSSMAN, A. S. (1968a). In *The Bacterial Spore*. London: Academic Press. (In the Press.)
SUSSMAN, A. S. (1968b). In *The Fungi*. New York: Academic Press. (In the Press.)
SUSSMAN, A. S., DISTLER, J. R. & KRAKOW, J. S. (1956). *Pl. Physiol., Lancaster* **31**, 126.
SUSSMAN, A. S. & HALVORSON, H. O. (1966). *Spores: Their Dormancy and Germination*. New York: Harper and Row.
SUSSMAN, A. S., HOLTON, R. & VON BÖVENTER-HEIDENHAIN, B. (1958). *Arch. Mikrobiol.* **39**, 38.
SUSSMAN, A. S. & LOWRY, R. J. (1955). *J. Bact.* **70**, 675.
SUSSMAN, A. S., LOWRY, R. J. & TYRRELL, E. (1959). *Mycologia* **51**, 237.
SWARTZ, M. N., KAPLAN, N. O. & FRECH, M. E. (1956). *Science, N.Y.* **123**, 50.
WENT, F. A. F. C. (1901). *Zentbl. Bakt. ParasitKde.* (Abt. II), **7**, 544.

SURVIVAL OF ALGAE UNDER ADVERSE CONDITIONS

By G. E. FOGG

Department of Botany, Westfield College, London, N.W. 3

INTRODUCTION

Among the phylogenetically diverse organisms classified as algae are some with remarkable ability to survive and even grow under extreme conditions. Snow algae, mostly members of the Chlorophyceae, appear capable of continued existence at temperatures which never exceed 0° C. by more than a fraction of a degree. On the other hand the Cyanophyceae of hot springs are capable of growth at temperatures of at least 75° C. Other Cyanophyceae grow in waters saturated with sodium chloride or carbonate and both Chlorophyceae and Cyanophyceae resist prolonged desiccation in desert soils. The Phaeophyceae characteristic of the upper part of the intertidal zone withstand desiccation and rapid change between extremes of temperature and salinity. Many oceanic plankton algae, members of the Bacillariophyceae, Chrysophyceae and Haptophyceae, are so delicate and exacting that there has so far been little success in keeping them alive for long after collection, yet it seems that they must survive prolonged periods of darkness when the water masses which contain them move out of the photic zone. Algae, being generally small and morphologically simple, must accept the environmental conditions in their immediate vicinity and cannot alter them or exploit the favourable characteristics of spatially separated phases, such as soil and the aerial environment, as a higher plant can. The ability to survive extreme conditions is undoubtedly of particular importance in this group.

This paper is not intended to be a comprehensive review. Some topics, such as the remarkable ability of some algae to survive heavy doses of ionizing radiation (Godward, 1962) and survival of marine and Antarctic algae in darkness (Fogg, 1969; Fogg & Horne, 1969), will not be considered. The object will be to discuss the physiological aspects of the resistance of algae to adverse conditions but, as will become apparent, precise investigations of the effects of individual factors and the mechanisms of resistance to them are scanty.

TEMPERATURE

Survival at low temperatures

Obviously, a great many different algae can survive temperatures around freezing point. For example, lists of Antarctic benthic marine algae (Papenfuss, 1964) and freshwater algae (Hirano, 1965) contain many hundreds of species names and suggest that no major groups, with the possible exception of the Haptophyceae and Dinophyceae, are characterized by cold-sensitiveness.

More precise information about the effects of exposure to low temperatures on microscopic algae has come from work on the preservation of cultures. Daily & McGuire (1954) investigated the viability of 32 species of algae freeze-dried after suspension in horse serum and freezing at $-55°$ C. Out of 22 species of Chlorophyceae 15 proved viable when tested 24 hr. after freeze-drying, some, but not all, remaining alive in the freeze-dried condition for up to 10 months, but the proportion of viable cells was low, ranging from 3·2% for *Bracteococcus cinnabarinus* to 0·013% for *Chlamydomonas pseudococcum*. All seven species of blue-green alga and both species of Chrysophyceae tested remained viable, no percentage survival values being obtained, but *Navicula minima* was killed. Watanabe (1959) found that the blue-green algae *Tolypothrix tenuis* and *Calothrix brevissima* could be preserved for 2 years in the freeze-dried state with human serum albumen as protective substance, about half the number of cells proving viable after this period. Whitton (1962) tested 18 strains of blue-green algae for ability to grow after portions of ordinary growing cultures had been frozen at $-15°$ C. for 2 months and found that 11 of them withstood the treatment. One species which did not, *Anacystis nidulans*, was found to have a detectable proportion of viable cells, about 1 in 10^6, up to 4 hr. after freezing; there thus appear to be distinct 'initial' and 'storage' effects such as are found in bacteria (Ingraham, 1962). Exponentially growing *Anabaena cylindrica* was killed rapidly and completely by freezing but its spores survived. The ability of blue-green algae to survive freezing is not improved by adjuvants such as horse serum or glycerol. This is in sharp contrast to Chlorophyceae, the survival of which is much improved by such additions (Holm-Hansen, 1967a). Hwang & Horneland (1965) found that *Euglena gracilis*, as well as various Chlorophyceae, could be preserved by freeze-drying and remain viable for periods of years if stored at the temperature of liquid nitrogen. Blue-green algae were found by Holm-Hansen (1967b) to survive when freeze-dried to a much better degree than Chlorophyceae.

Survival of freezing depends on the physiological condition of the cells and the exact circumstances under which freezing and revival takes place. Duthie (1964) found that desmids of the genera *Cosmarium* and *Staurastrum* survived freezing at $-5°$ C. for 7–42 days much better if the cells came from cultures grown at 4 °C. than if they had been grown at 20 °C. However, attempts to adapt seaweeds to cold have met with little success (Biebl, 1962). With the blue-green alga *Nostoc muscorum* Holm-Hansen (1963, 1967b) found no significant difference in viability between samples freeze-dried at $-25°$ C. or at room temperature and between those dried for 12 hr. or for periods up to 101 hr. However, *Chlorella pyrenoidosa* did not survive freeze-drying when the cells were at $-25°$ C. throughout the drying period, whereas it showed good survival when the frozen cells were dried rapidly in a desiccator at room temperature (Holm-Hansen, 1967b). Controlled freezing to $-30°$ C. at the rate of about $1°$ C./min. results in better survival of freeze-dried algae than does uncontrolled freezing, within about 25 sec., to the same temperature (Hwang & Horneland, 1965).

Observations on the survival of low temperatures by marine algae have been summarized by Biebl (1962). He stresses the point that death is dependent on time of exposure as well as temperature and suggests that an exposure period of 12 hr. is a practical and ecologically relevant period to use in experiments. *Fucus vesiculosus* in the Arctic survives temperatures of $-40°$ C. for many months but there is experimental evidence of seasonal variation in the lower limit of tolerance of this species, from $-45°$ C. in winter to $-30°$ C. in summer. As in higher plants the effect of freezing is complex, involving desiccation, through abstraction of water from the protoplasm in the formation of ice crystals, as well as low temperature. *Ulva pertusa* shows frost plasmolysis but not intracellular freezing at $-10°$ C. and may survive this for at least 24 hr. *Fucus* and *Ulva* are characteristic intertidal algae and contrast with sublittoral algae, which are much more sensitive to low temperatures. For example, five species of intertidal algae at Naples (40° 50′ N.) survived the lowest temperature employed, $-8°$ C., but of 13 sublittoral species from the same locality none survived freezing although about half tolerated cooling to $-2°$ C. if ice formation did not occur. Intertidal algae are also less affected by abrupt temperature changes, such as occur when they are submerged by the rising tide, than are sublittoral species.

Growth at low temperatures

Algae may not merely survive at temperatures around 0° C. but show active growth. This seems to be so for the phytoplankton of Arctic and Antarctic seas, which give the impression of being among the most productive of any sea areas. Determinations using the radiocarbon technique show that rates of photosynthesis per unit area of sea surface in such situations, while generally lower, are sometimes as high as those in fertile temperate waters (e.g. Horne, Fogg & Eagle, 1969). The best comparison is provided by data obtained by Steemann Nielsen & Hansen (1959), who showed that at light saturation the rate of photosynthesis per unit amount of chlorophyll was about the same (between 3 and 4 mg. carbon.mg.$^{-1}$. chlorophyll.hr.$^{-1}$) in cold-water plankton at about 5° C. as in plankton from temperate seas at 15° C. This they explained on the assumption that in the cold-water algae higher concentrations of enzymes compensated for the effect of the lower temperature on the rate of the non-photochemical reactions of photosynthesis. The relation between rate of photosynthesis at light saturation and chlorophyll content is highly variable, being affected considerably by availability of nutrients and the physiological state of the cells, so that by itself this evidence is not conclusive. However, Steemann Nielsen & Jørgensen (1968) and Jørgensen (1968) have shown that although under otherwise similar conditions rates of photosynthesis are nearly the same in cells of the marine diatom *Skeletonema costatum* adapted to grow at 7° and 20° C. the relative growth-rate is less than half at the lower temperature of what it is at the higher. In accordance with this it is found that cells grown at 7° C. are larger and contain more protein than those adapted to 20° C. The next step should be to test this hypothesis directly by assay of an enzyme such as ribulose diphosphate carboxylase. Some further support for it comes from determinations of the relative growth-rate of a cold-water planktonic *Chlorella* which showed no light saturation even at 15,000 lux at 20° C. (Fogg & Belcher, 1961). Since similar algae usually show light saturation at about 4000 lux this suggests a relatively high capacity of this *Chlorella* to carry out the non-photochemical reactions of photosynthesis. It may be noted that the Q_{10} of the growth of this *Chlorella* is low (Fig. 1), so that although its optimum temperature is as high as 22° C. its relative growth-rate is then much less than that of an ordinary *Chlorella* strain at the same temperature. This feature is also shown by psychrophilic bacteria but the explanation of the phenomenon is unknown (Ingraham, 1962). Other means of moderating the effect of low ambient temperatures probably operate as well as possible increase in concentration of particular enzymes. It is generally found that compensation occurs at lower light intensities at low temperatures (Rabinowitch, 1956), a circumstance which will favour growth under these conditions. In a study by mass spectrometry of the oxygen exchanges by pure cultures of a diatom, *Fragilaris sublinearis*, which grows in Antarctic sea ice at temperatures as low as $-2°$ C., Bunt, Owens & Hoch (1966) found complicated effects of light intensity, wavelength and temperature on oxygen uptake in the light. The outcome seems to be that whereas the optimum for photosynthetic oxygen evolution by this diatom occurs at around 7° C., oxygen uptake either in the light or in the dark is at a maximum between 10° and 24° C., so that in this temperature range net oxygen production cannot occur and the organism is thus obligately psychrophilic. Another feature which contributes to its inability to grow at temperatures as high as 24° C is that at such temperatures it is unable to compensate

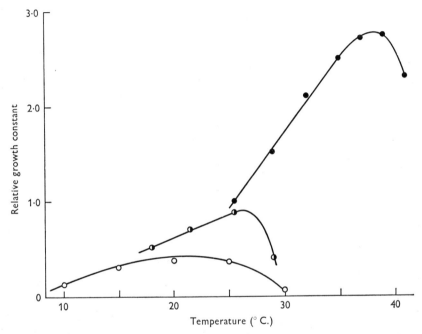

Fig. 1. Relative growth rates ($\log_{10}$ day^{-1} units) at about 15,000 lux of three strains of *Chlorella* as a function of temperature. ○, *C. pyrenoidosa*, cold-water strain isolated from Torneträsk, Sweden (data of Fogg & Belcher, 1961); ◐, *C. pyrenoidosa*, Emerson strain; ●, high-temperature strain, *C. sorokiniana* (data of Sorokin & Myers, 1953).

for photo-destruction of its pigments. Algal felts, which are more efficient in absorbing radiant energy and retaining heat than planktonic forms, are most successful in Antarctic fresh waters (Fogg & Horne, 1969).

It seems that most of the growth of terrestrial algae in frigid situations is achieved in short periods when their micro-environment heats up well above the ambient temperature. Holdgate (1964) has shown that soil or moss in the maritime Antarctic may achieve in sunshine temperatures 10 or 15° C. above that of the air, and Fogg & Stewart (1968) found that although *Nostoc commune* in the same area fixes atmospheric nitrogen at an appreciable rate at about freezing point, the rate accelerates enormously, with a Q_{10} of about 6, with rise in temperature. Apparently rapid growth at temperatures which cannot be greatly in excess of freezing point is, nevertheless, shown by snow algae. These algae, which belong to species of Chlorophyceae and Chrysophyceae almost exclusively confined to this habitat, occur in regions with permanent snow fields and appear, sometimes with great suddenness, as red, green or yellow patches on the snow surface during periods of thaw. Investigations both by direct counts and by *in situ* determinations of rates of photosynthesis using a radiocarbon technique have, however, shown that their rate of growth is slow, the doubling time being estimated as about 23 days (Fogg, 1967). The sudden appearance of patches seems to be due merely to mechanical accumulation at the surface of cells, previously distributed sparsely through its thickness, as ablation of the snow proceeds. The snow of Signy Island, on which snow algae are at times conspicuous, is relatively rich in mineral nutrients derived from the sea and rock debris. It seems likely that a major factor, apart from low temperature, limiting the growth of these algae is desiccation.

SURVIVAL AND GROWTH AT HIGH TEMPERATURES

At the other extreme, certain algae are able to survive at physiologically high temperatures. Most striking are those found in hot springs. The two characteristic algal groups in this habitat are the diatoms (mostly species of *Navicula* and *Nitzschia*) reported as occurring at temperatures up to $50.7°$ C., and the blue-green algae, reported as surviving in water at $86.6°$ C. (Gessner, 1955; Mann & Schlichting, 1967). It should be noted that investigators may not have always realized how steep the temperature gradients in thermal situations may be, and it must also be said that the taxonomy of the organisms found at the highest temperatures is unsatisfactory. Brock (1967) puts the upper limit for growth of blue-green algae in Yellowstone Park at $75°$ C., although he found bacteria, including 'flexibacteria' which seem to be apochlorotic blue-green algae, in the same situation at $91°$ C.

Blue-green algae are found in alkaline but not in acid hot springs. The number of species and varieties found decreases from about 90 in waters at $30°$ C. to one or two at temperatures in the neighbourhood of $80°$ C. (Gessner, 1955). Similarly, the biomass at the highest temperatures is small, the maximum occurring between $50°$ and $55°$ C. (Brock, 1967). *Cyanidium caldarium*, a coccoid alga with a blue-green chromatophore, is found in acid hot springs, having been isolated by Allen (1959), for example, from a spring containing about 0.1 N sulphuric acid and with a temperature of $70-75°$ C. This alga is more likely to be a member of the Cryptophyceae than of the Cyanophyceae (Fogg, 1956; Lewin, 1961). The algae of arid soils and rock surfaces, which are principally members of the Cyanophyceae and coccoid Chlorophyceae, reach high temperatures in full sun but there appear to be no adequate records of the temperatures which can be tolerated. Apart from Cyanophyceae most freshwater and marine phytoplankton and soil algae of temperate regions do not tolerate temperatures much in excess of 30° C. Intertidal algae at Naples (40° 50′ N.) mostly have an upper temperature limit for survival of about 35° C., whereas at Roscoff (48° 43′ N.) seven species tested died at temperatures of more than 30° C., only *Enteromorpha intestinalis* surviving a 12 hr. period at 35° C. Less difference was observed between sublittoral algae from these two places, temperatures from 27–30° C. being tolerated and all species being killed after 12 hr. at 35° C. (Biebl, 1962). Both for intertidal algae and algae of arid situations the effects of high temperatures are complicated by those of desiccation, a matter which will be discussed later.

Upper temperature limits for laboratory-grown algae are reported frequently in the literature. Such values are dependent, to some extent, not only on exposure times but on the levels of other factors. It might be expected that strains would be adaptable, by repeated subculture at progressively higher temperatures, to survive at higher temperatures, but Sorokin & Myers (1953) were unable to train *Chlorella pyrenoidosa* to maintain a stable growth-rate at temperatures higher than 30° C. On the other hand Löwenstein (1903) found that the blue-green alga *Mastigocladus laminosus*, which grows naturally at 52° C., lost its capacity to survive at such temperatures on prolonged culture at room temperature. Working on the hypothesis that inability to survive at high temperatures is due to impairment of particular biosynthetic systems, Baker, Hutner & Sobotka (1955) attempted to grow different strains of *Euglena gracilis* at high temperatures in media supplemented with various metabolites but could not raise the upper temperature limit in this way although similar methods employed with thermophilic bacteria were successful. However, Hutner *et al.* (1957) found that the flagellate *Ochromonas malhamensis* can be grown above 35° C., the 'natural' maximum temperature tolerated in the normally adequate basal medium, if thiamine and vitamin B_{12} were supplied at 300 times the normal concentrations. Nevertheless, it appears that the temperature limits of algae are mainly determined by genetically fixed characters. Peary & Castenholz (1964) from an examination of numerous isolates of the blue-green alga *Synechococcus*, some of which survived at 75° C., concluded that there were at least four strains characterized by different temperature optima and lethal temperatures as well as by morphological features.

McLean (1967) found that resistance of dried cells of the chlorococcalean alga *Spongiochloris typica* to heating at 100–105° C. for 1 hr. was dependent on the age of the culture from which the cells were derived. Maximum resistance, with about 8% survival as compared with 1% or less for older or younger material, was shown after 3 weeks. In medium diluted to one-tenth of its usual concentration the peak of resistance was reached after 2 weeks; this may be attributed to the equivalent physiological state being reached more quickly in deficient cultures than in ones with ample nutrient supply. The state of hydration of the cells is clearly important. Field observations suggest that desiccated algal cells are able to withstand higher temperatures than fully hydrated ones and laboratory experiments bear this out. For example, after freezing, samples of *Nostoc muscorum* and *Chlorella pyrenoidosa* were dried for varying periods of time, then heated to 100° C. for 10 min. *Chlorella* dried for 46 hr. or less failed to survive this treatment but samples dried for longer contained viable cells, the proportion rising from 0.11×10^{-6}% after drying for 59 hr. to 0.23×10^{-3}% after drying for 101 hr. *Nostoc*, however, behaved quite differently, surviving the heat treatment but showing no increase in viability with increase in length of the drying period (Holm-Hansen, 1963). Possibly only a certain level of dehydration is necessary to confer resistance and this was exceeded in these experiments. Glade (1914) found that cells of another blue-green alga, *Cylindrospermum majus*, could survive temperatures of 95–100° C. when dry. Intertidal seaweeds also withstand high temperatures better if desiccated. Dried plants of *Bangia fuscopurpurea* survived at 42° C. for 24 hr. whereas in sea water they could not tolerate 35° C. for a similar period (Biebl, 1962). A factor which may contribute to this increased resistance is the reduction in respiration rate which occurs in intertidal seaweeds on desiccation (Bergquist, 1957, and W. E. Isaac (personal communication), both quoted by Newell & Pye, 1968).

Thermosensitivity may be induced in cells of a variety of different kinds of plants by previous heat-shock (Fries & Söderström, 1964). *Chlorella pyrenoidosa* heated to 48° C. for 2 min. showed a reduction in viability to less than 50% of that of untreated control material when subsequently cultured at 28° C.

Spores, when produced, are perhaps generally more resistant to high temperatures than vegetative cells but there is little published evidence supporting this idea. Glade (1914) showed that spores of *Cylindrospermum* are much more resistant than its vegetative cells. From data given by Fay (1969) it can be calculated that the spores of *Anabaena cylindrica* contain only 0.15 dry matter per unit volume as compared with vegetative cells, so that resistance in this case is not conferred by reduced hydration.

Detailed information about growth characteristics and rates of photosynthesis exist for a thermophilic strain of *Chlorella* (Tx 71105 (Sorokin & Myers, 1953) now designated as *C. sorokiniana* (Shihira & Krauss, 1963)). The relation of its relative growth-rate to temperature is shown in Fig. 1, with corresponding curves for an ordinary strain of *Chlorella* and the cold-water *Chlorella* for comparison. The temperature optimum lies close to 39° C., at which it shows 9·2 doublings per day when light-saturated. Stable growth could not be obtained above 41·2° C. (Sorokin & Myers, 1953). Other data have been tabulated by Sorokin (1959). At 25° C. the rates of growth, photosynthesis and respiration are the same or slightly higher than those for the Emerson strain of *C. pyrenoidosa*. At its optimum temperature 14,000 metre-candles (lux) are required to saturate growth as compared with 5000 at 25° C. The rate of apparent photosynthesis at light saturation increases 3·62-fold between 25° and 39° C. as compared with a 2·75-fold increase in the rate of endogenous respiration. Less extensive studies have been made on *Cyanidium caldarium* by Fukuda (1958). At its optimum temperature of 55° C. the saturating light intensity was in excess of 100,000 m.-candles when 3 % carbon dioxide in air was supplied. The optimum temperature for the Hill reaction by this organism was found also to be near 55° C., the reaction being much retarded at 30° C. and not detectable at 25° C. (Fukuda, 1962). As with the thermophilic *Chlorella* the respiration rate was much less influenced by rise in temperature, unless inhibitory temperatures were reached, than was photosynthesis (Fukuda, 1958). This state of affairs, which is presumably of great importance for the maintenance of growth at high temperatures, was envisaged but not convincingly demonstrated (see Fogg, 1956) by Bünning & Herdtle (1946). Similar relations are also seen in non-thermophilic algae which are subject to abrupt changes of temperature. Intertidal algae belonging to the genera *Enteromorpha, Ulva, Fucus, Porphyra, Chondrus* and *Griffithsia* show, in the summer months, relatively slight increases in respiration rates, with Q_{10} less than 1·25, when the temperature is raised from 10 to 20° C. (Newell & Pye, 1968). Respiration–temperature curves obtained with material collected in autumn and winter similarly show a region of little change, approximately corresponding to the temperature of the shore at the particular time. It thus appears that the response of respiration to temperature in these algae is modifiable so that the temperature fluctuations in the habitat have a minimal effect. The mechanism of this adaptation is unknown but it does not involve perceptible change in the temperature at which maximum rate of respiration is shown, which agrees with the idea that the upper limit of temperature tolerance is set by genetic factors.

The two thermophilic algae studied in culture, *Chlorella sorokiniana* and *Cyanidium caldarium*, both show maximum photosynthetic activity at temperatures close to the highest tolerated. Marré & Servettaz (1956, quoted in Marré, 1962) found that *Aphanocapsa thermalis* was adapted to photosynthesize at maximum rate at the temperature at which it had been growing, both increase and decrease of temperature diminishing the rate. Temperatures a few degrees above that of acclimatization produced irreversible declines in photosynthesis and respiration, the margin before this inactivation being very narrow when the alga was acclimatized to the highest temperatures. Using a ^{14}C-tracer method for measuring rates of photosynthesis *in situ*, so as to eliminate the possibility of adaptive changes, Brock (1967) showed that individual samples of thermal algae which had grown at relatively constant temperatures had optimum temperatures for photosynthesis extraordinarily close to the actual temperatures in the habitat in which they were growing. Correspondingly rapid growth, with 2–10 doublings per day, was observed. It may be noted that the most abundant algal growth is in hot springs in unshaded positions and that the spring waters are usually rich in carbon dioxide, so that the conditions necessary for maintaining high rates of photosynthesis are met. Brock concluded that the upper temperature limit for the growth of blue-green algae, which he puts at 75° C., is set by some inherent characteristic of their photosynthetic apparatus rather than of protoplasmic functioning, since non-photosynthetic bacteria (and colourless blue-green algae, since 'flexibacteria' are probably of this nature) are found at higher temperatures. The finding that these thermophilic algae are actively growing and not merely surviving with a minimal rate of growth under these extreme conditions is borne out by other observations. Stewart (1969) found by *in situ* determinations by a ^{15}N-tracer technique that maximum rates of nitrogen fixation were achieved by *Mastigocladus laminosus* at 42·5° C. and that detectable fixation still occurred at 54° C. Baker, Hutner & Sobotka (1955) found that high-temperature strains of *Euglena gracilis* are characterized by particularly high rates of utilization of exogenous nutrients.

It thus appears that the thermophilic algae are characterized by having high rates of anabolism and low rates of catabolism. The idea that thermophilic algae are slow growing, often put forward in the older literature, is a myth. Adjustment of the balance of metabolic reactions, although essential for the maintenance of active life at high temperatures, is probably secondary and not the basic feature which confers resistance to high temperatures. Nor is it likely that accumulation of particular components could be the cause. McLean (1967) found no correlation between heat resistance and total lipid content in *Spongiochloris typica*. It seems probable that, as in bacteria (Ingraham 1962), the basic feature on which heat resistance depends is exceptional thermostability of proteins. Marré & Servettaz

(1957, quoted by Marré, 1962) found that NADPH-cytochrome *c* reductase extracted from the thermal alga *Aphanocapsa thermalis* showed greater heat stability than the corresponding enzyme from *Anabaena cylindrica*, which shows optimum growth at about 32·5° C. The enzyme from the former showed unimpaired activity after heating to 85° C. for 5 min., whereas that from the latter was completely inactivated by similar treatment. It was shown that the thermostability of the *Aphanocapsa* cytochrome reductase was not due to a dialysable substance with unspecific protective action, since *Aphanocapsa* extracts did not confer heat stability on the *Anabaena* enzyme. It was also found that cytochrome reductase and catalase from *Aphanocapsa* were not so easily inactivated by solutions of acetamide or urea, reagents which dissociate hydrogen bonds, as were the corresponding enzymes from *Anabaena*. As will be seen later, resistance to the dissociating effects of urea is characteristic of the protoplasm of blue-green algae in general. Kubín (1959) studied the thermal inactivation of catalase extracted from *Oscillatoria* spp. and found in Arrhenius plots that there was no difference between the characteristics of the inactivation and those reported for catalases extracted from beef liver and from various seaweeds. However, the algal material used, although from a hot spring, was stated to have an optimum temperature for growth of 35° C., so that it may not have been particularly temperature-resistant. The ability of thermal blue-green algae to photosynthesize at high temperatures evidently does not depend on any special properties of phycocyanin, since Moyse & Guyan (1963) found that in *Aphanocapsa thermalis* exposed to temperatures of 50° C. the phycocyanin-dependent photochemical activity decreased before that dependent on chlorophyll *a*.

SURVIVAL OF DESICCATION

Many different kinds of algae survive desiccation well and some show great longevity in the dry condition. Blue-green and chlorococcalean algae predominate in the flora of arid deserts (Cameron & Blank, 1966) and algae belonging to these groups are found to be more viable than others in samples stored in the dry condition for long periods. The record for survival is held by *Nostoc commune*, which has been revived from a dried herbarium specimen 107 years old (Cameron & Blank, 1966). Previously this same specimen had been shown to be viable after 87 years (Lipman, 1941). Other blue-green algae and the green alga *Protosiphon cinnamomeus* have been revived after air-dry storage of 65–85 years as museum specimens (Cameron & Blank, 1966). While resistance to desiccation is clearly a general characteristic of the group, species of blue-green algae vary markedly in their degree of resistance, this showing a very good correlation with the dryness of the habitat (Hess, 1962). Algae of other groups, while less resistant than blue-green algae and Chlorococcaceae may yet show surprising survival of desiccation. Evans (1958, 1959), in an investigation of pond algae belonging to various families of Chlorophyceae, Xanthophyceae, Cryptophyceae, Dinophyceae, Bacillariophyceae and Cyanophyceae found that, although both the number of species and of individuals decreased during drought, all species investigated showed some degree of survival of drying, except perhaps *Pandorina morum* and *Euglena mutabilis*. Intertidal algae belonging to the Chlorophyceae, Phaeophyceae and Rhodophyceae show a considerable resistance to desiccation, reviving after air-drying for periods of 1–14 days, their resistance showing a general correlation with the exposure to which they are accustomed (Biebl, 1962; Jeník & Lawson, 1967). Related sublittoral algae, on the other hand, die after only 1 or 2 hr. exposure to air. On the meagre evidence available it may be surmised that members of the Chrysophyceae, Haptophyceae and Xanthophyceae are most sensitive to drying. Diatoms, too, show little resistance as a group, although *Nitszchia palea* is an exception, having been revived from soils preserved air-dry for 50 years (Bristol, 1919) and having been shown experimentally to resist prolonged periods of drying (Evans, 1958, 1959).

Survival depends on the physiological state of the organism and conditions during drying. Fritsch & Haines (1923) noted that there was great variation between cells of *Pleurococcus*, *Prasiola* and *Zygogonium* in permeability to stains and resistance to plasmolysis and desiccation. Those cells which were not permeable to stains but resistant to plasmolysis seemed to survive desiccation best. The desiccation-resistance of the green alga *Spongiochloris typica* was found to reach a peak in young cultures (Fig. 2) and be favoured by dilution of the nutrient medium (McLean, 1967) and by growth on agar rather than in liquid media (McLean, 1968). More of the species studied by Evans (1959) survived slow drying (air-dryness reached in 110 hr.) than when drying was accomplished quickly (within 40 hr.). On the other hand rate of drying made little difference to the survival of various species of Oscillatoriaceae studied by Hess (1962). Hess also found that algae of this family grown in alternating periods of light and dark survived desiccation better if dried in the dark phase than in the light phase. Drying in the dark rather than in the light also enabled better survival. A slight increase in the resistance of *Tolypothrix distorta* grown at 99·1% relative humidity as compared with that grown at 99·7% relative humidity parallels McLean's (1968) observation that agar cultured material survives desiccation better than that grown in liquid medium. Whereas the vegetative cells of some species of Oscillatoriaceae survived desiccation well, those of species of *Nostoc*, *Calothrix* and *Scytonema* were sensitive, only the spores or hormocysts of these species showing high resistance.

Certain features are characteristic of algal cells which are successfully surviving desiccation. Fritsch & Haines (1923) found that such cells had a diminished tendency to plasmolyse in hypertonic solutions, more rigid and viscous protoplasm, and more abundant granules than cells with adequate moisture. Evans (1959) concluded that cells which survive pro-

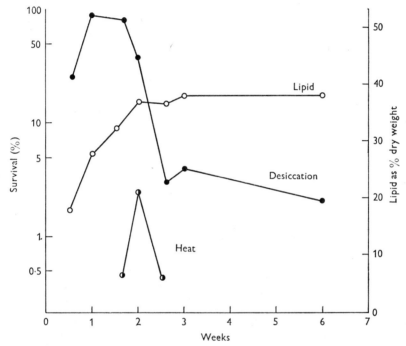

Fig. 2. Changes in resistance to desiccation (●), resistance to high temperature (◐) and total lipid content (○) of *Spongiochloris typica* during growth in culture in one-tenth concentration basal medium (data of McLean, 1967).

longed periods of drought usually show one or more of the following modifications: accumulation of fat, thickening of the cell wall, and production of mucilage. Desiccation is known to cause accumulation of fats in various micro-organisms and algae seem to conform to this rule (Collyer & Fogg, 1955).

Metabolism is presumably greatly retarded in desiccated algal cells but there seems to be little published information on this. Although some species of blue-green algae show great resistance to desiccation, nevertheless growth of these species does not occur at relative humidities less than 80% (Hess, 1962). The rate of photosynthesis of seaweeds is progressively reduced as they dry and it seems commonly the case that it is zero for a large part of the time for which intertidal species are exposed. Photosynthesis is resumed immediately on re-immersion of the more resistant forms growing higher on the shore (Biebl, 1962; Chapman, 1966).

The factors contributing to drought resistance in algae may be various. A lipid pellicle is sometimes produced around cells of micro-algae on drying (Evans, 1959) and Russell-Wells (1932) found the proportion of fat in different species of littoral marine algae to be correlated with the degree of exposure to which they were subjected. Nevertheless, it does not appear that this has any marked effect. The rate of loss of water from littoral seaweeds is determined solely by the physical factors of the environment and algae of similar morphological structure show no differences in rate of drying (Biebl, 1962; Jeník & Lawson, 1967). Low surface/volume ratio and mucilaginous sheaths and thick walls acting as reservoirs will effectively reduce the rate at which water is lost from the cells themselves and possibly give opportunity for physiological adaptation. Protoplasmic characteristics are presumably most important. Here again the expectation that lipid content should be important seems not to be justified, for McLean (1967) found no relation between desiccation-resistance and total lipid or carotenoid contents in *Spongiochloris typica* grown for various periods in media of different concentration (Fig. 2). Nevertheless, there may be a particular lipid component concerned in desiccation resistance.

HALOPHILISM

Algae occur in situations, such as salt marshes, salt ponds and rock pools, in which sea water becomes concentrated by evaporation, and also in saline springs and salt-lakes. Blue-green algae are usually abundant in all these habitats. Many species, such as *Microcoleus chthonoplastes*, are perhaps more correctly described as halotolerant than halophilic but others, *Spirulina subsalsa*, *Phormidium tenue* and *Aphanotheca halophytica*, for example, grow in strong brines and are truly halophilic (Hof & Frémy, 1933). Strains of the chlorophycean flagellate *Dunaliella* are particularly characteristic of the most concentrated brines. *D. viridis* is common in the Dead Sea, growing in media containing 15–27% sodium chloride. This species and *D. salina*, which is normally red in colour due to accumulation of carotenoids, are common in seawater evaporation ponds, the latter growing best at 10–12% sodium chloride but surviving at concentrations near saturation (Larsen, 1962). Both blue-green algae and *Dunaliella* sp. remain viable in dry salt, but the latter appears to survive longer, up to 7 years, under these conditions (Hof & Frémy, 1933). For the blue-green algae, at least, the chemical nature of the salt is not critical, for they are found in abundance in waters containing high concentrations of sodium carbonate or sulphate as well as in sodium chloride brines. The ability of algae other than those just mentioned to withstand high salt concentrations is less marked. Most intertidal algae can tolerate 3- or 4-fold concentrations of sea water for short periods but few of those near and below low-water mark can survive in sea water at more than 1·5 times normal concentration (Biebl, 1962). Intertidal and estuarine species clearly must possess an ability to resist rapid fluctuations in the salt content of the water in which they grow.

Few experimental studies on halophilic algae have been reported. Marré and his co-workers (quoted by Guillard, 1962) found that the sodium content of *Dunaliella salina* increased with the concentration of the medium in which it had been grown, the osmotic excess of the cells being 9 atm. when the osmotic value of the medium was 76 atm., and 25 atm. when the medium had an osmotic value of 170 atm. This situation is similar to that found in halophytic bacteria, and presumably, like the bacteria, these algae contain enzymes resistant to inhibition by salt (Larsen, 1962).

DISCUSSION

Among the algae the Cyanophyceae possess an outstanding capacity, second only to that of bacteria, to resist the adverse effects of extreme temperatures, drought and salt concentration. Since both Cyanophyceae and bacteria are procaryotic and have many biochemical features in common, this further resemblance is not surprising. The highly anomalous organism *Cyanidium caldarium* is the only eucaryotic species known to grow above 47° C. Otherwise among the Eucaryota, certain unicellular Chlorophyceae are characterized by more than ordinary resistance to these particular adverse conditions. Only the Cyanophyceae and Chlorophyceae, of course, have achieved success in the colonization of terrestrial habitats. Although a few species belonging to the other groups may show particular resistance to extreme temperatures or desiccation and although these groups show resistance to other adverse conditions, such as prolonged darkness, they have by and large remained confined to aquatic environments in which they are not exposed to extremes of temperature and desiccation.

Resistances to freezing, high temperatures, desiccation and high salt concentrations are often said to be associated. This association may partly result from the circumstance that these adverse factors usually occur in combination, e.g. high temperature, desiccation and high salt concentration in the hot desert and freezing and desiccation in the cold desert. From field observations alone it is difficult to separate their effects but there is evidence suggesting that the same basic feature is involved in resistance to these different factors. The most telling point is that algae showing the greatest resistance to cold, heat, desiccation and high salt concentration nearly all occur in the same two taxonomic groups. Furthermore, it is well established that desiccated cells are better able to withstand high temperatures than those with ample water, and Baker *et al.* (1955) found osmo-resistance in *Euglena* correlated with thermophily. On the other hand, resistances to different adverse conditions are not always present together in the individual species. A blue-green alga which is both thermophilic and halophilic has been reported by Anagnostidis (1961), but as a rule thermal algae do not survive desiccation (Castenholz, 1967). McLean (1967) found the resistance maxima of *Spongiochloris typica* to desiccation and heat developed at different times during growth in culture (Fig. 2), and whereas resistance to desiccation was favoured by dilute nutrient medium that to heat was best developed in concentrated medium.

Resistance to a given adverse condition by algae is doubtless dependent on a complex of factors as, from the much greater amount of information relating to bacteria (Ingraham, 1962; Larsen, 1962) and higher plants (Henckel, 1964; Olien, 1967), it evidently is in other forms. Low temperatures clearly have at least two distinct effects, namely the physical damage caused by freezing and the slowing down of chemical reactions. At high temperatures besides denaturation of proteins there is imbalance of metabolism resulting from the differential acceleration of its component reactions (as is evident from the work on *Ochromonas* (p. 129)). It is also clear that growth under adverse conditions calls for more adaptations than mere survival. Nevertheless a common feature essential for survival against frost, heat, desiccation and high salt concentrations must be stability of enzymic and protoplasmic structure against disruption, whether caused by mechanical, ionic or thermal factors. The necessary stability might be conferred by a higher proportion of main-valency bonds, as opposed to hydrogen or other homopolar cohesive bonds sensitive to heat and mechanical forces, linking the polypeptide chains of the proteins (Fogg, 1956). This idea is supported by the finding that the protoplasm of blue-green algae is rendered more resistant to mechanical disruption by a strong urea solution, which breaks hydrogen bonds, whereas treatment with thioglycollate, which breaks the main valence bond –S–S–, renders it more easily disrupted (Fig. 3). This is in contrast to what happens with a eucaryotic organism. Isolated enzymes from blue-green algae likewise, as we have seen (p. 133), are more stable towards urea treatment and heat than their counterparts from eucaryotic organisms. It is to be noted, however, that the bulked proteins of blue-green algae evidently do not contain more than the usual proportion of cysteine (Fowden, 1962). If resistance is dependent on some such feature it would be expected to be a genotypic rather than phenotypic character—in accordance with the finding that organisms apparently cannot be trained to withstand higher temperatures. The theory that resistance is basically dependent on greater molecular

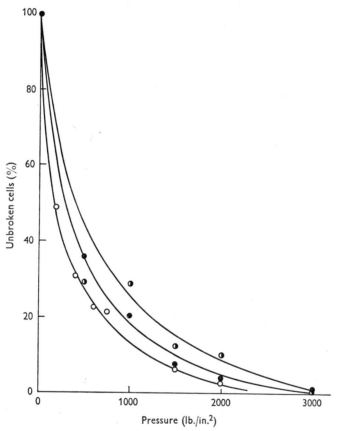

Fig. 3. Cell breakage of *Anabaena* sp. subjected to different pressures in the French press. ●, Cells suspended in distilled water; ◐, in 10 M urea; ○, in 1 % sodium thioglycollate.

stability conferred by a greater proportion of main valency linkages between polypeptide chains is of course distinct from, though not necessarily incompatible with, the theory of Levitt that frost desiccation causes formation of disulphide bonds and subsequent denaturation of proteins (see Olien, 1967). The variation in resistance to cold, heat and desiccation which is dependent on pretreatment of the cells must be attributable to other secondary factors which are phenotypic rather than genetically determined.

Finally, some comment should be made on the role of akinetes, zygospores and other 'resting spores' in the survival of adverse conditions by algae. Few experimental studies on this appear to have been made, but the meagre published information (Glade, 1914; Hess, 1962; Whitton, 1962) supports the expectation that such structures are more resistant to cold, heat, desiccation and perhaps other extreme conditions than are the corresponding vegetative cells. Nevertheless there are some curious anomalies. Evans (1958, 1959) found that survival of desiccation by pond algae bore little relation to spore formation. It is even more remarkable that spore-forming blue-green algae are generally absent from the floras of deserts (Cameron & Blank, 1966) and hot springs above 55° C. (see, for example, Anagnostidis, 1961), whereas they are conspicuous in the plankton of lakes. The little available information on the metabolism of spores does not suggest that this is greatly different from that of vegetative cells. Fogg (1967) found that *in situ* rates of photosynthesis per unit volume of cell material of red snow algae, mainly spores of *Chlamydomonas nivalis* and *Chlorosphaera antarctica*, was about the same as that of green snow algae, mainly vegetative cells of *Raphidonema nivale*. In studies on suspensions of isolated spores of *Anabaena cylindrica*, Fay (1969) found that, although the capacity for photosynthesis was reduced in spores and they showed no nitrogen fixation, yet their rate of respiration expressed on a dry-weight basis might be as much as twice that of vegetative cells. Clearly more investigation is required, but the impression at present is that the formation of spores is a relatively unimportant means of survival of adverse conditions by most algae. So far no elaborate mechanisms for ensuring dormancy during unfavourable periods such as are found in higher plants have been reported in this group.

REFERENCES

ALLEN, M. B. (1959). *Arch. Mikrobiol.* **32**, 270.
ANAGNOSTIDIS, K. (1961). *Untersuchungen über die Cyanophyceen einiger Thermen in Griechenland.* University of Thessalonika.
BAKER, H., HUTNER, S. H. & SOBOTKA, H. (1955). *Ann. N.Y. Acad. Sci.* **62**, 349.
BIEBL, R. (1962). In *Physiology and Biochemistry of Algae*, p. 799. Ed. R. A. Lewin. New York: Academic Press.
BRISTOL, B. M. (1919). *New Phytol.* **18**, 92.
BROCK, T. D. (1967). *Nature, Lond.* **214**, 882.
BÜNNING, E. & HERDTLE, H. (1946). *Z. Naturf.* **1**, 93.
BUNT, J. S., OWENS, O. VAN H. & HOCH, G. (1966). *J. Phycol.* **2**, 96.
CAMERON, R. E. & BLANK, G. B. (1966). *Desert Algae: Soil Crusts and Diaphanous Substrata as Algal Habitats.* Jet Propulsion Laboratory, Tech. Rep. no. 32-971, Pasadena, California.
CASTENHOLZ, R. W. (1967). In *Environmental Requirements of Blue-green Algae*, p. 55. U.S. Department of the Interior, Federal Water Pollution Control Administration.

CHAPMAN, V. J. (1966). *Proc. Vth Int. Seaweed Symp.* p. 29. London: Pergamon Press.
COLLYER, D. M. & FOGG, G. E. (1956). *J. exp. Bot.* **6**, 256.
DAILY, W. A. & McGUIRE, J. M. (1954). *Butler Univ. bot. Stud.* **11**, 139.
DUTHIE, H. C. (1964). *Br. phycol. Bull.* **2**, 376.
EVANS, J. H. (1958). *J. Ecol.* **46**, 149.
EVANS, J. H. (1959). *J. Ecol.* **47**, 55.
FAY, P. *J. exp. Bot.* (1969) (in the press).
FOGG, G. E. (1956). *Bact. Rev.* **20**, 148.
FOGG, G. E. (1967). *Phil. Trans. R. Soc.* B **252**, 279.
FOGG, G. E. (1969). In *Marine Ecology*, ed. O. Kinne. Interscience Publishers.
FOGG, G. E. & BELCHER, J. H. (1961). *Verh. int. Verein theor. angew. Limnol.* **14**, 893.
FOGG, G. E. & HORNE, A. J. (1969). *S.C.A.R. Symposium on Antarctic Ecology.* Cambridge.
FOGG, G. E. & STEWART, W. D. P. (1968). *Br. Antarct. Surv. Bull.* no. 15, 39.
FOWDEN, L. (1962). In *Physiology and Biochemistry of Algae*, p. 189. Ed. R. A. Lewin. New York: Academic Press.
FRIES, N. & SÖDERSTRÖM, I. (1964). *Expl Cell Res.* **32**, 199.
FRITSCH, F. E. & HAINES, F. M. (1923). *Ann. Bot.* **37**, 683.
FUKUDA, I. (1958). *Bot. Mag., Tokyo* **71**, 79.
FUKUDA, I. (1962). *Bot. Mag., Tokyo* **75**, 349.
GESSNER, F. (1955). *Hydrobotanik*, vol. 1. Berlin: Deutscher Verlag der Wissenschaft.
GLADE, R. (1914). *Beitr. Biol. Pfl.* **12**, 295.
GODWARD, M. B. E. (1962). In *Physiology and Biochemistry of Algae*, p. 551. Ed. R. A. Lewin. New York: Academic Press.
GUILLARD, R. R. L. (1962). In *Physiology and Biochemistry of Algae*, p. 529. Ed. R. A. Lewin. New York: Academic Press.
HESS, U. (1962). *Arch. Mikrobiol.* **44**, 189.
HENCKEL, P. A. (1964). *A. Rev. Pl. Physiol.* **15**, 363.
HIRANO, M. (1965). In *Biography and Ecology in Antarctica. Monographiae biol.* **15**, 127.
HOF, T. & FRÉMY, P. (1933). *Recl Trav. bot. néerl.* **30**, 140.
HOLDGATE, M. (1964). In *Antarctic Biology*, p. 181. Ed. R. Carrick, M. Holdgate and J. Prévost. Paris: Hermann.
HOLM-HANSEN, O. (1963). *Nature, Lond.* **198**, 1014.
HOLM-HANSEN, O. (1967a). In *Environmental Requirements of Blue-green Algae*, p. 87. U.S. Department of the Interior, Federal Water Pollution Control Administration.
HOLM-HANSEN, O. (1967b). *Cryobiology* **4**, 17.
HORNE, A. J., FOGG, G. E. & EAGLE, D. J. (1969). *J. mar. biol. Ass. U.K.* **49**, 409.
HUTNER, S. H., BAKER, H., AARONSON, S., NATHAN, H. A., RODRIGUEZ, E., LOCKWOOD, S., SANDERS, M. & PETERSEN, R. A. (1957). *J. Protozool.* **4**, 259.
HWANG, S.-W. & HORNELAND, W. (1965). *Cryobiology* **1**, 305.
INGRAHAM, J. L. (1962). In *The Bacteria*, vol. IV, p. 265. Ed. I. C. Gunsalus and R. Y. Stanier. New York: Academic Press.
JENÍK, J. & LAWSON, G. W. (1967). *J. Phycol.* **3**, 113.
JØRGENSEN, E. G. (1968). *Physiologia Pl.* **21**, 423.
KUBÍN, Š. (1959). *Biologia Pl., Prague* **1**, 3.
LARSEN, H. (1962). In *The Bacteria*, vol. IV, p. 297. I. C. Gunsalus and R. Y. Stanier. New York: Academic Press.
LEWIN, R. A. (1961). *Phycol. News Bull.* **14**, 6.

LIPMAN, C. B. (1941). *Bull. Torrey bot. Club* **68**, 664.
LÖWENSTEIN, A. (1903). *Ber. dt. bot. Ges.* **21**, 317.
MANN, J. E. & SCHLICHTING, H. E., Jr. (1967). *Trans. Am. microsc. Soc.* **86**, 2.
MARRÉ, E. (1962). In *Physiology and Biochemistry of Algae*, p. 541. Ed. R. A. Lewin. New York: Academic Press.
MCLEAN, R. J. (1967). *Can. J. Bot.* **45**, 1933.
MCLEAN, R. J. (1968). *J. Phycol.* **4**, 73.
MOYSE, A. & GUYON, D. (1963). In *Microalgae and Photosynthetic Bacteria*, p. 253. Tokyo: Japanese Society of Plant Physiologists.
NEWELL, R. C. & PYE, V. I. (1968). *J. mar. biol. Ass. U.K.* **48**, 341.
OLIEN, C. R. (1967). *A. Rev. Pl. Physiol.* **18**, 387.
PAPENFUSS, G. F. (1964). *Antarct. Res. Ser.* **1**, 1.
PEARY, J. A. & CASTENHOLTZ, R. W. (1964). *Nature, Lond.* **202**, 720.
RABINOWITCH, E. I. (1956). *Photosynthesis and Related Processes*, vol. II (2). Interscience, New York.
RUSSELL-WELLS, B. (1932). *Nature, Lond.* **129**, 654.
SHIHIRA, I. & KRAUSS, R. W. (1963). *Chlorella: Physiology and Taxonomy of Forty-one Isolates.* University of Maryland.
SOROKIN, C. (1959). *Nature, Lond.* **184**, 613.
SOROKIN, C. & MYERS, J. (1953). *Science, N.Y.* **117**, 330.
STEEMANN NIELSEN, E. & HANSEN, V. K. (1959). *Physiologia Pl.* **12**, 353.
STEEMANN NIELSEN, E. & JØRGENSEN, E. G. (1968). *Physiologia Pl.* **21**, 401.
STEWART, W. D. P. (1969). *Phycologia* (in the press).
WATANABE, A. (1959). *J. gen. appl. Microbiol., Tokyo* **5**, 153.
WHITTON, B. A. (1962). *Br. phycol. Bull.* **2**, 177.

SEED GERMINATION AND THE CAPACITY FOR PROTEIN SYNTHESIS

By A. MARCUS

The Institute for Cancer Research, Philadelphia, Pa. 19111

INTRODUCTION

Initiation of germination in many seeds requires only contact with water at the appropriate temperature. Subsequently, an entire array of metabolic phenomena are set into motion. Considering that the ability to make new protein is quite likely an early requirement during germination, we set out several years ago to study protein synthesis in dry and imbibed seeds.

NUCLEIC ACID METABOLISM OF GERMINATING PEANUT COTYLEDONS

Initial studies were prompted by the report of Oota, Fugii & Osawa (1953) indicating a very rapid loss of RNA from cotyledons of *Vigna sesquipedalis* early in germination. Assuming a similar situation in the peanut, and armed with earlier observations (Kornberg & Beevers, 1957; Marcus & Velasco, 1960; Cherry, 1963a) that germination is accompanied by the formation of specific enzymes, it appeared that the peanut cotyledon system would be analogous to the 'stepdown' culture of Hayashi & Spiegelman (1961), and would presumably provide an excellent source of messenger RNA (mRNA). The experimental findings (Marcus & Feeley, 1962) turned out to be just the reverse of these expectations. The cotyledons of germinating peanuts synthesized considerable ribosomal and transfer RNA (rRNA and tRNA), in fact resulting in 30–40% net increase in cotyledon RNA in 6 days of germination. Similar results were obtained independently by Cherry (1963b). Since there is no cell division in the cotyledon during germination, and since the dry cotyledon contains ample quantities of rRNA and tRNA, the purpose of such RNA synthesis in an essentially quiescent tissue remains an enigma.

AMINO ACID INCORPORATION
IN VITRO AND IN VIVO

We next turned our attention to amino acid incorporation into protein *in vitro* with ribosomal preparations. Initially it was noted that ribosomal preparations from dry peanut cotyledons showed a low level of incorpora-

Table 1. [^{14}C]*Phenylalanine incorporation with wheat embryo preparations*

	Ribosomes	Counts/min. in protein	
Conditions	(mg. RNA)	0 day	1 day
No poly U	0·17	2	120
	0·85	9	881
Poly U	0·02	240	408
	0·16	3681	3522

The complete incubation mixture contained 50 μmoles tris buffer pH 7·9, 5 μmoles $MgCl_2$, 50 μmoles KCl, 20 μmoles 2-mercaptoethanol, 1 μm ATP, 10 μmoles creatine phosphate, 50 μg. creatine phosphate kinase, 120 μg. peanut cotyledon tRNA, 2·1 mμmoles [^{14}C]phenylalanine (101,000 counts/min.), and 0·27 ml. 0 day, dialysed 100,000g supernatant, in a total volume of 1·0 ml. Incubation was for 40 min. at 30° C. and 100 μg. poly U were added, where indicated. 0 day and 1 day refer respectively to dry and 16 hr. germinated embryos.

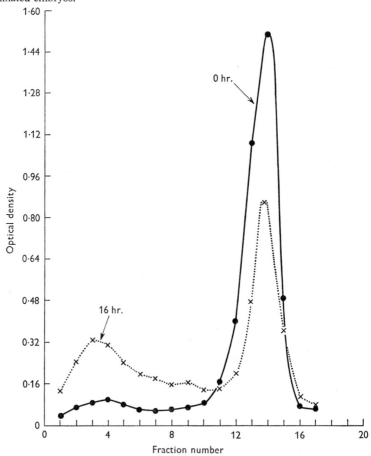

Fig. 1. Sucrose density-gradient profile of ribosomes from dry and imbibed wheat embryos. Ribosomes were centrifuged through 4·0 ml. of a linear 5–20 % sucrose gradient (0·01 M tris-acetate buffer (pH 7·8) + 0·001 M $MgCl_2$ + 0·02 M KCl) over 0·8 ml. of 50 % sucrose for 40 min. at 37,000 rev./min. in a Spinco SW 39 rotor. Fractions were collected from the bottom of the tube.

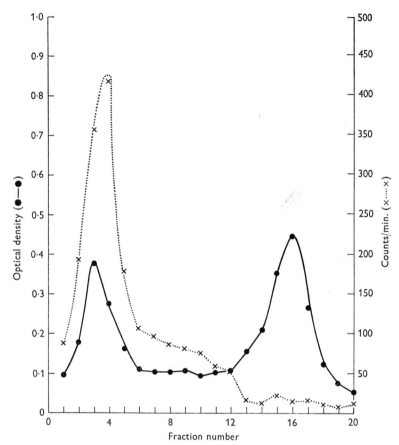

Fig. 2. Amino acid-incorporating capacity of wheat embryo ribosomal components. Ribosomes were centrifuged through the sucrose gradient as in Fig. 1. After dilution with 0·01 M tris-acetate buffer (pH 7·6) + 0·001 M $MgCl_2$ + 0·02 M KCl, one set of aliquots was analysed for absorbance at 260 mμ while a second set of aliquots was assayed for amino acid incorporation in a system similar to that of Table 1.

Table 2. *Correlation of ribosomal activity, polysome content and* in vivo *incorporation*

Imbibition	Ribosome activity (counts/min/mg. RNA)	Polysome content O.D. units	*In vivo* activity units
	288	0·01	—
15 min.	6,680	0·10	—
30 min.	23,200	1·61	6·0
1·5 hr.	31,900	2·42	11·0
6 hr.	50,300	3·66	16·9

Ribosomal activity is obtained by isolating ribosomes after various periods of imbibition and assaying [^{14}C]leucine incorporation. Polysome content is determined by the difference in absorbance of the polysome region of sucrose gradients with two aliquots of each ribosomal sample, one treated with deoxycholate alone, and one treated with RNA-ase and then with deoxycholate. *In vivo* activity is a measure of counts incorporated into protein during a 19 min. exposure of the intact embryo to a radioactive amino acid.

tion increasing about tenfold by 36 hr. after imbibition. This observation was extended to peanut axis and wheat embryo (Johnston & Stern, 1957). In contrast to the ribosomes, both tRNA and 100,000g supernatant were equally active in extracts of dry or imbibed seeds. The ribosomal situation was examined more closely in experiments with polyuridylic acid (poly U) as 'messenger' and the incorporation of [^{14}C]phenylalanine (Table 1). In the presence of poly U, ribosomes of dry seed were as active as those from imbibed seed. These results (Marcus & Feeley, 1964a) suggested that perhaps the transition occurring in imbibition involved 'activation' or formation of mRNA. Later experiments involving sucrose density-gradient analysis (Marcus & Feeley, 1965) indicated that imbibed seed contains a significant polysome fraction in contrast to the dry seed (Fig. 1). These observations were subsequently confirmed by Allende & Bravo (1966), Wolfe & Kay (1967) and more recently by van Huystee, Jachymczyk, Tester & Cherry (1968). That the polysome fraction was reponsible for the increased amino acid incorporation could be shown by direct analysis (Fig. 2). Furthermore, when polysome content, ribosomal activity, and *in vivo* incorporation were compared in time-course experiments (Marcus & Feeley, 1966), good correlation was obtained (Table 2). Together, these data indicated that the ribosome was probably the site of transition and that the major change occurring during imbibition was the formation of polysomes.

MECHANISM OF POLYSOME FORMATION

In considering mechanisms that might result in such formation of polysomes, three possibilities were apparent: (*a*) that hydration allows the combination of messenger and ribosome (forming a polysome), both being functional in the dry seed, but separated spatially, (*b*) that messenger is made available to the ribosome (by synthesis or liberation from an inhibitor), or (*c*) that the ribosome is made available to messenger. Monroy, Maggio & Rinaldi (1965), studying a somewhat similar situation in sea-urchin eggs where fertilization results in polysome formation, have supported the concept of an inactive ribosome, on the basis of the ability to increase ribosomal activity (of unfertilized eggs) by trypsin treatment. Several unpublished attempts have been made to 'activate' ribosomes of dry seed by similar trypsin treatment, all unsuccessful. With regard to messenger synthesis, several studies (Marcus & Feeley, 1964b; Dure & Waters, 1964) in which inhibitors of RNA synthesis failed to inhibit germination, suggested the pre-existence of messenger. We therefore considered as the most likely possibility that messenger and ribosomes preexist in an active form in the dry seed and that some sort of complexing occurs after imbibition. Since simple homogenization (which would presumably bring messenger and ribosome into direct contact) does not result in the formation of active ribosomes, it was necessary to postulate that some metabolic reaction is necessary.

Polysome formation in vitro

The approach was a straightforward one. We reasoned that since the embryo itself can convert its ribosomes to polysomes, perhaps it could be induced to do this in an incubated homogenate. The approach and results are indicated in Fig. 3. Dry-wheat embryos were ground as described under Fig. 3 and 'homogenate' was obtained as the supernatant after removing cell debris in a low-speed centrifugation. One aliquot of such a homogenate

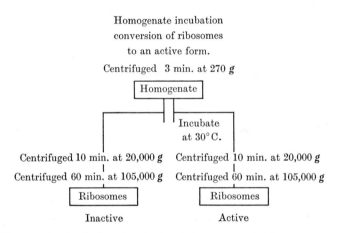

Fig. 3. Ribosome 'activation' by incubation of homogenate. Embryos were ground by hand in a mortar in 0·02 M KCl+0·001 M Mg-acetate. After centrifuging for 3 min. at 270 g one aliquot was incubated for 12 min. at 30° C. with ATP and an ATP-generating system while another aliquot was kept in ice. 0·5 M sucrose+0·025 M KCl+0·001 M Mg-acetate+0·05 M tris (pH 7·6)+0·005 M mercaptoethanol was added to both samples and the ribosomes were isolated and assayed for incoporating activity. Specific data are given in Table 3.

Table 3. *Homogenate incubation and formation of active ribosomes*

Conditions	Ribosomal activity
Complete, 30° C.	5140
Complete, 0° C.	490
'ATP'-omitted	285
Act D (12 μg.)	5100
DNA-ase (10 μg.)	5800
RNA-ase (0·005 μg.)	2070
RNA-ase (0·01 μg.)	830

The experimental details are given in Fig. 3. The data are for an assay similar to Table 1 and are given in counts/min. per mg. of rRNA.

was incubated at 30° C. under specified conditions while another aliquot was kept in ice. Subsequently, ribosomes were isolated from both samples and assayed for incorporating activity in a standard amino acid incorporating system. Since only polysomes catalyse incorporation, this assay is essentially a quantitation of the polysome content in the ribosome population

Table 4. *Inhibition of ribosomal activation by cycloheximide*

Preincubation additions	Ribosomal activity (counts/min/mg. RNA)
o-ribosomes	4,990
o-ribosomes + 15 μg. cycloheximide	1,600
90 min. ribosomes	11,680
90 min. ribosomes + 15 μg. cycloheximide	14,200
90 min. ribosomes (no preincubation)	15,200

The experimental details are those of Fig. 3 and Table 3. 0 and 90 refer to dry embryos and embryos imbibed for 90 min., respectively.

being tested. As shown in Fig. 3, ribosomes isolated from a homogenate that had been incubated are active. In contrast, control ribosomes isolated from a non-incubated homogenate are inactive. Table 3 presents some salient data pertaining to the incubation of the homogenate. The only

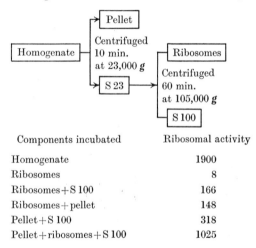

Fig. 4. Cell fractions required for ribosomal activation. Embryos were ground as in Fig. 3 and the particular fractions were obtained as indicated. Pellets were suspended in 0·02 M KCl + 0·001 M Mg-acetate and the various components were mixed as shown and incubated. Subsequently, ribosomes were isolated as in Fig. 3 and assayed for activity. The data shown are counts/min. incorporated by the total ribosomal fraction.

exogenous addition required is ATP and an ATP-generating system. Presence of DNA-ase or actinomycin D is without effect. In contrast, the presence of minute quantities of RNA-ase during the homogenate incubation strongly inhibits ribosome 'activation'. In similar experiments addition of cycloheximide to the homogenate incubation was markedly inhibitory (Table 4). When one fractionates the homogenate prior to incubation in order to determine which cellular components must be present during the incubation to activate the ribosomes, data such as those of

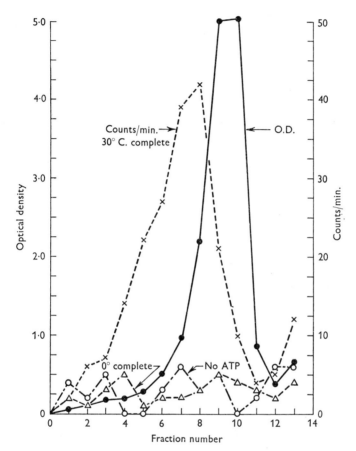

Fig. 5. Polysome formation by homogenate incubation. Homogenate was incubated as in Fig. 3 except that a radioactive leucine pulse was included for the last 2 min. The isolated ribosomes were then centrifuged through a sucrose gradient as in Fig. 1. After determining the absorbance at 260 mμ, TCA-insoluble material was collected on membrane filters and counted. In one of the controls the incubation was at 0° C. and the 2 min. pulse was at 30° C. In the second control, both treatments were at 30° C. with 'ATP' omitted during the initial incubation and added just prior to the [^{14}C]leucine pulse.

Fig. 4 are obtained. Ribosomes, 'pellet fraction', and S 100 must be present during incubation for conversion of ribosomes from the inactive to the active state. Finally, if the ribosomes isolated from an incubated homogenate are examined on a sucrose gradient after [^{14}C]leucine pulse-labelling (Fig. 5), a radioactive peak is obtained distinct from that of the absorbance, indicating that incorporation is indeed occurring on polysomes. Our interpretation of these data (Marcus & Feeley, 1966) is that the unimbibed embryo contains a pre-formed messenger spatially separated from the soluble cytoplasm (Fig. 6). During incubation of the homogenate, this messenger is released to the cytoplasm (reaction 1), where it combines with ribosomes (R) to form a polysome (reactions 2 and 3), the latter process requiring the simultaneous transfer of aminoacyl tRNA to the ribo-

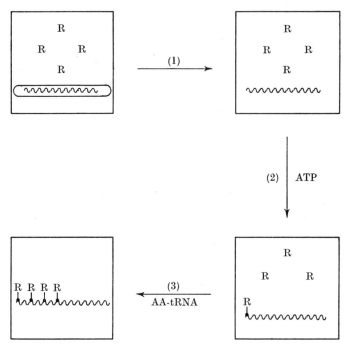

Fig. 6. Polysome formation from a pre-formed messenger. See text for explanation.

somes. Initially, it appeared that the ATP requirement might be ascribed to aminoacyl tRNA formation; however, recent studies have indicated a more specific function of ATP (see below).

Pre-formed messenger

The hypothesis was made that a pre-formed messenger in the dry seed explained the insensitivity of protein-synthesizing preparations from the seed to DNA-ase and actinomycin D as well as the high sensitivity to RNA-ase. Furthermore, existence of such a 'pre-formed messenger' would explain the requirement for 'pellet fraction'. Attempts were therefore made to isolate and identify such a messenger. At this point, we simplified our assay for messenger function using an *in situ* assay whereby both polysome formation and function are allowed to occur in one incubation. This is contrasted to the earlier studies in which the polysomes were first formed in one incubation, isolated, and assayed in a second incubation. (That the *in situ* process does indeed occur in two stages may be seen either by examining products of pulse-labelling on sucrose gradients or more simply by following early kinetics of incorporation.) Figure 7 presents the application of such an assay for messenger activity in the 'pellet fraction'. Three arbitrary fractions (A–C) were obtained by differential centrifugation and

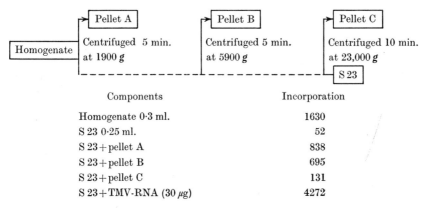

Components	Incorporation
Homogenate 0·3 ml.	1630
S 23 0·25 ml.	52
S 23 + pellet A	838
S 23 + pellet B	695
S 23 + pellet C	131
S 23 + TMV-RNA (30 μg)	4272

Fig. 7. 750 mg. wheat embryo were ground in 8·6 ml. 0·001 M Mg-acetate + 0·002 M-CaCl$_2$ + 0·05 M KCl. The 'homogenate' obtained after centrifuging for 3 min. at 270g was subfractionated by differential centrifugation as shown, yielding pellets A, B and C and S 23. The pellets obtained from each 1·5 ml. of homogenate were resuspended in 0·45 ml. of S 23. Thus each 0·1 ml. of suspended pellet was equivalent to 0·3 ml. of initial pellet and contained 0·1 ml. S 23. The incubation components were added to give final concentrations of Ca^{2+}, Mg^{2+} and K$^+$ of 0·0007 M 0·0025 M and 0·027 M respectively. All vessels contained 0·25 ml. S 23, 0·06 ml. of 0·5 M tris (pH 7·6), 0·01 M ATP, 0·08 M creatine phosphate containing 24 μg. creatine phosphate, and 0·02 ml. [^{14}C]-L-leucine (1·5 mμmoles + 0·16 μC.), to a final volume of 0·7 ml. The mixtures were incubated for 30 min. at 30° C. with 0·1 ml. of the specific pellet fractions. With 0·05 ml. quantities of pellets A and B the counts/min. incorporated were 425 and 355 respectively, demonstrating at least a limited degree of linearity.

assayed for ability to stimulate incorporation in a 30 min. assay. Stimulatory activity was found primarily in the pellet sedimenting in 5 min. at 5900g. If TMV-RNA was added in place of pellet, considerable stimulation was obtained, further suggesting that the pellet is providing the source of messenger. However, in contrast to TMV-RNA, when the total RNA of the pellet fraction was assayed, only a 2- to 4-fold stimulation was obtained. A variety of fractionation procedures attempted on this RNA failed to produce a fraction enriched in a stimulatory activity. More recently, Dr D. P. Weeks, in our laboratory, has succeeded in 'solubilizing' the stimulatory activity of the pellet and has obtained an RNA fraction giving

10- to 15-fold stimulation. This activity is still considerably below that of the 'solubilized' pellet and Dr Weeks has obtained preliminary evidence that a protein fraction may be involved.

Polysome formation with TMV-RNA

The other aspect of our current studies has taken advantage of the observation (Fig. 7) that TMV-RNA was highly stimulatory in catalysing amino acid incorporation by cell-free preparations from wheat embryo. We con-

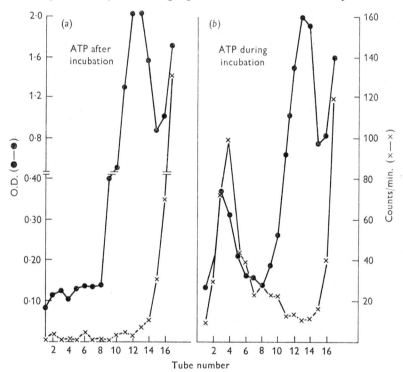

Fig. 8. Both incubations contained 0·38 ml. S 23, 0·02 ml. TMV-RNA (20 μg.) and 0·02 ml. 0·05 M Mg-acetate. In addition (a) contained 0·04 ml. of 0·5 M tris buffer (pH 7·6) and (b) 0·04 ml. 0·5 M tris buffer (pH 7·6), 0·01 M ATP, 0·08 M creatine phosphate and creatine phosphate kinase at 400 μg./ml. After 20 min. incubation at 30°, 0·04 ml. of the ATP-creatine phosphate-creatine phosphate kinase (without tris pH 7·6) was added to (a) and 0·04 ml. H_2O to (b). Both were charged with 0·02 ml. [H^3]leucine (2·7 μC.; 1·34 mμmoles) and reincubated 2 min. at 30° C. Deoxycholate (DOC) was added to a concentration of 0·75 % and 0·45 ml. aliquots were centrifuged through a 5–25 % sucrose gradient with an underlayer of 60 % sucrose. Sixteen-drop fractions were collected for determining absorbance and TCA-insoluble radioactivity.

sidered that if TMV-RNA was serving as a messenger in a manner analogous to the endogenous wheat-embryo messenger, then one could study the second part of the problem, namely the *formation* of the polysome, with a clearly defined messenger. Experimentally, the TMV-RNA system has proved very useful (Marcus, Luginbill & Feeley, 1968). Incubation of TMV-RNA with a ribosome-supernatant system (S 23) results in net polysome synthesis (Fig. 8b), demonstrable either by increased absorbance in the polysome region of a sucrose gradient or by the pulse-label technique. The requirements for the process are those established previously for the endogenous messenger. Ribosomes, ATP, S 100 and incubation at 30° C. are necessary. If cycloheximide is added to the incubation, polysome

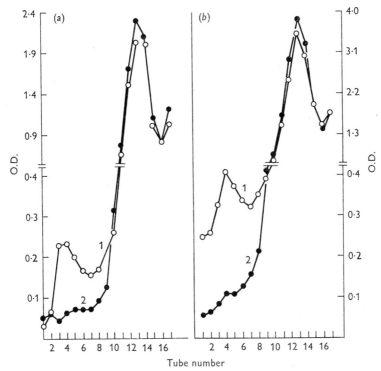

Fig. 9. Cycloheximide inhibition of polysome formation. All incubations included S 23, TMV-RNA and 'ATP' as in Fig. 8 B except that S 23 for (b) was prepared in 0·02 M-KCl+0·002 M Mg-acetate+0·01 M KHCO$_3$. In each case 10 μg. cycloheximide were added to incubation 2.

formation is strongly inhibited (Fig. 9). The time course of polysome formation is shown in Fig. 10. The process proceeds by sequential increase of both polysome content and size (i.e. number of TMV-RNA-ribosome units and the number of ribosomes per TMV-RNA strand). Close examination of a series of time-course experiments (similar to those of Fig. 10) suggested that the initial rate of polysome formation was slower than that obtained later in the incubation. Additional indications of the same phenomenon were seen when the incubations were carried out in the absence of ATP, and ATP was then added after the incubation together with the radioactive pulse (Fig. 8a). Since formation and transfer of the radioactive aminoacyl-tRNA is not limiting under these conditions, it would appear that another step, not involving aminoacyl-tRNA, is rate-limiting. Direct evidence for such a rate-limiting step was obtained by a careful kinetic

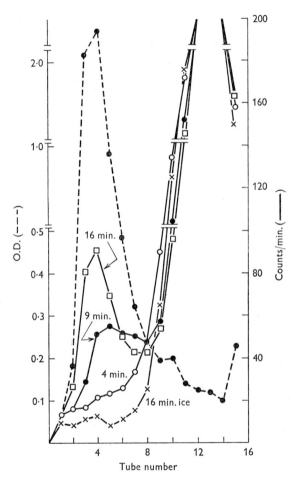

Fig. 10. Time course of polysome formation. Incubation contents as in Fig. 8. The times shown are the duration of a 30° C. incubation. [H³]Leucine was added at 12 min. to the vessel incubated for 16 min. (4 min. pulse).

analysis of incorporation (Fig. 11). The system used in these experiments employed washed ribosomes and a supernatant passed through DEAE-cellulose (at 0·3 M KCl), so that exogenous addition of tRNA was required. The reaction in each case was started by the addition of tRNA and radioactive leucine. Curve 1 presents the normal situation. Incorporation occurs at a linear rate only after 5–8 min. In curve 2 all components were preincubated for 5 min. at 30° C. and then the reaction was begun. The lag is essentially removed. In curve 3 all components were preincubated as in curve 2 except that 'ATP' was added after the preincubation. The lag is retained.

Analysis of this system allows the definition of the components required to carry out the initial rate-limiting reaction. Thus, curve 3 establishes that

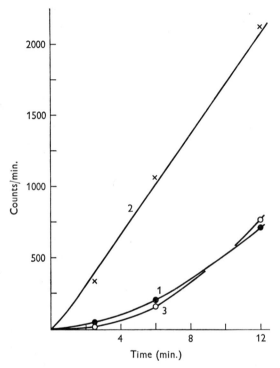

Fig. 11. Preincubation and the time course of incorporation. See text for explanation.

ATP is required. In a similar manner it was shown (Marcus et al. 1968) that ribosomes, S 100 and TMV-RNA are involved and that tRNA* and GTP are required only for subsequent reactions. These data have led us to conclude that the first product of the messenger–ribosome interaction is a monosome, i.e. a ribosome attached at the initiation site of the messenger, and that the attachment is enzymic, requiring ATP and not requiring

* The fact that the incorporation rate is not markedly increased by the addition of tRNA during preincubation suggests that once the initial ribosome attachment to messenger has occurred, subsequent attachment of other ribosomes is facilitated. Alternatively, it is possible that the polysomes obtained with the DEAE supernatant are products involving one ribosome per cistronic area. This would result from a much reduced rate of ribosome attachment relative to ribosome movement so that even when movement is made possible (presence of tRNA), polysomes of no greater size accumulate.

aminoacyl-tRNA. This is shown schematically in Fig. 6 as reaction 2. Subsequently there would occur movement of the ribosome along the messenger strand (Fig. 6, reaction 3) in a process requiring GTP and aminoacyl-tRNA transfer.

Further evidence that ATP is functioning in a reaction other than the synthesis of aminoacyl-tRNA or its transfer comes from two types of experiments: (*a*) incorporation experiments in which radioactive leucyl-tRNA was added in the presence of 19 unlabelled aminoacyl-tRNA's and an excess of non-radioactive leucine. ATP was still an absolute requirement. (*b*) Experiments with polyuridylic acid and phenylanine incorporation. When the incorporation of free phenylalanine was measured, ATP was an absolute requirement whether the system was primed either with poly U or with TMV-RNA. With the poly U system, however, there was only a slight lag, which could be removed best by preincubation in the *absence* of ATP. This is in marked contrast to the TMV-RNA system where, as shown above, considerable lag is obtained and pre-incubation to remove the lag must be carried out in the presence of ATP. In addition, in the poly U system when phenylalanyl-tRNA incorporation was followed, there was no requirement for ATP, again in marked contrast to the TMV-RNA-catalysed system. These studies would appear to rule out the function of ATP at the level of aminoacyl-tRNA formation. Our current endeavours are directed at the demonstration of the initial messenger-ribosome complex and the ascertaining of the specific role of ATP.

GERMINATION AND PROTEIN SYNTHESIS

Having established that protein synthesis is an active process early in germination, it would seem appropriate to ascertain a bit more closely the link with germination. To do this, one would first like to define germination. Generally, it is implied that germination means the protrusion of the embryo root through its surrounding coats. However, since this protrusion is brought about by normal growth processes (cell division and cell elongation), it cannot actually be accepted as a special physiological entity distinct from growth. A partial solution to this problem appears to be afforded by the isolated embryo.

Figure 12 shows the water uptake pattern of isolated wheat embryos (Marcus, Feeley & Volcani, 1966) as measured by increased fresh weight. There is a rapid initial increase to approximately twice the initial weight. The process is complete within 30 min. During the next 5·5 hr. there is little change in fresh weight. This quiescent period, which is followed by a slow sustained secondary increase in fresh weight, has been observed in

Raphanus embryos (Fujisawa, 1965), as well as in lima bean axes (Pollock & Toole, 1966). It is a physiological response distinct from growth and we therefore suggest that it be defined as the 'germination phase'.

If one accepts this concept, it becomes of specific importance to learn what is transpiring in the embryo during this period of time. As we have indicated earlier, protein synthesis appears to be one of the earliest processes activated. Figure 13 shows that the rapid increase in ribosomal

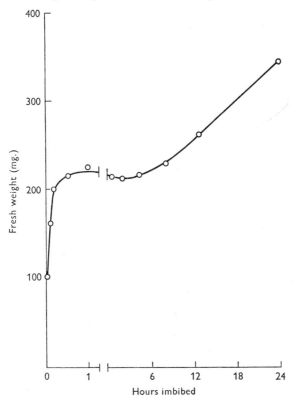

Fig. 12. Time course of water uptake. One hundred mg. samples of embryos were imbibed for the times shown, blotted thoroughly and weighed.

capacity begins already in the imbibition phase. It therefore appears that protein synthesis is a major function during 'germination'. One would obviously like to know what types of proteins are made. Hoping to make a start at this question, Dr Ray Huffaker decided to examine the enzyme, 3'-nucleotidase. This enzyme activity had previously been shown by Shuster & Gifford (1962) to increase 80-fold in the wheat embryo after 48 hr. of germination. Dr Huffaker first looked at a more detailed analysis of the early time period. The question we were concerned with was would the increase in enzyme activity begin during 'germination' or only later with cell extension. Disappointingly the answer was clear-cut (Fig. 14). Enzyme increase did not begin until cell extension. In further studies designed to ascertain whether or not the 3′-nucleotidase is synthesized *de novo*, Dr Huffaker made the interesting observation that if embryos were transferred to cycloheximide after the 'germination period', enzyme

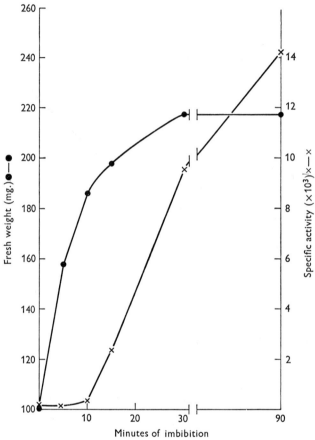

Fig. 13. Comparison of water uptake and ribosomal activity in the first phase of imbibition. One hundred mg. embryo samples were analysed for fresh weight as in Fig. 12 and for ribosomal activity as in Table 1.

activity increased for about 2 hr. despite the immediate shut-off of incorporation of radioactive amino acids. If, however, cycloheximide was present during the 'germination' period enzyme formation was completely halted. Our working hypothesis is that an 'activator' is synthesized during the 'germination' period whose subsequent action results in 3′-nucleotidase formation. If this is indeed so, understanding the process may provide an identifiable system which is synthesized specifically in the 'germination' period.

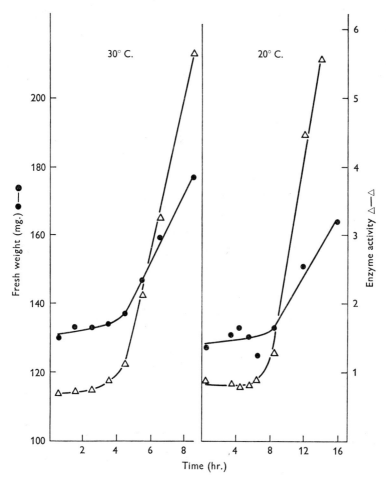

Fig. 14. Comparison of water uptake and the appearance of 3'-nucleotidase. Sixty mg. embryo samples were imbibed for various periods of time at either 20° or 30° C. Fresh weight was determined as in Fig. 12 and 3'-nucleotidase activity according to Shuster & Gifford (1962).

SUMMARY

Early in imbibition the ribosomes of the wheat embryo are converted from a non-functional to a functional form; more specifically to polysomes. Such a conversion can be performed *in vitro* under conditions where transcription cannot occur, suggesting that the conversion involves a 'preformed' endogenous messenger. The attachment of ribosomes to messenger can also be carried out with TMV-RNA and an analysis of the process indicates that such an 'initiation' reaction is enzymic, requiring ATP in a hitherto undescribed process. Participation of aminoacyl-tRNA occurs subsequent to the 'initiation' reaction and results in the *in vitro* formation of polysomes.

This work was supported in part by grants GM-15122, CA-06927 and FR-05539 from the U.S. Public Health Service, grant IN-49 from the American Cancer Society, and by an appropriation from the Commonwealth of Pennsylvania.

REFERENCES

ALLENDE, J. E. & BRAVO, M. (1966). *J. biol. Chem.* **241**, 5813.
CHERRY, J. H. (1963a). *Pl. Physiol., Lancaster* **38**, 440.
CHERRY, J. H. (1963b). *Biochim. biophys. Acta* **68**, 193.
DURE, L. & WATERS, L. (1964). *Science, N.Y.* **147**, 410.
FUJISAWA, H. (1965). *Mem. Coll. Sci. Kyoto Univ.* **32**, 9.
HAYASHI, M. & SPIEGELMAN, S. (1961). *Proc. natn. Acad. Sci. U.S.A.* **47**, 1564.
VAN HUYSTEE, R. B., JACHYMCZYK, W., TESTER, C. F. & CHERRY, J. H. (1968). *J. biol. Chem.* **243**, 2315.
JOHNSTON, F. B. & STERN, H. (1957). *Nature, Lond.* **179**, 160.
KORNBERG, H. L. & BEEVERS, H. (1957). *Biochim. biophys. Acta* **26**, 531.
MARCUS, A. & FEELEY, J. (1962). *Biochim. biophys. Acta* **61**, 830.
MARCUS, A. & FEELEY, J. (1964a). *Proc. natn. Acad. Sci. U.S.A.* **51**, 1075.
MARCUS, A. & FEELEY, J. (1964b). *Biochim. biophys. Acta* **89**, 170.
MARCUS, A. & FEELEY, J. (1965). *J. biol. Chem.* **240**, 1675.
MARCUS, A. & FEELEY, J. (1966). *Proc. natn. Acad. Sci. U.S.A.* **56**, 1770.
MARCUS, A., FEELEY, J. & VOLCANI, T. (1966). *Pl. Physiol., Lancaster* **41**, 1167.
MARCUS, A., LUGINBILL, B. & FEELEY, J. (1968). *Proc. natn. Acad. Sci. U.S.A.* **59**, 1243.
MARCUS, A. & VELASCO, J. (1960). *J. biol. Chem.* **235**, 563.
MONROY, A., MAGGIO, R. & RINALDI, A. M. (1965). *Proc. natn. Acad. Sci. U.S.A.* **54**, 107.
OOTA, Y., FUGII, R. & OSAWA, S. (1953). *J. Biochem., Tokyo* **40**, 649.
POLLOCK, B. M. & TOOLE, V. K. (1966). *Pl. Physiol., Lancaster* **41**, 221.
SHUSTER, L. & GIFFORD, R. H. (1962). *Archs Biochem. Biophys.* **96**, 534.
WOLFE, F. H. & KAY, C. M. (1967). *Biochemistry* **6**, 2853.

SEED DORMANCY AND OXIDATION PROCESSES

By E. H. ROBERTS

Department of Agriculture, University of Reading

INTRODUCTION

The evolutionary advantages of seed dormancy are largely self-evident. In contrast to the growing plant the seed is a resting body which is very resistant to adverse conditions such as drought, frost and darkness. A dormancy mechanism which temporarily prevents germination provides a means of ensuring that the species can survive periods of adverse conditions. Very often the mechanism is nicely adapted so that the period of dormancy ends in response to signals which suggest that conditions are suitable for plant growth. The ecological adaptations of seed-dormancy responses have been discussed in a number of recent reviews (e.g. Harper, 1957; Mayer & Poljakoff-Mayber, 1963; Koller, 1964, and this symposium). In cultivated plants seed dormancy is generally less pronounced, though seldom completely absent since a certain degree of dormancy is necessary to prevent vivipary. In breeding cereals a certain degree of seed dormancy is often deliberately selected for in order to prevent sprouting in the head during a wet harvest season.

In view of the importance of seed dormancy it is not surprising that different species often appear to have evolved different mechanisms to achieve it; thus, quite properly, it is often the differences between species which have been emphasized (Crocker & Barton, 1953). Nevertheless, in spite of the differences, many species show a number of features in common. One such feature is that loss of seed dormancy often seems to involve some oxidation: a corollary of this is that the induction of secondary dormancy is often associated with anaerobic conditions (Vegis, 1964).

THE INFLUENCE OF SEED-COVERING STRUCTURES

In some cases seeds show true embryo dormancy; in others dormancy is maintained only if the covering structures (testa, pericarp, etc.) remain intact. Even in cases of true embryo dormancy, the dormancy is usually reinforced by the presence of covering structures. Seed coats are generally quite permeable to water, except in the case of 'hard' seeds, for example in some of the legumes (Owen, 1956). Even so they are often relatively impermeable to gases (Mayer & Poljakoff-Mayber, 1963; Brown, 1965). It is not surprising then that it has often been suggested that seed coats enforce dormancy by restricting respiration. The implication is that dormancy is regulated by the availability of an energy supply. The view that restriction of respiration may be important in the maintenance of dormancy has been reinforced by the repeated finding with seeds of different species that removal or puncture of covering structures increases the rate of respiration (see reviews by Stiles, 1960; Carr, 1961; and Nyman, 1963). Nutile & Woodstock (1967), for example, reported that drying treatments which induce dormancy in *Sorghum vulgare* lead to a decrease in the oxygen uptake on rehydration when compared with non-dormant seeds. It was suggested that this dormancy may be caused by physical changes brought about in the integumentary membrane resulting in a restriction of gas exchange. Treatment of intact seeds with increased oxygen partial pressures often accelerates the breaking of dormancy and at the same time increases the rate of oxygen uptake, e.g. in Scots Pine (Nyman, 1963).

Barley seeds show similar responses, high oxygen tensions accelerate the breaking of dormancy and also increase the rate of oxygen uptake (Major & Roberts, 1968b). In this species removal of the husk also reduces dormancy (Pollock, 1958) and increases oxygen uptake (Merry & Goddard, 1941; Belderok, 1963). Nevertheless the seed dormancy of barley or rice cannot be explained entirely on the impermeability of covering structures to oxygen and the suggestion (Caldwell, 1959) that loss of dormancy results from an increased permeability caused by micro-organisms now appears unwarranted. The evidence for this is that the gas exchange during the early imbibition of intact barley and rice is *greater* in dormant seeds than in non-dormant seeds (Major & Roberts, 1968b). Thus the inability of seeds to germinate cannot simply be related to a restricted respiration.

THE ROLE OF GERMINATION INHIBITORS

It was thought formerly that the upper seed of *Xanthium* was dormant because of the impermeability of the seed coat to oxygen, but it now appears that the coat acts mainly by preventing the leaching of two water-soluble germination inhibitors from the embryo. Increased oxygen tensions accelerate the enzymic oxidation of these inhibitors to inactive forms (Wareing & Foda, 1957; Wareing, 1963). As will be seen shortly, at least one component of dormancy in wild oats (*Avena fatua*) appears to be caused by a similar mechanism. Thus in some cases there is evidence that the oxidation reaction involved in loss of dormancy is the oxidation of germination inhibitors to inactive forms. In the examples mentioned the inhibitory effect of covering structures has been assigned more to their prevention of an outward leaching of inhibitors from the embryo than to an inward diffusion of oxygen.

Oxygen may also be involved in the inactivation of an inhibitor in birch seeds. On the basis of their work on *Betula pubescens* and *B. verrucosa*, Black & Wareing (1959) suggested the following hypothesis. The seed coat impedes oxygen entry and also contains an inhibitor. Lack of oxygen itself does not prevent germination but lack of oxygen in the presence of inhibitor does. It may be said that either the inhibitor increases the oxygen requirement of the embryo or that high oxygen concentrations are necessary to overcome the effect of the inhibitor. Light and photoperiodic treatments also help to overcome the inhibitor. It is not known whether light or oxygen destroy the inhibitor but it may be that oxygen destroys the inhibitor while light increases the concentration of a promoter.

Generally speaking, the role of germination inhibitors in seed dormancy has not been as satisfactorily demonstrated in cultivated species as in wild species. Since the review of germination inhibitors cannot claim to be comprehensive it is proposed to concentrate now on discussing the evidence for inhibitors in cereal seeds. The reason for this is that the last part of this paper will be concerned with discussing an alternative dormancy hypothesis for which most of the evidence is derived from work on cereals. In wheat, Miyamoto, Tolbert & Everson (1961) isolated four types of water-soluble inhibitors from the seed coats which inhibited the germination of isolated wheat embryos as follows: (*a*) about 38% of the total inhibitory activity was chloroform-insoluble and heat-labile; (*b*) 12% was also chloroform-insoluble but heat-stable; the isolation procedure suggested that both this and fraction *a* were catechin and tannin-like materials; (*c*) 30% was said to be alkaloid-like, since it was more soluble in chloroform than in alkaline aqueous solution, and more soluble in acidic water than in ether; (*d*) 20% was ether-soluble and obtained as an unknown crystalline material. In 3 weeks after harvest it was shown that there was a loss of both chloroform-soluble and chloroform-insoluble inhibitors, and since dormancy also decreases with period of storage after harvest, it could be argued that these inhibitors play a role in the dormancy mechanism. However, although this evidence is suggestive, it is difficult to assess the role of these inhibitors in the intact seed. The bioassay used depended on the delay in germination of isolated embryos for a period of 24 hr. This method was adopted since the inhibition of non-dormant intact seeds 'was found to be a very insensitive assay'. The authors concluded: 'How many of these inhibitory compounds in the wheat seed coat may actually be involved in dormancy is not known.'

Ching & Foote (1961) also found water- and ethanol-soluble growth inhibitors in extracts of dormant wheat seeds and it was postulated that loss of dormancy may be due to the oxidation of these inhibitors—but there was no evidence of this. The significance of these inhibitors is obscure since no attempt was made to examine extracts from non-dormant seeds or to see if the inhibitor content decreased with loss of dormancy.

To emphasize that the dormancy mechanism in wheat is still an open question, it should be mentioned that in the same year in which the two groups above published their work on inhibitors, Durham & Wellington (1961), on the basis of their own experiments, suggested that differences in dormancy between wheat varieties could be ascribed to differences in the resistance of the covering layers to the expansion of the embryo. Belderok (1961), on the other hand, concluded that there is a close relationship between the degree of dormancy in wheat and the content of bound disulphide linkages, particularly in the covering layers of the grain, and implied that these linkages may be involved in the imposition of dormancy.

In cultivated oats (*Avena sativa*) dormancy of the intact grain has been associated with the presence of a water-soluble inhibitor in the hulls (Elliot & Leopold, 1953). The evidence suggested that the inhibitor is a polypeptide of high molecular weight which might act by complexing with sulphydryl groups. However, this work needs confirming and extending before it can be concluded that the inhibitor is functional *in vivo*. For example, it would be interesting to compare the inhibitor content of dormant and non-dormant seeds.

Wild oats (*Avena fatua*) show a much deeper and more prolonged dormancy than the corresponding cultivated species. The mechanism seems to be very complicated and the following is a brief outline which has been presented in a series of papers by Black (1959), Naylor & Simpson (1960), Simpson & Naylor (1962) and Simpson (1965). The hull acts as a barrier to the leaching of an inhibitor. High oxygen tensions reduce the level of inhibitor and conversely a high inhibitor level is induced under anaerobic conditions. Exogenous gibberellic acid (GA) can overcome two blocks to germination. At high concentrations it activates enzymes in the aleurone layer, which make reserves from the endosperm available to the embryo; the release of maltase seems to be particularly important in this respect. After the requirement for high concentrations of GA has disappeared, percentage germination can still be improved by low concentrations of GA, although low concentrations have no effect during early dormancy, presumably because the first block is then limiting. The second block, which can be overcome by a low concentration of GA, is thought to be concerned with the suppression of synthesis of enzymes within the embryo, which consequently is unable to utilize the nutrients derived from the endosperm. During the period of after-ripening these blocks disappear. Originally it was not clear whether the disappearance is due to a decrease in inhibitor level or an increase in GA. Black (1959) could find no differences in inhibitor level between dormant and after-ripened seeds and therefore suggested that loss of dormancy is probably due to an increase in level of a promoter. Naylor & Simpson (1960) found circumstantial evidence that there was no increase in endogenous GA level during after-ripening since it was the block which responded to low concentrations of GA which was the last component of dormancy to disappear; it was therefore suggested, in contrast to Black (1959), that loss of dormancy is due to a decrease in inhibitor content. Later, however (Simpson & Naylor, 1962), dialysis experiments on the ability of excised embryos to induce isolated endosperms to digest potato starch indicated that there was an increase in a gibberellin-like factor during after-ripening.

Hay (1962) found inhibitors in wild oats with similar characteristics to those described by Black. These were bioassayed on embryos but never caused complete inhibition of growth, even when used at high concentration. Furthermore, Hay could not find any quantitative or qualitative difference in inhibitor content between dormant and non-dormant seeds. Weeks (1967) also confirmed the presence of inhibitors with similar characteristics but efforts at further concentration and purification resulted in loss of activity.

Chen & Varner (1967) have recently questioned the notion that one aspect of dormancy is concerned with inability of the seeds to utilize endosperm reserves since they found that the maltose content of dormant and non-dormant seeds is the same and labelling experiments showed that maltose is *more* readily used by dormant seeds than by non-dormant ones.

Khan, Tolbert & Mitchel (1964) reported that they had partially isolated a germination inhibitor from barley (*Hordeum distichon*). Later Bruin & Tolbert (1965) showed that this inhibitor decreased the amount of α-amylase secreted by the aleurone in the excised endosperm under the influence of GA; it also inhibited the release of amino acids. Although this 'dormancy factor' prevented the germination of wheat and lettuce seeds and inhibited the subsequent growth of barley seedlings, it did not affect the germination of barley seeds. Nevertheless it was implied that the dormancy factor may function *in vivo* by inhibiting the release of amino acids from their site of formation and thus cut off the nutrient supply to the developing embryo. However, there is some evidence that the release of nutrients from the aleurone layer and endosperm in barley follows the onset of germination and therefore lack of nutrients from these regions could not be a cause of dormancy (MacLeod & Palmer, 1967). Likewise the development of amylase activity in various *Avena* species is said to follow the onset of germination (Drennan & Berrie, 1962) and lack of amylase in the endosperm was dismissed as a potential cause of dormancy. Major (1966) was unable to find evidence of an inhibitor in *H. distichon* which could be related specifically to the dormant condition. In *Hordeum spontaneum*, a species which shows a more pronounced dormancy, Ogowara & Hayashi (1964) did partially isolate a water-soluble inhibitor of 'low activity' from dormant seeds, but no attempt was made to compare this with extracts of non-dormant seeds and in any case the authors were not inclined to ascribe any importance to this substance in the dormancy mechanism.

Mikkelsen & Sinah (1961) demonstrated the presence of a number of water-soluble inhibitory substances in the hulls of rice (*Oryza sativa*) (e.g. vanillic, ferrulic, p-hydroxybenzoic, p-coumaric and indole-acetic acids). However, the evidence presented showed that although these substances delayed germination they did not prevent it. It is not known whether these substances play a part in the dormancy mechanism since dormant and non-dormant seeds were not compared. The same year a specific attempt was made to investigate the possible association of endogenous inhibitors with dormancy in rice seeds, but no evidence was found which would relate the dormancy mechanism with extractable inhibitors (Roberts, 1961).

This brief review of the relationship between seed dormancy and endogenous germination inhibitors shows that in some species there is good evidence that dormancy is largely controlled by germination inhibitors. In many of the cereals, however, even though a variety of inhibitors have been partially characterized, their role in dormancy mechanisms is obscure; this is particularly true of the cultivated cereals.

We will now consider various aspects of seed dormancy in rice and barley to show how a consideration of these has led to a different approach to the dormancy problem.

Dormancy in rice seeds

In rice it was shown that increase in the partial pressure of oxygen surrounding the 'dry' seeds during storage shortened the dormancy period (Roberts, 1962). Increasing the oxygen tension during the germination test or applying hydrogen peroxide also increased the germination of a partially dormant seed population (Roberts, 1962, 1964*a*). High storage temperatures increased the rate of breaking of dormancy, at least over the range 3–57° C.; this response was investigated in detail over the range 27–47° C., where the relationship was shown to be

$$\log d = K_{\bar{d}} - C_{\bar{d}} t,$$

where d = mean dormancy period, t = temperature, and $K_{\bar{d}}$ and $C_{\bar{d}}$ are constants (Roberts 1962, 1965). There was evidence that the slope constant, $C_{\bar{d}}$, was identical for all species of *Oryza sativa* and that varietal differences could be resolved entirely into differences in the value of the intercept constant, $K_{\bar{d}}$. The Q_{10} for rate of loss of dormancy in all cases had a value of about 3·1 (Roberts, 1965).

The effect of temperature on the imbibed seeds, i.e. during the germination test, was quite different: in this case the lower the temperature the higher the percentage germination of a partially dormant population of seeds (Roberts, 1962). The lower limit to this relationship is set by the fact that non-dormant seeds are incapable of germinating below about 16° C. (Jones, 1926). Nevertheless a continuation of the same phenomenon below this temperature can be inferred from the fact that dormancy may be removed from many seeds by giving the imbibed seeds a cold treatment (say 3° C.), though the seeds then have to be removed to a higher temperature for germination to take place. This response, which is the equivalent of the horticultural practice of stratification, is very common in temperate species and in such cases can be related to ecological behaviour in that it ensures that seeds do not germinate until after the unfavourable conditions of winter. It is interesting that the response is also shown by a tropical crop in which there is no apparent ecological or evolutionary significance.

The responses mentioned so far, together with experiments which showed that removal or partial removal of covering structures stimulated the germination of dormant seeds (Roberts, 1961), suggested that loss of dormancy is associated with some oxidation reaction. It could be postulated that high temperatures during storage of the 'dry' seed increase the rate of this oxidation reaction. On the other hand, when seed is wet it could be envisaged that oxygen enters mainly in the dissolved state; and the greater solubility of oxygen at lower temperatures would ensure a greater supply of oxygen (other explanations are possible, of course, but these will be discussed later). This hypothesis was supported by an investigation into the effects of nitrate (Roberts, 1963a), which is one of the best-known agents for stimulating the breaking of seed dormancy in a wide range of species (Toole, Hendricks, Borthwick & Toole, 1956). It was shown that nitrate was effective in removing dormancy of rice seeds and nitrite was even more effective; on the other hand, ammonium ions, urea and various amino acids had no effect (Roberts, 1963a, b). These results, together with the fact that methylene blue is also stimulatory (Roberts, 1964a), led to the suggestion that nitrate and nitrite are effective in their capacity as alternative electron acceptors to oxygen rather than as nitrogen sources (Roberts, 1964b).

If an oxidation is involved in loss of dormancy it was felt that this might well be respiration. If the effect of temperature on loss of dormancy during dry storage were due to its effect on respiration the Q_{10} was somewhat higher than would be generally anticipated for reactions limited by respiration, but on the other hand, as we have discussed previously (Roberts, 1965), the kinetics of 'dry' systems may be peculiar.

It was thought that a simple test of whether respiration is involved in the dormancy mechanism would be to inhibit it with respiratory inhibitors. It was postulated that, if respiration were involved, respiratory inhibitors should slow down the rate of loss of dormancy. On the other hand, it could be that the oxidation mechanism was unconnected with respiratory enzymes but, for example, be some kind of auto-oxidation (which would not be entirely unexpected in view of the relatively high Q_{10} value). In this case, unless they were toxic, it was expected that respiratory inhibitors would have no effect on dormancy. It was thus a serendipitous finding that many respiratory inhibitors turned out to be extremely potent dormancy-breaking agents; millimolar cyanide or azide, for example, which were amongst the most potent, typically raised germination from less than 20% to more than 80% (Roberts, 1964a). Twenty-two substances were investigated at this stage, including inhibitors of glycolysis, Krebs cycle and terminal oxidation. Only terminal-oxidase inhibitors were found to break dormancy: carbon monoxide, cyanide, azide, hydrogen sulphide and hydroxylamine. Because of the positive results obtained with these inhibitors and the fact that salts of heavy metals, p-chloromercuribenzoate, 8-hydroxyquinoline, diethyldithiocarbamate (DIECA) and 2,3-dimercapto-1-propanol (BAL) did not stimulate germination, it was suggested that it was the inhibition of cytochrome oxidase which was important. Against this interpretation, however, was the fact that filtered light, with maximum transmission at 420–440 mμ (the main absorption band of the CO-cytochrome-oxidase complex), not only failed to reverse the stimulatory effect of CO, but actually enhanced it. In spite of this complication, however, the balance of evidence still favoured the idea that inhibition of cytochrome oxidase leads to the breaking of dormancy. Growth following germination (in contrast to germination itself) was, as expected, inhibited by these substances and, in the case of CO, the inhibition was light-reversible, as would be predicted if the energy for growth depends on the operation of cytochrome oxidase (Roberts, 1964a).

From all these results the following dormancy hypothesis was postulated (Roberts, 1964b). It is necessary for some oxidation reaction to occur before germination can take place. The conditions within the seed are relatively anaerobic and the seed-covering structures act as a barrier to the inward diffusion of oxygen; consequently the loss of dormancy is normally a slow process. Any treatment which leads to an increase in the rate of the oxidation reaction results in a more rapid loss of dormancy—for example, puncturing the covering structures, storage in oxygen, application of hydrogen peroxide or storage at high temperature. Conventional respiration is a strong competitor for the available oxygen since the oxygen affinity of cytochrome oxidase is extremely high. Consequently any treatment which decreases this competition for oxygen also speeds up the breaking of dormancy—for example, the application of inhibitors of cytochrome oxidase or the provision of alternative hydrogen acceptors which can be used in respiratory processes, such as nitrate, nitrite or methylene blue.

Water-sensitivity in barley seeds

In the following section dormancy in barley (*Hordeum distichon*) will be considered, but first it is important to consider the phenomenon of 'water-sensitivity' in barley, which has many attributes in common with dormancy and, unless the experiments are carefully designed, is likely to be confused with dormancy. Water-sensitivity is not shown by rice, but there is evidence that it may exist to some extent in other seeds, e.g. in wheat (Belderok, 1961), in *Beta vulgaris* (Chetram & Heydecker, 1967) and in *Spinacea oleracea* and *Phaseolus vulgaris* (Orphanos & Heydecker, 1967). Whether or not similar mechanisms are involved, however, is not yet clear.

In some samples of barley it has been shown that after they have lost what may be termed normal dormancy they are still unable to germinate if set to imbibe in slightly supra-optimal amounts of water (say 8 ml. or more in a 9 cm. Petri dish), although germination is quite satisfactory under optimal conditions (4 ml. water per 9 cm. dish). This phenomenon, which was termed 'water-sensitivity', was exposed by Essery, Kirsop & Pollock (1954) and subsequently investigated in greater detail and more clearly defined by the same workers (Pollock, Kirsop & Essery, 1955a, b).

The major similarities between normal dormancy and water-sensitivity are that the removal of both is stimulated to various extents by high temperature and low moisture content during storage, by stratification, low germination temperature, removal of covering structures, increased oxygen pressure, hydrogen peroxide, sodium hypochlorite, gibberellic acid, cyanide, azide, and low pH. The major differences are revealed by responses to nitrogenous compounds: nitrate and nitrite are very effective agents for the removal of normal dormancy while urea has no effect; in contrast, nitrate and urea accentuate water-sensitivity while nitrite has no effect (Roberts & Gaber, 1969).

Blum & Gilbert (1957) and Kudo & Yashida (1958) suggested that the failure of water-sensitive seeds to germinate in supra-optimal amounts of water is caused by the harmful activities of micro-organisms. However, Jansson, Kirsop & Pollock (1959) rejected this idea and Jansson (1962 a–d) postulated that the failure is caused by the production under these conditions of a 'lag' factor which leads to a lag phase in the induction of indole-acetic acid oxidase; thus IAA builds up to a toxic level and, it was suggested, inhibits germination by uncoupling oxidative phosphorylation. More recent work has shown that neither respiratory uncouplers nor auxins specifically increase water-sensitivity and thus Jansson's hypothesis has been criticized (Gaber & Roberts, 1969 a), and the most recent work (Gaber & Roberts, 1969 b) strongly supports the earlier micro-organism hypotheses. The main evidence for the involvement of micro-organisms in water-sensitivity depends on the fact that germination of water-sensitive seeds in supra-optimal amounts of liquid is markedly stimulated by a combination of a fungal and bacterial antibiotic (Gaber & Roberts, 1969 b). For example, a combination of any one of the fungal antibiotics nystatin (500 p.p.m.), amphatericin B (50 p.p.m.) or pimafucin (400 p.p.m.) with either of the bacterial antibiotics streptomycin sulphate (800 p.p.m.) or crystapen benzylpenicillin is effective. None of these substances on their own has any effect. Traditional seed-sterilizing agents (e.g. mercuric chloride and sodium hypochlorite), although occasionally stimulating the germination of water-sensitive seed (Blum & Gilbert, 1957; Kudo & Yashida, 1958), usually have little effect (Jansson et al. 1959). This suggests that these sterilizing agents are comparatively ineffective in inhibiting the seed's microflora, and in a large measure the ineffectiveness of these agents accounts for the previous criticism of the micro-organism hypothesis. Even one of the more potent combinations of antibiotics (500 p.p.m. nyastin plus 800 p.p.m. streptomycin) may be ineffective on freshly harvested seeds and it then may be necessary to use a wider range of antibiotics at higher doses to stimulate germination in supra-optimal amounts of liquid. For example, the following combination was found stimulatory to freshly harvested seeds where two antibiotics were without effect: streptomycin (1600 p.p.m.), penicillin (1600 p.p.m.), polymyxin B (400 p.p.m.), and nystatin (1600 p.p.m.). It is suggested that nitrate and urea increase water-sensitivity by acting as a nitrogen source for micro-organism activity which then causes water-sensitivity; nitrite is less effective as a nitrogen source and therefore does not increase water-sensitivity.

It was shown that about 80% of the water-sensitive seeds which do not germinate in supra-optimal liquid become non-viable after 6 days. Largely on the basis of this, measurements of gas exchange and the evidence described above, the following hypothesis has been put forward to explain water-sensitivity in barley (Gaber & Roberts, 1969b). Bacteria and fungi often infect the covering structures of the seed during anthesis. In humid conditions the degree of infection may be severe. A severe infection results in water-sensitivity, i.e. the inability to germinate under conditions which reduce the rate of entry of oxygen. When water-sensitive seeds are set to germinate in excess water, the entry of oxygen is reduced and furthermore is competed for by the high populations of the microflora in the covering structures. As a result, the rate of germination is decreased and some of the micro-organisms attack the embryos and kill most of the seeds. If the rate of germination is accelerated by the use of gibberellic acid, or by increasing the rate of oxygen entry by increasing the oxygen concentration, removal of covering structures, or supplying hydrogen peroxide, the seeds germinate sufficiently rapidly to enable growth to take place before the rapidly increasing microflora in the covering structures can cause loss of viability. Alternatively, water-sensitivity is not expressed (i.e. germination may take place) in excess water if the activity of micro-organisms is inhibited. However, both fungi and bacteria play a part in water-sensitivity and it is important to suppress the activity of both groups of organisms. If only bacteria are inhibited, for example, the activity of fungi is visibly greater than when bacteria are allowed to be active (and presumably the same argument would apply to the suppression of fungi only, although the effects are not immediately visible).

Oxygen-availability, then, plays a part in water-sensitivity, and water-sensitive seeds react to a number of dormancy-breaking agents in a similar way to dormant seeds. Nevertheless it is suggested that water-sensitivity should not be considered as part of the general dormancy phenomenon.

Dormancy in barley seeds

It is evident that in investigations of the normal dormancy of barley care has to be taken not to confuse effects due to micro-organisms (water-sensitivity) with the intrinsic dormancy mechanism of the seed. Effects of micro-organisms on germination are minimized, and apparently for the most part are negligible, if the germination tests are carried out in optimal amounts of liquid. When measuring gas exchange, however, it is important to take precautions to avoid interference by the respiration of micro-organisms and our own measurements are always carried out in the presence of streptomycin and nystatin.

Dormancy in barley shows many features in common with rice. For example, as reviewed previously (Roberts, 1962), high temperatures during storage of 'dry' seed accelerate the loss of dormancy. Low temperatures during the germination test on a dormant sample result in a greater percentage germination. This rule holds right down to typical stratification temperatures (namely 3° C.). Unlike rice, barley is capable of germinating at this temperature and does not necessarily have to be moved to a higher temperature for germination to take place (Gaber, 1967). Removal of the covering structures stimulates the germination of dormant seeds. Nitrate and particularly nitrite are potent dormancy-breaking agents, whereas ammonium ions and other nitrogen sources have no effect (Major, 1966). As in rice, methylene blue can break dormancy and in barley it has been shown that this property is shared by another hydrogen-acceptor, 2,6-dichlorophenol-indo-phenol (Major, 1966). Because of these similarities with rice, and because there had also been reports that two terminal-oxidase inhibitors, cyanide and carbon monoxide, stimulate the germination of dormant seeds (Fischnich, Thielbein & Grahl, 1962), barley (*Hordeum distichon*) was chosen for a further investigation of the paradox that dormancy could be broken by either increased oxygen tensions or respiratory inhibitors.

The effect of oxygen in barley, however, is not entirely clear. Although 'dry' storage in oxygen shortens the period of dormancy (Caldwell, 1959) and hydrogen peroxide applied during the germination test increases the percentage germination of a dormant sample (Major, 1966; Pollock *et al.* 1955*b*), it has been reported (Pollock, Kirsop & Essery, 1956) that treatment of dormant barley with oxygen as compared with air during the germination test has no effect. However, it has now been shown that although it is true that treatment with oxygen for 3 days (the period used by Pollock *et al.*) has no effect, treatment for 1 day is stimulatory and treatment for 6 days is inhibitory; it appears as though oxygen has two opposing effects which tend to cancel each other during the 3-day application (Major & Roberts, 1968*a*). The mechanism of the inhibitory effect is obscure; it is specific to a partially dormant population since a similar 6-day treatment on non-dormant seeds has no deleterious effects. Parallel experiments on both dormant and non-dormant seeds showed that 100% oxygen, in comparison with air, caused a considerable increase in oxygen uptake during the initial stages of imbibition (Major & Roberts, 1968*b*).

The results of experiments on the effects of respiratory inhibitors on the germination of dormant barley gave somewhat variable results, depending on the cultivar used; nevertheless an increase in germination was repeatedly achieved with the terminal oxidase inhibitors carbon monoxide, potassium cyanide, sodium azide and hydroxylamine (Major & Roberts, 1968a). In addition, Pollock & Kirsop (1956) reported that hydrogen sulphide stimulates the germination of dormant barley. Unlike rice, however, it was found that an additional terminal oxidase inhibitor, DIECA, was also effective (Major & Roberts, 1968a). DIECA is supposed to be a specific inhibitor or ascorbic acid oxidase (James & Boulter, 1955); but it would be unwise to place too much confidence in its reputed specificity since, as we have discussed (Major & Roberts, 1968a), it is relatively unstable and some of its breakdown products could inhibit other enzymes.

In addition to DIECA, some other inhibitors which did not affect rice were found to stimulate the germination of dormant barley seeds. These included two which specifically affect the Krebs cycle (malonate, an inhibitor of succinic dehydrogenase, and monofluoroacetate, which combines with oxaloacetate) and two which inhibit glycolysis (iodoacetate, an inhibitor of triosephosphate dehydrogenase, and sodium fluoride, an inhibitor of enolase). Thus although barley is similar to rice in that respiratory inhibitors can break dormancy it has less specific requirements and it can no longer be assumed that the inhibition of a terminal oxidase is an essential aspect of the dormancy-breaking mechanism. It is interesting to note that in another species of barley which shows greater dormancy, *Hordeum spontaneum*, Ogowara & Hayashi (1964) showed that the two respiration inhibitors they tried, potassium cyanide and malonic acid, both increased the percentage germination of dormant seeds.

Since the products of fermentation—ethanol, acetaldehyde and lactic acid—can inhibit respiration in some tissues, the effect of these compounds on the germination of dormant barley was investigated. All were found to be stimulatory (Major & Roberts, 1968b).

As part of the attempt to elucidate the stimulatory effect of respiratory inhibitors on germination, an investigation has been carried out on the effects of a number of the respiratory inhibitors on gas exchange during the initial phase of imbibition (Major & Roberts, 1968a). The main finding relevant to the present argument was that, of those inhibitors investigated which break dormancy—cyanide, azide, malonate and monofluoroacetate in barley, and cyanide and azide in rice—all inhibit oxygen uptake of dormant seeds. In contrast, the oxygen uptake of rice was not depressed by fluoride—an inhibitor which has no effect on dormancy in this cereal. Consequently it was concluded that the ability of a respiratory inhibitor

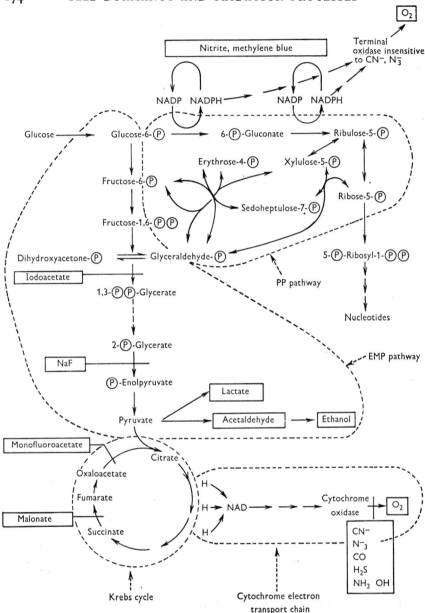

Fig. 1. Diagram of respiratory pathways showing the reactions affected by respiratory inhibitors. Substances in boxes are stimulatory to the germination of dormant barley seeds.

to break dormancy is associated with its ability to inhibit oxygen uptake: thus the dormancy-breaking activity is related more to the activity of the inhibitor *in vivo* than to the particular enzyme which it is supposed to inhibit. Providing they are effective in their inhibitory role, substances which block the Krebs cycle or glycolysis, in addition to those affecting terminal oxidation, are capable of breaking dormancy. Most of the substances which affect both dormancy in barley and respiration are shown in their proposed sites of action in Fig. 1.

Evidence that seeds of other species have some similar dormancy characteristics to rice and barley

It has already been mentioned that the property of seed dormancy appears to have been achieved in a number of different ways by different species. Nevertheless there are many points of similarity; for example, it is rare to find a seed whose dormancy is not reduced by gibberellic acid. However, since the application of gibberellic acid can produce so many different biochemical results, the reaction to gibberellic acid will not be considered here. In this discussion attention will be drawn to the similar responses to nitrates, nitrites and respiratory inhibitors.

In addition to the more intensive work on rice and barley, a preliminary survey has been carried out on the effects of respiratory inhibitors on dormant seeds of oats (*Avena sativa*) and ryegrass (*Lolium perenne*) (Major & Roberts, 1968a). In oats increases in the germination of dormant seeds were obtained with cyanide, azide, hydrogen sulphide, hydroxylamine, DIECA, and malonate. In ryegrass the effect was only obtained with cyanide (though azide and hydroxylamine were also tried). In a previous brief survey of the effects of cyanide, azide and hydroxylamine on the seeds of nine miscellaneous species (Roberts, 1964a) it was found that all these inhibitors caused small but significant increases in the germination of dormant seeds of the cultivated cereal *Digitaria exilis*; hydroxylamine caused a large increase in *Zea mays*; azide caused an increase in *Neomarica gracilis* (Iridaceae); and cyanide caused an increase in *Amaranthus caudatus* (Amaranthaceae).

Apart from these specific investigations, observations incidental to other experiments point to a similar response in lettuce seed. Mayer, Poljakoff-Mayber & Appleman (1957) showed that potassium cyanide could be stimulatory to the germination of lettuce seed and, although the results may not be statistically significant, their graphs suggest a very slight stimulation of germination at the lowest concentrations (which were still quite high, i.e. a little above 10^{-3} M) of hydroxylamine and sodium sulphide.

Potassium cyanide has been shown to increase percentage germination in *Silene noctiflora* and *Melandrium noctiflorum* (Borriss, 1967). The uncouplers 2,4-dinitrophenol (DNP) (10^{-4}–10^{-5} M) and sodium arsenate (10^{-2}–10^{-3} M) inhibited germination.

Recently Ballard & Lipp (1967) have reported that the germination of dormant seeds of *Trifolium subterraneum* is stimulated by respiratory uncouplers. They obtained significant increases in germination using 0·2 mM DNP, 0·8 mM sodium azide, 1·0 mM sodium arsenate, 4·0 mM sodium salicylate and 0·2 mM dicoumarol. Much the greatest stimulation was obtained with sodium azide, which it may be recalled is also stimulatory to the germination of dormant seeds of rice, barley, oats, *Digitaria exilis* and *Neomarica gracilis*. The interpretation of the results on subterranean clover is still open to question however, since although sodium azide generally acts as an uncoupler at lower concentrations, it usually acts as a terminal-oxidase inhibitor at millimolar concentration. Furthermore a number of the other substances can also inhibit oxygen uptake at higher concentrations in a number of tissues. For example, although DNP generally uncouples at 10^{-5} M, at 10^{-4} M it can inhibit oxidation and at 10^{-3} M it can inhibit fermentation (Simon, 1953). In rice millimolar DNP caused an increase in germination which was significant only at $P = 0·1$ and it had no effect at lower concentrations; in barley a slight but significant effect was obtained on one variety at 10^{-6} M (Major & Roberts, 1968a). Apart from azide, other uncouplers which we have tried are sodium arsenate and phloridzin. Sodium arsenate had no stimulatory effect on rice at 10^{-2}–10^{-4} M (Roberts, 1964a) nor on barley at 10^{-2}–10^{-6} M (Gaber & Roberts, 1969a). Phloridzin from 10^{-3}–10^{-6} M, tried on barley only, also had no stimulatory effect (Gaber & Roberts, 1969a).

In view of this it is felt that to establish that the stimulations obtained in *Trifolium subterraneum* are in fact due to uncoupling activity it would be necessary to obtain information on the effect of the agents on oxygen uptake and also data on the germination of these seeds when subjected to treatments with other respiratory inhibitors which do not show uncoupling activity.

Nitrate has for long been recognized as one of the major dormancy-breaking agents for the seeds of a wide range of species (Toole *et al*. 1956). Steinbauer & Grigsby (1957) studied the germination of 85 species of weed plants in 15 families and half showed more germination in the presence of nitrate. In spite of this, little is known about the nitrate stimulation of germination (Koller, Mayer, Poljakoff-Mayber & Klein, 1962). In a number of species various other nitrogen sources have the same effect; Crocker & Barton (1953) have mentioned nitrites, ammonium salts and urea. In addition to nitrate, Ogowara & Ono (1955) showed that nitrite, hydroxylamine and a wide range of amino acids were effective in tobacco seeds. In such cases it would seem that what is required is a source of nitrogen. However, in barley and rice the evidence is that only nitrate or nitrite will do: reduced forms of nitrogen have no effect. This restriction has also been shown to apply to cultivated oats since Schwendiman & Shands (1943) showed that nitrate is stimulatory whereas ammonium ions are not. In non-cereals a similar response has also been found. For example, it has been shown that nitrate and nitrite are stimulatory to the germination of the seeds of *Capsella bursa-pastoris*, whereas ammonium ions have no effect (Popay & Roberts, 1969a). Williams & Harper (1965) reported that nitrate, nitrite and hydroxylamine were equally effective as dormancy-breaking agents in *Chenopodium album* and that a lesser but positive effect was shown by methylene blue; ammonium ions and glycine were said to be less effective, but no figures were given.

Although in some cases (e.g. tobacco) it seems that the nitrate and other nitrogenous compounds are effective by virtue of their ability to act as nitrogen sources, there is circumstantial evidence in other species, including at least three cultivated cereals, that the function of nitrate and nitrite is bound up with their ability to act as hydrogen acceptors. In passing it is interesting to note that in those cases where the germination response is restricted to nitrate or nitrite, this may be of ecological significance. It was shown that in *Capsella bursa-pastoris* alternating temperatures and nitrate act synergistically in stimulating germination (Popay & Roberts, 1969a). And in a study of the intermittent flushes of germination in the field which are characteristic of this and many other annual weeds it was shown that some of the flushes could plausibly be accounted for by a combination of the changes in temperature and nitrate concentration in the soil (Popay & Roberts, 1969b). In the cases where seeds also respond to ammonium ions the possibility of nicely timed responses in relation to changes in the environment would be less since there is less seasonal change in the concentration of ammonium ions in the soil (Russell, 1961).

OXIDATION HYPOTHESES TO ACCOUNT FOR DORMANCY IN CEREALS

Recent work on seed dormancy in *Phalaris arundinacea* shows that it has a number of features in common with other cereals (Vose, 1962). The following is an outline of the hypothesis which was suggested to explain dormancy in this species. The husk limits gas exchange and the resulting anaerobic conditions within the seed lead to the formation of a germination inhibitor which is a side product of fermentation. Oxidation of the inhibitor leads to loss of dormancy. It was suggested that deactivation occurs through an aerobic dehydrogenase system.

Vose's hypothesis is very similar to a general theory of dormancy due to Thornton (1945), who suggested:

Secondary dormancy, and even primary dormancy, has its inception, it is believed, in the accumulation of intermediate products, formed by partial anaerobic respiration, that act as inhibitors because the oxidation system has been temporarily impaired through an insufficient supply of oxygen. The system is therefore unable to function in the normal manner in the removal of many products, among which may be mentioned acetaldehyde, reducing sugar, polypeptides and no doubt many other substances...

More recently Vegis (1964) has also proposed a general theory of dormancy. Vegis suggested that under conditions of oxygen deficiency such as may occur in seeds and buds, particularly at high temperature, the oxidative breakdown of acetyl-CoA is limited and hence the hydrogen in respiration can only partly be transferred to atmospheric oxygen. As a result, part of certain intermediates and products can no longer be broken down oxidatively and are diverted to other pathways, particularly to the formation of fatty acids and fats. This theory is very similar to that of Thornton's except it is envisaged that it is the diversion of metabolites under anaerobic conditions which is important rather than the toxicity of the products which was implied by Thornton.

To explain secondary dormancy in *Avenu fatua*, Hay (1962) suggested the following mechanism:

One might assume that, during the induction of dormancy, electrons, instead of going to oxygen, are passed to another acceptor which, in its reduced form, then blocks the electron transport system at some specific locus where there are no alternative pathways, or possibly stops metabolism in general. If dormancy is due to an inhibitor which is active in its reduced form and inactive when oxidized, this could account for the apparent reversibility of dormancy and afterripening. Furthermore, if the 'reduced' inhibitor is destroyed by oxidases that are activated by wounding, this could explain the promotion of germination when seeds are punctured.

Hay introduced the concept of activation of enzymes by wounding since his results suggested that the stimulatory effect of puncturing the testa could not be attributed to allowing increased entry of oxygen. It will be seen that Hay's hypothesis, although elaborated in greater detail, shows a basic resemblance to that of Vose. It would be difficult to apply any of these hypotheses to rice, barley, oats and a number of other species since they would seem to demand that terminal-oxidase inhibitors should increase rather than decrease dormancy. The stimulatory effects of the products of fermentation, ethanol, acetaldehyde and lactic acid in barley would also reduce the attraction of these hypotheses.

As a general theory of seed dormancy Gadd (1939) suggested that seed coats are initially alive and deprive the embryos of oxygen until sufficient time has elapsed for the seed coat to dry out. A more specific theory for cereal grains had already been proposed by Hiltner (1910), who suggested that inner covering layers of dormant grains contain an oxygen-absorbing substance, probably enzymic, which would prevent oxygen reaching the embryo. This type of theory was developed still further for barley by Pollock & Kirsop (1956), who noticed that a number of agents which were able to stimulate the germination of dormant barley seeds were also known inhibitors of phenol oxidases; these substances included hydrogen sulphide, various mercaptans, thiourea and hydroquinone. It was tentatively suggested therefore that this inhibitory function is intimately bound up with dormancy and they concluded: 'It may be supposed that the presence of phenol oxidase enzymes in the pericarp of barley may prevent the access of oxygen to the embryo and, as a consequence, impair the growth of the latter.' In this case phenol oxidases—which are absent from the embryo (James, 1953)—are seen as a kind of chemical filter in the pericarp which prevents access of oxygen to the embryo. Mayer (1960), referring to lettuce, made a different tentative suggestion, but still linking the effect of thiourea to copper-containing enzymes (presumably including phenol oxidases). Mayer pointed out: 'The copper-containing enzymes are all inhibited to a greater or lesser extent by the germination-stimulating thiourea. It is therefore tempting to relate the action of this chemical to its inhibition of alternative oxidative systems and to the resulting more rapid entry of the energy-yielding Krebs cycle mechanism into operation. However, it could be premature to draw so definite a conclusion from this effect.' This suggestion is not easy to reconcile with the stimulatory effect of cyanide, which presumably would also affect cytochrome oxidase.

An hypothesis of the type suggested by Pollock & Kirsop, on the other hand, does have some attractions. Belderok (1961) considered this hypothesis to explain dormancy in wheat but rejected it because he found that a decrease in the catechol-oxidase activity in the pericarp was confined to the period before harvest-ripeness; after this the level remained practically constant, and thus there was no change in catechol-oxidase activity which coincided with the period of most rapid loss of dormancy which occurred several weeks after harvest. However, there may not be sufficient evidence to reject the hypothesis entirely. In addition to the polyphenol oxidases in the covering structures one should perhaps also consider cytochrome oxidase, since it was shown that in rice (Roberts, 1964a) and in barley (Major & Roberts, 1968b) the aleurone layer and a layer associated with the testa gave a strong cytochrome-oxidase reaction. If one modified Pollock and Kirsop's hypothesis and suggested that the activity of cytochrome oxidase in the covering structures might also lead to oxygen starvation of the embryo, and if it is assumed that the activity of this oxidase will depend on a supply of electrons from previous dehydrogenations, then the hypothesis would now explain the paradox that it is possible to break dormancy either by increased oxygen concentrations or by the application of a wide spectrum of respiratory inhibitors.

In our work on barley, however, we have recently been investigating an alternative hypothesis to explain the effects of the various dormancy-breaking agents. Because of the similarities which have already been referred to, this hypothesis might also apply to other species. It is postulated that as a prerequisite of germination, seeds need to operate a considerable proportion of their initial respiratory metabolism via the pentose phosphate (PP) pathway relative to the Embden–Meyerhof–Parnas (EMP) pathway and that in dormant seeds this ability is restricted. This is a development of our earlier hypothesis, which has already been mentioned, that in dormant seeds normal respiration competes with some dormancy-breaking oxidation. It is now suggested that the dormancy-breaking oxidation is that concerned with the operation of the PP pathway. The main clues which led to the development of this hypothesis are as follows.

(1) Nitrite and methylene blue are effective dormancy-breaking agents. Methylene blue increases the participation of the PP pathway in potato disks and carrot slices (ap Rees & Beevers, 1960 a, b), and nitrite, methylene blue and other electron acceptors were shown to have the same effect in maize roots (Butt & Beevers, 1960, 1961). This stimulation is assumed to be associated with the fact that nitrite and methylene blue preferentially oxidize NADPH (rather than NADH) and that the dehydrogenases of the PP pathway reduce NADP rather than NAD. Nitrate, which is also a dormancy-breaking agent, appears to oxidize NADH in plants rather than NADPH (Beevers, Flescher & Hageman, 1964); but the nitrate is reduced to nitrite and nitrite reductase utilizes NADPH. In this connection it may be relevant that in barley and rice nitrite is a more efficient dormancy-breaking agent than nitrate.

(2) Terminal-oxidase, Krebs-cycle and glycolysis inhibitors are to various extents stimulatory to the germination of dormant seeds. There is some evidence that oxidations associated with the PP pathway may not operate via the conventional oxidases. For example, Gibbs (1954) showed that in pea seedlings cyanide or azide does not inhibit the oxidation of glucose-6-phosphate and cyanide may accelerate it. It has also been concluded that the dehydrogenases of the PP pathway are not sensitive to fluoride or iodoacetate (Gibbs, 1954), which are dehydrogenase inhibitors which tend to stimulate the breaking of dormancy. If this is a general phenomenon, then terminal-oxidase inhibitors, Krebs-cycle inhibitors and

glycolysis inhibitors would be expected to divert respiration towards the PP pathway.

(3) During the germination of non-dormant 'Grand Rapids' lettuce seed there is some evidence that the PP pathway is important and NADP is the major electron acceptor in oxidative processes during the early stages (Poljakoff-Mayber, 1952; Poljakoff-Mayber & Mayer, 1958). During this early phase of germination the Krebs cycle and conventional electron-transfer system are either not working or of minor importance (Poljakoff-Mayber, 1955; Mayer et al. 1957; Poljakoff-Mayber & Evenari, 1958): it is only after growth has started that metabolism shifts from the PP pathway to glycolysis, Krebs cycle and the cytochrome-oxidase system (Poljakoff-Mayber & Evanari, 1958). During the early stages of germination oxygen uptake is hardly affected by carbon monoxide, but sensitivity gradually develops (Harel & Mayer, 1963). Cyanide (2×10^{-3} M) does not inhibit germination and in fact stimulates it slightly (Mayer et al. 1957), whereas $3\cdot3 \times 10^{-5}$ M cyanide inhibits 90% of the activity of isolated mitochondria. Accordingly it was suggested (Mayer & Poljakoff-Mayber, 1962) that some alternative electron-transport system, as yet unknown, may be functioning in the seeds. Presumably from this work it may be implied that during early germination most, if not all, respiratory metabolism is via the PP pathway and this is associated with a terminal oxidase system which is insensitive to typical terminal-oxidase inhibitors (e.g. CN^- and CO).

There is also evidence that the PP pathway is important during the early germination of other seeds but then disappears. Chakravorty & Burma (1959), working on mung beans (*Phaseolus radiatus*), demonstrated the presence of all the enzymes necessary for the oxidation of glucose-6-phosphate to ribulose phosphate, which was then further metabolized. The system could only be operative for 3 days, since the activity of phosphogluconate dehydrogenase fell markedly and that of glucose-6-phosphate dehydrogenase virtually disappeared within this period. In buckwheat (*Fagopyrum esculentum*) it has been shown that the Bloom and Stetten ratios (C-6/C-1) obtained in labelled-glucose experiments on 2-day-old seedlings are very low (about 0·2), suggesting considerable activity of the PP pathway (Effer & Ranson, 1967a).

(4) It could be argued that storage-root and tuber tissue have properties similar to dormant embryos in that cell enlargement, cell division and metabolism are held in check. Slicing and washing such tissue leads to a stimulation of metabolism and 'induced respiration' (perhaps the equivalent of breaking dormancy?). When respiration is induced in this way, there is evidence that the PP pathway increases to become a major component (ap Rees & Beevers, 1960a, b). Furthermore there is evidence that sensi-

tivity to cyanide, azide and carbon monoxide decreases after induction until oxygen uptake may be hardly affected by the inhibitor (Thimann, Yocum & Hackett, 1954; MacDonald, 1959). Thus again the operation of the PP pathway appears to be associated with oxidations which are relatively insensitive to an inhibitor of the well-known terminal oxidases.

It has been sometimes suggested that the breaking of dormancy caused by puncturing seeds cannot be explained entirely on the resulting increased accessibility of oxygen, but some more specific wound-stimulus is involved; for example, in wheat (Belderok, 1961), in wild oats (Hay, 1962), in *Hordeum spontaneum* (Ogowara & Hayashi, 1964) and in birch seeds (Black & Wareing, 1959). It is tempting to speculate that such a stimulus may cause a similar series of changes to those that occur on slicing and washing storage tissue, and one of these changes would be an increase in pentose phosphate metabolism.

(5) An increase in oxygen tension tends to break dormancy. There is also some evidence that increasing the oxygen tension may increase the relative proportion of respiration occurring via the PP pathway. For example, Effer & Ranson (1967b) suggest that the results they obtain in seedlings of *Fagopyrum esculentum* can be explained on two alternative hypotheses: either (a) there is a decreased rate of glycolysis in anoxia; or (b) the contribution of the PP pathway to total respiratory catabolism decreases as the oxygen concentration is reduced to zero and glycolysis is faster in anoxia than in air.

It may also be relevant that in potato tubers high oxygen concentrations can lead to a blockage of the Krebs cycle between citrate and α-ketoglutarate (Barker & Mapson, 1955). Such 'oxygen poisoning' eventually leads to a shift to the more oxidized state of the ascorbic acid–glutathione redox system. It has often been suggested that the oxidation of NADPH (the co-enzyme of the PP-pathway dehydrogenases) occurs via a redox system of this type (see reviews by Mapson, 1958, and Beevers, 1961).

(6) Low-temperature treatments of imbibed seeds break dormancy. Seeds of *Prunus cerasus* require a period of chilling to overcome dormancy. LaCroix & Jaswal (1967) reported: 'During the seventh week of after-ripening (at low temperature) a striking increase in respiration at 25° C. of the embryonic axis occurred along with a sharp drop in the dormant to non-dormant state of the seed. On the basis of C-6/C-1 ratios (a change from about 0·95 to about 0·6) this change may be related to an increased activity of the pentose phosphate cycle.' (At first it may seem that the results of Bradbeer & Colman (1967) on *Corylus avellana* are contrary to the above results in that these workers reported an increase in C-6/C-1 ratios from 0·3 to 0·5 after 50 days' chilling. But these results refer to

cotyledon slices and not to the embryonic axis. In the case of *Prunus*, LaCroix and Jaswal said that the C-6/C-1 ratio of cotyledons varied from 0·82 to 0·89 but showed no obvious relationship to after-ripening time.) Thus the work of LaCroix and Jaswal suggests that germination is correlated with an increased ability of the embryonic axis to operate the PP pathway. In addition these authors found that during the period of after-ripening at 5° C. there was a decrease in the DNP-stimulated respiration of the embryonic axis. This suggests the possibility, at least, that the operation of the PP pathway is not associated with the typical energy-yielding electron-transport of the cytochrome system. Pollock & Olney (1959), working on the same species, had previously shown that the percentage stimulation of oxygen uptake in the embryonic leaf primordia caused by DNP declined with period of after-ripening at 5° C. But if the seeds were held at 25° C. (when dormancy was not broken) then the percentage stimulation by DNP rose to a high level, then gradually declined, though it was always much higher than in seeds held at 5° C. From this Pollock and Olney suggested that the rate-limiting step in the respiration of cells which showed a marked DNP-stimulated respiration is the supply of phosphatic acceptors or their turnover rate. Thus they concluded that their results indicated that one cause of the breaking of dormancy may be the increased availability of energy to the embryo, possibly by an increased supply of phosphate acceptors. On the other hand their results could also be interpreted on the basis that the proportion of conventional respiration (EMP, Krebs, energy-coupled electron transport) decreases as the embryos become capable of germination, and that there is an increase in the proportion of respiration occurring through some other oxidizing system. On this interpretation the results of Pollock and Olney would agree with those of LaCroix and Jaswal.

(7) Bruinsma (1962) has reviewed the Japanese work on dormancy in the potato tuber. In this case there is a decrease in sensitivity of respiration to terminal-oxidase inhibitors during the change from the dormant to non-dormant state: the tissue becomes less sensitive to cyanide and DIECA. However, in this case it is suggested that terminal oxidation changes from phenol oxidase rather than from cytochrome oxidase to some other system. The evidence for this is that extracts of phenol oxidase show decreases in activity with time and that antimycin A (which specifically inhibits the electron transport of the cytochrome system) retains its same inhibitory properties before and after the breaking of dormancy. It is suggested that the changeover occurs to glycollate oxidase because acriflavine (a specific inhibitor of this system) has no effect on dormant tissue but inhibits the respiration of non-dormant tissue. However, since the acriflavine inhibition

is the sole evidence here, the suggestion of the glycollate system is at present only tentative.

(8) It has been shown (Major & Roberts, 1968a) that the depression of the rate of oxygen uptake by cyanide (the most effective respiratory inhibitor for breaking dormancy) increases with duration of imbibition. This suggests that much of the initial respiration during germination occurs via a cyanide-insensitive terminal oxidase: the participation of a cyanide-sensitive terminal oxidase increases with time. The extent of the cyanide inhibition of oxygen uptake during the early stages (the first 6 hr. of imbibition) is greater in dormant than in non-dormant seeds (Table 1). This suggests that the operation of a cyanide-insensitive component is initially greater in non-dormant seeds than in dormant seeds. If the oxidations connected with the PP pathway operate via a cyanide-insensitive terminal oxidase (as the work on lettuce seed and storage tissue suggests), then the results now discussed imply that the operation of the PP pathway during the initial stages of imbibition is relatively more active in non-dormant seeds than in dormant seeds.

Table 1. *The rate of oxygen uptake of dormant and non-dormant seeds of barley and rice in the presence of millimolar KCN expressed as a percentage of oxygen uptake in water (from original data of Major & Roberts, 1968a)*

	Period of imbibition			
	0–6 hr.		22–24 hr.	
	Dormant	Non-dormant	Dormant	Non-dormant
Barley	62	85	7	7
Rice	75	92	48	24

(9) It may be no more than a coincidence but it is interesting that the change in inhibitor-sensitivity of terminal oxidase systems which appear to operate in lettuce, barley and mung bean seeds mentioned above appears to have a parallel in the germination of bacterial spores. Sussman & Halvorson (1966) have reviewed the evidence that in *Bacillus cereus* the main pathway of electron transport in vegetative cells proceeds by a particulate cytochrome system while activated spores (i.e. those which are about to germinate) primarily utilize a soluble flavoprotein oxidase for terminal oxidation. These differences are associated with differences in cytochrome content and cyanide-sensitivity. It is suggested that terminal oxidation during dormancy may occur through pathways which involve intermediates of the PP pathway. Spore extracts contain a complete hexosemonophosphate shunt leading to triose formation but there is doubt as to the existence of the EMP pathway in dormant spores. Experiments based on CO_2

evolution derived from different labelled carbon atoms of glucose suggest that vegetative cells operate almost exclusively via the glycolytic pathway whereas in spores only 75% of glucose is metabolized this way. Completely germinated spores utilized about 20% of glucose via the PP pathway and this dropped to 10% in the swollen elongated spore. By the first cell division the percentage of glucose metabolized by this pathway is close to that found in cells in the logarithmic stage of growth. In addition to glucose, intermediates of the PP pathway and related compounds break spore dormancy.

It seems then that an initial high operation of the PP pathway associated with cyanide-insensitive oxygen uptake may be typical of resting bodies which are about to begin growth and that during growth the EMP pathway and cyanide-sensitive oxygen uptake become important.

THE POSSIBILITY THAT LOSS OF DORMANCY IN BARLEY IS ASSOCIATED WITH CHANGES IN RESPIRATORY PATHWAYS

So far only circumstantial evidence has been presented to suggest that breaking of dormancy may be related to the stimulation of the PP pathway. The arguments were based on the stimulatory effects of oxygen, respiratory inhibitors, and electron acceptors such as nitrite and methylene blue. The possibility that the PP pathway may play a part in early germination is supported by the demonstration that the embryo and aleurone layer of barley contain glucose-6-phosphate dehydrogenase and 6-phosphogluconate dehydrogenase (Major & Roberts, 1968b). More recently we have attempted to obtain information concerning the operation of the PP pathway by feeding either glucose-6-^{14}C or glucose-1-^{14}C for 3 hr. periods to seeds which have been set to imbibe in the presence of antibiotics and examining the C-6/C-1 ratios of the $^{14}CO_2$ evolved. This is the technique first devised by Bloom & Stetten (1953). There have been many criticisms of the interpretation of such experiments and they cannot be used to give quantitative information concerning the relative activities of the PP and EMP pathways. Nevertheless such experiments can give useful qualitative indications of changes in the balance between these two respiratory pathways. A C-6/C-1 ratio of unity would be consistent with the proposition that breakdown of glucose is entirely by the EMP sequence, whereas a ratio of less than one would show that some unknown fraction of glucose is diverted to the PP pathway. Details of such experiments on barley seeds which have been carried out by R. D. Smith will be published in due course; the following is an outline of the results obtained so far.

Table 2 shows the C-6/C-1 ratios which were generated when the labelled glucose was added for 3 hr. periods at various times from the beginning of imbibition. The results show that initially the C-6/C-1 ratios are remarkably low, which suggests that the PP pathway may play an important part in respiration during the early stages; but later the values rise, suggesting that the participation of the EMP pathway increases. The change seems to coincide more or less with the time when visible signs of germination appear—say after about 33 hr. in this sample of non-dormant seeds. It is also evident that the ratio in non-dormant seeds starts lower, and remains lower for a longer period, than in the dormant seed. Thus dormant barley seed not only shows a greater rate of gas exchange (Major & Roberts, 1968b), but the C-6/C-1 ratios suggest that the proportion of respiration occurring via the PP pathway is less than in non-dormant seeds.

Table 2. *The evolution of $^{14}CO_2$ from glucose-6-^{14}C and from glucose-1-^{14}C (C-6/C-1 ratios) when applied for 3 hr. periods to Pallas barley seeds at different times from the beginning of imbibition*

Period from start of imbibition (hr.)	C-6/C-1 ratio	
	Dormant	Non-dormant
0·5–3·5	0·18	0·13
3·5–6·5	0·29	0·15
6·5–9·5	0·28	0·13
12·5–15·5	0·42	0·14
21·5–24·5	0·33	0·12
33·5–36·5*	0·35	0·28

* Radicle appeared during this period in non-dormant seeds.

Preliminary experiments have also been carried out on the effects of a number of dormancy-breaking agents on the C-6/C-1 ratios. From what has been said previously it would be expected that both nitrite and cyanide would increase the amount of catabolism via the PP pathway relative to the EMP pathway. But in addition to these two agents it was thought that it might be profitable to investigate the effects of two other dormancy-breaking agents whose effects on the C-6/C-1 ratios could not be rationally forecast. The first of these was GA, which seems to be almost a universal dormancy-breaking agent in plant material. The second was chloramphenicol, which was first shown by Black & Richardson (1965, 1967) to be a dormancy-breaking agent in either the D-threo or L-threo isomeric form for light-sensitive lettuce seed. It has since been shown that D-threo-chloramphenicol can also cause considerable increases in the germination of dormant barley seeds (Major, 1966). R. D. Smith has recently compared the effects on dormant barley seeds of D-threo-chloramphenicol, which has antibiotic properties, with the L-threo isomer, which does not. It was thought that this comparison might shed some light on the mode of action of chloramphenicol since they have different biochemical effects. The D-isomer is said to inhibit protein synthesis at low concentration while at higher concentrations it inhibits oxidation and phosphorylation of mitochondria; the L-isomer is even more effective at inhibiting oxidation and phosphorylation and can do this at a relatively low concentration, but it only inhibits protein synthesis at higher concentrations (MacDonald, Bacon, Vaughan & Ellis, 1966; Hanson & Krueger, 1966). In other words at a given concentration D-threo-chloramphenicol affects protein synthesis more, but oxidation and phosphorylation less, than L-threo-chloramphenicol.

Unfortunately, unlike the dormant barley sample used by Major (1966), D-threo-chloramphenicol has shown very little dormancy-breaking activity on the sample used by Smith; it is interesting to note, however, that, as in lettuce, the L-threo isomer appears to have had a comparable effect (Table 3).

Table 3. *The effect of dormancy-breaking agents on the evolution of $^{14}CO_2$ from glucose-6-^{14}C and glucose-1-^{14}C in Proctor barley applied during the period from 3 to 6 hr. after the start of imbibition*

Treatment	Germination %	C-6/C-1 ratio	C-1 treated / C-1 untreated	C-6 treated / C-6 untreated
Dormant				
KCN, 10^{-3} M	26	0·16	0·26	0·22
Untreated	10	0·19		
NaNO$_2$, 10^{-2} M	32	0·20	1·39	1·25
Untreated	10	0·22		
GA, 10^{-3} M	100	0·11	0·98	0·45
Untreated	10	0·24		
D-Threo-chloramphenicol, 1000 p.p.m.	12	0·11	0·81	0·35
Untreated	10	0·25		
L-Threo-chloramphenicol, 1000 p.p.m.	22	0·13	0·74	0·32
Untreated	14	0·30		
Non-dormant				
KCN, 10^{-3} M	100	0·09	0·27	0·17
Untreated	96	0·14		
NaNO$_2$, 10^{-2} M	100	0·11	0·68	0·68
Untreated	96	0·11		
GA, 10^{-3} M	100	0·14	0·94	1·01
Untreated	100	0·13		
D-Threo-chloramphenicol, 1000 p.p.m.	79	0·16	0·67	0·99
Untreated	96	0·11		
L-Threo-chloramphenicol, 1000 p.p.m.	100	0·11	0·46	0·45
Untreated	97	0·11		

This particular barley sample has also shown only small responses to cyanide and nitrite. Nevertheless, in spite of these rather disappointing germination responses, all the dormancy-breaking agents tried decreased the C-6/C-1 ratio of dormant seeds, thus suggesting that they were tending to shift glucose catabolism in the direction of the PP pathway and make the seeds behave more like non-dormant ones in this respect. Somewhat surprisingly, nitrite had the least effect, but although the shift in the ratio is small, it can be reproduced consistently. The effect of chloramphenicol is particularly striking. The L-threo isomer appears to have a greater effect in decreasing the C-6/C-1 ratio. This would be expected if the L-threo isomer has a greater effect than the D-threo isomer in inhibiting oxidation via the conventional electron transport chain. It could be suggested that it is this inhibitory property of chloramphenicol which leads to its dormancy-breaking property, rather than its ability to inhibit protein synthesis. This suggestion is supported by the fact that other protein-synthesis inhibitors do not break dormancy in lettuce seed (Frankland & Smith, 1967).

The effects of the dormancy-breaking agents on non-dormant seeds are much more variable. This might be related to the indication that the PP pathway is already operating at a high level in these seeds and consequently further decreases in the ratio are difficult to produce.

The possible significance of the pentose phosphate pathway during early germination

The function of the PP pathway during early germination at present remains obscure. Degradation of glucose via this pathway leads to the production of intermediates which are essential to the synthesis of many important substances (Neish, 1960). For example, erythrose-4-phosphate is required for the synthesis of shikimic acid, which in turn appears to be a key intermediate in the formation of the aromatic amino acids, phenolic compounds, flavones, lignin and alkaloids. Then 5-ribosyl-1-pyrophosphate, a key intermediate in the formation of nucleotides and nucleic acids, is formed from ribose-5-phosphate. In addition to the importance of the intermediates it is now generally accepted that the dehydrogenations of the pathway supply NADPH for many reductive synthetic reactions (Conn & Stumpf, 1966). Thus if it is the restriction of the activity of the PP pathway which is important in dormancy, it may not be easy to disentangle the ultimate limiting factor for germination. Nevertheless it would seem possible at this stage to suggest some aspects of pentose phosphate metabolism which are unlikely to be important. For example, it seems unlikely that the production of reduced NADP is important, since the dormancy-breaking agents nitrite and methylene blue would be expected to return NADPH to the oxidized state.

Another possibility which should be considered is that dormancy is associated with a limitation of nucleic acid synthesis. Tuan & Bonner (1964) have shown that chromatin extracted from dormant potato-tuber buds is incapable of acting as a template for RNA synthesis when supplied with nucleoside triphosphates and appropriate co-factors, whereas the chromatin from non-dormant buds is capable of RNA synthesis. However, recent work has shown that dormancy in barley seeds cannot be explained on genome repression since the chromatin extracted from both dormant and non-dormant seeds is equally capable of RNA synthesis (Brenda Higham & E. H. Roberts, unpublished). Furthermore Chen & Varner (1967) have recently shown that dormant seeds of *Avena fatua* are *more* active than non-dormant seeds in the *in vivo* synthesis of RNA. Fractionation of nucleic acids into the components of MAK-column chromatography after incorporation of ^{32}P or glucose-^{14}C-U during the early stages of imbibition have also indicated that dormant seeds of *Avena fatua* are at least as capable as non-dormant seeds of synthesizing all fractions (Diana Madden & E. H. Roberts, unpublished). Thus it seems unlikely that dormancy in cereal seeds is connected with an inability to synthesize nucleic acids and the significance of the PP pathway must be sought elsewhere.

CONCLUDING REMARKS

There is evidence that different species have evolved different mechanisms of seed dormancy. In spite of the differences, oxidation mechanisms frequently seem to be involved. At one time it was commonly thought that in many species covering structures of the seed impose a limitation on oxygen uptake, and germination is prevented by a restriction of respiration. In most cases, however, there is little evidence for this view. Recently attention has been focused on endogenous germination inhibitors and there seems little doubt that in a number of species these are important. In some cases the inhibitors increase the requirement for oxygen and in others there is evidence that the function of oxygen is to oxidize the inhibitor to an inactive form. In many cultivated species the involvement of specific germination inhibitors is less obvious. In this paper attention has been drawn to the fact that in a number of species dormancy is decreased by oxygen, nitrate, nitrite and other electron acceptors but not by reduced nitrogenous compounds. In such cases dormancy is often also reduced by a wide spectrum of respiratory inhibitors. There is some evidence that the pentose phosphate pathway may be important during the early stages of germination and that the operation of this pathway may be restricted in dormant seeds. The hypothesis is advanced for future investigation that a number of dormancy-breaking agents stimulate germination by increasing the operation of the PP pathway.

REMARKS

AP REES, T. & BEEVERS, H. (1960a). *Pl. Physiol., Lancaster* **35**, 830.
AP REES, T. & BEEVERS, H. (1960b). *Pl. Physiol., Lancaster* **35**, 839.
BALLARD, L. A. T. & LIPP, A. E. G. (1967). *Science, N.Y.* **156**, 398.
BARKER, J. & MAPSON, L. W. (1955). *Proc. R. Soc.* B **143**, 523.
BEEVERS, H. (1961). *Respiratory Metabolism in Plants*. Illinois. New York: Row, Peterson and Co.
BEEVERS, L., FLESCHER, D. & HAGEMAN, R. H. (1964). *Biochim. biophys. Acta* **89**, 453.
BELDEROK, B. (1961). *Proc. int. Seed Test Ass.* **26**, 697.
BELDEROK, B. (1963). [*Versl. Tienjarenplan Graanonderz*, 1962, **9**, 63.] *Fld Crop Abstr.* 1966, **19**, abstr. no. 124.
BLACK, M. (1959). *Can. J. Bot.* **37**, 393.
BLACK, M. & RICHARDSON, M. (1965). *Nature, Lond.* **208**, 1114.
BLACK, M. & RICHARDSON, M. (1967). *Planta* **73**, 344.
BLACK, M. & WAREING, P. F. (1959). *J. exp. Bot.* **10**, 134.
BLOOM, G. & STETTEN, D. (1953). *J. Am. chem. Soc.* **75**, 5446.
BLUM, P. H. & GILBERT, S. G. (1957). *Echo de la Brasserie* **48**, 1155.
BORRISS, H. (1967). In *Physiology, Ecology and Biochemistry of Germination*, vol. I, p. 155. Ed. H. Borriss. Greifswald: Ernst-Moritz-Arndt-Universität.
BRADBEER, J. W. & COLMAN, B. (1967). In *Physiology, Ecology and Biochemistry of Germination*, ed. H. Borriss. Greifswald: Ernst-Moritz-Arndt-Universität.
BROWN, R. (1965). *Handb. PflPhysiol.* **15** (2), 894.
BRUIN, W. J. & TOLBERT, N. E. (1965). *Pl. Physiol., Lancaster* **40**, xxxvii.
BRUINSMA, J. (1962). *Eur. Potato J.* **5**, 195.
BUTT, V. S. & BEEVERS, H. (1960). *Biochem. J.* **76**, 51 P.
BUTT, V. S. & BEEVERS, H. (1961). *Biochem. J.* **80**, 21.
CALDWELL, F. (1959). *Agric. hort. Engng* (Abstr.) **9**, 189.
CARR, D. J. (1961). *Handb. PflPhysiol.* **16**, 737.
CHAKRAVORTY, M. & BURMA, D. P. (1959). *Biochem. J.* **73**, 48.
CHEN, S. S. C. & VARNER, J. E. (1967). *MSU/AEC Plant Res. Lab. Ann. Rep.*, 1967. East Lansing: Michigan State University.
CHETRAM, R. S. & HEYDECKER, W. (1967). *Nature, Lond.* **215**, 210.
CHING, T. M. & FOOTE, W. H. (1961). *Agron. J.* **53**, 183.
CONN, E. E. & STUMPF, P. K. (1966). *Outlines of Biochemistry*, 2nd ed. New York: Wiley.
CROCKER, W. & BARTON, L. V. (1953). *Physiology of Seeds*. Waltham, Mass: Chronica Botanica.
DRENNAN, D. S. H. & BERRIE, A. M. M. (1962). *New Phytol.* **61**, 1.
DURHAM, V. M. & WELLINGTON, P. S. (1961). *Ann. Bot.* **25**, 197–205.
EFFER, W. R. & RANSON, S. L. (1967a). *Pl. Physiol., Lancaster* **42**, 1042.
EFFER, W. R. & RANSON, S. L. (1967b). *Pl. Physiol., Lancaster* **42**, 1053.
ELLIOT, B. B. & LEOPOLD, A. C. (1953). *Physiologia Pl.* **6**, 65.
ESSERY, R. E., KIRSOP, B. H. & POLLOCK, J. R. A. (1954). *J. Inst. Brew.* **60**, 473.

FISCHNICH, O., THIELBEIN, M. & GRAHL, A. (1962). *Saatgut-Wirt.* **14**, 12.
FRANKLAND, B. & SMITH, H. (1967). *Planta* **77**, 354.
GABER, S. D. (1967). Investigations into factors controlling water-sensitivity in barley seeds. Ph.D. thesis, University of Manchester.
GABER, S. D. & ROBERTS, E. H. (1969a). *J. Inst. Brew.* (In the Press.)
GABER, S. D. & ROBERTS, E. H. (1969b). *J. Inst. Brew.* (In the Press.)
GADD, I. (1939). *Proc. Int. Seed Test Assoc.* **11**, 108.
GIBBS, M. (1954). *Pl. Physiol., Lancaster* **29**, 34.
HANSON, J. B. & KRUEGER, W. A. (1966). *Nature, Lond.* **211**, 1322.
HAREL, E. & MAYER, A. M. (1963). *Physiologia Pl.* **16**, 804.
HARPER, J. L. (1957). *Proc. IVth Int. Congr. Crop Protection,* Hamburg, vol. 1, p. 415.
HAY, J. R. (1962). *Can. J. Bot.* **40**, 191.
HILTNER, L. (1910). *Jber. Verein. angew. Bot.* **8**, 219.
JAMES, W. O. (1953). *Proc. R. Soc.* B **141**, 289.
JAMES, W. O. & BOULTER, D. (1955). *New Phytol.* **54**, 1.
JANSSON, G. (1962a). *Ark. Kemi.* **19**, 141.
JANSSON, G. (1962b). *Ark. Kemi.* **19**, 149.
JANSSON, G. (1962c). *Ark. Kemi.* **19**, 161.
JANSSON, G. (1962d). *Svensk kem. Tidskr.* **74**, 181.
JANSSON, G., KIRSOP, B. H. & POLLOCK, J. R. A. (1959). *J. Inst. Brew.* **65**, 165.
JONES, J. W. (1926). *J. Am. Soc. Agron.* **18**, 576.
KHAN, A. A., TOLBERT, N. E. & MITCHEL, E. D. (1964). *Pl. Physiol., Lancaster* **39**, xxviii.
KOLLER, D. (1964). *Herb. Abstr.* **34**, 1.
KOLLER, D., MAYER, A. M., POLJAKOFF-MAYBER, A. & KLEIN, S. (1962). *A. Rev. Pl. Physiol.* **13**, 437.
KUDO, S. & YASHIDA, T. (1958). *Rep. Kirin Brewing Co.* **1**, 39.
LACROIX, L. J. & JASWAL, A. S. (1967). *Pl. Physiol., Lancaster* **42**, 479.
MACDONALD, I. R. (1959). *Ann. Bot.* **23**, 241.
MACDONALD, I. R., BACON, J. S. D., VAUGHAN, D. & ELLIS, R. J. (1966). *J. exp. Bot.* **17**, 822.
MACLEOD, A. M. & PALMER, G. H. (1967). *Nature, Lond.* **216**, 1342.
MAJOR, W. (1966). Investigations into the physiology of dormancy in cereal seeds. Ph.D. thesis, University of Manchester.
MAJOR, W. & ROBERTS, E. H. (1968a). *J. exp. Bot.* **19**, 77.
MAJOR, W. & ROBERTS, E. H. (1968b). *J. exp. Bot.* **19**, 90.
MAPSON, L. W. (1958). *A. Rev. Pl. Physiol.* **9**, 119.
MAYER, A. M. (1960). *Indian J. Pl. Physiol.* **3**, 13.
MAYER, A. M. & POLJAKOFF-MAYBER, A. (1962). *Pl. Cell Physiol., Tokyo* **3**, 309.
MAYER, A. M. & POLJAKOFF-MAYBER, A. (1963). *The Germination of Seeds.* Oxford: Pergamon.
MAYER, A. M., POLJAKOFF-MAYBER, A. & APPLEMAN, W. (1957). *Physiologia Pl.* **10**, 1.
MERRY, J. & GODDARD, D. R. (1941). *Proc. Rochester Acad. Sci.* **8**, 28.
MIKKELSEN, D. S. & SINAH, M. N. (1961). *Crop Science* **1**, 332.
MIYAMOTO, T., TOLBERT, N. E. & EVERSON, E. H. (1961). *Pl. Physiol., Lancaster* **36**, 739.
NAYLOR, J. M. & SIMPSON, G. M. (1960). *Can. J. Bot.* **39**, 281.
NEISH, A. C. (1960). *A. Rev. Pl. Physiol.* **11**, 55.
NUTILE, G. E. & WOODSTOCK, L. W. (1967). *Physiologia Pl.* **20**, 554.
NYMAN, B. (1963). *Studies on the Germination in Seeds of Scots Pine.* Stockholm: Skogshögskolan.

OGOWARA, K. & HAYASHI, J. (1964). *Ber. Ohara Inst. landw. Biol.* **12**, 159.
OGOWARA, K. & ONO, K. (1955). *Bull. School Educ., Okayama Univ.* no. 1, p. 97.
ORPHANOS, P. I. & HEYDECKER, W. (1967). *Univ. Nottingham School of Agric. Rep.* 1966–7, p. 73.
OWEN, E. B. (1956). *The Storage of Seeds for the Maintenance of Viability.* Hurley: Commonwealth Agricultural Bureaux.
POLJAKOF-MAYBER, A. (1952). *Palest. J. Bot. Jerusalem Ser.* **5**, 180.
POLJAKOFF-MAYBER, A. (1955). *J. exp. Bot.* **6**, 313.
POLJAKOFF-MAYBER, A. & EVENARI, M. (1958). *Physiologia Pl.* **11**, 84.
POLJAKOFF-MAYBER, A. & MAYER, A. M. (1958). *Bull. Res. Coun. Israel.* **6**D, 86.
POLLOCK, B. M. & OLNEY, H. O. (1959). *Pl. Physiol., Lancaster* **34**, 131.
POLLOCK, J. R. A. (1958). *Chemy Ind.* p. 387.
POLLOCK, J. R. A. & KIRSOP, B. H. (1956). *J. Inst. Brew.* **62**, 323.
POLLOCK, J. R. A., KIRSOP, B. H. & ESSERY, R. E. (1955*a*). *European Brewery Conv. Congr.* p. 203.
POLLOCK, J. R. A., KIRSOP, B. H. & ESSERY, R. E. (1955*b*). *J. Inst. Brew.* **61**, 301.
POLLOCK, J. R. A., KIRSOP, B. H. & ESSERY, R. E. (1956). *J. Inst. Brew.* **62**, 181.
POPAY, A. I. & ROBERTS, E. H. (1969*a*). *J. Ecol.* (In the Press.)
POPAY, A. I. & ROBERTS, E. H. (1969*b*). *J. Ecol.* (In the Press.)
ROBERTS, E. H. (1961). *J. exp. Bot.* **12**, 430.
ROBERTS, E. H. (1962). *J. exp. Bot.* **13**, 75.
ROBERTS, E. H. (1963*a*). *Physiologia Pl.* **16**, 732.
ROBERTS, E. H. (1963*b*). *Physiologia Pl.* **16**, 745.
ROBERTS, E. H. (1964*a*). *Physiologia Pl.* **17**, 14.
ROBERTS, E. H. (1964*b*). *Physiologia Pl.* **17**, 30.
ROBERTS, E. H. (1965). *J. exp. Bot.* **16**, 341.
ROBERTS, E. H. & GABER, S. D. (1969). *J. Inst. Brew.* (In the Press.)
RUSSELL, E. W. (1961). *Soil Conditions and Plant Growth*, 9th ed., p. 303. London: Longmans.
SCHWENDIMAN, A. & SHANDS, H. L. (1943). *J. Am. Soc. Agron.* **35**, 681.
SIMON, E. W. (1953). *J. exp. Bot.* **4**, 377.
SIMPSON, G. M. (1965). *Can. J. Bot.* **43**, 793.
SIMPSON, G. M. & NAYLOR, J. M. (1962). *Can. J. Bot.* **40**, 1659.
STEINBAUER, G. P. & GRIGSBY, B. (1957). *Weeds* **5**, 175.
STILES, W. (1960). *Handb. PflPhysiol.* **12** (2), 465.
SUSSMAN, A. S. & HALVORSON, H. O. (1966). *Spores, Their Dormancy and Germination.* New York and London: Harper and Row.
THIMANN, K. V., YOCUM, C. S. & HACKETT, D. P. (1954). *Archs Biochem. Biophys.* **53**, 239.
THORNTON, N. C. (1945). *Contr. Boyce Thompson Inst. Pl. Res.* **13**, 487.
TOOLE, E. H., HENDRICKS, S. B., BORTHWICK, H. A. & TOOLE, V. K. (1956). *Ann. Rev. Pl. Physiol.* **7**, 299.
TUAN, D. Y. H. & BONNER, J. (1964). *Pl. Physiol., Lancaster* **39**, 768.
VEGIS, A. (1964). *A. Rev. Pl. Physiol.* **15**, 185.
VOSE, P. B. (1962). *Ann. Bot.* **26**, 197.
WAREING, P. F. (1963). In *Vistas in Botany*, vol. III, pp. 195–227. Ed. W. B. Turrill. Oxford: Pergamon.
WAREING, P. F. & FODA, H. A. (1957). *Physiologia Pl.* **10**, 266.
WEEKS, D. P. (1967). Characteristics of polyribosomes in germinating seeds and growing root tip tissue. Ph.D. thesis, University of Illinois.
WILLIAMS, J. T. & HARPER, J. L. (1965). *Weed Res.* **5**, 141.

LIGHT-CONTROLLED GERMINATION OF SEEDS

By M. BLACK

Department of Biology, Queen Elizabeth College,
University of London

INTRODUCTION

The first recorded references to a possible effect of light on seed germination appeared towards the end of the eighteenth century in the writings of Ingenhousz and Senebier. Their suspicions were confirmed by Caspary in 1860 and later by Gassner and other workers about 50–60 years ago. Kinzel (1908), for example, records over 600 species of seed which were stated to be light-sensitive. But light-sensitivity of seeds did not become firmly established as part of the photomorphogenetic lore until the work of Flint & McAlister (1935), Evenari and his colleagues (e.g. Evenari, 1965b) and the Beltsville group (Borthwick, Hendricks, Parker, Toole & Toole, 1952). Perhaps it may be claimed that this group ensured for lettuce seeds what Morgan did for *Drosophila*—a place in the biological hall of fame. Both of these organisms were used to help lay down the foundations of important biological concepts, and it was using lettuce seeds that the phytochrome system was discovered (Borthwick *et al.* 1952). Following these basic discoveries research was intensified, and it has become apparent that the operation of the phytochrome system in seeds is very complex. Moreover, even the notion that light-sensitivity depends solely upon phytochrome may have to be abandoned as it has been found that the other photomorphogenetic system, the high energy reaction, is commonly present in seeds (Mohr & Appuhn, 1963; Rollin, 1963).

Comprehensive accounts of light-controlled germination have been published in the last few years (Evenari, 1965a, b) in which the details of the many types of behaviour are discussed. Thanks to these full reviews it is necessary here to consider only some important general aspects of the phenomenon in order to set the background to the problem.

THE LIGHT-SENSITIVITY OF SEEDS

The majority of seeds seem to be insensitive to light but many are stimulated or inhibited by exposure to continuous or short periods of illumination. Imbibed seeds which do not germinate at all in darkness (or show very low percentages) may be fully promoted by only a few seconds or minutes of white light. The phytochrome system is operative in all these cases as shown by the facts that red light (approx. 660 mμ) is the most effective waveband, and that its inductive action can be reversed by far-red (approx. 730 mμ), both at relatively low intensities. Red light causes the formation of the active form of phytochrome (P_{fr}) which then induces germination. The conversion P_r to P_{fr} requires about one-quarter the amount of energy than does the reverse reaction, which is why white light has the same action as red.

The Grand Rapids variety of lettuce is a well-known example of control by the phytochrome system, which has been the subject of numerous investigations (e.g. Borthwick, Hendricks, Toole & Toole, 1954). Reversal by far-red of red-light induction gradually diminishes as the period of darkness between red and far-red increases; during this time the seeds escape from phytochrome control as P_{fr} initiates the reactions leading to germination. In this variety the temperature conditions determine the response to light. There is little germination in darkness at 25° C. but this increases as the temperature is lowered, until at 10° C. it may be almost complete. Thus, Grand Rapids lettuce has a light-requirement only at supra-optimal temperatures. Other cases are known in which light-sensitivity decreases as the temperature is increased, or if daily temperature alternations are experienced, e.g. *Nicotiana* (Toole, Toole, Hendricks & Borthwick, 1957). These particular characteristics generally hold for any one species but one frequently encounters wide variation. For example, the germination levels in darkness of different batches of Grand Rapids may range from 0 to 30 or 40 %, the sensitivity to light and the escape times may differ, and so might the response to variations in temperature. Some of these differences are no doubt due to after-ripening, but others may be determined by the environmental conditions under which the seeds ripen on the mother plant (Koller, 1962).

Many seeds exhibit more complex light requirements than Grand Rapids lettuce. *Plantago* seeds must be illuminated for 1 hr., *Epilobium* seeds for 24 hr. (Isikawa & Yokohama, 1962) and *Paulownia* seeds for 48 hr., either continuously or intermittently (Toole, 1961). Some of the seeds for which repeated daily exposures are necessary show typical photoperiodic responses, many having long-day requirements (Isikawa, 1954; Black & Wareing, 1955; Black, 1967). All of these are cases where the active form of phytochrome (P_{fr}) must be repeatedly renewed, either because P_{fr} itself fairly quickly reverts or is destroyed, or because subsequent reactions which are initiated by P_{fr} proceed rather slowly.

Many of the different kinds of responses to light can be understood on the basis of the various properties of phytochrome which have been elucidated over the past few years (e.g. Butler, Lane & Siegelman, 1963), since the extraction and *in vitro* study of the pigment has been possible. The P_r form exhibits peak absorption at 660 mμ but also absorbs over the whole visible spectrum as well as in the far-red, so that even under the latter irradiation some 1–4 % P_r is converted to P_{fr}. The small amounts of P_{fr} which are formed in blue, yellow and far-red light are apparently sufficient to promote germination in *Artemisia* (Koller, Sachs & Negbi,

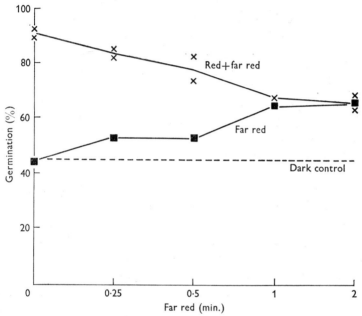

Fig. 1. Germination of *Eragrostis* induced by far-red, and by red followed by far-red (after Toole & Borthwick, 1968).

1964); it seems that in this seed the 'efficiency' of P_{fr} is extraordinarily high. Conversion of some P_r to P_{fr} even by far-red explains results obtained by Downs (1964) with *Wittrockia superba*, in which germination is fully promoted by a short exposure to red light or by continuous far-red, and partially promoted by a few minutes of far-red. In fact, the response of these seeds to these various light treatments is very much like that of Grand Rapids lettuce which have been treated with kinetin (Miller, 1958). Similarly, in *Eragrostis*, germination above the level in darkness is obtained after a short exposure to far-red or with red followed by far-red (Fig. 1). Evidently, in all these seeds much less pigment conversion is required than the 50 % thought to be necessary to achieve the 50 % germination level in *Lactuca* (Hendricks, Borthwick & Downs, 1956).

The amount of P_{fr} which is required for the promotion of germination depends strongly on the conditions of stress imposed upon the embryo; stress is used here to mean anything which impedes embryo growth. Normally, this is achieved by the tissues surrounding the embryo, especially the intact endosperm. For example, isolated embryos of Grand Rapids germinate in darkness (Evenari & Neumann, 1952), as do half-seeds, but a light requirement can be reinstated by solutions having relatively high osmotic pressures (Scheibe & Lang, 1965; Kahn, 1960a). The germination which is permitted by these 'surgical' operations evidently still depends on P_{fr}, since it is significantly lowered by a short exposure to far-red. The low level of P_{fr} present in darkness is presumably inadequate to stimulate embryo growth in the presence of the endosperm but can do so when the restraining influence of this tissue is removed. It is not known how the enclosing structures succeed in inhibiting embryo growth. Some authors suggest that it is primarily a mechanical effect (Ikuma & Thimann, 1963c) while others have proposed that inhibitors are involved (Black & Wareing, 1959). In any case, it seems likely that the growth potential of the embryo must be increased in order to overcome the block.

We see clearly how important stress conditions are in those cases where a light-requirement is induced by treatment with chemical inhibitors (Nutile, 1945). In the Progress variety of lettuce, for example, there is no evidence of the existence of the phytochrome system until normal dark germination is prevented by coumarin; then germination requires red light and red-far-red reversibility can be demonstrated. In the presence of the inhibitor, a high level of P_{fr} becomes necessary.

The requirement for light can in many cases be circumvented by various chemical treatments, as well as by the low temperature or surgical treatments discussed above. Grand Rapids seeds germinate in darkness when treated with gibberellin (Kahn, Goss & Smith, 1957), thiourea (Thompson & Kosar, 1938) and chloramphenicol (Black & Richardson, 1965). There is evidence, though, that the low level of P_{fr} present in darkness contributes or even causes all these apparent by-passes of the light-requirement (e.g. Scheibe & Lang, 1965; Roth-Bejerano, Koller & Negbi, 1966). P_{fr} is certainly required for kinetin-promoted germination of Grand Rapids (Miller, 1958). Moreover, a far-red-sensitive pigment (which may not be the 'classical' phytochrome) is essential for both gibberellin- and thiourea-induced germination, a point which we shall discuss in some detail below.

Clearly, there is much to be learned about the properties of phytochrome in seeds. Unfortunately, although seeds were used for the physiological demonstration of the existence of this photoreceptor, the pigment has only very recently been detected spectrophotometrically in them (Mancinelli &

Tolkowsky, 1968; Rollin, 1968 (personal communication)). Before we can understand the complex positive responses to the quality and quantity of light we must discover more about the amount of pigment conversion necessary to promote germination; about the rates of destruction and reversion of phytochrome in seeds; and, of course, about the biochemical processes which phytochrome controls.

We have so far considered the positive responses to light. Many seeds are, however, inhibited when held imbibed under continuous white light or long daily photo-periods (Isikawa, 1954; Black & Wareing, 1960). The inhibition is apparently due mainly to blue and far-red light, the latter being much more effective than the former. In these seeds (e.g. *Nemophila*, *Nigella*, *Phacelia*) illumination for several hours is necessary to achieve full effect, and this must be given at certain times after the beginning of imbibition. It is interesting that the effect of light on these seeds is again determined by stress conditions imposed upon the embryo. Many seeds are not inhibited by light if the structures enclosing the embryo are removed. Conversely, some seeds which normally show no response whatsoever to light (e.g. radish) are converted into light-inhibited types when treated with solutions having relatively high osmotic pressures (McDonough, 1967).

White light is effective in these seeds because the actions of the blue and far-red parts of the spectrum are dominant. But in some other seeds inhibition is achieved only by light which contains a high proportion of far-red. Some varieties of tomato and lettuce, which germinate in darkness, are inhibited by several hours irradiation with far-red. In tomato, irradiation between the 6th and 18th hr. of imbibition is necessary (Mancinelli, Borthwick & Hendricks, 1966). This is apparently due to the continued synthesis of a far-red absorbing form of pigment which is fully reversed only by continual or intermittent far-red extending over several hours. After this far-red treatment, seeds are rendered sensitive to typical red–far-red reversible control.

Another example of the inhibitory action of long periods of far-red is seen in the lettuce varieties Grand Rapids and Reine de Mai. Short far-red irradiation does not inhibit germination of these light-stimulated seeds after the escape time has passed; but if the far-red is extended to 3 or 4 hr., complete inhibition is achieved (Mohr & Appuhn, 1963; Rollin, 1963). How this phenomenon is related to the previously described effects of prolonged far-red is still unclear.

LIGHT IN THE NATURAL CONTROL OF DORMANCY

What is the importance of light in the control of dormancy? Various experiments carried out in the laboratory and field indicate that dormancy of a wide range of species is governed by light. Wesson & Wareing in 1967 studied seeds of 23 species which commonly occur in arable land and found that 20 of these fail to germinate when buried, because of the absence of light. Evenari (1965a) refers to some desert plants whose distribution and survival depend upon their light-sensitivity. Seeds of *Juncus maritimus* var. *arabicus*, for example, have an absolute requirement over a wide range of temperatures and thus they do not germinate under dense vegetation or in deep water. Some of the fruits of *Salsola* germinate only when uncovered, even after lying buried for several years. Conversely, because *Calligonum comosum* seeds are light-inhibited they germinate only when well covered, at depths where the water content is, incidentally, sufficient to support subsequent growth.

Light undoubtedly plays an extremely important role in restricting germination to certain times of the year. Photoperiodic control is especially important in this respect, birch seed being a good example. Seeds which have not been chilled in the imbibed condition require long days for germination and therefore will remain dormant when shed from the mother plant in autumn, but will germinate when the days lengthen in the following spring. Chilled seeds, on the other hand, germinate even in complete darkness when the temperature is suitable (Black & Wareing, 1955). Light and temperature therefore interact to control germination and to confine it to the most suitable time of year.

We have mentioned above that in many seeds control by light can be by-passed either by low temperatures (including chilling) or by after-ripening, as in lettuce and numerous other examples. One might wonder therefore how important light really is as a natural control and how often it exerts its regulatory role. Even in birch seed, the photoperiod would fully regulate germination only in the very rare years when chilling is not experienced. However, rigorous control of germination might well be exerted by prolonged far-red, which would also affect many seeds which are apparently insensitive to white light. Far-red light is found under leaf canopies (Vezina & Boulter, 1966); Cumming (1963) has already drawn attention to its possible ecological role in regulating germination of *Chenopodium*. In a simple experiment we have shown that two leaves of *Tilia* act as a perfect far-red filter which can be used to demonstrate red–far-red reversibility in Grand Rapids lettuce; peak transmission through these leaves is at 730 mμ (Fig. 2). All the treatments which normally serve to release lettuce seed from control by white light are quite ineffective when seeds are held under far-red for 12 hr. per day (Table 1). Further, one 6 hr. exposure to far-red filtered through leaves completely prevents germination of half-seeds, isolated embryos and of seeds later transferred

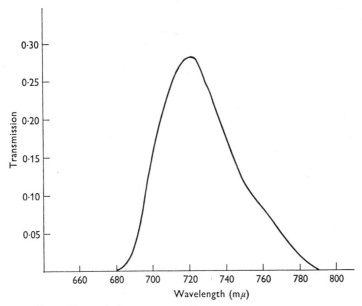

Fig. 2. Transmission spectrum of leaves of *Tilia europea* from 640 mμ.

Table 1. *The effect of 12 hr. far-red per day on germination of Grand Rapids lettuce given various germination-stimulating treatments*

Treatment	Germination (%)	
	Darkness	Far-red
Intact	6	0
Thiourea (4 mg./ml.)	67	0
GA$_3$ (50 μg./ml.)	87	9
Embryos	100	12
Half-seeds	100	15
'Low' temperature (16° C.)	79	0
5 min. red light	97	0

to the low temperatures (10–15° C.) which favour dark germination. Similarly, chilled birch seeds are inhibited by prolonged far-red. This aspect of seed dormancy has so far received little attention from the ecological standpoint, but it is most likely to be important in regulating germination under foliage such as in forest communities.

MECHANISM OF PHYTOCHROME ACTION IN GERMINATION

The question as to the mechanism of action of phytochrome is a basic problem in developmental physiology which still remains unanswered; some interesting clues which may lead to its solution have, however, recently been found.

If we examine a relatively simple case of phytochrome-controlled germination—the familiar Grand Rapids lettuce seed at 25° C.—we see that

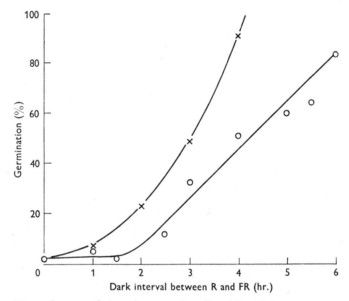

Fig. 3. 'Escape' curves of phytochrome action. Grand Rapids lettuce were exposed to red light (R) (180 μW..cm.$^{-2}$) for 30 sec. and, after different periods in darkness, were irradiated with far-red (FR) (70 μW..cm.$^{-2}$) for 5 ($\times$) or 15 min. (O). 'Escape' depends on the amount of far-red.

there is little or no germination in total darkness, but when the seeds are exposed to red light (100 μW..cm.$^{-2}$ for 30 sec.) germination begins about 7 hr. later and all the seeds have germinated after 12 hr. The conversion of P_r to P_{fr} has led to profound changes in the seeds which culminate in protrusion of the radicle. The 'escape' curve for P_{fr} action indicates that the pigment must act for at least 4 hr. to ensure germination of the whole population; action for 1 hr. is largely ineffectual (Fig. 3). The fundamental problem is: what does P_{fr} do during those few hours? It might help us to find the answer if we knew whether cell division or cell enlargement is the result of P_{fr} action. Unfortunately, it does not seem that we can come down squarely on the side of either of these. In light-induced lettuce, mitosis and cell enlargement coincide at the time of protrusion of the radicle (Evenari, Klein, Anchori & Feinbrun, 1957), although it is quite possible to achieve germination without any mitoses (Haber & Luippold, 1960). On the other hand, mitoses are always found before enlargement in light-stimulated seeds of *Pinus sylvestris* (Nyman, 1961).

The ideas concerning phytochrome action in plants can be placed under five headings:

(a) P_{fr} is a key enzyme possibly involved in fat metabolism.
(b) P_{fr} stimulates respiration.
(c) P_{fr} is involved in gene activation, hence enzyme synthesis.
(d) P_{fr} causes gibberellin biosynthesis.
(e) P_{fr} is involved in membrane permeability.

Unfortunately, as far as seeds are concerned, most of the evidence relating to these ideas is of a negative kind, and we know more about what phytochrome apparently does not do than what it does. However, as we shall see below, the picture is not entirely bleak and some statements of a positive sort can be made.

P_{fr} as an enzyme in fat metabolism

It is an interesting fact that many of the light-requiring seeds store fat. There have been occasional reports (Tietz, 1953; Rimon, 1957) of changed lipolytic activity in seeds following illumination, but no firm favourable evidence has yet emerged. On the contrary, Nyman (1966) carried out a thorough examination of fat metabolism in the light-sensitive *Pinus sylvestris* and found no indication of any effect of light on lipolytic activity before visible germination had occurred.

Respiration

Changes in respiration in response to light have sometimes been observed. In *Pinus sylvestris*, for example, anaerobic respiration increases after irradiation (Nyman, 1961). However, it is frequently found in other seeds that respiration is unaffected by light; there is therefore no strong evidence that P_{fr} generally controls respiration in seeds. It appears, though, that seeds must be undergoing active respiration in order for P_{fr} to act, since anaerobic conditions imposed after the conversion of P_r to P_{fr} inhibit germination (Ikuma & Thimann, 1964).

Nucleic acid and protein synthesis

Mohr and his co-workers have argued that P_{fr} is involved in the activation of 'potentially active' genes (Mohr, 1967; Rissland & Mohr, 1967). Some of this evidence is direct and concerns specific enzymes involved in anthocyanin synthesis. However, there is no support for a similar action of P_{fr} in the morphological responses which the pigment controls, of which germination is one. We have found (as has Khan, 1967) that various inhibitors which affect nucleic acid and protein synthesis (6-azauracil, cycloheximide and 6-methyl purine) all inhibit the germination of lettuce seeds which have been given a saturating red-light dose. On the other hand, actinomycin D, puromycin and chloramphenicol have no inhibitory action; in fact, chloramphenicol promotes germination in darkness even though total protein synthesis is depressed (Black & Richardson, 1967). It seems likely on the basis of these studies with inhibitors that protein synthesis (at least the fraction inhibited by cycloheximide) is necessary for germination to occur, but we can conclude nothing about the protein synthesis being phytochrome-controlled. Protein synthesis occurs in darkness in any case (Black & Richardson, 1967) and this is sensitive to cycloheximide.

We have so far been unable to detect any effect of P_{fr} on incorporation of [^{3}H]leucine into protein or [^{3}H]orotate into RNA in the whole seed or in the axis alone, during the time between illumination and the start of germination. These results do not, however, rule out the possibility that protein synthesis is promoted by P_{fr}. If one key enzyme is induced our methods would not detect it and it is difficult to know what enzyme to look for. Moreover, since the response to light occurs only in the radicle and probably only in a relatively few cells of this organ, any possible changes in protein synthesis might well be much too small to measure.

Gibberellin synthesis

Since gibberellin often simulates the action of phytochrome in plant development it has been suggested that P_{fr} causes gibberellin biosynthesis (Brian, 1958); but on various grounds several workers have rejected this proposal as far as germination is concerned (Haber & Tolbert, 1959; Ikuma & Thimann, 1963b; Scheibe & Lang, 1965). However, favourable experimental evidence comes from the work of Yamaki, Hashimoto, Ishii & Yamada (1958) and Reid, Clements & Carr (1968) on leaves, and Köhler (1966a, b) using etiolated peas and the light-sensitive variety of lettuce Schreibers Tenax. Köhler reported that in these seeds there was a steadily increasing level of extractable gibberellin after the illumination. A considerable rise occurred in the first 2 hr., i.e. even over the period when seeds cannot be expected to have escaped from control by P_{fr} (see Fig. 3). Presumably the level of gibberellin produced during this time is too low to induce germination.

We have performed a series of experiments to investigate this point, reasoning that the supply of further but still subthreshold amounts of gibberellin might magnify the germination-promoting potential of the endogenous hormone (Bewley, Negbi & Black, 1968). Grand Rapids seeds were exposed to 2 min. red light and returned to darkness for periods up to 4 hr., after which they were irradiated with far-red for 5 min., i.e. P_{fr} was allowed to act for varying periods. Immediately after the far-red they were transferred to GA_3 (5 μg./ml.) for a further 24 hr., when germination was counted. Now, 1 hr. of P_{fr} activity induces about 7% germination when seeds are held on water; and the GA_3 dark control shows about 25% germination. However, transfer to GA_3 strikingly enhances the 1 hr. of P_{fr} activity, to give the high germination figure of 71%. There is clearly a strong synergism between P_{fr} action for 1 hr. and GA_3 supplied externally.

In further experiments a different light regime was used, consisting of 30 sec. red light and 5 min. far-red. This duration of far-red causes virtually complete reversion of P_{fr} to P_r, so that the treatment in which red was followed immediately by far-red before transferring the seeds to GA_3 gave almost the same germination value as the GA_3 dark control (Fig. 4). The times of P_{fr} action allowed before supplying GA_3 ranged from 5 to 30 min. Figure 4 shows the observed results and, for comparison, a curve calculated for an additive effect between GA_3 and P_{fr} action. Again the strong synergism is evident, but even more remarkable is the effect of only 5 min. of P_{fr} activity. This technique of supplying a subthreshold level of GA_3 to seeds in which P_{fr} has acted for some time therefore reveals that the action of P_{fr} is rapid and has caused significant changes in 5 min., even though these changes alone are insufficient to induce germination. We have never failed to obtain this synergism; but there are reports both for and against synergism between light and gibberellin (Kahn, 1960b; Ikuma & Thimann, 1963a). Furthermore, we have investigated several other gibberellins and they all give similar results to GA_3 (Table 2). Our results therefore throw considerable doubt on the possibility that gibberellins are produced by the rapid action of phytochrome. Since these experiments were carried out, Dr Köhler has informed us that his earlier findings are unrepeatable, and so the contradiction seems to be resolved.

In spite of the above results we still cannot dismiss the question of gibberellin production in light-sensitive seeds. McDonough (1965) inhibited light-induced *Verbascum* seeds by Phosphon D, which could mean

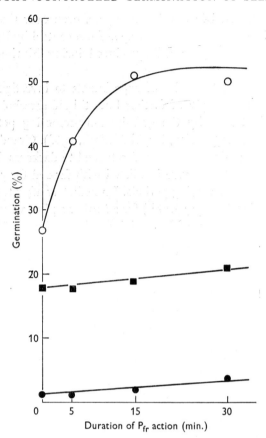

Fig. 4. Interaction between P_{fr} and subthreshold GA_3 concentration. Seeds were irradiated with 30 sec. red light and then given 0, 5, 15 or 30 min. in darkness before exposure to 5 min. far-red. They were then transferred to water (●—●) or 5 μg./ml. GA_3 (○—○). The calculated curve for an additive effect between GA_3 and P_{fr} (■—■) is obtained from the sum of the water control and the germination in darkness on GA_3 alone. (From Bewley, Negbi & Black, 1968.)

Table 2. *The effect of adding different gibberellins to seeds in which P_{fr} has been allowed to act for 5 min.*

Gibberellin	Concentration (μg./ml.)	Germination (%)		
		$-P_{fr}$ action	$+P_{fr}$ action	Calculated*
GA_1	10	7±1	50±1	15
GA_4	0·25	21±2	82±1	29
GA_5	25	2±2	38±5	10
GA_7	0·25	35±2	94±3	43
GA_9	25	8±3	44±4	16
GA_{13}	30	4±1	60±10	12
None	—	2	10	—

* Calculated value is the sum of gibberellin dark control and the P_{fr} action on water alone, from which the dark control on water has been subtracted (from Bewley *et al.* 1968).

that gibberellin production is essential for their germination. We have found that the promotive effect of light on Grand Rapids lettuce is nullified by AMO 1618 and Phosphon D, but chloro-choline chloride (CCC) is inactive (Bewley, 1968); application of GA_3 overcomes the effect of AMO and Phosphon (Table 3). These results suggest that gibberellin production may well be essential for light-controlled germination to occur, but they do not necessarily indicate that light actually causes this production. In fact, very similar results are obtained with half-seeds which can normally germinate in complete darkness (Table 4); thus, gibberellin biosynthesis might occur even in the absence of light. A ready interpretation of these results is that light-induced germination does indeed require gibberellin biosynthesis, but this occurs quite independently of any light action. Since we have already shown that P_{fr} can synergize with gibberellin the interesting possibility is now raised that normal light-stimulated germination is caused by P_{fr} action for 3–4 hr. together with gibberellin production which proceeds in darkness but at a level too low to induce germination directly.

Table 3. *Effect of AMO, Phosphon D and CCC on germination of light-stimulated Grand Rapids lettuce seed*

Substance	mg./ml.	No. of seeds germinated out of 20		
		Red light	Dark	$+GA_3$
AMO 1618	1·2	2	1	18
CCC	1·25	19	2	18
CCC	2·5	17	2	—
CCC	5	17	2	—
Phosphon D	0·5	0	0	18
Water	—	20	2	19

Seeds were punctured through the seed coat and cotyledons and kept on the appropriate solution for 12 hr. in darkness before exposure to light (from Bewley, 1968).

Table 4. *Effect of AMO 1618 and Phosphon D on germination in darkness of half-seeds of Grand Rapids lettuce*

	Germination (%)	
	$-GA_3$	$+GA_3$
AMO 1618 (1·2 mg./ml.)	45	75
Phosphon D (0·5 mg./ml.)	26	62
Water	100	—

This hints at the idea that both gibberellin and P_{fr} may be necessary for germination. The behaviour of some seeds does suggest that this is indeed a real possibility, as gibberellin sometimes has no action in the dark, but only in the light (Alcorn & Kurtz, 1959). In both *Kalanchoë* and *Begonia evansiana*, GA_3 does not stimulate germination in darkness but it strongly reduces the critical day-length (Bunsow & von Bredow, 1958; Nagao *et al.*

1959). Gibberellic acid apparently sensitizes seeds of *Lythrum salicaria* to light (Rollin & Bidault, 1963). Seeds of various species of *Pinus* do not respond to light unless they have been chilled at 5° C. for several weeks (Toole, Toole, Hendricks & Borthwick, 1961; Toole, Toole, Borthwick & Snow, 1962); there is evidence that chilling increases the endogenous gibberellin levels in seeds (Frankland & Wareing, 1966).

Whether or not this interpretation is correct, we still have to deal with the problem of the action of P_{fr}. Our results demonstrate that P_{fr} has an extremely rapid action. This was an unexpected finding as the visible response to P_{fr} formation is not evident until several hours later in the case of Grand Rapids, and perhaps days later in other species. Evidence has been accumulating recently that P_{fr} acts very quickly in many other systems which are controlled by the pigment. The work of Tanada (1968) reveals the most dramatic rapidity; here, the 'stickiness' of root tips is determined by only 30 sec. of P_{fr} action and is reversible by far-red. Another effect measurable in seconds is that of the action of P_{fr} in the suppression of flowering in *Kalanchoë*. The results of Fredericq (1965) suggest that P_{fr} acts within less than 180 sec. A visible quick response to P_{fr} is found in the nyctinastic movements of certain legumes, in which the pigment induces closing of the leaflets. In *Mimosa pudica* and *Albizzia julibrissin* the response is detectable after about 5 min. (Fondeville, Borthwick & Hendricks, 1966; Hillman & Koukkari, 1967; Jaffe & Galston, 1967). The action of P_{fr} is apparently in the pulvini themselves (Koukkari & Hillman, 1968). These findings give us the best clue to P_{fr} action as they point to the ability of P_{fr} to regulate turgidity, and the suggestion has been put forward that it does so by acting upon membrane permeability (Hendricks & Borthwick, 1967). The phytochrome-controlled root stickiness (Tanada, 1968) and chloroplast orientation (Haupt, 1959) would support the idea of changes in cell membranes (accompanied perhaps by alterations in charge) brought about by P_{fr}. As stated by Hendricks & Borthwick (1967): 'An effect of P_{fr} on cell permeability could well lead to a variety of displays.' Quick changes in permeability in radicle cells of Grand Rapids seeds might possibly lead to the occurrence of reactions when enzymes and substrates are allowed to come into contact. Some of these reactions apparently take place in lettuce seeds in at least 5 min., and can be magnified in the presence of gibberellin. We see, then, how these recent findings might help us to understand the basis of phytochrome-controlled germination and these important developments suggest new and interesting lines of attack.

THE ACTION OF PROLONGED FAR-RED LIGHT ON GERMINATION

Several examples of the inhibition of germination by prolonged far-red have previously been mentioned. Far-red treatment can convert dark-germinating seeds into those which display typical red–far-red reversibility, or it can inhibit germination of light-stimulated seeds after the escape time has passed. Since rigorous control of germination might be exerted by prolonged far-red it could be of considerable ecological importance. Prolonged far-red also inhibits the action of gibberellic acid, especially if this is at subthreshold levels (Kahn, 1960b; Toole & Cathey, 1961), but it is not clear from their work if far-red acts on a gibberellin-induced process or quite independently of gibberellin. The report of Ikuma & Thimann (1960) suggests that the latter may be the case, since the sensitivity of lettuce seeds to GA_3 was lowered when they were irradiated for 50 min. before the hormone was supplied.

We have investigated the effect of prolonged far-red on gibberellin action, and some interesting possibilities concerning the role of this far-red have arisen (Negbi, Black & Bewley, 1968). Continuous far-red light (between 700 and 1000 mμ at 70 μW..cm.$^{-2}$) inhibits germination of Grand Rapids lettuce seeds kept on GA_3 (50 μg./ml.). It is not strictly necessary to expose seeds continuously, as far-red from the 6th to the 12th hr. of imbibition on GA_3 also prevents germination, at least during the following 24 hr., though some seeds start to germinate after this time. Thus, the timing and duration of prolonged far-red required to inhibit gibberellin action are similar to those found to be effective for the inhibition of non-light-requiring seeds, and also light-stimulated seeds (Mancinelli & Borthwick, 1964; Mohr & Appuhn, 1963).

By giving prolonged far-red before supplying GA_3 it is possible to determine if a GA_3-induced process or one independent of GA_3 is inhibited by the radiation. From Fig. 5 we can clearly see that prolonged far-red prevents subsequent gibberellin action and therefore cannot act on a gibberellin-induced process. Irradiation during the first 6 hr. of imbibition is less effective than at later times; during the first 3 hr. far-red is found to have no action at all. This inhibition by far-red is apparently irrevocable in darkness, for even when seeds are kept on GA_3 for 72 hr. after the far-red treatment no germination occurs. Thus, if we assume the existence of a far-red-sensitive pigment which is necessary for GA_3 action, most of this pigment seems to be present only after approximately the 6th hr. of imbibition, and once it is changed by far-red light it apparently does not reform nor is any more synthesized. A short exposure to red light, however, reverses the action of prolonged far-red. One cannot tell, though, if this is a real reversion or if red light is only acting on P_r and is stimulating germination this way.

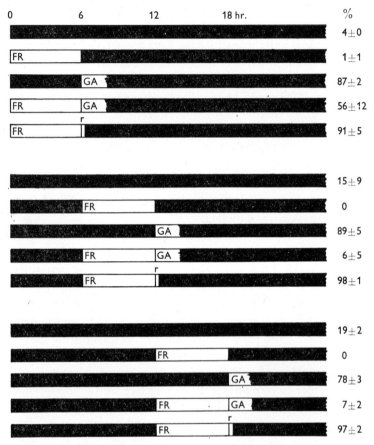

Fig. 5. Desensitization by prolonged far-red. Grand Rapids lettuce seeds were irradiated with 6 hr. far-red at different times after the beginning of imbibition, and then transferred to GA_3 (50 µg./ml.). Germination in this experiment was counted 24 hr. after transfer to GA_3, but in other experiments it was found not to increase even up to 72 hr. (from Negbi, Black & Bewley, 1968).

The conclusion that the action of far-red is on a process quite independent of gibberellin action is fully supported by the finding that the promotive action of thiourea on germination is also completely prevented by 6 hr. far-red prior to the supply of this chemical. The concentration of GA_3 commonly used (50 µg./ml.) is completely inactivated by preceding far-red treatment for 6 hr. However, with higher concentrations (100–500 µg./ml.) longer durations of far-red are necessary to achieve desensitization; 6 hr. far-red is ineffective but 18 hr. prevents the action of 100 µg./ml. GA_3.

We can conclude from this that the small amount of pigment remaining after 6 hr. of far-red can support the action of gibberellin only if the concentration of the latter is relatively high.

Figure 6 illustrates that, in fact, 3–4 hr. far-red are sufficient for full desensitization to GA_3 (50 μg./ml.), no matter what the previous imbibition time is. This indicates that the pigment which begins forming at approximately the 6th hr. does not accumulate as a far-red absorbing form. If it

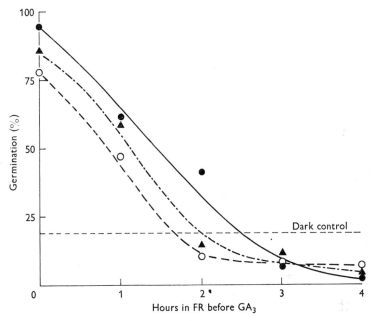

Fig. 6. The duration of far-red (FR) required for desensitization to GA_3. Far-red was given after different times of imbibition in darkness (D), then seeds were transferred to GA_3 (●—●, 6 hr. D; ○--○, 2 hr. D; ▲-.-▲, 18 hr. D). (From Negbi et al. 1968.)

did accumulate in this form, shorter irradiation with far-red would be more effective after the longer imbibition times. The amount of pigment therefore seems to remain constant.

In these experiments, however, there are in fact two variables: one being the duration of far-red and the other being the time before GA_3 is applied to the seeds. When we keep the duration of far-red constant, but vary the time before GA_3 is supplied, by intercalating different periods of darkness, we find that relatively short periods of far-red are appreciably effective, provided there are some hours delay in darkness before adding the hormone (Fig. 7). These treatments succeed in inhibiting germination over the 24 hr. after adding GA_3, but there is gradual recovery during the next 24 hr. Evidently, these regimes are not as effective as 4–6 hr. continuous far-red;

none the less they indicate that there is appreciable pigment conversion in 15 min. The inhibitory effect of delaying gibberellin application after pigment conversion suggests the slow loss, in darkness, of some condition or substance necessary for gibberellin action which has been produced earlier when the pigment was still present.

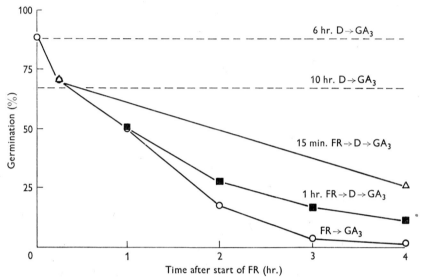

Fig. 7. The light and dark components of the 'prolonged' far-red effect. Seeds were imbibed on water for 6 hr. in darkness before irradiation. GA_3 (50 µg./ml.) was supplied to the seeds immediately after 15 min. 1, 2, 3 or 4 hr. far-red (○—○). In the other treatments, seeds were given 15 min. (△—△) or 1 hr. (■—■) of far-red and then kept in darkness for different periods before transfer to GA_3 (50 µg./ml.). In these cases the total length of time preceding the GA_3 (i.e. far-red plus darkness) was equivalent to the various periods of continuous far-red in curve ○—○ (D = darkness). (From Negbi et al. 1968.)

The finding that 15 min. far-red can cause considerable pigment conversion can be used to determine the relative importance of intensity and duration of far-red. Results of such an experiment are set out in Table 5, and these show that the duration of far-red is more important than intensity.

Table 5. *The effect of high and low intensities of far-red*

Treatment before GA_3	Germination ± S.D. (%)	
	300 µW..cm.$^{-2}$	70 µW..cm.$^{-2}$
6 hr. dark, 15 min. FR	65 ± 5	68 ± 3
6 hr. dark, 15 min. FR, 3·75 hr. dark	28 ± 6	27 ± 3
6 hr. dark, 60 min. FR	52 ± 2	45 ± 3
Dark controls		
GA_3 from 6th hr.	78 ± 8	
GA_3 from 10th hr.	72 ± 3	

We can postulate, on the basis of these experiments, that processes necessary for gibberellin action and which are dependent upon a far-red-sensitive pigment occur in Grand Rapids seeds from approximately the 6th hr. of imbibition. If this is true, gibberellin-induced germination can only begin when these processes have taken place. Now, at 25° C. the first seeds germinate in darkness 9 hr. after they have started imbibition on GA_3, and during part of these 9 hr. GA_3 has been ineffective because the postulated preparatory dark processes have not yet occurred. But when seeds are allowed to imbibe on water for 12 hr. before the supply of GA_3, germination then begins after 5 hr., i.e. the lag period is almost halved (Fig. 8). A similar situation exists with regard to light-stimulated germination (Fig. 9). The fact that the accelerative effect of long imbibition times can be completely cancelled by the presence of far-red indicates that water uptake is not the decisive factor. These results support the idea of preparatory far-red-sensitive processes occurring in darkness; the accelerative effect of a long imbibition period has also recently been found by Karssen (1967) using *Chenopodium* seeds. These far-red-sensitive processes are probably the ones affected when seeds, released from control by red light (Table 1), are nevertheless inhibited by prolonged far-red. As already mentioned above, 6 hr. far-red given at the appropriate time completely prevents germination on water of isolated embryos, half-seeds, and intact seeds which are afterwards transferred to a relatively low temperature.

The question now arises as to the nature of this pigment system. The facts that duration and not intensity of far-red is important, and that 4 hr. irradiation must be given to secure full inhibition, suggest that the far-red-sensitive pigment is slowly formed over a period of a few hours. However, only a small amount of it is present at any one time, otherwise considerably less than 4 hr. far-red would be necessary, particularly after 12 hr. imbibition in darkness. We can explain these observations if we assume that the pigment is formed by thermal reactions from a precursor, but the equilibrium between these two is very much on the side of the precursor. Far-red removes the pigment and more forms from the precursor to restore equilibrium, but again the pigment is removed by far-red. Eventually, prolonged far-red drains off all the pigment, and the precursor. The precursor is apparently not resynthesized; if it were, inhibition by far-red would not be permanent. This hypothesis is similar in some respects to one proposed by Mancinelli & Borthwick (1964), Mancinelli *et al.* (1966) and Mancinelli, Yaniv & Smith (1967) to explain the requirement for prolonged or intermittent far-red on the dark-germinating varieties of lettuce and tomato. In cucumber seeds it has been demonstrated that total phytochrome does indeed increase during darkness (Mancinelli & Tolkowsky, 1968).

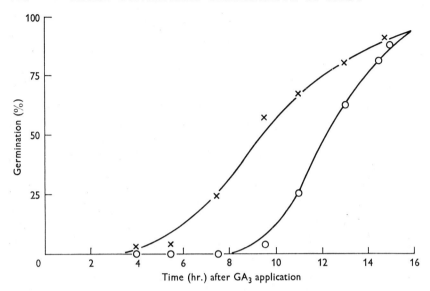

Fig. 8. The accelerative effect of the preparatory processes on the time course of GA_3-induced germination. Seeds were imbibed in darkness for 1·5 (O—O) and 12 hr. (×—×) before transfer to GA_3 (250 μg./ml.). (From Negbi et al. 1968.)

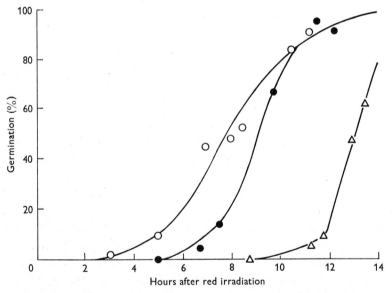

Fig. 9. The accelerative effect of far-red-sensitive preparatory processes on the time course of germination. Seeds were imbibed in darkness for 1·5 hr. (●—●) and 12 hr. (O—O) or in far-red (70 μW..cm.$^{-2}$) for 12 hr. (△—△) before irradiation with red light (180 μW..cm.$^{-2}$ for 5 min.). (From Negbi et al. 1968.)

Rollin (personal communication) has also recently succeeded in detecting the production of phytochrome in lettuce seed.

The question remains as to the identity of this far-red-absorbing pigment which is produced in Grand Rapids only after about 6 hr. imbibition. Is it in fact the P_{fr} form of the 'classical' phytochrome? A low level of P_{fr} does exist in lettuce seeds (e.g. Scheibe & Lang, 1965) and this might conceivably be the pigment sensitive to prolonged far-red. However, this seems to be

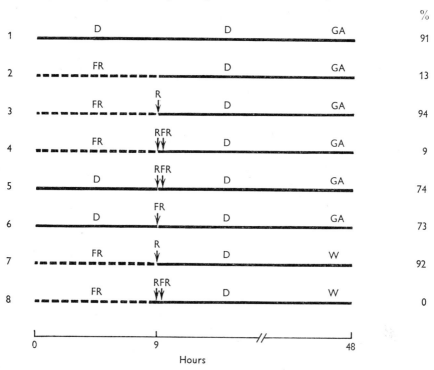

Fig. 10. Reversion of prolonged far-red by a short red-light treatment. (Seeds were exposed to far-red (90 μW..cm.$^{-2}$) for 9 hr. before transfer to GA$_3$ (50 μg./ml.).) In some treatments the prolonged far-red was followed by 3 min. red light (180 μW..cm.$^{-2}$), or 3 min. red light then 10 min. far-red (90 μW..cm.$^{-2}$); after these treatments, seeds were transferred to GA$_3$ (50 μg./ml.) or water. D = darkness; FR = far-red, R = red, W = water.

unlikely as prolonged far-red is not really effective until approximately the 6th hr. of imbibition; P_{fr} is certainly present before that, since inhibition of germination by short far-red is achieved 2 hr. or so after the start of imbibition (Scheibe & Lang, 1965; Roth-Bejerano et al. 1966). Furthermore, if P_{fr} itself is the pigment in question it is presumably converted back to P_r during the far-red irradiation. This P_r should then act later as a source of more P_{fr} (though only a small amount) so that inhibition by prolonged far-red should not be irrevocable. Further evidence that the pigment required for gibberellin action is not likely to come from a red-absorbing form is given by the experiment shown in Fig. 10. This experiment illustrates that a reversible pigment, controlled by short periods of red and far-red, is present after the prolonged far-red treatment. When all the pigment is driven over into the red-absorbing form, this time by only a short exposure to far-red, the gibberellin-treated seeds fail to germinate.

The pigment system described above clearly has properties reminiscent of 'classical' phytochrome, but for various reasons does not seem to be identical to it. Thus, one could suggest as a working hypothesis that a second phytochrome-like pigment system makes its appearance in the far-red absorbing form in Grand Rapids lettuce after about 6 hr. of imbibition. This pigment seems to cause the formation of a labile product (see Fig. 7).

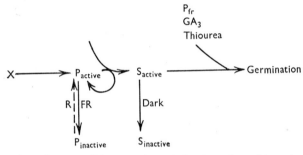

Fig. 11. Model to show possible operation of the far-red-sensitive pigment system. X = precursor of pigment; P = pigment; S = substrate; R, FR = red, far-red. (From Negbi et al. 1968.)

The pigment is necessary for both gibberellin- and thiourea-induced germination, and even possibly for germination promoted by 'classical' phytochrome (Fig. 11). The possibility that more than one type of phytochrome exists has, incidentally, been raised recently by Hillman (1967).

How far the phenomena described above are related to the high-energy reaction in other seeds is not yet clear. It is possible, but by no means established, that the high-energy reaction described by Mohr & Appuhn (1963) and Rollin (1963), which operates in light-stimulated seeds after the escape time has passed, is the same as the one we have described. Hartmann (1966) has argued that the effects of prolonged far-red given after the escape time are connected with the destruction of P_{fr}. This is a debatable point which raises some interesting questions about the possible multiple roles of phytochrome, for it suggests that P_{fr} is required at two stages—an early one to initiate the germination processes, and one about 5 hr. later, just before protrusion of the radicle. Moreover, if phytochrome were destroyed by prolonged far-red, it is difficult to understand why the effect is so easily reversible by a short exposure to red light. We should note, furthermore, that the experimental evidence obtained by Clarkson & Hillman (1967) is not in agreement with the idea of P_{fr} destruction in prolonged far-red. Whatever conclusions we may care to make about the identity of the 'second' phytochrome one fact seems clear—that this pigment is necessary for all the treatments which by-pass the requirement for red light (Table 1). How far similar systems are present in other seeds, so as to render them sensitive to the prolonged far-red experienced in certain environments, is worthy of investigation.

CONCLUSIONS

Although many of the diverse light-regimes required for the stimulation of germination can be understood on the basis of some of the known properties of phytochrome, we are still unable to answer the fundamental question: what does phytochrome do? Our evidence shows that even in lettuce seeds, where the visible response occurs several hours after P_{fr} formation, the pigment does, in fact, appear to act within a few minutes. Gibberellin production is not promoted by P_{fr}, but the endogenous synthesis of the hormone might nevertheless be necessary for germination to take place. These possibilities raise the question of interactions between phytochrome and gibberellin. Certainly a far-red-sensitive pigment affected by prolonged irradiation is essential for gibberellin to act and the same pigment is necessary even for germination in which control by P_{fr} is by-passed. The whole question of the role of phytochrome, whether the 'classical' one or additional forms of the pigment, is clearly much more complex than appeared a few years ago. Nevertheless, through the operation of the pigment(s), light, day-lengths and the far-red predominating under foliage exert their control over germination.

I should like to thank Dr M. Negbi of the Hebrew University, Jerusalem, who participated in some of the work described here, for many useful and enjoyable discussions about seeds, and for permission to use some of his results. My thanks are due, no less, to Dr J. D. Bewley who, similarly, took part extensively in these investigations and whose results I have drawn upon.

Some of this work was made possible by a grant from the Agricultural Research Council, whose help is gratefully acknowledged.

REFERENCES

ALCORN, S. M. & KURTZ, E. B. (1959). *Am. J. Bot.* **46**, 526.
BEWLEY, J. D. (1968). Ph.D. Thesis, University of London.
BEWLEY, J. D., NEGBI, M. & BLACK, M. (1968). *Planta* **78**, 351.
BLACK, M. (1967). In *Physiology, Ecology and Biochemistry of Germination*, ed. H. Borriss. Greifswald: Ernst-Moritz-Arndt Universität.
BLACK, M. & RICHARDSON, M. (1965). *Nature, Lond.* **208**, 1114.
BLACK, M. & RICHARDSON, M. (1967). *Planta* **73**, 344.
BLACK, M. & WAREING, P. F. (1955). *Physiologia Pl.* **8**, 300.
BLACK, M. & WAREING, P. F. (1959). *J. exp. Bot.* **10**, 134.
BLACK, M. & WAREING, P. F. (1960). *J. exp. Bot.* **11**, 28.
BORTHWICK, H. A., HENDRICKS, S. B., PARKER, M. W., TOOLE, E. H. & TOOLE, V. K. (1952). *Proc. natn. Acad. Sci. U.S.A.* **38**, 662.
BORTHWICK, H. A., HENDRICKS, S. B., TOOLE, E. H. & TOOLE, V. K. (1954). *Bot. Gaz.* **115**, 205.
BRIAN, P. W. (1958). *Nature, Lond.* **181**, 1122.
BUNSOW, R. & VON BREDOW, K. (1958). *Naturwissenschaften* **45**, 95.
BUTLER, W. L., LANE, W. C. & SIEGELMAN, H. W. (1963). *Pl. Physiol., Lancaster* **38**, 514.
CASPARY, R. (1860). *Schr. Kgl. phys.-oekon. Ges.* Koenisberg, **1**, 66.
CLARKSON, D. T. & HILLMAN, W. S. (1967). *Planta* **75**, 286.
CUMMING, B. G. (1963). *Can. J. Bot.* **41**, 1211.
DOWNS, R. J. (1964). *Phyton, B. Aires* **21**, 1.
EVENARI, M. (1965a). In *Recent Progress in Photobiology*, ed. E. J. Bowen. Oxford: Blackwell Scientific Publications.
EVENARI, M. (1965b). In *Encyclopedia of Plant Physiology*, vol. XV (2). Ed. Ruhland. Berlin: Springer-Verlag.
EVENARI, M. & NEUMANN, G. (1952). *Bull. Res. Coun. Israel* **2**, 15.
EVENARI, M., KLEIN, S., ANCHORI, S. & FEINBRUN, N. (1957). *Bull. Res. Coun. Israel* **6D**, 33.
FLINT, L. H. & MCALISTER, E. D. (1935). *Smithson misc. Collns.* **94** (5), 1.
FONDEVILLE, J. C., BORTHWICK, H. A. & HENDRICKS, S. B. (1966). *Planta* **69**, 359.
FRANKLAND, B. & WAREING, P. F. (1966). *J. exp. Bot.* **17**, 596.
FREDERICQ, H. (1965). *Biol. Jaarb. Dodonaea* **33**, 66.
GASSNER, G. (1911). *Ber. dt. bot. Ges.* **29**, 708.
HABER, A. H. & LUIPPOLD, H. J. (1960). *Pl. Physiol. Lancaster* **35**, 168.
HABER, A. H. & TOLBERT, N. E. (1959). In *Photoperiodism and Related Phenomena in Plants and Animals*, ed. Withrow. Washington: American Association for the Advancement of Science.
HARTMANN, K. M. (1966). *Photochem. Photobiol.* **5**, 349.
HAUPT, W. (1959). *Planta* **53**, 484.
HENDRICKS, S. B. & BORTHWICK, H. A. (1967). *Proc. natn. Acad. Sci. U.S.A.* **58**, 2125.
HENDRICKS, S. B., BORTHWICK, H. A. & DOWNS, R. J. (1956). *Proc. natn. Acad. Sci. U.S.A.* **42**, 19.
HILLMAN, W. S. (1967). *A. Rev. Pl. Physiol.* **18**, 301.
HILLMAN, W. S. & KOUKKARI, W. L. (1967). *Pl. Physiol., Lancaster* **42**, 1413.
IKUMA, H. & THIMANN, K. V. (1960). *Pl. Physiol., Lancaster* **35**, 557.
IKUMA, H. & THIMANN, K. V. (1963a). *Pl. Cell Physiol., Tokyo* **4**, 113.
IKUMA, H. & THIMANN, K. V. (1963b). *Nature, Lond.* **197**, 1313.
IKUMA, H. & THIMANN, K. V. (1963c). *Pl. Cell Physiol., Tokyo* **4**, 169.

Ikuma, H. & Thimann, K. V. (1964). *Pl. Physiol., Lancaster* **39**, 756.
Isikawa, S. (1954). *Bot. Mag., Tokyo* **67**, 51.
Isikawa, S. & Yokohama, Y. (1962). *Bot. Mag., Tokyo* **75**, 127.
Jaffe, M. J. & Galston, A. W. (1967). *Planta* **77**, 135.
Kahn, A. (1960a). *Pl. Physiol., Lancaster* **35**, 1.
Kahn, A. (1960b). *Pl. Physiol., Lancaster* **35**, 333.
Kahn, A., Goss, J. A. & Smith, D. E. (1957). *Science, N.Y.* **125**, 645.
Karssen, C. M. (1967). *Acta bot. neerl.* **16**, 156.
Khan, A. A. (1967). *Physiologia Pl.* **20**, 1039.
Kinzel, W. (1908). *Ber. dt. bot. Ges.* **26a**, 105.
Köhler, D. (1966a). *Planta* **69**, 27.
Köhler, D. (1966b). *Planta* **70**, 42.
Koller, D. (1962). *Am. J. Bot.* **49**, 841.
Koller, D., Sachs, M. & Negbi, M. (1964). *Pl. Cell Physiol., Tokyo* **5**, 85.
Koukkari, W. L. & Hillman, W. S. (1968). *Pl. Physiol., Lancaster* **43**, 698.
Mancinelli, A. L. & Borthwick, H. A. (1964). *Annali Bot.* **28**, 9.
Mancinelli, A. L., Borthwick, H. A. & Hendricks, S. B. (1966). *Bot. Gaz.* **127**, 1.
Mancinelli, A. L. & Tolkowsky, A. (1968). *Pl. Physiol., Lancaster* **43**, 489.
Mancinelli, A. L., Yaniv, Z. & Smith, P. (1967). *Pl. Physiol., Lancaster* **42**, 333.
McDonough, W. T. (1965). *Pl. Physiol., Lancaster* **40**, 575.
McDonough, W. T. (1967). *Nature, Lond.* **214**, 1147.
Miller, C. O. (1958). *Pl. Physiol., Lancaster* **33**, 115.
Mohr, H. (1967). *Photochem. Photobiol.* **5**, 469.
Mohr, H. & Appuhn, U. (1963). *Planta* **60**, 274.
Nagao, M., Esashi, Y., Tanaka, T., Kumagai, T. & Fukumoto, S. (1959). *Pl. Cell Physiol., Tokyo* **1**, 39.
Negbi, M., Black, M. & Bewley, J. D. (1968). *Pl. Physiol., Lancaster* **43**, 35.
Nutile, G. E. (1945). *Pl. Physiol., Lancaster* **20**, 433.
Nyman, B. (1961). *Nature, Lond.* **191**, 1219.
Nyman, B. (1966). *Physiologia Pl.* **19**, 63.
Reid, D. M., Clements, J. B. & Carr, D. J. (1968). *Nature, Lond.* **217**, 580.
Rimon, D. (1957). *Bull. Res. Coun. Israel* **6D**, 53.
Rissland, I. & Mohr, H. (1967). *Planta* **77**, 239.
Rollin, P. (1963). *C. r. hebd. Séanc. Acad. Sci., Paris* **257**, 3642.
Rollin, P. & Bidault, Y. (1963). *Photochem. Photobiol.* **2**, 59.
Roth-Bejerano, N., Koller, D. & Negbi, M. (1966). *Pl. Physiol., Lancaster* **41**, 962.
Scheibe, J. & Lang, A. (1965). *Pl. Physiol., Lancaster* **40**, 485.
Tanada, T. (1968). *Proc. natn. Acad. Sci. U.S.A.* **59**, 376.
Thompson, R. C. & Kosar, W. F. (1938). *Science, N.Y.* **87**, 218.
Tietz, N. (1953). *Biochem. Z.* **324**, 517.
Toole, E. H. (1961). *Proc. int. Seed Test. Ass.* **26**, 659.
Toole, V. K. & Borthwick, H. A. (1968). *Pl. Cell Physiol., Tokyo* **9**, 125.
Toole, V. K. & Cathey, H. M. (1961). *Pl. Physiol., Lancaster* **36**, 663.
Toole, V. K., Toole, E. H., Borthwick, H. A. & Snow, A. G. (1962). *Pl. Physiol., Lancaster* **37**, 228.
Toole, E. H., Toole, V. K., Hendricks, S. B. & Borthwick, H. A. (1957). *Proc. int. Seed Test. Ass.* **22**, 1.
Toole, V. K., Toole, E. H., Hendricks, S. B. & Borthwick, H. A. (1961). *Pl. Physiol., Lancaster* **36**, 285.
Vezina, P. E. & Boulter, D. W. K. (1966). *Can. J. Bot.* **44**, 1267.
Wesson, G. & Wareing, P. F. (1967). *Nature, Lond.* **213**, 600.
Yamaki, T., Hashimoto, T., Ishii, T. & Yamada, M. (1958). *Abstr. 2nd Meeting Jap. Gibb. Res. Ass.* p. 55.

THE PROBLEM OF DORMANCY IN POTATO TUBERS AND RELATED STRUCTURES

By L. RAPPAPORT and N. WOLF

Department of Vegetable Crops, University of California, Davis

INTRODUCTION

Interest in the regulation of dormancy in storage organs is attributable in part to the commercial importance of tubers, corms, and bulbs either as propagules or as foodstuffs. Thus, horticulturists have worked to develop techniques either to shorten or to prolong the dormant period. In addition to its practical implications, dormancy has major biological interest and an understanding of its control by environmental and hormonal or other biochemical factors has been sought for years.

Of those storage organs that may be characterized as unextended stems, the potato tuber has been studied most extensively. Therefore, the focal point of this paper will be arrested growth in the potato tuber in relation to hormonal regulation. Since a great deal has been written about environmental effects on such storage organs as onion, *Allium cepa* (Abdalla & Mann, 1963), *Begonia* (Nagao & Okagami, 1966), garlic, *Allium sativum* (Mann & Lewis, 1956), and *Gladiolus* (Denny, 1936; Tsukamoto & Yagi, 1960), less attention will be devoted to this subject. Comprehensive reviews of dormancy with heavy emphasis on environmental relationships were published by Lang (1965), Vegis (1961, 1964) and Hartsema (1961). Where appropriate, information concerning other modified stem structures will be introduced. Amen (1968) recently published an interesting review of the mechanism of dormancy in seeds and an excellent volume of papers on the growth of the potato was edited by Ivins & Milthorpe (1963).

CONCEPTS OF REST AND DORMANCY

There is no general agreement on the meaning of the term 'dormancy'. It should be understood that the bud is the morphological structure of central concern in this discussion, whether it is in the 'eye' of a potato, in a corm, in a rhizome, or in a bulb. What is observed during ontogeny of the organ is the broadening of the bud through periclinal divisions with a long period when there is little elongation. This period is followed by accelerated and sustained elongation.

Much of the research on the subject of bud dormancy has been characterized by the search for a factor, or factors, that regulate growth. Thus, in keeping with fashion, dormancy has at various times been studied in relation to the starch–sugar relationships (Müller-Thurgau, 1882), concentrations of ascorbic acid and glutathione (Emilsson, 1949), excess or deficiency of CO_2 or O_2 (Thornton, 1939; Burton, 1963), volatile inhibitors (Burton, 1952, 1963), dehydration (Henkel & Sitnikova, 1953), chemical inhibitors and promoters (Hemberg, 1949, 1952; Madison & Rappaport, 1968), enzyme activity (Emilsson & Lindbloom, 1963), and lack of protein synthesis (Levitt, 1954; Cotrufo & Levitt, 1958).

That there is no universal agreement on terminology is understandable, because it is difficult to characterize either induction or termination of arrested growth in unextended stems by a distinct morphological change, such as occurs during conversion from a vegetative to a reproductive form in plants. Thus there are at least two major points of view in defining arrested growth. One view, advocated by Burton (1963), is that dormancy is any restriction to elongation of the bud. This is a broad, generalized concept which considers the tuber as being in a 'state of continual flux' with the external environment intimately affecting internal processes. It is a condition heralded by the onset of tuber formation and terminated by acceleration of sprout growth.

In contrast, Emilsson (1949) and others (Pollock, 1953; Samish, 1954) distinguished between 'rest' and 'dormant' periods. They considered the rest period to be a specialized form of dormancy in which buds fail to elongate even under environmental conditions suitable for growth. The condition of rest is sometimes referred to as internal dormancy. The dormant period, according to Emilsson, is that time when buds fail to grow precisely because environmental conditions, particularly temperature, are unfavourable (i.e. external dormancy). When temperature is favourable for termination of rest, dormant and rest periods coincide. Burton's definition simply requires that extension growth be accelerated. Clearly, a more specific definition will have to await a better understanding of the phenomenon of arrested growth. For purposes of this discussion Emilsson's terminology (1949) will be utilized but the question of when rest begins and ends will be considered under Discussion.

HORMONAL CONTROL OF BUD REST

The concept of hormonal control of bud rest in potato is an old one, dating back to Appleman (1918). Michener (1942) attributed accelerated termination of bud rest in potato by ethylene chlorhydrin to ability of the compound to lower auxin levels. Auxins, however, have never been serious contenders for the role of controlling agents in rest because little consistent relationship has been found between auxin concentration and the timing of bud break (Samish, 1954; Hemberg, 1952). Moreover, concentration of auxin is not sufficient to account for inhibition even when correlated with bud break. In potatoes, Hemberg (1952) was unable to show that auxin levels increased as sprouting of tubers began, although Szalai (1959) provided evidence for such a change. Attempts to cause termination of rest by auxins have failed; indeed, one auxin, methyl ester of naphthaleneacetic acid, is used commercially to prolong the rest period in potatoes. The effect of auxin may be one of controlling apical dominance rather than of prolonging the rest period *per se* (Goodwin, 1963).

The involvement of inhibitors in controlling bud break in trees was postulated as early as 1893 by Reinitzer, who discussed *Ermudungsstoffen* or fatigue substances. However, it was not until Hemberg's reports (1949, 1952, 1954) that research on the concept of inhibitor control of bud rest was initiated seriously. Hemberg prepared acidic ether extracts of potato peelings and interior tissues which he found inhibited elongation of oat coleoptiles. Measurement of inhibitor activity during storage of resting potatoes revealed that inhibitor concentration remained high during rest period and decreased prior to sprouting. Treatment of tubers with ethylene chlorhydrin, which accelerated sprouting, lowered the inhibitor level. Bennet-Clark & Kefford (1953) chromatographed the inhibitor extract on paper and detected a zone, dubbed inhibitor β-complex, hereafter referred to as (β), that strongly inhibited elongation of coleoptiles. Continuing study by Hemberg (1958b) led him to conclude that β was specific for inhibition of bud growth, i.e. was the factor controlling rest period in potato buds. Other studies of endogenous inhibitors in relation to rest in potatoes have supported Hemberg's findings (Varga & Ferenczy, 1957; Blommaert, 1954; Walker, 1968). Evidence for presence of growth inhibitors controlling bud rest in trees that require either low temperature or long day for elongation has been provided by Allen (1960), Hemberg (1958a), Waxman (1957), Blommaert (1955), Eagles & Wareing (1963, 1964), Phillips & Wareing (1959), Kawase (1961) and many other workers.

The characterization of β as the controlling agent in rest did not go unchallenged since there was no evidence that a single component of β controlled rest, and there was no assurance of specificity of action (Smith & Kefford, 1964). A major concern was a lack of evidence indicating that β inhibited elongation of the tissue from which it was extracted. Most work on inhibitors had utilized coleoptile elongation or seed germination as assays for inhibitor activity. Wareing & Villiers (1961) stated this problem clearly in their studies of seed dormancy in *Fraxinus excelsior*, and the point was emphasized by Rappaport, Blumenthal-Goldschmidt & Hayashi (1965b). Inhibitor β-complex applied to potato buds failed to inhibit their growth, a result which Hemberg (1949) attributed to enzymic destruction. Buch & Smith (1959) were unable to show inhibition of bud growth in potato plugs placed on chromatographic strips containing β, although the extracts from the treated plugs could inhibit the elongation of coleoptiles. They concluded, therefore, that β had nothing to do with rest period. In addition, Burton (1956), using a coleoptile assay, was unable to show a correlation between inhibitor concentration and time of sprouting in all of the potato varieties tested.

In Bennet-Clark & Kefford's research (1953) components of β occurred over a broad zone on the paper chromatogram. In a chromatographic study of potato-tuber extracts, Varga & Ferenczy (1957) used organic compounds as standards to identify tentatively substances, some phenolic in nature, capable of causing inhibition of coleoptiles. Thus the question became one of identifying the compound and showing that it could substitute for the activity of β on buds. A part of the problem of physiological action of β in rest period was solved by Blumenthal-Goldschmidt & Rappaport (1965), working with an excised potato-plug bioassay (Rappaport et al. 1965a). Inhibition of elongation of buds in plugs resulted only after repeated application of the chromatographic eluates from the inhibitor β zone. These eluates, one of which lost activity in storage, also inhibited wheat coleoptile growth. Failure of β to cause inhibition (Hemberg, 1952; Buch & Smith 1959) was attributed to possible loss of activity or to application to buds that had already emerged from rest and therefore were no longer sensitive. Even this result, while helpful, was not satisfying, since the identity of the inhibitor was unknown. The picture was clarified by the nearly simultaneous isolation of the same inhibiting compound, 'abscisin II', from cotton bolls (Ohkuma, Addicott, Smith & Thiessen, 1965) and of 'dormin' from sycamore leaves (Cornforth, Milborrow, Ryback & Wareing, 1965). This inhibitor, now referred to as abscisic acid (ABA), was isolated from many tissues, including potatoes (Cornforth, Milborrow & Ryback, 1966) and shown to inhibit elongation of potato buds effectively (Blumenthal-Goldschmidt & Rappaport, 1965; El Antably, Wareing & Hillman, 1967; Madison & Rappaport, 1968). At the present time the physiological role of ABA in prolonging bud rest is not understood.

Although there is mounting evidence indicating the involvement of inhibitors in the rest period, it is unlikely that the mechanism is simple. Total control of bud growth by inhibitors would require that their levels decrease prior to termination of rest. Since this result has not been obtained with all species studied, growth-promoting substances as well as inhibitors have been implicated in the explanation of bud rest (Nitsch, 1957; Wareing, Eagles & Robinson, 1964; Eagles & Wareing, 1963, 1964; Blumenthal-Goldschmidt & Rappaport, 1965; Madison & Rappaport, 1968).

The possibility that endogenous gibberellins play a prominent role as promoters in reducing rest in potato tubers was enhanced by the observations that gibberellin treatment terminated rest period in potatoes (Brian, Hemming & Radley, 1955; Rappaport, 1956; Rappaport, Lippert & Timm, 1957), and that gibberellins occur naturally in potato tubers (Okazawa, 1959, 1960; Smith & Rappaport, 1960) and increase in concentration near the termination of the rest period (Smith & Rappaport, 1960; Rappaport & Smith, 1962). In other studies the observations that gibberellin A_3 was active at extremely low concentrations in terminating rest of tubers on or off the plant (Lippert, Rappaport & Timm, 1958; Rappaport & Smith, 1962), in specifically promoting elongation of cultured potato buds (Oshima & Livingston, 1963; Morel & Müller, 1964) and buds in excised plugs (Rappaport & Blumenthal-Goldschmidt, 1961; Rappaport et al. 1965 a), in promoting stolon elongation while preventing tuber formation (Okazawa, 1960; Smith, 1962), and in reducing level of β while promoting sprouting, have all sustained the view that gibberellins are important factors in the control of rest in potato buds. However, agreement on the action of gibberellins in terminating rest period is not unanimous. Doorenbos (1958) raised some doubts about the action of applied gibberellin in terminating rest period in potato tubers, and Boo (1962) was unable to show that gibberellin concentration increased as rest terminated. Possibly the method of extraction used by Boo was not adequate (Rappaport, Blumenthal-Goldschmidt & Hayashi, 1965 a), especially since the results of Smith & Rappaport (1960) were recently confirmed (Rappaport, Sachs, & Schuster, unpublished). Applied gibberellin terminated physiological dwarfism in tree peony, *Paeonia suffruticosa* (Barton & Chandler, 1957), induced bud break in peach, *Prunus persica* (Donoho & Walker, 1957) and promoted growth in dormant rhubarb crowns (Tompkins, 1965). Studies by Nitsch (1957), Eagles & Wareing (1963, 1964) and Kawase (1961) supported the view that gibberellins are involved in photoperiodically controlled rest period. It became clear, however, for buds requiring low temperature for termination of rest, that gibberellins only reduced the length of low temperature needed to induce normal bud growth (Brown, Griggs & Iwakiri, 1960; Donoho & Walker, 1957). In addition, gibberellin application seems to have little or no effect on duration of rest (or on shoot growth) in onion, *Allium cepa* (Rappaport, 1956; Thomas, 1969). Indeed, in aerial tubers of *Begonia evansiana* which require low temperature for termination of rest (Esashi & Nagao, 1959; Nagao & Mitsui, 1959; Nagao & Okagami, 1966), and grapevine, *Vitis vinifera* (Weaver, 1959), gibberellin actually prolongs the rest period. Thus, while gibberellins appear to shorten rest in many species, they may have no effect or may even extend rest in others.

Despite certain inconsistencies regarding the role of inhibitors and gibberellins in governing rest period, results derived from two lines of research have served to intensify interest in this area. One of these is the spectacular induction of enzyme synthesis in the barley aleurone layer by gibberellins (Paleg, 1965) and its inhibition by ABA (Chrispeels & Varner, 1967). In addition, gibberellin A_3 induced invertase production in Jerusalem artichoke (*Helianthus tuberosus*) tuber tissue (Edelman & Hall, 1964). The other line of evidence concerns the combined action of gibberellins and ABA in regulating germination in peach seeds (Lipe & Crane, 1966), birch, *Betula pubescens* (Eagles & Wareing, 1963, 1964) and in potato buds (Boo, 1961; Blumenthal-Goldschmidt & Rappaport, 1965; El Antably, Wareing & Hillman, 1967; Madison & Rappaport, 1968). Evidence indicating inhibition of gibberellin synthesis by ABA was recently presented by Wareing, El Antably and Good (1967). In addition, papers by Van Overbeek & Loeffler (1967) and Van Overbeek, Loeffler & Mason (1967) revealed that ABA, which induced bud dormancy in duckweed (*Lemna*), also markedly lowered levels of DNA and RNA. Restoration of growth by treatment with benzyladenine, a cytokinin, resulted in speedy enhancement of DNA and RNA syntheses. Interestingly, benzyladenine, in contrast to gibberellin, also shortened rest period in grapevines (Weaver, 1963). Thus it appears that for a number of species rest period is under hormonal control, and in potato at least, the hormones that presently appear to be most involved are gibberellins and abscisic acid.

CELLULAR EVENTS ASSOCIATED WITH TERMINATION OF REST IN POTATO BUDS

From the foregoing section it is apparent that the duration of rest period of buds may be controlled by photoperiod (as in *Betula*), by temperature (as in aerial tubers of *Begonia*), or may not be controlled qualitatively by environment as in potato and onion. Rest can be shortened variously by treatment with ethylene chlorhydrin, glutathione, thiourea, cytokinins, gibberellins or even by wounding. The mode of action of these substances in regulating rest is not understood. Temperature (Pollock, 1953; Pollock & Olney, 1959) and photoperiod (Nitsch, 1957, 1963) appear to activate metabolic processes and hormonal changes in species that are sensitive to such environmental effects. However, in potato, in which there is no clear-cut environmental control, it appears that there is an internal 'clock' that requires primarily time for it to trigger cellular events leading to renewed sprout growth. The importance of the clock is underlined by the knowledge that the duration of the rest period of different cultivars is similar over the generations.

Some insights into the nature of the chemical changes in buds, with special emphasis on nucleic acids and proteins, were reported by Barskaya & Okinina (1959), Clegg (1967), Clegg & Rappaport (1966), Kupila-Ahvenniemi (1966), Rappaport (1964), Rappaport & Stahl (1965), Rappaport & Wolf (1968a, b), Satarova (1950) and Tuan & Bonner (1964). Tuan & Bonner showed that increase in RNA and DNA in excised buds from potatoes, treated with ethylene chlorhydrin to shorten rest, occurred prior to increase in fresh weight of the buds. The RNA synthesis was inhibited by actinomycin D. Moreover, they showed that chromatin material from sprouted buds could support RNA synthesis, *in vitro*, whereas chromatin material from resting buds was unable to do so. Tuan and Bonner concluded that the genome regulating bud rest in potato tubers is repressed, and that emergence from rest is signalled by derepression of the genetic material. Attempts by Tuan and Bonner to activate the chromatin material by treatment with GA_3 failed (private communication).

The initial studies on hormonal control of potato-bud rest, plus the increasing evidence for the involvement of nucleic acid and protein synthesis in germination and bud growth, led us to increase efforts to identify the mechanism of hormonal control of events at the level of nucleic acid synthesis. This required new techniques and approaches, not the least of which was development of a good bioassay for dormancy of potato buds.

A considerable portion of the early work was done with intact tubers; however, there are many difficulties in working with them. Accordingly, Rappaport & Blumenthal-Goldschmidt (1961) and Rappaport, Blumenthal-Goldschmidt, Clegg & Smith (1965a) devised a simple bioassay involving plugs with buds removed with a 'de-eyer' from freshly harvested, resting tubers. Plugs of different dimensions may be obtained by using de-eyers of different diameters. Plugs, usually 0·5 × 0·8 cm., but varying in size from as large as 2·3 × 0·5 cm. (Rappaport *et al.* 1965a) to 0·2 × 0·2 cm. (Rappaport & Wolf, 1968b), have been used in different experiments. Ten plugs are ordinarily placed in each of four Petri dish replications on Whatman no. 1 filter paper moistened with 1 ml. of water. The advantages of the assay are as follows. The size of experiments can be reduced, and uniformity of test material can be approached since the apical bud, or only buds in the region of the apex, are used. The buds in the excised plugs overtly behave similarly to those in intact tubers and are appropriately responsive to such substances as naphthaleneacetic acid, which inhibits sprouting, and various gibberellins which promote sprouting (Rappaport et al. 1965a). Rate of sprouting varies with plug size, just as it does with the size of potato 'seed-pieces' used for commercial planting.

The information that nucleic acid and protein synthesis are essential processes associated with termination of rest in potato buds was supported by the observations of Madison & Rappaport (1968), Rappaport et al. (1965a), and Rappaport & Wolf (1968a). They noted that deoxyadenosine, actinomycin D and puromycin, inhibitors of DNA, RNA and protein syntheses, respectively, could reduce sprouting of buds in excised potato plugs. Interestingly, ABA was equally effective in these studies (Madison & Rappaport, 1968). Rappaport & Wolf (1968a) found that buds in potatoes treated 3 days with ethylene chlorhydrin incorporated [^{3}H]thymidine in the first 24 hr. We have similar evidence for [^{3}H]uridine. The speed of activation of elongation and mitotic activity by such treatment is indicated by the data in Table 1. Cell elongation and division increased significantly in the first 24 hr. in buds that had been treated with ethylene chlorhydrin. Control buds incorporated very little of the precursor and did not change measurably in mitotic activity or elongation in the same period.

Table 1. *The effect of ethylene chlorhydrin on mitotic activity and elongation of buds in White Rose potato tubers*

(Tubers were placed in respiration jars containing a beaker with sufficient ethylene chlorhydrin to accelerate sprouting. The potatoes were transferred to air and plugs were excised and fixed 1 or 5 days later. Data represent the average number of mitotic figures per 6 μ section and of average length (mm.) of 15 buds (from Rappaport & Wolf, 1968a).

	Control (no ethylene chlorhydrin)	Days after transfer from ethylene chlorhydrin	
		1	5
Mitoses	0.0	0.57 ± 0.14	2.00 ± 0.2
Bud length	0.2	0.34 ± 0.05	0.58 ± 0.8

Even with some understanding of hormonal relationships in the rest period, the questions of why hormonal concentrations change and how hormones act in renewing elongation in the resting bud remain unanswered. We have studied the participation of hormones in terminating rest in potato tubers for some years. In particular, the apparent significance of gibberellins in controlling rest period (Smith & Rappaport, 1960; Rappaport & Smith, 1962) led to an extensive study of the mode of action of this hormone in promoting growth and biochemical changes in potato tubers. Other studies (Hashimoto & Rappaport, 1966a, b; Hayashi, Blumenthal-Goldschmidt & Rappaport, 1962; Hayashi & Rappaport, 1965; Rappaport, Hsu, Thompson & Yang, 1967) have concerned the binding of gibberellins as a means of regulating their supply. This binding might have implications for inhibition and promotion of bud growth.

In many studies rate of visible sprouting has been used as an index of response to treatment. However, it is obvious that emergence of the bud apex through the bracts is the final link in the chain of events associated with termination of bud rest. Thus, in order to study the nature of bud rest in potato, it was recognized that a two-pronged approach involving both biochemical and historadioautographic analyses of the buds was needed. The former would give insight into biochemical changes taking place shortly after excision and during the course of sprouting, while the latter would permit analysis of the rate and site of nucleic acid synthesis in relation to the timing of cell division and enlargement. Moreover, in view of the effects of wounding on nucleic acid synthesis in potato buds (Rappaport & Wolf, 1968b) the method assured the use of 'normal' and uninjured buds. The techniques were described earlier (Rappaport & Wolf, 1968a, b).

In the radioautographic studies, only qualitative changes in RNA and DNA syntheses were of interest since there is no measurable synthesis of new nucleic acid in cells of buds of the intact resting tubers (Rappaport & Wolf, 1968a). Therefore no attempt was made to count silver grains. Appearance of a plug (0·5 × 0·8 cm.) with a bud that had been treated with [^{3}H]thymidine and held at 20° C. for 48 hr. is shown in longitudinal section in Plate 1. Labelled nuclei are seen in cells of the apex, leaf bases, procambial and vascular system which extends to the periphery of the plug. New cambial development is also seen coincident with labelling in cells along the periphery. An enlarged view of peripheral labelling after 24 and 48 hr. is seen in Plate 2 (a, b). Cambial development was not apparent in the first 24 hr.

The effect of temperature on incorporation of precursors

If, as suggested, duration of rest period is essentially independent of temperature, it would be expected that nucleic acid synthesis would occur both at low and at optimal temperatures for sprouting. Therefore, excised plugs (0·5 × 0·8 cm.) from freshly harvested potatoes were placed in Petri dishes at 6° or at 22° C. One lot at each temperature was treated immediately with [^{3}H]uridine or [^{3}H]thymidine and fixed after 24 hr. The second lot was treated 24 hr. after excision and killed 24 hr. later. The first lot had no incorporation of [^{3}H] thymidine and only moderate incorporation of [^{3}H]uridine, whereas heavy labelling by both precursors was evident after 48 hr. Labelling at 6° and 22° C. is shown in Plates 3(a, b) and 4(a, b). Most significantly, the timing of incorporation of each precursor was similar, although the amount of labelling was less at 6° than at 22° C.

Thus nucleic acid synthesis occurs whether or not temperature conditions favour bud elongation, supporting the view that changes associated with the onset of bud growth are not qualitatively temperature-dependent.

Timing of nucleic acid synthesis in relation to plug size

It is well known that cutting potato tubers results in accelerated sprouting (Appleman, 1916). There is a relationship between size of seed-piece and sprouting: the smaller the seed-piece, the faster the rate of sprouting. Similarly, rate of bud emergence is related to size of the plug. Treatment with GA_3 of buds in 'large' (2·3 × 0·5 cm.) plugs causes them to sprout as quickly as those in smaller (0·8 × 0·5 cm.) plugs (Rappaport *et al.* 1965*a*). To determine the relationship between plug size and the onset of nucleic acid synthesis in the buds, plugs of three sizes were placed in Petri dishes and treated either with [^{3}H]uridine or [^{3}H]thymidine (10 μC./ml.). They were fixed after 6, 12, 24 or 48 hr. From Table 2 it is evident that plug size decidedly influenced the timing of synthesis of RNA and DNA in the buds. A slight increase in DNA had occurred in buds of the smallest plugs (0·2 × 0·2 cm.) by 12 hr. after excision. Enhanced RNA synthesis was evident only in the smallest plugs in the first 6 hr. Typically, incorporation of [^{3}H]thymidine into buds of the largest plugs occurred after 24 hr. and labelling indicating RNA synthesis was observed only after 12 hr. In these studies, as in all, RNA synthesis preceded DNA synthesis, and both preceded cell enlargement and cell division.

Table 2. *Nucleic acid synthesis in buds in potato plugs of three sizes*

(Buds in excised plugs were treated with microdroplets of [^{3}H]thymidine or [^{3}H]uridine and lots were killed after 6, 12, 24 or 48 hr. (from Rappaport & Wolf (1968*b*).)

Size of plugs (cm.)	Treatment	Duration of treatment (hr.)			
		6	12	24	48
0·2 × 0·2	[^{3}H]uridine	+	+ +	−	−
	[^{3}H]thymidine	o	+	−	−
0·3 × 0·3	[^{3}H]uridine	−	+	+ +	−
	[^{3}H]thymidine	−	−	+	+ +
0·5 × 0·8	[^{3}H]uridine	−	−	+	+ +
	[^{3}H]thymidine	−	−	o	+ +

+, Low intensity; + +, moderate to high intensity of labelling; o, no labelling; −, treatment not run.

Influence of GA_3 on incorporation

The effect of GA_3 on rate of synthesis of nucleic acids was studied by applying the hormone separately and together with labelled precursors to excised buds. In one experiment 0·2 × 0·2 cm. plugs from freshly harvested

White Rose and Russet Burbank potatoes were treated with [^{3}H]uridine or [^{3}H]thymidine, with or without GA$_3$ (10^{-6} M). Plugs were killed in formalin acetic acid 12, 24 or 48 hr. after excision. Both cultivars behaved similarly. Incorporation of [^{3}H]thymidine was undetectable in the first 12 hr. (Plate 5a) unless the buds were simultaneously treated with GA$_3$ (Plate 5b). As expected, DNA synthesis in the buds not treated with GA$_3$ increased in the next 12 hr. Because of the nature of the stain used (fast green) suitable photographs of graining indicating RNA synthesis were not obtained for all time exposures; however, the promoting effect of GA$_3$ on RNA synthesis 12 hr. after excision was shown earlier (Rappaport & Wolf, 1968b).

To determine more precisely when GA$_3$ enhances nucleic acid synthesis, another experiment was done in which 0·2 × 0·2 cm. buds were fixed 3, 6 or 12 hr. after excision and treatment with the precursors and GA$_3$. It was clear that DNA synthesis, which proceeded at a slow rate throughout this experiment, was accelerated only after 12 hr., confirming the results shown in Plate 5(a, b). Plate 6(a–f) shows labelling attributable to incorporation of [^{3}H]uridine in the presence and absence of GA$_3$ after 3, 6 and 12 hr. It was of interest that labelling was very slight and uniform both in buds treated and not treated with GA$_3$ after 3 hr. (Plate 6a, b). After 6 hr., labelling intensified slightly both in the presence and absence of the hormone (Plate 6c, d). However, in the next 6 hr. there was a marked increase in labelling in the buds treated with GA$_3$ (Plate 6e, f). Thus it appears that RNA synthesis increases slightly even without additional GA$_3$, but is then accelerated by the hormone. DNA synthesis, on the other hand, which is minimal and equal despite GA$_3$-treatment in the first 6 hr., is accelerated markedly by the hormone in the following 6 hr. period.

It is not known whether there is resistance to penetration of the precursor or the hormone; however, it is noteworthy that the results were reproducible over three separate experiments involving a total of 30 buds per treatment.

In addition to, or as part of, the process of induction of renewed growth in potato buds, are certain other changes associated with excision and treatment with GA$_3$. Appleman (1916) observed that the reducing sugar content of wounded potatoes increases with time, as it does as tubers emerge from rest (Emilsson, 1949). Excision of plugs from freshly harvested tubers also results in an increase in amount of reducing sugars, and GA$_3$ accelerates the process markedly (Fig. 1). Characteristically, as in the control, the reducing sugar level decreases slowly and then increases. Treatment with GA$_3$ accelerates both the decrease and the increase. The relationship between concentration of GA$_3$ and level of reducing sugar in plugs that

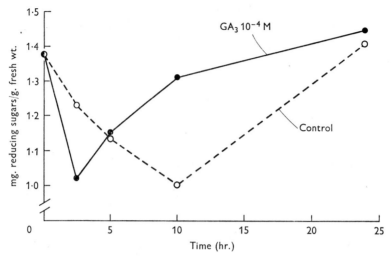

Fig. 1. Reducing sugar content of excised plugs (0·5 × 0·8 cm.) after treatment with 10^{-5} M GA_3 or water over a 25 hr. period. Note that both control and treated plugs undergo similar patterns of change; however, increase in concentration of reducing sugars is accelerated by GA_3. (From Clegg, 1967.)

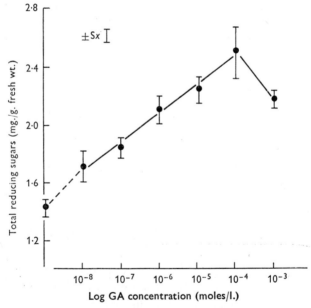

Fig. 2. Effect of concentration of GA_3 on reducing sugar content of excised 0·8 × 0·5 cm. plugs incubated 48 hr. at 20° C. Each point is the average of four samples. (From Clegg, 1967.)

were incubated 48 hr. is seen in Fig. 2. This reaction accelerates as temperature is increased above 0° C. (Clegg, 1967). Attempts to identify the mechanism of the reaction have failed. Whereas protein synthesis in excised plugs was detected within the first hours after excision, total content remained constant in the plugs, whether treated with GA_3 or not, over a period of 48 hr. Edelman & Singh (1968) measured protein in disks from the interior of sprouted and non-sprouted tubers and found that protein increased only in the latter. Levitt (1954) found that protein increased near the end of rest period of potato tubers.

Clegg (1967) showed that both invertase and amylase increased in GA_3-treated plugs, but only after 20 hr. at 20° C. Phosphorylase was not affected. Thus, while sugar level is increased rapidly by GA_3, an explanation of the mechanism awaits further investigation. Nevertheless, the rapid and reproducible increase in sugar is highly reminiscent of what happens, although at a slower rate, in the wounded tuber. This effect suggests the development of a system, either induced or accelerated by GA_3, capable of causing the increase. It should be kept in mind that gibberellin concentration in wounded tissue increased within 12 hr. (Rappaport & Sachs, 1967). More recently it was shown to increase within 6 hr. (unpublished). The timing of the increase in sugar is also of interest in relation to the already described increase in RNA soon after plug excision.

DISCUSSION

The available evidence supports the view that rest period in potato buds is strongly influenced by hormones. Whether the primary control of the rest period is at the level of the gene remains to be seen. In buds of potato tubers, as well as in other storage organs, it appears that the path leads inevitably to a quest for knowledge about the role of gene action in regulating rest.

There are problems in this quest that may turn out to be insurmountable. A major one is that there are several kinds of tissues in a bud. These are not uniform except that to a certain extent they are synchronously quiescent, or nearly so, during rest. That there are occasional cell divisions adds to the problem. The acceleration of mitotic activity in potato buds is confined primarily to meristematic tissues. This is not true in the developing tuber, according to Artschwager (1924) and Plaisted (1957). In such tubers it is possible to find parenchymatous cells undergoing division, an observation we have confirmed. Thus, one aspect of an understanding of the nature of rest is to determine the primary site of arrested growth, if there is one. There is reason to expect that mitotic activity leading to elongation is initiated in the subapical meristem. Sachs, Bretz & Lang (1959) found, in vegetative rosette plants, that cell division in this region proceeded at a very slow rate. After photoperiodic or low temperature induction, or treatment with GA_3, mitotic activity in the subapical region increased dramatically within 24 hr. There is an indication, though not always consistent, that the elongation in potato buds is affected in the same way (Plates 6 a–f).

Another difficult problem is the characterization of when rest begins and ends, since even after tuber formation bud primordia continue to develop (Rosa, 1928; Davidson, 1958). In our studies, mitotic activity in apical buds of photoperiodically induced tubers doubled. That mitotic activity in the bud is prolonged during tuber enlargement is indicated by the increased size of buds in the 'mature' tuber. However, it is evident that cell divisions in the apex during tuber development are primarily periclinal, resulting in a broadening rather than a lengthening of the bud. Despite the mitotic activity in buds during growth of the tuber, mitotic activity was not detected at harvest (Rappaport & Wolf, 1968 a). As already indicated, such activity was greatly enhanced by treatment of the tuber with ethylene chlorhydrin or, merely, by excision of the bud. Davidson (1958) detected greater mitotic activity in buds of harvested potatoes stored at a high temperature (30° C.) and Abdalla & Mann (1963) found that mitotic activity in onion apical buds never really stopped even in the 1-week interval they designated as the rest period. While there may be some question about degree of mitotic activity in such meristems, it appears that there is a time, as yet undisclosed, when mitotic activity slows down very significantly. Whether this is a true and definite physiological stage prior to harvest or is simply a reduction in cell division following death of the tops, as suggested by Emilsson & Lindbloom (1963) and Slater (1963), is not known. Although there appears to be a consistent sequence of responses in potato buds, and probably in other species—RNA and DNA synthesis always appear to precede cell enlargement and cell division—this response is not consistent for all species. Indeed, Haber & Luippold (1960) showed that the early phases of seed germination in lettuce, *Lactuca sativa*, including radicle and epicotyl elongation, could proceed in the absence of cell division. This was supported by the observation of Feinbrun & Klein (1962) that radicle protrusion from lettuce seed anticipates labelling of nuclei with [^{3}H]thymidine. Nevertheless, the observations of Tuan & Bonner (1964) and those we report here foster the conclusion that in the buds of freshly harvested tubers there is a depression in nucleic acid and protein syntheses associated with extremely slow elongation and little or no mitotic activity.

The duration of rest may vary for a given cultivar from year to year. In fact, it is possible to find that rest had already terminated at the time of harvest. Nevertheless, the most remarkable observation in studies of the rest period of numerous cultivars is the consistency of response over generations (Schippers, 1956; Burton, 1966). This fact adds to the view that, while many factors may interact at different levels to influence duration of rest, the basic mechanism is to be found at the level of the gene. While admitting to the complexities and recognizing that environment may exert partial effects on duration of rest (Vegis, 1964; Davidson, 1958), it appears fruitful to seek a solution to the problem of rest by studying the consequences of gene action.

The simplest interpretation of what is happening would require that protein synthesis be blocked at some level during rest period. Tuan & Bonner, working with potato buds, proposed that repression of the genome resulted in inhibition of RNA synthesis. In our own studies, it was clear that the cells of the resting bud produced little new RNA or DNA, and that synthesis of these nucleic acids increased with time. Our results do not exclude the possibility of control at levels other than the genome. The nature of derepression in potato buds is not understood. In view of the multiple effects of gibberellins on rest, it was plausible to speculate that they play a functional role in governing early events associated with termination of rest. Indeed, the parallel between action of gibberellin on rest and ecdysone on moulting of certain insects (Berendes, 1967) was unavoidable at the start of the radioautographic research. However, results presented in this paper cast some doubt on the role of gibberellin as a primary effector in termination of rest. Within the limitations of the radioautographic technique, it appeared that RNA synthesis early after excision increased equally, albeit slightly, in excised buds whether or not they were treated with GA_3 (Plate 6). Conceivably, analysis of the RNA species present in the resting buds and study of their changes in response to GA_3 will reveal significant early effects of the hormone. While changes in total protein are difficult to distinguish, characterization of the proteins during emergence from rest as well as study of the appearance of messenger and other RNA species would be of great interest.

It is emphasized that our study heretofore has centred primarily on the response of excised plugs with buds. Whether the effects of gibberellin will be shown to be identical for buds in intact tubers still remains to be seen. There is ample evidence to lead to the conclusion that in excised buds gibberellins play a functional role in processes related to bud growth. It appears that gibberellins produced in the superficial cell layers of the plug (Rappaport & Sachs, 1967; Madison & Rappaport, 1968) move inward to the bud, there enhancing a number of processes essential to bud elongation. The relation between size of plug and rate of sprouting could be related to the production or release of the hormone in cells along the periphery of the plug. Presumably the slower sprouting in the larger plug could be attributed to the greater distance over which the hormone has to travel. The effects of GA_3 on sprouting, sugar production and nucleic acid synthesis, as well as initial translocation experiments with $[^3H]GA_1$, strongly indicate that this happens.

That gibberellin should affect a number of possibly unrelated processes is not surprising. Tata (1968), in discussing animal hormone systems, made it clear that such hormones typically have not been shown to direct processes in cell-free systems. This is also true in plants, with the possible exception of Johri & Varner's (1968) observation that GA_3 enhances RNA synthesis in nuclei isolated from pea, *Pisum sativum*, seedlings. Tata indicates that hormones may affect a single process, as, for example, regulating supply of a particular metabolite, or may affect multiple processes, such as readjustment of permeability barriers. The former mechanism could account for rapid, hormone-specific effects, whereas the latter would affect other processes that, in turn, regulate the synthesis of hormone-specific constituents. The consequences of both kinds of action ultimately may be the same. Thus, gibberellin may play a direct and primordial role in regulating duration of rest, but conceivably it also has other indirect effects which ultimately feed back to contribute to the process of bud elongation. The well-established effect of gibberellin in inducing enzyme synthesis cannot be overlooked. It may well be that this role may be the significant one resulting in release of stored food materials for the bud elongation.

If GA_3 is not the only factor in bud elongation, and surely it is not, what other explanation can be offered for acceleration of nucleic acid synthesis and the subsequent morphogenetic events in excised buds? At least two alternative hypotheses were considered early in the study: (1) a volatile inhibitor is liberated as a result of wounding, and (2) a chemical inhibitor is destroyed. We have not explored these hypotheses and both remain distinct possibilities.

Burton (1952) found that volatile compounds, including N-amyl alcohol and ethylene, were evolved by potatoes in storage and inhibited sprouting. This work, which triggered an important line of research into metabolic control of respiration (Laties, 1963; MacDonald, 1967), is significant in that it suggested a possible mechanism for control of the rest period. Burton (1952, 1963, 1966) advanced the hypothesis that such volatiles might act as internal regulators, decreasing in concentration prior to initiation of accelerated growth in storage. However, there are some difficulties connected with this alternative. For example, how would such a mechanism regulate induction of rest period? There is especially a problem if, as

Burton believes, the onset of rest is considered to occur at the time of tuber formation when periderm development has not yet started. It would be necessary to postulate a metabolic control of the supply of the volatile. It is also difficult to relate the presumed role of a volatile to the activity in cells of the buds in developing tubers and those at harvest. The volatile would have to depress elongation and transverse, but not longitudinal, mitoses during tuber formation, and depress both elongation and essentially all mitotic activity after harvest. In the same vein, it is not apparent how changes in the tubers during rest would account for release of the inhibitor, especially since, during storage, periderm development continues.

Despite these problems with the concept of volatile inhibitors, the growing importance of ethylene as a regulator of growth requires some discussion. Ethylene has been detected in the gas released from potato tubers at exceedingly low concentration and its concentration is increased in stored potatoes by GA_3 (Poapst, Durkee, McGugan & Johnston, 1968). The effect of ethylene on sprouting is not clear since both stimulation and inhibition of sprouting by ethylene have been reported (see Poapst *et al.* 1968). Most of the evidence favours the action of ethylene as an inhibitor of sprouting, which would conflict with the fact that GA_3 accelerates termination of rest. It appears that a thorough investigation of the role of ethylene in sprouting is necessary.

The case for soluble inhibitors of sprouting, particularly ABA, was made earlier. How excision of a plug would result in a reduction in level of the inhibitor in the bud is not immediately clear, although there is evidence that ABA is not stable (Milborrow, 1967). The metabolism of ABA in plant tissues has not yet been described, but there is information that its effect is short-lived, at least in *Lolium* (Evans, 1966). This is also indicated by the results of Hemberg (1949) and Blumenthal-Goldschmidt & Rappaport (1965) with β. Since ABA is an effective inhibitor of bud elongation in potatoes and other species, an investigation of its effects on nucleic acid synthesis and cell division and enlargement in potato buds is indicated. There is evidence that it does not act as a competitive inhibitor of gibberellin action (Milborrow, 1966), but it is clear, at least in the barley, *Hordeum vulgare*, endosperm (Chrispeels & Varner, 1967) and *Lemna* (Van Overbeek and Loeffler, 1967; Van Overbeek *et al.* 1967) systems that it can block or reduce protein synthesis. Thus, conceivably, ABA would block a step prior to the one affected by GA_3. Its destruction or inactivation could explain the gradual increase in RNA synthesis in excised buds in the first 6 hr., before GA_3 is seen to enhance such synthesis (Plate 6). If shown to be effective, the role of ABA in regulating dormancy of potato buds, the first source from which it was isolated, may be characterized.

GENERAL CONCLUSIONS

An explanation of the great variety of expressions of the rest period in plants is not immediately forthcoming. In addition, the response to numerous agents that decrease the length of rest is difficult to explain (see Samish, 1954). The approach we have taken has unquestionably ignored certain important aspects while stressing others that we believe to be salient features of the rest period in potatoes and other organs. While it is a truism to say that rest is under genetic control, there is little reason to assume that an understanding of the rest period in all species will be found by studying their capacities for DNA and RNA syntheses. There are several types of dormancy (see Amen, 1968; Vegis, 1961, 1964) and the explanation for limitation to growth may easily be found in reasons as widespread as limitation of respiration (Pollock, 1953), inhibition by chemical or volatile inhibitors (Burton, 1952, 1963; Wareing *et al.* 1964) or of capacity for RNA synthesis (Marcus & Feeley, 1964; Tuan & Bonner, 1964). Recognizing the disparity of requirements for onset of growth, however, helps provide a scheme, no more than a working hypothesis, to integrate existing knowledge into an understanding of the many facets of bud dormancy. If, indeed, ultimate control of rest is at the level of the gene, it is conceivable that in some species the ability to germinate is indeed blocked by a repression of DNA synthesis. In other species there may be a block at the level of translation or transcription. In still others, the limitation may be insufficient water supply, a problem perhaps associated with membrane permeability. The knowledge that light and temperature markedly influence levels of growth substances (Nitsch, 1963) leads to an explanation of environmentally controlled growth. The interesting results of Waxman (1957), Nitsch (1957) and Wareing & Villiers (1961) certainly support the view that development and growth of certain embryos and buds is linked to hormonal control by way of effects of light or temperature, or both. The picture that emerges therefore is that of the basic morphogenetic scheme: DNA → RNA → protein, with hormonal factors impinging on the normal progression of events. Thus, while the rationale of the scheme is in harmony with present thought about gene action, the complexities are such that a general-purpose 'unified field theory' is still a dream. It appears that an understanding of the rest period in buds will require a detailed study of one or a few species with similar requirements.

We acknowledge the generous assistance of Professors M. Evenari, D. Koller, N. Feinbrun, M. Negbi, and A. Fahn, Department of Botany, Hebrew University, Jerusalem, where one of us (L. R.) was on sabbatical leave in 1963-4 with financial assistance through Guggenheim Foundation and Fulbright awards. Thanks are also due to Professor W. Jensen and Miss Mary Ashton, Department of Botany, University of California, Berkeley, for advice and assistance. The research at Davis was assisted by grants from the U.S. Public Health Service (GM 12885) and National Science Foundation (GB 1335).

REFERENCES

ABDALLA, A. A. & MANN, L. K. (1963). *Hilgardia* **35**, 85.
ALLEN, R. M. (1960). *Physiologia Pl.* **13**, 555.
AMEN, R. D. (1968). *Bot. Rev.* **34**, 1.
APPLEMAN, C. O. (1916). *Bot. Gaz.* **61**, 265.
APPLEMAN, C. O. (1918). *Science, N.Y.* **48**, 319.
ARTSCHWAGER, E. F. (1924). *J. agric. Res.* **27**, 809.
BARSKAYA, E. N. & OKININA, E. Z. (1959). *Fiziologiya Rast.* **6**, 470.
BARTON, L. V. & CHANDLER, C. (1957). *Contr. Boyce Thompson Inst. Pl. Res.* **19**, 201.
BENNET-CLARK, T. A. & KEFFORD, N. P. (1953). *Nature, Lond.* **171**, 645.
BERENDES, H. D. (1967). *Chromosoma* **22**, 274.
BLOMMAERT, K. L. J. (1954). *Nature, Lond.* **174**, 970.
BLOMMAERT, K. L. J. (1955). *Sci. Bull. Dep. Agric. For. Un. S. Afr.* no. 368.
BLUMENTHAL-GOLDSCHMIDT, S. & RAPPAPORT, L. (1965). *Pl. Cell Physiol., Tokyo* **6**, 601.
BOO, L. (1961). *Physiologia Pl.* **14**, 676.
BOO, L. (1962). *Svensk bot. Tidskr.* **56**, 193.
BRIAN, P. W., HEMMING, H. G. & RADLEY, M. (1955). *Physiologia Pl.* **8**, 899.
BROWN, D. S., GRIGGS, W. H. & IWAKIRI, B. T. (1960). *Proc. Am. Soc. hort. Sci.* **76**, 52.
BUCH, M. L. & SMITH, O. (1959). *Physiologia Pl.* **12**, 706.
BURTON, W. G. (1952). *Nature, Lond.* **169**, 117.
BURTON, W. G. (1956). *Physiologia Pl.* **9**, 567.
BURTON, W. G. (1963). In *The Growth of the Potato*, pp. 17-41. Ed. J. D. Ivins and F. L. Milthorpe. London: Butterworth.
BURTON, W. G. (1966). *The Potato*, p. 382. Wageningen: H. Veenman and Zonen, N.V.
CHRISPEELS, M. J. & VARNER, J. E. (1967). *Pl. Physiol., Lancaster* **42**, 1008.
CLEGG, M. D. (1967). Ph.D. thesis, University of California. Davis.
CLEGG, M. D. & RAPPAPORT, L. (1966). *Proc. XVIIth Int. Hort. Congr.* vol. 1, p. 75.
CORNFORTH, J. W., MILBORROW, B. V. & RYBACK, G. (1966). *Nature, Lond.* **210**, 627.
CORNFORTH, J. W., MILBORROW, B. V., RYBACK, G. & WAREING, P. F. (1965). *Nature, Lond.* **205**, 1269.
COTRUFO, C. & LEVITT, J. (1958). *Physiologia Pl.* **11**, 240.
DAVIDSON, T. M. W. (1958). *Am. Potato J.* **25**, 451.
DENNY, F. E. (1936). *Contr. Boyce Thompson Inst. Pl. Res.* **8**, 137.
DONOHO, C. W. & WALKER, D. R. (1957). *Science, N.Y.* **126**, 1178.
DOORENBOS, J. (1958). *Neth. J. agric. Sci.* **6**, 267.
EAGLES, C. F. & WAREING, P. F. (1963). *Nature, Lond.* **199**, 874.
EAGLES, C. F. & WAREING, P. F. (1964). *Physiologia Pl.* **17**, 697.

EDELMAN, J. & HALL, M. A. (1964). *Nature, Lond.* **201**, 296.
EDELMAN, J. & SINGH, S. P. (1968). *J. exp. Bot.* **19**, 288.
EL ANTABLY, H. M. M., WAREING, P. F. & HILLMAN, J. (1967). *Planta* **73**, 75.
EMILSSON, B. (1949). *Acta agric. suec.* **3**, 189.
EMILSSON, B. & LINDBLOOM, H. (1963). In *The Growth of the Potato*, pp. 45–62. Ed. J. D. Ivins and F. L. Milthorpe. London: Butterworth.
EVANS, L. T. (1966). *Science, N.Y.* **151**, 107.
ESASHI, Y. & NAGAO, M. (1959). *Sci. Rep. Tôhoku Univ.* 4th ser. (*Biol.*) **25**, 191.
FEINBRUN, N. & KLEIN, S. (1962). *Pl. Cell Physiol., Tokyo* **3**, 407.
GOODWIN, P. B. (1963). In *The Growth of the Potato*, pp. 63–71. Ed. J. D. Ivins and F. L. Milthorpe. London: Butterworth.
HABER, A. H. & LUIPPOLD, H. J. (1960). *Pl. Physiol., Lancaster* **35**, 486.
HARTSEMA, A. (1961). *Encyclopedia of Plant Physiol.* vol. XVI, 123.
HASHIMOTO, T. & RAPPAPORT, L. (1966a). *Pl. Physiol., Lancaster* **41**, 623.
HASHIMOTO, T. & RAPPAPORT, L. (1966b). *Pl. Physiol., Lancaster* **41**, 629.
HAYASHI, F., BLUMENTHAL-GOLDSCHMIDT, S. & RAPPAPORT, L. (1962). *Pl. Physiol., Lancaster* **37**, 774.
HAYASHI, F. & RAPPAPORT, L. (1965). *Nature, Lond.* **205**, 414.
HEMBERG, T. (1949). *Physiologia Pl.* **1**, 24.
HEMBERG, T. (1952). *Physiologia Pl.* **5**, 115.
HEMBERG, T. (1954). *Physiologia Pl.* **7**, 312.
HEMBERG, T. (1958a). *Physiologia Pl.* **11**, 610.
HEMBERG, T. (1958b). *Physiologia Pl.* **11**, 615.
HENKEL, P. A. & SITNIKOVA, O. A. (1953). *Trudȳ Inst. Fiziol. Rast.* **8**, 276.
IVINS, J. D. & MILTHORPE, F. L. (1963). *The Growth of the Potato.* London: Butterworth.
JOHRI, M. M. & VARNER, J. E. (1968). *Proc. natn. Acad. Sci. U.S.A.* **59**, 269.
KAWASE, M. (1961). *Pl. Physiol., Lancaster* **36**, 643.
KUPILA-AHVENNIEMI, S. (1966). *Aquilo*, Ser. Botanica, **4**, 59. Societas Amicorum Naturae Ouluensis.
LANG, A. (1965). *Encyclopedia of Plant Physiology*, vol. XV, p. 848.
LATIES, G. G. (1963). In *Control Mechanisms in Respiration and Fermentation*, p. 129. Ed. B. Wright. New York: Ronald Press.
LEVITT, J. (1954). *Physiologia Pl.* **7**, 597.
LIPPERT, L. F., RAPPAPORT, L. & TIMM, H. (1958). *Pl. Physiol., Lancaster* **33**, 132.
LIPE, W. N. & CRANE, J. C. (1966). *Science, N.Y.* **153**, 541.
MACDONALD, I. R. (1967). *Pl. Physiol., Lancaster* **42**, 227.
MADISON, M. & RAPPAPORT, L. (1968). *Pl. Cell Physiol., Tokyo* **9**, 147.
MANN, L. K. & LEWIS, D. A. (1956). *Hilgardia* **26**, 161.
MARCUS, A. & FEELEY, J. (1964). *Proc. natn. Acad. Sci. U.S.A.* **51**, 1075.
MICHENER, H. D. (1942). *Am. J. Bot.* **29**, 558.
MILBORROW, V. T. (1966). *Planta* **70**, 155.
MILBORROW, V. T. (1967). *Planta* **76**, 93.
MOREL, G. & MÜLLER, J. (1964). *C. r. hebd. Séanc. Acad. Sci., Paris* **258**, 5250.
MÜLLER-THURGAU, H. (1882). *Landw. Jbr.* **11**, 751.
NAGAO, M. & MITSUI, E. (1959). *Sci. Rep. Tôhoku University*, 4th. ser. (*Biol.*) **25**, 199.
NAGAO, M. & OKAGAMI, N. (1966). *Bot. Mag., Tokyo* **79**, 687.
NITSCH, J. P. (1957). *Proc. Am. Soc. hort. Sci.* **70**, 512.
NITSCH, J. P. (1963). In *Env. Control Plant Growth*, pp. 175–193. Ed. L. T. Evans. New York: Academic Press.
OKAZAWA, Y. (1959). *Proc. Crop Sci. Soc. Japan* **28**, 129.
OKAZAWA, Y. (1960). *Proc. Crop Sci. Soc. Japan* **29**, 121.

OHKUMA, K., ADDICOTT, F. T., SMITH, O. E. & THIESSEN, W. E. (1965). *Tetrahedron Lett.* **29**, 2529.
OSHIMA, N. & LIVINGSTON, C. H. (1963). *Am. Potato J.* **40**, 9.
PALEG, L. G. (1965). *A. Rev. Pl. Physiol.* **16**, 291.
PHILLIPS, I. D. J. & WAREING, P. F. (1959). *J. exp. Bot.* **10**, 504.
PLAISTED, P. H. (1957). *Pl. Physiol., Lancaster* **32**, 445.
POAPST, P. A., DURKEE, A. B., MCGUGAN, W. A. & JOHNSTON, F. B. (1968). *J. Sci. Fd Agric.* **19**, 325.
POLLOCK, B. M. (1953). *Physiologia Pl.* **6**, 47.
POLLOCK, B. M. & OLNEY, H. O. (1959). *Pl. Physiol., Lancaster* **34**, 131.
RAPPAPORT, L. (1956). *Calif. Agric.* **10**, 4.
RAPPAPORT, L. (1964). *Abstr. Xth Int. Bot. Congr.*, Edinburgh, p. 6.
RAPPAPORT, L. & BLUMENTHAL-GOLDSCHMIDT, S. (1961). *Pl. Physiol., Lancaster* (Suppl.) **36**, viii.
RAPPAPORT, L., BLUMENTHAL-GOLDSCHMIDT, S., CLEGG, M. D. & SMITH, O. E. (1965a). *Pl. Cell Physiol., Tokyo* **6**, 587.
RAPPAPORT, L., BLUMENTHAL-GOLDSCHMIDT, S. & HAYASHI, F. (1965b). *Pl. Cell Physiol., Tokyo* **6**, 609.
RAPPAPORT, L., HSU, A., THOMPSON, R. H. & YANG, S. F. (1967). *Ann. N.Y. Acad. Sci.* **144**, 211.
RAPPAPORT, L., LIPPERT, L. F. & TIMM, H. (1957). *Am. Potato J.* **34**, 254.
RAPPAPORT, L. & SACHS, M. (1967). *Nature, Lond.* **214**, 1149.
RAPPAPORT, L. & SMITH, O. E. (1962). In *Eigenschaften und Wirkungen der Gibberelline*, pp. 37–45. Ed. R. Knapp. Heidelberg: Springer-Verlag.
RAPPAPORT, L. & STAHL, N. (1965). *Am. J. Bot.* **52**, 623.
RAPPAPORT, L. & WOLF, N. (1968a). *Int. Symp. Growth Substances*, Calcutta, India. (In the Press.)
RAPPAPORT, L. & WOLF, N. (1968b). Int. Symp. Plant Biochem. Regulation in Viral and other Diseases or injury. *Phytopath. Soc. Japan*, Tokyo, Proc. 204.
REINITZER, F. (1893). *Ber. dt. bot. Ges.* **11**, 531.
ROSA, J. T. (1928). *Hilgardia* **3**, 99.
SACHS, R. M., BRETZ, C. F. & LANG, A. (1959). *Am. J. Bot.* **46**, 376.
SAMISH, R. M. (1954). *A. Rev. Pl. Physiol.* **5**, 183.
SATAROVA, N. A. (1950). *Trudȳ Inst. Fiziol. Rast.* **7**, 67.
SCHIPPERS, A. A. (1956). *Publ. Stichting voor Aardappelbew*, ser. A, **108**, 13. Wageningen.
SLATER, J. W. (1963). In *The Growth of the Potato*, p. 114. Ed. J. D. Ivins and F. L. Milthorpe. London: Butterworth.
SMITH, H. & KEFFORD, N. P. (1964). *Am. J. Bot.* **51**, 1002.
SMITH, O. E. (1962). Ph.D. thesis, University of California. Davis.
SMITH, O. E. & RAPPAPORT, L. (1960). *Adv. Chem. Ser.* **28**, 42.
SZALAI, I. (1959). *Physiologia Pl.* **12**, 237.
TATA, J. R. (1968). *Nature, Lond.* **219**, 331.
THOMAS, T. H. (1969). *J. Expt. Bot.* **20**, 124.
THORNTON, N. C. (1939). *Contr. Boyce Thompson Inst. Pl. Res.* **10**, 339.
TOMPKINS, D. R. (1965). *Proc. Am. Soc. hort. Sci.* **87**, 371.
TSUKAMOTO, Y. & YAGI, M. (1960). *Pl. Cell Physiol., Tokyo* **1**, 221.
TUAN, D. Y. H. & BONNER, J. (1964). *Pl. Physiol., Lancaster* **39**, 768.
VAN OVERBEEK, J. & LOEFFLER, J. E. (1967). *Abstr. VIth Int. Growth Reg. Symp.*, Ottawa, p. 66.
VAN OVERBEEK, J., LOEFFLER, J. E. & MASON, M. I. R. (1967). *Science, N.Y.* **156**, 1497.
VARGA, M. B. & FERENCZY, L. (1957). *Acta bot. hung.* **13**, 111.

VEGIS, A. (1961). *Encyclopedia of Plant Physiology*, vol. XVI, p. 168.
VEGIS, A. (1964). *A. Rev. Pl. Physiol. U.S.A.* **15**, 185.
WALKER, M. G. (1968). *Nature, Lond.* **218**, 878.
WAREING, P. F., EAGLES, C. F. & ROBINSON, P. M. (1964). In *Regulateurs naturels de la croissance végétale*, p. 377.
WAREING, P. F., EL ANTABLY, H. M. M. & GOOD, N. (1967). *Abstr. VIth Int. Growth Reg. Symp.*, Ottawa, p. 98.
WAREING, P. F. & VILLIERS, T. A. (1961). In *Plant Growth Regulation*, p. 95. Ed. R. M. Klein. Iowa State University Press.
WAXMAN, S. (1957). Ph.D. thesis, Cornell University, Ithaca, New York.
WEAVER, R. J. (1959). *Nature, Lond.* **183**, 1198.
WEAVER, R. J. (1963). *Nature, Lond.* **198**, 207.

PLATE I

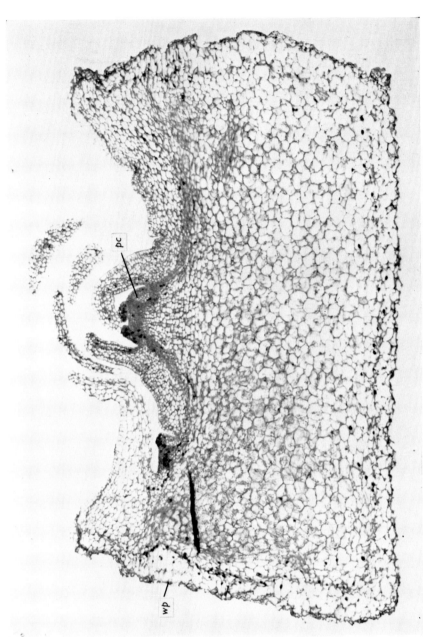

Localization of DNA synthesis in cells of the apical bud, procambial (*pc*) and vascular system, leaf bases and along the wounded periphery (*wp*) of the plug 48 hr. after excision and treatment. Note new cambial development in the same zone as new DNA synthesis and the absence of such synthesis in the parenchymatous interior of the plug. The section was prepared from a cylinder of tissue incubated in Petri dishes on filter paper and [³H]thymidine (10 μC./ml.) was applied to the bud apex and the filter paper. Stained with safranin-fast green. (From Rappaport & Wolf, 1968a.)

(*Facing p.* 240)

PLATE 2

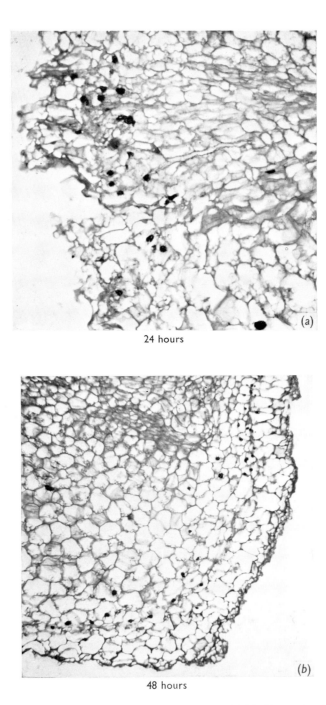

24 hours

48 hours

Enlarged photo of the peripheral region of a plug, showing (*a*) labelling 24 hr. after excision in the region of the vascular system and (*b*) peripheral labelling 48 hr. after excision and treatment. Labelling in (*a*) was confined to the region of the severed vascular system. (*b* from Rappaport & Wolf, 1968*a*.)

PLATE 3

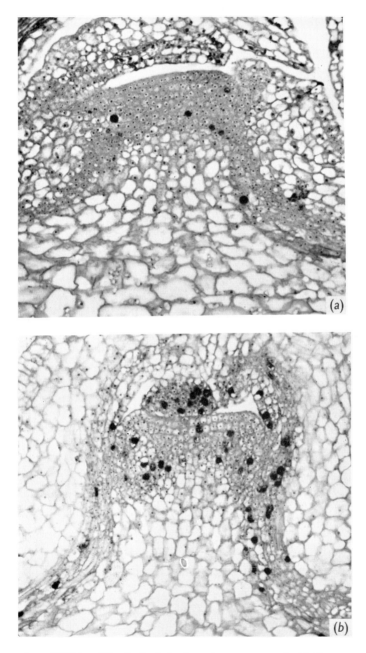

Incorporation of [³H]thymidine by buds in plugs that were treated with the precursor for a period of 24 hr., beginning 1 day after excision and storage at (a) 6° and (b) 22° C. Note that the bud at 22° C. is considerably elongated. Stained with safranin-fast green. The nucleoli are stained deeply with safranin.

PLATE 4

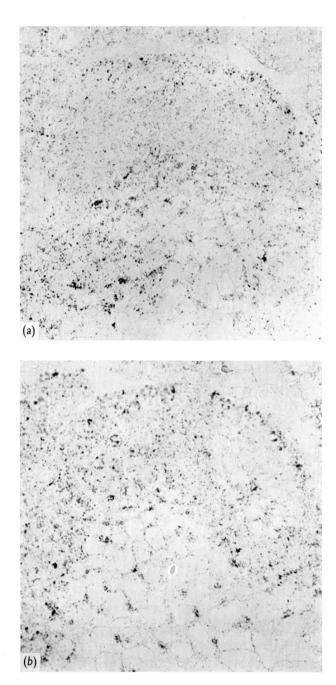

Incorporation of [³H]uridine 24 hr. after excision and treatment with precursor. Plugs were excised, treated and held at (a) 6° and (b) 22° C. No stain.

PLATE 5

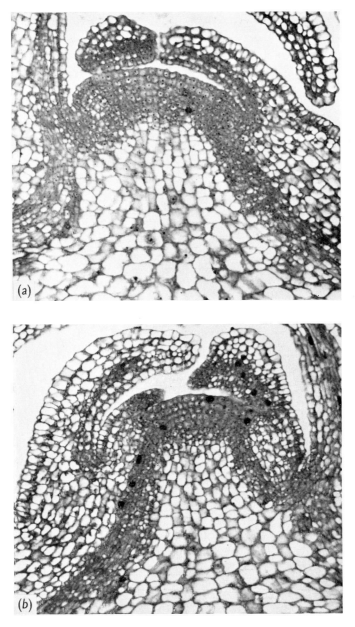

Effect of GA$_3$ (0·009 µg./bud) on incorporation of [^{3}H]thymidine 12 hr. after excision. (*a*) Water control and (*b*) treated with GA$_3$.

PLATE 6

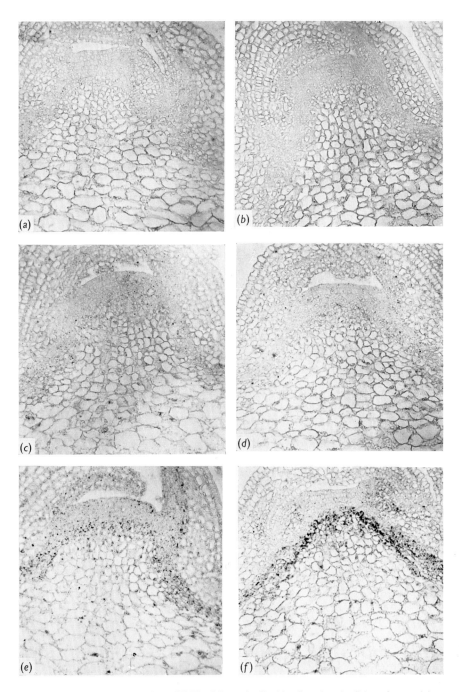

Effect of GA_3 on incorporation of [^{3}H]uridine 3 (a, b), 6 (c, d) and 12 (e, f) hr. after excision and treatment of 0·2 × 0·2 cm. plugs. (a), (c) and (e), water control. (b), (d) and (f), treated with 0·009 μg. GA_3/bud. Stained with fast green.

THE CONTROL OF BUD DORMANCY IN SEED PLANTS

By P. F. WAREING

Botany Department, University College of Wales, Aberystwyth

INTRODUCTION

The term 'dormancy' will be used here as a convenient 'shorthand' form for 'innate dormancy' or 'rest', in which the causes of dormancy lie within the bud itself; we shall not be concerned with cases where the dormancy is due to some influence arising outside the bud itself, as in correlative inhibition or where low temperature or some other external factor is preventing growth. We speak of innate dormancy or rest where a bud or other organ fails to grow even under external conditions which, at other phases of the life-cycle, are quite favourable for active growth; for example, we frequently find in northern temperate regions that the buds of woody plants fail to grow in September, even though the conditions of temperature, light and water supply prevailing at that time may be no less favourable than in, say, May, when growth is normally very active.

The phenomenon of bud dormancy is widespread in seed plants and involves not only the winter-resting buds of woody plants but a variety of other organs, including stem tubers (*Solanum* spp., *Begonia evansiana*), bulbs (*Allium*, *Tulipa*), corms (*Gladiolus*), rhizomes (*Convallaria*) and the turions of aquatic plants (*Hydrocharis*, *Stratiotes*).

Resting buds are essentially adaptations for survival during unfavourable seasons, when climatic conditions present special hazards, such as low temperature, associated with water stress due to the freezing soil water, or hot, dry conditions, as in arid regions. Adaptations which make possible survival during such periods of stress are of several types. In some cases, as with underground resting organs such as bulbs, corms, rhizomes and tubers, some protection from excessive heat or cold, or excessive desiccation, is brought about by the burial of the organ below ground, or below water in the case of aquatic plants. Species showing this type of overwintering are included in the life-forms referred to as cryptophytes by Raunkiaer (1934). On the other hand, the buds of the typical woody plants (phanaerophytes) are exposed to the air during the unfavourable season. Desiccation is no doubt reduced by enclosure of the meristematic tissue of the buds within protecting bud scales in many woody plants. However, cold and heat resistance in both cryptophytes and chamaephytes is also dependent upon certain physiological and biochemical changes within the tissues, including not only mature, differentiated tissue, but also the meristematic tissues of the shoot apices and the cambium. A discussion here of the physiological basis of frost and heat resistance would take us too far afield, but it is apparent that the change from the actively growing to the dormant condition involves a number of changes in the tissues, particularly in the protein fractions (Levitt, this symposium, p. 393).

In the following account we shall consider, first, the environmental factors controlling the onset and disappearance of dormancy in buds and other organs, and secondly, the present state of knowledge regarding the hormonal and biochemical control of dormancy.

THE DEVELOPMENT OF BUD DORMANCY IN WOODY PLANTS

The majority of temperate woody plants, including both coniferous and dicotyledonous species, show a well-marked dormancy or resting phase during the annual growth cycle and this is usually, but by no means always, characterized by changes at the shoot apices, with the development of resting buds. The typical resting bud involves the 'telescoping' of the bud scales and leaf primordia in the apical region, due to the arrest of normal internode extension. In some species which have stipules (e.g. *Betula, Fagus*) this telescoping of the shoot apical region leads to the formation of a resting bud, since the overlapping stipules in this region form the bud scales. In other species the protective scales represent leaves, which may be only slightly modified, as in *Viburnum* spp., or more highly modified, so that they comprise only the leaf base with the lamina underveloped. In such 'foliar' buds the scales can be regarded as leaf primordia which have been arrested at an early stage of development. Thus, in stipular buds inhibition of normal internode extension is involved, while in foliar buds there is also inhibition of lamina development. In some trees a terminal bud is not formed, since growth of the shoot is terminated by the death and abscission of the apical region, and growth is later continued from the uppermost axillary bud. Such species (which include *Tilia, Ulmus, Castanea, Robinia, Ailanthus*) are said to show a sympodial growth habit.

When terminal buds are first formed they can frequently be induced to resume growth by various treatments including defoliation, whether by hand or by insects (Büsgen & Münch, 1929). It would appear, therefore, that at this stage the terminal buds are not themselves innately dormant, but their growth is apparently inhibited by the mature leaves on the shoot. Similarly, lateral buds may be inhibited by the leaves, or by the main apical region in actively growing shoots, and are held in check by correlative inhibition rather than by innate dormancy. This phase of bud development is referred to as *summer dormancy* or *pre-dormancy*.

Later, the buds are, in many species, found to have entered into a state variously referred to as *deep dormancy, true dormancy, winter dormancy* or *rest*. When they have entered this condition the buds will no longer resume growth if the shoots are defoliated, so that they are now innately dormant and not simply held in check by environmental conditions or inhibitory influences within the plant itself, as is the case during pre-dormancy. After an appropriate period of true dormancy the buds become capable, in the later part of the winter or early spring, of resuming growth when external conditions, particularly temperature, are favourable for growth. Thus, at this stage the buds are no longer innately dormant, but nevertheless for some time they may fail to grow because of low outdoor temperatures. This phase is referred to as *post-dormancy* or *after-rest*.

We shall now consider some of the environmental factors controlling the induction and breaking of dormancy in temperate woody plants, of which day-length and temperature are among the most important.

FACTORS CONTROLLING THE INDUCTION OF BUD DORMANCY IN WOODY PLANTS

One of the most important factors affecting and controlling the induction of dormancy in woody plants is day-length. This subject has been dealt with in a number of review articles (Wareing, 1956; Nitsch, 1957; Vegis, 1965) and only the main features of photoperiodism in woody plants need be summarized here.

For obvious reasons of experimental convenience, nearly all studies have been carried out with seedlings of woody plants and it has been found that in the majority of species so far studied long days (LD's) promote vegetative growth and short days (SD's) bring about the cessation of extension growth and the formation of resting buds. However, a number of common cultivated fruit trees (*Pyrus, Malus, Prunus*) and certain other species, including the family *Oleaceae*, appear to be relatively insensitive to day-length changes.

The seedlings of some species, e.g. *Robinia pseudacacia, Betula pubescens* and *Larix decidua*, can be maintained in continuous growth for at least 18 months under LD conditions in a warm greenhouse, whereas under SD they cease growth within 10–14 days. On the other hand, other species such as *Acer pseudoplatanus, Aesculus hippocastanum*, and *Liquidambar styraciflua* (Borthwick & Downs, 1956) show delayed dormancy under LD, but they cannot be maintained in growth indefinitely under these conditions. For those species which can be maintained in continuous growth under LD, there appears to be a certain critical day-length, below which dormancy is induced, and above which dormancy does not occur. In other species there appears to be a graded response, with an increasing number of leaves being produced, up to a certain maximum, with increasing day-length (Wareing, 1950; Downs & Borthwick, 1956).

As in herbaceous plants, the photoperiodic responses of woody seedlings appears to depend upon the length of the dark period, rather than of the photoperiod, and if a long dark period is interrupted by a short light interruption ('light-break') the effect of the dark period is largely nullified and dormancy is delayed (Wareing, 1950; Zahner, 1955). The most effective region of the spectrum for this light-break effect lies in the red (Nitsch, 1957), suggesting that phytochrome is involved, and clear-red/far-red reversibility has been demonstrated for seedlings of *Larix europaea* (R. van der Veen, personal communication).

The response of woody seedlings depends on the day-length conditions to which the *leaves* are exposed. In *Acer pseudoplatanus*, as in herbaceous species, it is the young, fully expanded leaves which are the most sensitive to day-length, but in birch seedlings even quite young leaves in the apical region show sensitivity to photoperiod (Wareing, 1954).

The question arises as to how important are these photoperiodic responses in determining the formation of resting buds and the onset of dormancy in woody plants in nature. There seems little doubt that the seasonal decline in day-length is important in determining the onset of dormancy in seedlings of species which normally continue active growth into the autumn, e.g. *Larix decidua*, *Betula pubescens* and *Robinia pseudacacia* (Wareing, 1956).

We have little information as to how far dormancy is controlled by day-length in older trees, but it is very common to find that older trees show a very much shorter period of extension growth than seedlings of the same species, and they frequently cease growth in June or July, while the natural photoperiods are still long (Wareing, 1948). In such cases it seems doubtful whether declining day-length is important in determining the formation of resting buds, and it seems more likely that some change in the internal conditions, either in nutrients or hormonal balance, within the tree itself, determines the period of growth and the onset of dormancy. However, resting buds formed under LD appear to be still in a state of summer dormancy or pre-dormancy, since they can frequently be induced to expand and resume growth if the tree is defoliated. It is only later in the season that such buds enter a state of true dormancy, and since this deepening of dormancy is correlated with declining day-lengths in the autumn it may well be that it is affected by photoperiod. Thus, buds may be formed in older trees under LD, but the entry of such buds into dormancy may be an SD response. Some evidence in support of this hypothesis is given by observations on the effect of day-length on the axillary buds of birch seedlings (Wareing & Black, 1958). When the seedlings are actively growing under LD the axillary buds are not dormant but are held in check by correlative inhibition, as shown by the fact that decapitation of the seedlings causes them to grow out. But after exposing the seedlings to 14 SD cycles the axillary buds do not grow out when transferred to LD at 15° C. following decapitation; they have thus become dormant in response to SD, although externally they are indistinguishable from non-dormant buds, apart from the fact that they may be a little larger than the latter. These experimental results appear to be paralleled by the transition from pre-dormancy to true dormancy in nature, since if shoots of *Ribes nigrum* are decapitated on successive dates the axillary buds show a progressive decline in the ability to grow out (Nasr & Wareing, 1961).

In addition to the effects of day-length on bud formation and dormancy, SD's also have an effect on leaf-fall and cambial activity in some species. Garner & Allard (1923) first reported the effects of day-length on leaf-fall in *Liriodendron tulipifera*, and since then there have been a number of similar reports for other species (Nitsch, 1957). However, photoperiodic control of leaf-fall does not appear to be universal in all woody species; for example, if seedlings of *Acer pseudoplatanus* or *Betula pubescens* are maintained under SD at warm temperatures in a greenhouse their leaves remain green and healthy for many weeks. It seems likely that, in nature, low temperatures, and possibly low light intensity, are at least as important as day-length in determining the onset of leaf senescence and abscission.

In broad-leaved trees the duration of cambial activity varies greatly between species. In 'diffuse-porous' species cambial activity depends upon continued extension growth, and when the latter ceases xylem differentiation normally ceases also. By contrast, in 'ring-porous' broad-leaved trees cambial activity may continue long after extension growth has ceased, although there is a change from 'early-wood' to 'late-wood' formation at the time that extension growth ceases (Wareing, 1958). In the ring-porous species *Robinia pseudacacia* and *Ailanthus glandulosa* continued 'late-wood' formation, after extension growth has ceased, is dependent upon LD, and under SD xylem formation ceases although phloem formation may continue (Digby & Wareing, 1966). In *Pinus* spp. also, cambial activity may continue after extension growth has ceased under LD, but the cambium becomes dormant under SD (Wareing, 1951).

In addition to the various observable morphological and anatomical changes associated with the induction of dormancy by SD's there are also chemical changes which are reflected in increased cold resistance. Moshkov (1935) observed that seedlings of *Robinia pseudacia* and *Ribes nigrum* which had been exposed to SD's were markedly more frost-resistant than seedlings which had been grown only under LD's. Recently, van Huystee, Weiser & Li, (1968) have shown that cold acclimation in *Cornus stolonifera* depends upon exposure both to SD's and to decreasing temperatures. Plants subjected to SD's alone did not show cold resistance, while those exposed to low temperature alone showed only slightly increased cold resistance. When exposure to SD's preceded exposure to gradually decreasing temperatures the cold acclimation was most rapid and effective.

It is now well established that wide-ranging woody species such as *Pinus sylvestris* and *Picea abies* show marked ecotypic differences in photoperiodic responses in relation to both the latitude and the altitude at which they occur naturally (Vaartaja, 1959). Northern races or provenances are frequently found to require longer photoperiods for active extension growth than do more southern races, adapted to shorter natural photoperiods. This fact suggests that woody plants are rather closely adapted to natural day-length conditions, and that the latter probably play an important controlling role in the seasonal cycle of growth and dormancy.

FACTORS CONTROLLING EMERGENCE FROM DORMANCY

As we have seen, SD's play an important role in the induction of bud dormancy in many, though by no means all, woody plants. What controls emergence from dormancy? Do naturally increasing photoperiods in the spring determine the resumption of growth in woody plants?

In the majority of woody species so far studied it is found that once the buds have entered true dormancy in response to SD's they cannot be induced to resume growth by exposing them to long photoperiods. It is well known that in many species the buds require to be exposed to chilling treatments for several weeks before they become non-dormant. In general, the most effective temperatures are just above freezing (0–5° C.). The total period of chilling required ranges from 260 to over 1000 hr. Frequently the chilling requirements are met by January, but in many species the buds fail to resume growth when they have become non-dormant, because the temperatures still remain too low for growth and they are then in the phase of post-dormancy (p. 243).

Although in the majority of north-temperate woody plants dormancy can be overcome by chilling but not by LD's, in a number of species the unchilled buds *can* be induced to expand by exposure to LD's. This appears to be true, for example, for *Fagus sylvatica, Betula pubescens* (Wareing, 1953) and *Larix decidua* (Wareing, unpublished). It was first shown by Klebs (1914) that the buds on leafless shoots of *F. sylvatica* will expand if placed under continuous light in a warm greenhouse in the winter, and this response appears to be dependent on day-length (Wareing, 1953).

At first sight it would seem that the response of dormant buds to photoperiod contradicts the rule that it is the leaves which are the organs of photoperiodic 'perception', and that the apical meristematic region is insensitive to photoperiod. However, it should be remembered that resting buds contain well-developed leaf primordia and that therefore the difference between species such as *F. sylvatica* and other species relates primarily to a difference in age at which the leaves become sensitive to photoperiod.

If the buds of *B. pubescens* can be induced to resume growth by exposing them to long photoperiods, does the presence of mature leaves affect the responses of dormant seedlings of this species? This problem was studied by first inducing seedlings of *Betula pubescens* to become dormant by SD treatment and then exposing the buds and leaves separately to LD's or SD's (Wareing, 1954) (Fig. 1). The results may be summarized as follows: (1) the buds themselves must be directly exposed to LD's, and they will expand if either (*a*) no leaves are present or (*b*) if the leaves as well as the buds are exposed to LD's; (2) if the buds are exposed to LD's and the leaves to SD's, the buds cannot expand. Thus, *when the leaves are exposed to SD's they exert an inhibitory effect upon the buds*. Similar results were obtained by Waxman (1957). Moreover, Downs & Borthwick (1956) have observed that plants of *Weigela* rendered dormant by SD's will resume growth even under SD's, if defoliated.

It is not clear whether the photoperiodic control of bud-break is important in nature, but there is evidence that bud-break in *Fagus sylvatica* may be dependent upon lengthening photoperiods in the spring (Wareing, 1953), although in many regions temperature is also likely to be a limiting factor for this species. In *Rhododendron*, also, bud-break appears to be determined by day-length (Doorenbos, 1955).

Photoperiodic control of bud growth appears to be important in species such as *Quercus robur*, which normally make more than one flush of growth per growing season. In this latter species the number of nodes in each flush of growth is already laid down in the winter-resting bud, and the spring flush is limited to the extension of internodes predetermined in the bud. Growth then normally ceases for a time from early June until mid-July, during which time a number of new primordia are formed in the apical region of the shoot. A second flush ('lammas-shoot') is then made during July or early August. Normally, further flushing is prevented by the onset of SD's in the autumn, but if the natural day-length is extended artificially as many as 3-4 flushes of growth may be obtained in a single season (Wareing, 1954). The buds formed during the summer are evidently not

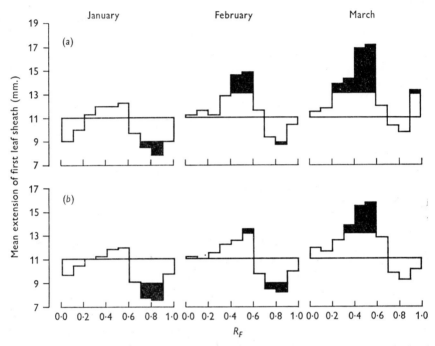

Fig. 1. Effect of winter chilling on endogenous gibberellin and inhibitor levels in buds of blackcurrant (var. Wellington XXX). (*a*) Shoots collected from out-of-doors in January and kept in cold room at 2° C. (*b*) Buds collected from out-of-doors at different dates, and hence subjected to natural chilling temperatures. Extract from equal dry weights (10 g.) bud tissue were loaded on chromatograms, which were developed in isopropanol: ammonia:water (10:1:1). Activity was assayed by dwarf maize (mutant d_1) leaf sections. Points above and below the horizontal lines indicate gibberellin activity and growth inhibition, respectively (from El-Antably, 1965).

in a state of dormancy, but show summer dormancy and are easily stimulated to grow by LD's. On the other hand, the bud formed in the autumn under SD conditions ultimately becomes fully dormant and its dormancy can then only be overcome by chilling. Conifers, such as *Pinus* spp., show similar responses (Kramer, 1936; Wareing, 1951).

DORMANCY IN VARIOUS ORGANS

As already stated above (p. 241), dormancy is shown by a wide variety of organs of perennial plants. The dormancy of the winter resting buds ('turions') of the aquatic plants *Utricularia*, *Stratiotes* and *Hydrocharis* has been studied in some detail by Vegis (1961, 1965). In *Stratiotes aloides* the dormancy of the turions is induced by SD's in association with high temperature. The length of the dormancy period is affected by temperature, and is shorter at 10° than at 20-25° C. SD's have also been found to promote the formation of resting buds ('hibernaculae') in the insectivorous plant *Pinguicula grandiflora* (Heslop-Harrison, 1962), and the dormancy of the buds is overcome by chilling. By contrast, dormancy in bulbs of onion (*Allium cepa*) is promoted by long photoperiods, and the period of dormancy is shortest when the bulbs are stored in cool temperatures (Mann & Lewis, 1956). Similarly, dormancy in *Lunularia cruciata*, a desert liverwort from Israel, is promoted by LD's and overcome by SD's (Schwabe & Nachmony-Bascombe, 1963). This latter response appears to be an adaptation to the hot, dry conditions prevailing during the period of summer dormancy in the natural habitat of *L. cruciata*.

Dormancy of the corms of *Gladiolus*, the rhizomes of *Convallaria* and the tubers of *Helianthus tuberosus* is most rapidly overcome by storage at chilling temperatures. On the other hand, the tubers of most varieties of potato emerge from dormancy more rapidly when stored at 22° than at 10° C. (Burton, 1963).

Theories concerning the mechanism of bud dormancy

Theories concerning the biochemical and physiological basis of bud dormancy have been manifold, but at the present time there are three main hypotheses which require consideration:

(1) Several authors, of whom Vegis (1956, 1961, 1964b) has been the leading proponent, have stressed the importance of restricted gaseous exchange, especially of oxygen, arising from the presence of bud scales.

(2) A second group of theories is based upon the view that bud dormancy is under hormonal control and involves the interaction of both growth promoters and growth inhibitors.

(3) A recent approach to dormancy relates to the control of gene activity at the molecular level and postulates that dormancy control involves the repression and derepression of DNA (Bonner, 1965).

We shall consider briefly each of these three theories, which are not necessarily mutually incompatible.

Temperature and restricted oxygen uptake as factors in bud dormancy

Earlier workers, notably Davis (1930), Thornton (1945) and Crocker (1948), stressed the importance of restricted oxygen uptake, especially in association with high temperatures, as a factor in the induction dormancy in buds and seeds. Vegis has recently argued strongly in support of this theory (1956, 1961, 1964 *a, b*). Vegis points out that as the dormancy of buds develops (during the phase of pre-dormancy) the range of temperatures over which they are capable of growing becomes increasingly narrow, and that, conversely, as they emerge from dormancy after a period of chilling, they become capable of growing over an increasingly wide range of temperatures. In some species, the seeds or buds may lose the ability to germinate or grow at high temperatures, but may be capable of germinating at lower temperatures. In other species the freshly harvested seeds are not capable of germinating at low temperatures but will do so at higher temperatures. This type of response is shown also by tubers of potato, turions of *Utricularia* and buds of *Fagus* and *Betula* (under LD's).

A third type of response, shown apparently by the majority of species having dormant buds and seeds, involves a loss of ability to grow at both high and low temperatures. As dormancy is reduced, the seeds and buds become capable of growing over a gradually widening range of temperatures. Examples are provided by the buds of many tree species, the turions of *Stratiotes* and by many seeds. Vegis attempts to relate these various types of temperature response to the climatic conditions in which the species occur naturally.

Vegis has put forward a theory of the mechanism of dormancy in which temperature is held to play an important role. He points out that in many species the inability of the dormant seeds to germinate at certain temperatures is dependent on the presence of intact covering structures, such as testa, endosperm or pericarp, and that if the latter are damaged or removed, the embryos become capable of germinating. There is little evidence that the dormancy of buds is affected by the presence of enclosing scales, since in the fully dormant condition removal of the scales does not usually lead to growth, although in a few cases their removal during the phase of pre-dormancy may do so.

There is considerable evidence that the inhibitory effect of seed coats arises primarily from interference with oxygen uptake by the embryo (see Wareing, 1968; Roberts, p. 161 above). It appears also that the bud scales of *Acer* interfere with oxygen uptake (Pollock, 1953). Vegis notes that secondary dormancy may be induced in certain seeds by keeping them at high temperatures under conditions of oxygen deficiency. He maintains that such secondary dormancy is directly analogous to primary dormancy and postulates that 'under natural conditions the cause of the formation of the resting condition is a temperature too high for growth of the young, recently formed cells, which are surrounded by structures limiting oxygen diffusion'. He points out that the embryos of developing seeds are liable to experience oxygen deficiency because of the surrounding seed coats and maternal tissues, and postulates that under such conditions of partial anaerobiosis certain metabolic changes occur, leading to the formation of fatty acids and fats, which tend to accumulate in dormant tissues. He extends the theory also to buds, noting that high temperatures, in association with SD's, promote the formation of winter buds in *Hydrocharis*, and he suggests that the onset of dormancy in shoots of peach is brought about by high summer temperatures.

However, the application of this theory to bud dormancy raises a number of difficulties. Thus, although bud-scales may interfere with oxygen uptake by the meristematic tissue in a fully developed bud, until such scales are formed they do not constitute any barrier, and it is difficult to see, therefore, how interference with oxygen uptake can be important in the initial formation of a resting bud. On the other hand, we have seen that there is good evidence that dormancy induced by SD's depends upon some stimulus arising in the leaves. This latter observation is difficult to reconcile with the view that dormancy arises from conditions within the bud itself. To meet this difficulty Vegis (1964b) suggests that SD's only cause the cessation of growth and the formation of terminal buds and not the onset of dormancy, which he suggests is brought about by high temperatures and restricted oxygen supply after the buds have formed. However, we have seen (p. 245) that the state of dormancy of axillary buds, formed initially under LD's, is increased when the leaves are subsequently exposed to SD's.

Hormonal control of dormancy

In view of the well-established importance of hormones in many aspects of growth and development, it is reasonable to examine their possible role in the control of bud dormancy, especially since the growth-promoting substances, gibberellic acid and kinetin, will overcome bud dormancy in many species (Vegis, 1965).

It is clear that in true dormancy certain processes are blocked, so that growth cannot take place even though environmental conditions appear favourable. Assuming hormonal control, this block might be due to either (*a*) the absence of certain factors necessary for growth, or (*b*) the presence of growth inhibiting substances. The observation that leaves exposed to SD's have an inhibitory effect on the growth of buds is difficult to explain in terms of a lack of growth-promoting substances, unless one postulates that the leaves 'draw off' such substances from the buds, and it seems more likely that the inhibitory effect of leaves is due to the fact that they produce growth-inhibitory substances which are transported to other parts of the plant.

The suggestion that bud dormancy may be due to the presence of growth-inhibiting substances was first made by Hemberg (1949), and supporting evidence for this theory has continued to accumulate since then. Thus, Hemberg's (1949) original observation that there is a correlation between the state of dormancy and the levels of endogenous inhibitors in potato tubers and buds of *Fraxinus excelsior* have been confirmed by several authors for a range of species (Phillips & Wareing, 1958; von Guttenberg & Leike, 1958; Dörffling, 1963). In general, it is found that if samples of buds are taken at intervals throughout the winter, the levels of endogenous inhibitors are high during the period October–December, but thereafter the levels tend to decline and there is a parallel decline in the depth of dormancy of the buds. Moreover, it has been shown that when seedlings of various woody species are transferred from LD to SD conditions there is a rapid increase in the levels of endogenous inhibitors (Phillips & Wareing, 1959). Such increases in inhibitor levels may be detected after 2–5 SD cycles, long before there are any morphological changes. Similar increases in inhibitor levels may also be detected in the axillary buds of seedlings of *Betula pubescens* and *Ribes nigrum*, when transferred to SD's.

Thus, there is a good correlation between the endogenous levels of inhibitors in the leaves and buds of woody seedlings and their state of dormancy. The inhibitory effect of SD leaves can therefore possibly be explained in terms of increased production of inhibitor under these conditions.

These latter observations led to attempts to isolate the inhibitors present in the buds and leaves of woody seedlings. The inhibitors extracted from the leaves of certain woody species and of potato tubers closely resemble the 'β-inhibitor' present in a wide range of plant tissues (Hemberg, 1965), in their behaviour in paper chromatography. Varga (1957) found the β-inhibitor fraction to contain various phenolic substances, and concluded that the activity of the β-inhibitor was due to these substances. However, an intensive study of the β-inhibitor fraction in *Acer pseudoplatanus* showed that in certain chromatographic solvents it is possible to separate the phenolic material from the inhibitory activity and that the latter appeared to be due to the presence of a single highly active factor which was not an aromatic compound (Robinson & Wareing, 1964). Further studies by Cornforth, Milborrow, Ryback & Wareing (1965) led to the isolation of a crystalline fraction which was shown to be identical with the substance 'Abscisin II', isolated by Ohkuma, Addicott, Smith & Thiessen (1965) from young cotton fruits and now known as abscisic acid (ABA). Cornforth, Milborrow & Ryback (1966) achieved the successful synthesis of ABA, but the synthetic material is a racemic mixture of the (+) and (−) enantiomorphs (referred to as (RS)-ABA), whereas the naturally occurring ABA apparently consists only of the (+) form. Subsequently (+)-ABA has been shown to be present in a considerable number of other species, including *Betula pubescens*, *Salix viminalis* and *Fraxinus excelsior* (Milborrow, 1967).

ABA is a highly active growth inhibitor in a variety of growth tests and it antagonizes the effects of auxins, gibberellins and cytokinins (p. 256). The availability of synthetic (RS)-ABA has made it possible to study its effects in various physiological processes, including bud dormancy. Thus, it has been possible to study the effect of ABA on seedlings of woody plants growing under LD. In the first tests (Eagles & Wareing, 1963) with partially purified natural inhibitor it was possible to induce the formation of typical resting buds in seedlings of *Betula pubescens* and this result has been confirmed with synthetic ABA (El-Antably, Wareing & Hillman, 1967). However, as mentioned above (p. 242), the buds of *Betula* are stipular and are formed primarily by the arrest of internode elongation in the apical region, and application of a growth inhibitor such as ABA could therefore lead to the formation of a bud by inhibiting the activity of the apical meristem and internode extension. It became of interest therefore to study the effects of ABA on species which form foliar buds, involving the modification of leaf primordia to form bud scales. Application of (RS)-ABA to seedlings of *Acer pseudoplatanus* and *Ribes nigrum* leads to the cessation of growth and the formation of typical bud scales (El-Antably *et al.* 1967). Moreover, application of ABA to seedlings of *Ailanthus glandulosa*, a species which has a 'sympodial' growth habit, led to the arrest of growth, although the apical region was abscinded only at a high concentration of ABA.

The effects produced by exogenous ABA depend upon the method of application. It is found that daily spraying of ABA solutions on the leaves and apices of woody seedlings has little effect on growth, whereas if ABA solution is applied by keeping a leaf permanently immersed in the solution, or by feeding through a lip cut in the stem, the inhibitory effects are readily demonstrated.

Although it has been possible to induce the morphological changes involved in bud formation by application of exogenous ABA, it should be mentioned that the buds so formed resume growth when the treatment is discontinued. Thus, the successful induction of a state of true dormancy remains to be demonstrated. There are two possible explanations for this failure to induce a state of dormancy. First, even with induction of bud dormancy by SD, buds of birch resume growth when the seedlings are transferred back to LD. Secondly, in the case of *Acer pseudoplatanus* and *Ribes nigrum*, it requires more than 4 weeks of continuous SD treatment to bring about the formation of fully developed resting buds, and until this occurs, even in these species growth may be resumed on transfer back to LD if they are only partially induced. Furthermore, SD induction results in decreased levels of gibberellins, as well as increased inhibitor levels, whereas when exogenous ABA is applied to one or two leaves only it is possible that high levels of gibberellins continue to be produced in the remaining leaves or other parts of the plant, and these gibberellins may tend to counteract the effect of the exogenous ABA.

It was an essential part of the inhibitor hypothesis that inhibitor is formed in the leaves under SD and transported from there to the shoot apices. Strong evidence in support of this hypothesis has been provided by the experiments of Hoad (1967) using aphids feeding on willow (*Salix viminalis*) plants maintained under LD or SD. As is well known, the aphids 'tap' the sieve tubes and 'honeydew' excreted by them gives an indication of the sieve-tube contents, apart from possible modifications arising in the gut of the animal. It was found that the honeydew from aphids feeding on SD plants contained significantly higher levels of inhibitor than that of aphids on LD plants; moreover, this inhibitor was shown to be ABA. Thus, it appears from these results that ABA formed in the leaves in high amounts under SD is transported through the phloem to other parts of the plant.

Role of growth-promoting hormones in bud-break

The possibility that natural gibberellins may be involved in bud dormancy is suggested by the observation that application of exogenous gibberellic acid (GA_3) or kinetin will induce bud-break in a number of woody species (Vegis, 1965). Earlier workers studying changes in endogenous inhibitors during the rest period of buds observed that there is an increase in auxin levels at about the time of bud-break, but did not include studies on changes in gibberellins. More recent studies have shown, however, that there is a significant increase in gibberellin levels during the latter part of the rest period, while the inhibitor levels are declining and the buds are emerging from dormancy. When observations are made on changes in buds exposed to natural fluctuating temperatures it is not possible to determine whether the increased gibberellin levels occur during the period of exposure to chilling, or whether they are found during intervening periods when the temperatures are higher. However, observations carried out on shoots of blackcurrant maintained at a constant chilling temperature of $2°$ C. in a cold room has shown that there is an increase in gibberellin levels during the actual period of chilling (Fig. 1). If it had been found that the gibberellin levels only increased after the shoots had been transferred from the chilling treatment to warmer temperatures, it must have been concluded that the increase in gibberellin levels was a secondary *result* of the removal of dormancy, rather than its cause. The observed results, however, do not exclude the possibility that an increase in gibberellin levels may be an essential part of the removal of dormancy by chilling. Thus, it seems possible that the control of dormancy involves an interaction between ABA and gibberellins, and that induction of dormancy occurs with high ABA and low gibberellin levels, while the reverse is true for emergence from dormancy.

The interaction between ABA and gibberellins in bud dormancy has been studied in various ways. Thus, if birch seedlings which have been induced to form buds in response to application of ABA are treated with gibberellic acid they rapidly resume growth. If GA_3 and ABA are applied simultaneously to actively growing seedlings under LD they continue in active growth and the effect of the ABA is completely nullified. On the other hand, if shoots of birch are collected in the spring, after they have received their chilling treatment but have not yet commenced growth, the breaking of the buds under warm conditions is inhibited by standing the shoots in solutions of ABA, but this effect can be largely overcome by applying GA_3 together with ABA (Eagles & Wareing, 1963). Although budbreak is very effectively inhibited when shoots are allowed to stand in a solution of ABA, the latter is not effective if it is sprayed externally on to the buds, presumably because of the difficulty of penetration into and through the bud scales.

Effects of abscisic acid on gibberellin levels

Although there is much circumstantial evidence to support the view that endogenous inhibitors, such as ABA, play an important role in the regulation of dormancy, it will not be possible to state unequivocally whether this is so until we know the mode of action of the inhibitors at the molecular level. It was suggested above that the control of dormancy involves an interaction between ABA and gibberellins. This hypothesis is supported by the observed antagonism between ABA and gibberellic acid (GA_3) in the responses of birch and other woody species, both in the induction of dormancy in actively growing seedlings and in the emergence from dormancy and growth of seedlings or non-dormant shoots of older trees (see above). It becomes of importance therefore to study the nature of the interaction between ABA and growth-promoting hormones, especially gibberellins. A study of the interaction between ABA and the growth-promoters IAA, GA$_3$ and kinetin in various growth tests has given little evidence of direct competitive interaction between ABA and any of the other hormones. However, with sections of shoots from germinating maize seedlings it is possible to overcome the effects of ABA by GA$_3$ and this effect will be discussed in more detail below. In the germination of lettuce seed, on the other hand, the inhibitory effects of ABA can be reversed by kinetin but not by GA$_3$.

Studies on the levels of endogenous inhibitors and gibberellins in buds and leaves of woody plants and of potato plants and tubers, both in relation to the induction of dormancy and to emergence from dormancy, have

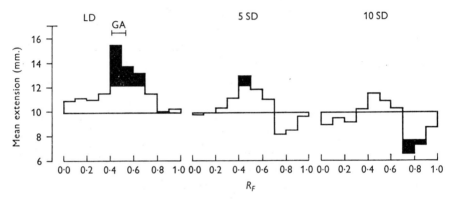

Fig. 2. Effects of day-length on endogenous gibberellin and inhibitor levels in shoot apical region of blackcurrant. Left: plants maintained throughout under 18 hr. days (LD). Centre and right: plants exposed to 5 and 10 short-day (SD) (9 hr.) cycles respectively. Extract from 10 g. dry weight of apical tissue (including young leaves) was used for each test (from El-Antably, 1965).

frequently revealed a striking reciprocal variation in the levels of these two types of growth substance (Libbert, 1964; Digby & Wareing, 1966). Thus, whereas there are high levels of gibberellins and relatively low levels of inhibitors in buds and shoot apices under LD's when the plants are transferred to SD's, there is a rapid decrease in gibberellins and an increase in inhibitors (Fig. 2). It is possible that these changes in the levels of gibberellins and inhibitors are independently controlled by day-length, but it is also possible that the two effects are interconnected, so that high levels of ABA result in reduced gibberellin levels, by inhibiting the biosynthesis of gibberellins or by bringing about their inactivation. In order to test this hypothesis, ABA was applied to the leaves of birch seedlings, and after 3 days the shoot apices were removed and placed on agar, using the method of Phillips & Jones (1964). Extraction and bioassay of the gibberellins which had diffused into the agar after 24 hr. showed that very much less gibberellin had diffused from the apices of the plants treated with ABA than of the control plants (Thomas, Wareing and Robinson, 1965).

Further studies have been carried out on the effects of ABA on endogenous gibberellin levels in etiolated shoots of maize and in green plants of spinach (*Spinacia oleracea*) (Wareing, Good & Manuel, 1968). Seedlings of maize (c.v. White Horse Tooth) were raised in vermiculite in dark at 25° C. for 6 days. The shoots were then severed 5 mm. below the first node and placed in tubes containing either (*a*) distilled water or (*b*) ABA solution

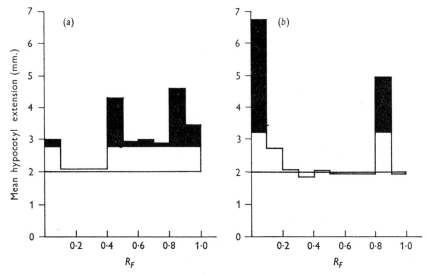

Fig. 3. Effect of abscisic acid (ABA) on gibberellin levels in maize shoots (var. White Horse Tooth). Maize shoots were placed in either (*a*) 5 p.p.m. ABA solution or (*b*) distilled water for 24 hr. prior to extraction. Thin-layer chromatograms were loaded into the eluate from R_F zone 0·1–0·6 of paper chromatograms of extracts. The TLC absorbent was silica gel, and the plates were developed in the solvent system di-isopropyl ether:acetic acid (95:5, v/v) (Good, 1967).

at a concentration of 5 p.p.m. The shoots were allowed to stand in the tubes for 24 hr. at 25° C. in the dark, after which gibberellins were extracted and purified by partition into ethyl acetate. After thin-layer chromatography the different fractions were assayed for gibberellin activity, using the lettuce hypocotyl test. It was found that pretreatment with ABA had markedly reduced the levels of endogenous gibberellins in the extracts (Fig. 3). It was found that the main peak of activity in the control shoots corresponded with GA_3 and this peak was greatly reduced in the extracts of shoots which had been pretreated with ABA. This peak was found to run at the same R_F as GA_3 in several thin-layer chromatographic systems. A second peak, corresponding to the marker spots of GA_4 and GA_7, appeared in both extracts and was *not* reduced by pretreatment with ABA.

Similar results have been obtained using plants of spinach (*Spinacia oleracea*.) It has been shown by Radley (1963) that if plants previously grown under SD conditions are exposed to a single LD cycle there is a very marked increase in the levels of gibberellins. When spinach plants were sprayed with ABA at 20 p.p.m. during the LD cycle, the increase in gibberellins was almost completely inhibited. In this experiment also the main peak of gibberellin activity was found to correspond to GA_3. We have also found that application of ABA to plants of blackcurrant (*Ribes nigrum*) and potato (*Solanum andigena*) leads to reduced levels of endogenous gibberellins. It would seem therefore that the effect is a rather general one.

The observed changes in gibberellin levels might be due to an inhibitory effect of ABA on an early stage in the pathway of gibberellin biosynthesis or to an interconversion between several closely related gibberellins, which is known to occur readily. If ABA blocks some step in gibberellin biosynthesis, then the observation that the effects of ABA can be overcome in maize by exogenous GA_3 is readily understood.

Effects of abscisic acid on RNA metabolism

The observation that ABA inhibits the gibberellin-stimulated synthesis of amylase in barley endosperm (Thomas, Wareing & Robinson, 1965) suggests that it may act by inhibiting some step in protein synthesis. Chrispeels & Varner (1966) were unable to detect any effect of ABA on total protein synthesis in barley endosperm, but the observation that ABA accelerates leaf senescence (Addicott, Ohkuma, Smith & Thiessen, 1966; El-Antably *et al.* 1967) suggests that it may have a more marked effect on total protein synthesis in this material, since it is known that protein synthesis declines during leaf senescence. It has, in fact, been shown that treatment with ABA causes a rapid decline in the over-all protein and RNA levels of leaves of *Lemna* (van Overbeek & Loeffler, 1967), and leaf disks of *Nasturtium* (Beevers, 1968), *Xanthium* (Osborne, 1967) and radish (El-Antably *et al.* 1967). Moreover, ABA markedly reduces the incorporation of [^{14}C]leucine and [^{14}C]cytidine into protein and RNA of radish leaves, following either 4 hr. or 20 hr. pretreatment with ABA at 10 p.p.m. (Table 1). Thus, it becomes of interest to determine at what step in protein synthesis ABA is acting, whether at the 'translation' stage or at some other stage in RNA synthesis. Studies have therefore been carried out with disks of radish leaves to determine whether ABA appears to inhibit the synthesis of specific fractions of RNA. Sucrose-density analysis of the RNA of radish-leaf disks showed that pretreatment with ABA for only 4 hr. causes a marked reduction in both the light and heavy ribosomal fractions and in the soluble RNA, with indications that the most marked effects are upon

Table 1. *Effect of abscisic acid on incorporation of [^{14}C]cytidine and [^3H]leucine into radish-leaf disks*

Labelled precursors were fed to the leaf disks for 4 hr., either (1) after 20 hr. pretreatment with ABA (10 p.p.m.) or (2) simultaneously with ABA (from Beevers, Pearson and Wareing, unpubl.)

	Water	ABA
[^{14}C]cytidine		
4 hr.		
Uptake	232,945 d.p.m.	209,572 d.p.m.
Incorporation	108,558 d.p.m.	74,772 d.p.m.
Incorporation (%)	46·6%	35·6%
20 hr.		
Uptake	300,399 d.p.m.	154,942 d.p.m.
Incorporation	91,799 d.p.m.	39,242 d.p.m.
Incorporation (%)	30·5%	25·3%
[^3H]leucine		
4 hr.		
Uptake	1,224,363 d.p.m.	1,195,395 d.p.m.
Incorporation	888,563 d.p.m.	743,295 d.p.m.
Incorporation (%)	72·0%	62·1%
20 hr.		
Uptake	1,142,876 d.p.m.	591,357 d.p.m.
Incorporation	660,666 d.p.m.	233,974 d.p.m.
Incorporation (%)	57·8%	39·5%

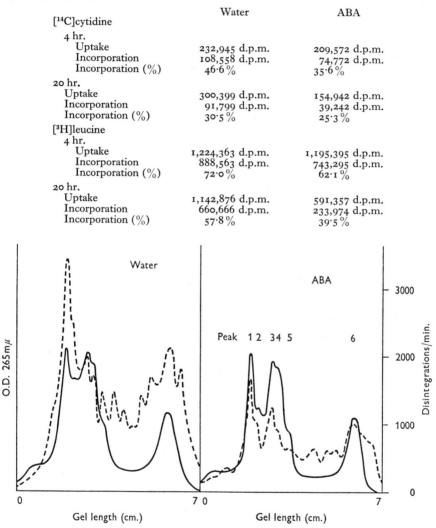

Fig. 4. Effect of abscisic acid (ABA) on RNA fractions of radish leaf. Sterile leaf-disks were treated with ABA for 4 hr. at 10 p.p.m., prior to incubating with [^3H]cytidine for 3 hr. in the dark. The optical density profile was measured on a Chromoscan densitometer with a 2 × 0 O.D. expander and a 3 : 1 gel-length expansion. Peaks numbered as follows: 1, heavy cytoplasmic ribosomal; 2, heavy chloroplastic ribosomal; 3, light cytoplasmic ribosomal; 4, light chloroplastic ribosomal; 5, breakdown product; 6, soluble components. (Continuous line: optical density, 260 mμ; broken line: radioactivity.)

the former fractions. The decreased capacity for RNA synthesis is even more apparent after 16–24 hr. In further experiments better separation of the RNA components was achieved by disk gel electrophoresis (Fig. 4). Here again it was found that there was reduced incorporation of [^{14}C]-cytidine into all fractions of RNA, but with some indications that incorporation into ribosomal RNA was more inhibited than into soluble RNA. Extraction procedures for the isolation of polyribosomes, followed by density-gradient fractionation, showed that both monoribosome and polyribosome contents were reduced by pretreatment with ABA, but the percentage of polyribosomes present was not affected by ABA as compared with the water controls, and both poly- and monoribosome fractions appeared to be equally reduced by ABA. The continued presence of polyribosomes in the ABA-treated disks indicates that messenger-RNA was still present and no evidence of a differential effect upon the synthesis of messenger-RNA was obtained.

The foregoing results clearly indicate a rapid and marked effect of ABA on RNA synthesis, even at relatively low concentrations. Although much further work will be required to determine the precise mode of action of ABA on RNA synthesis, the results so far obtained are consistent with the hypothesis that ABA may regulate dormancy through its effect upon RNA and protein synthesis. The possibility that dormancy may be controlled by the regulation of nucleic acid synthesis has also been suggested by Bonner (1965). Tuan & Bonner (1964) studied RNA synthesis in dormant and non-dormant buds of potato. In dormant buds there is only very limited RNA and DNA synthesis but the rates of synthesis are markedly increased by treatments, such as exposure to ethylene chlorhydrin, which break dormancy. RNA synthesis by non-dormant buds is inhibited by actinomycin D and hence is DNA-dependent. Preparations of 'chromatin' (i.e. DNA) of dormant potato buds are incapable of stimulating RNA synthesis *in vitro*, whereas chromatin from non-dormant buds is highly effective in supporting RNA synthesis by added RNA polymerase. They concluded that in dormant potato tissue the DNA is blocked in some way (in a repressed state) and hence cannot support RNA synthesis.

Evidence has recently been obtained that addition of ABA to radish hypocotyl tissue during extraction of chromatin reduces the RNA-polymerase activity of the latter (Pearson & Wareing, 1969).

REFERENCES

ADDICOTT, F. T., OHKUMA, K., SMITH, O. E. & THIESSEN, W. E. (1966). *Adv. Chem. Ser.* **53**, 47.
BEEVERS, L. (1968). *Proc. VIth Int. Conf. on Plant Growth Substances.* Ottawa: The Runge Press.
BONNER, J. (1965). *The Molecular Biology of Development.* Oxford: Clarendon Press.
BURTON, W. G. (1963). In *The Growth of the Potato,* p. 17. Ed. J. D. Ivins and F. L. Milthorpe. London: Butterworth.
BÜSGEN, M. & MÜNCH, E. (1929). *The Structure and Life of Forest Trees.* London: Chapman and Hall.
CHRISPEELS, M. J. & VARNER, J. E. (1966). *Nature, Lond.* **212**, 1066.
CORNFORTH, J. W., MILBORROW, B. V., RYBACK, G. & WAREING, P. F. (1965). *Nature, Lond.* **205**, 1269.
CORNFORTH, J. W., MILBORROW, B. V. & RYBACK, G. (1966). *Nature, Lond.* **206**, 715.
CROCKER, W. (1948). *Growth of Plants.* New York: Reinhold.
DAVIS, W. E. (1930). *Am. J. Bot.* **17**, 58, 77.
DIGBY, J. & WAREING, P. F. (1966). *Ann. Bot.* **30**, 607.
DOORENBOS, J. (1955). *Euphytica* **4**, 141.
DÖRFFLING, K. (1963). *Planta* **60**, 390.
DOWNS, R. J. & BORTHWICK, H. A. (1956). *Bot. Gaz.* **117**, 310.
EAGLES, C. F. & WAREING, P. F. (1963). *Nature, Lond.* **199**, 874.
EL-ANTABLY, H. M. M. (1965). Ph. D. thesis, Univ. of Wales.
EL-ANTABLY, H. M. M., WAREING, P. F. & HILLMAN, J. (1967). *Planta* **73**, 74.
GARNER, W. W. & ALLARD, H. A. (1923). *J. agric. Res.* **53**, 871.
GOOD, J. E. G., (1967). Ph. D. thesis, Univ. of Wales.
GUTTENBERG, H. VON & LEIKE, H. (1958). *Planta* **52**, 96–120.
HEMBERG, T. (1949). *Physiologia Pl.* **2**, 24, 37.
HEMBERG, T. (1965). *Handb. PflPhysiol.* **15**(2), 669.
HESLOP-HARRISON, Y. (1962). *Proc. R. Ir. Acad.* **62B**, 23.
HOAD, G. V. (1967). *Life Sci.* **6**, 1113.
KLEBS, G. (1914). *Abh. heidelb. Akad. Wiss.* (Math-Nat Kl.), 3.
KRAMER, P .J. (1936). *Plant Physiol.* **11**, 127.
LIBBERT, E. (1964). *Proc. Vth Int. Conf. on Plant Growth Substances,* C.N.R.S. Paris.
MANN, L. K. & LEWIS, D. A. (1956). *Hilgardia* **26**, 161.
MILBORROW, B. V. (1967). *Planta* **76**, 93.
MOSHKOV, B. S. (1935). *Planta* **23**, 774.
NASR, T. A. A. & WAREING, P. F. (1961). *Jour. hort. Sci.* **36**, 1.
NITSCH, J. P. (1957). *Proc. Am. Soc. hort. Sci.* **70**, 526.
OHKUMA, K., ADDICOTT, F. T., SMITH, O. E. & THIESSEN, W. E. (1965). *Tetrahedron Lett.* **29**, 2529.
OSBORNE, D. J. (1967). *Symp. Soc. exp. Biol.* **21**, 305.
OVERBEEK, J. VAN & LOEFFLER, J. E. (1967). *Science, N.Y.* **156**, 1497.
PEARSON, J. A. & WAREING, P. F. (1969). *Nature, Lond.* **221**, 672.
PHILLIPS, I. D. J. & JONES, R. L. (1964). *Nature, Lond.* **204**, 497.
PHILLIPS, I. D. J. & WAREING, P. F. (1958). *J. exp. Bot.* **9**, 350.
PHILLIPS, I. D. J. & WAREING, P. F. (1959). *J. exp. Bot.* **10**, 504.
POLLOCK, B. M. (1953). *Physiologia Pl.* **6**, 47.
RADLEY, M. (1963). *Ann. Bot.* **27**, 373.
RAUNKIAER, C. (1934). *The Life Forms of Plants.* Oxford: Clarendon Press.
ROBINSON, P. M. & WAREING, P. F. (1964). *Physiologia Pl.* **17**, 315.

SCHWABE, W. W. & NACHMONY-BASCOMBE, S. (1963). *J. exp. Bot.* **14**, 353.
THOMAS, T. H., WAREING, P. F. & ROBINSON, P. M. (1965). *Nature, Lond.* **205**, 1270.
THORNTON, N. C. (1945). *Contr. Boyce Thompson Inst. Pl. Res.* **13**, 487.
TUAN, D. Y. H. & BONNER, J. (1964). *Pl. Physiol., Lancaster* **39**, 768.
VAARTAJA, O. (1956). *Can. J. Bot.*, **34** 377.
VAN HUYSTEE, R. B., Weiser, C. J. and Li, P. H. (1968). *Bot. Gaz.* **128**, 200.
VARGA, M. (1957). *Acta biol., Szeged* **3**, 212.
VEGIS, A. (1956). *Experientia* **12**, 94–99.
VEGIS, A. (1961). *Handb. PflPhysiol.* **16**, 168.
VEGIS, A. (1964a). In *Environmental Control of Plant Growth*, ed. L. T. Evans. New York: Academic Press.
VEGIS, A. (1964b). *A. Rev. Pl. Physiol.* **15**, 185.
VEGIS, A. (1965). *Handb. PflPhysiol.* **15** (2), 499.
WAREING, P. F. (1948). *Forestry* **22**, 211.
WAREING, P. F. (1950). *Physiologia Pl.* **3**, 258.
WAREING, P. F. (1951). *Physiologia Pl.* **4**, 41.
WAREING, P. F. (1953). *Physiologia Pl.* **6**, 692.
WAREING, P. F. (1954). *Physiologia Pl.* **7**, 261.
WAREING, P. F. (1956). *A. Rev. Pl. Physiol.* **7**, 191.
WAREING, P. F. (1958). *J. Inst. Wood Sci.* **1**, 1.
WAREING, P. F. (1968). In *The Physiology of Plant Growth Development Responses*. London: McGraw-Hill.
WAREING, P. F. & BLACK, M. (1958). In *The Physiology of Forest Trees*, p. 643. Ed. K. V. Thimann. New York: Ronald Press.
WAREING, P. F., GOOD, J. & MANUEL, J. (1968). *Proc. VIth Int. Conf. on Plant Growth Substances.* Ottawa: The Runge Press.
WAXMAN, S. (1957). Ph.D. thesis, Cornell University. (Cited by Nitsch (1963). In *Environmental Control of Plant Growth*. New York: Academic Press.
ZAHNER, R. (1955). *Forest Sci.* **1**, 193.

DIAPAUSE AND SEASONAL SYNCHRONIZATION IN THE ADULT COLORADO BEETLE (*LEPTINOTARSA DECEMLINEATA* SAY)

By J. DE WILDE

Laboratory of Entomology, Agricultural University, Wageningen, Netherlands

INTRODUCTION

One of the outstanding problems of present-day ecophysiology is the processing of ecological information provided in the form of seasonal tokens. This 'language of nature', not unlike human language, has its structural and semantic aspects. The structure of seasonal tokens is in many cases rather simple, and may take the form of a critical photoperiod or temperature, or even a chemical compound. Complications may arise when a seasonal-adaptation phenomenon is governed by several environmental factors which may interact to form 'complex tokens' (Adkisson, Bell & Wellso, 1963).

The semantic significance of seasonal tokens depends on the way in which physiological processes are subordinated to the action of a critical environmental factor. A 10 hr. photoperiod and moderate temperatures are 'translated' in such a way as to induce diapause in a long-day (LD) insect such as the Colorado beetle. In a short-day (SD) insect, e.g. the bivoltine race of the commercial silkworm, the same conditions evoke the opposite response and allow for continued growth and reproduction. Similarly, a period at 5° C. breaks diapause in *Hyalophora cecropia* and prolongs it in *Leptinotarsa*.

Seasonal tokens are not necessarily sensory stimuli. In some cases the extrasensory nature of their action has been proven without doubt. Photoperiodic induction takes place by a direct effect of light on the brain (Benoit, 1937; Lees, 1964; Williams & Adkisson, 1964), and the same is true for low-temperature breaking diapause (Williams, 1956). In other cases, where nutrients are involved, the sensory pathway is unlikely (Adkisson, 1961; de Wilde & Ferket, 1967). It may well be that responses to seasonal tokens are phylogenetically 'old' responses which occurred already in the primitive Metazoa before the differentiation of specialized sense organs had taken place.

In many insect species diapause is a means of synchronization of the active state with the favourable season. Seasonal conditions may favour development and reproduction to a degree varying from year to year, and even the many ecological requisites for growth and nutrition (e.g. temperature, host-plant condition and a favourable site for pupation) may vary in a rather independent manner. Maintenance of the species is generally assured by a large degree of variability in response (e.g. depth and length of diapause) and by a system of 'multiple insurance', in which one ecological factor may prevent or reverse the effect of another factor. High temperatures generally annihilate the effect of short photoperiods.

While limitations are generally set to cold resistance in insects, and populations of one species obtained from various climatic zones often show surprisingly little differences in lethal minimum temperature (Danylievsky & Kuznetzova, 1968), 'token' adaptations protecting the species against exposure to adverse seasonal conditions are often highly developed.

In this paper diapause in the adult Colorado beetle (*Leptinotarsa decemlineata* Say) will be discussed along the above lines of thought, recapitulating the results obtained by authors in different countries, as well as those of our experiments during nearly 20 years in Amsterdam and Wageningen. We will try to follow the causal chains initiated by environmental tokens on the level of organs, tissues, and subcellular processes.

TEMPERATURE REQUIREMENTS AND INCIDENCE OF DIAPAUSE IN *LEPTINOTARSA*

The threshold temperature for embryonic development is 12° C. in *Leptinotarsa*, and the threshold of larval development in the field is around 17° C. Oviposition rarely takes place below 17° C. Optimum temperature for embryonic and postembryonic development is between 25° and 30° C. (de Wilde, 1950). Despite these rather high temperature requirements, the beetle is of minor importance in subtropical regions and most of its damage has been reported from the temperate zone. The zone of dispersal is shown in Fig. 1, and generally speaking is situated between the mean year isotherms of 7° and 20° C.

On the North American continent the beetle occurs between 30° and 46° N.L. and on the European continent between 40° and 55° N.L. Maximum length of photoperiod during the favourable season varies within the range of dispersal between 14–17 hr. in North America and 16–19 hr. in Europe.

Notwithstanding the more favourable temperature conditions in the southern part of the area, voltinism does not show an important increase in these regions. In Table 1 the prevailing number of generations per year is given. According to different authors, a second generation, whenever it develops, is usually partial. As the hibernating beetles usually emerge over a broad range of time, early beetles from regions with prevailing bivoltinism may give rise to a partial third generation, overlapping the second one. The same is true with regard to a second generation in regions with prevailing univoltinism.

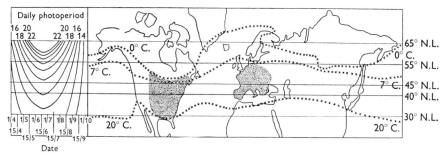

Fig. 1. Range of dispersal of the Colorado beetle (*Leptinotarsa decemlineata* Say) in relation to temperature and photoperiod. × × × = Mean year isotherm. (After Trouvelot, 1936, and own data.)

Table 1. *Voltinism of the Colorado beetle as reported from various geographical locations*

Location	Northern latitude	No. of generations	Author
Tuckson (Arizona, U.S.A.)	32°	1	Tower (1906)
Georgia (U.S.A.)	32–34°	2	Girault (1907)
Illinois (U.S.A.)	38–42°	2	Girault (1909)
Ohio (U.S.A.)	40°	2	Girault (1908)
Philadelphia (U.S.A.)	40°	1	Fink (1925)
Avignon (Fr.)	44°	2	Boczkowska (1944)
Ahun (Fr.)	46°	1	Grison (1938)
Paris (Fr.)	49°	1	Grison (1944)
Mukachevo (U.S.S.R.)	48°	2	Ushatinskaya (1966)
Netherlands	52°	1	de Wilde (1949)

As diapause in *Leptinotarsa* is started during summer, the beetles first aestivate and subsequently hibernate. Dormancy may show a varying degree of depth during both periods, depending on the environmental conditions. In Transcarpathia (U.S.S.R.), according to Minder & Petrova (1966), aestivation may comprise (1) summer sleep of short duration, without profound metabolic changes, and (2) summer diapause of longer duration, resembling winter diapause. In the same region, Ushatinskaya (1966) distinguishes between (1) normal winter diapause and (2) prolonged winter diapause, which may last up to 3 years.

In the Netherlands the hibernating beetles emerge from April to June and oviposit from May till August. The larva moults three times before entering the soil to pupate. Two weeks later, the adult of the first generation emerges, and is found active over a period extending from July till September.

Only during warm summers a small number of these beetles may start oviposition. The large majority enter the soil after about 2 weeks of intense food intake. Beetles may already be found underground in the beginning of August (de Wilde, 1954). Diapause is passed in the soil at a depth of 25–40 cm., depending on temperature and soil conditions. The minimum lethal temperature is $-7°$ C. This may be a limiting factor for dispersal in extremely continental climates, as prevailing in Canada and central U.S.S.R.

When taken indoors and kept at room temperature in moist sand, increasing numbers of beetles emerge over a period extending from 3 to 6 months after the beginning of diapause. It follows that 'diapause development' (*sensu* Andrewartha, 1952) does not require a period of chilling. In fact, it has been shown that chilling delays rather than promotes break of diapause (de Wilde, 1949).

When the beetles are not fed after post-diapause emergence, they repeatedly re-enter the soil and re-emerge over a period which may last for 2 months. A similar ambivalent behaviour may be shown by young adult beetles fed under illumination conditions suboptimal for reproduction: when placed upon soil, digging behaviour may be induced in the same beetles which, when placed upon potato foliage, would feed and gain weight (de Wilde, 1949, 1954).

ENVIRONMENTAL INDUCTION AND BREAK OF DIAPAUSE

Photoperiod is a primary factor controlling the state of activity in *Leptinotarsa*. The Colorado beetle is an LD insect with a critical photoperiod of 15 hr. (de Wilde, 1954, 1955; Jermy & Saringer, 1955; Goryshin, 1958).

Below this value, diapause is induced in 100% of the treated beetles. Photoperiods above 15 hr. never completely prevent diapause, but merely reduce the percentage. Natural day-length for the Colorado beetle largely surpasses the astronomical day-length but includes a considerable period of twilight. Threshold intensity of photoperiodic response is below 0·4 ergs/cm.^{2}/sec., but the system is saturated at 20 ergs/cm.^{2}/sec. (de Wilde & Bonga, 1958). Moonlight, which falls in the range of 0·4–2 ergs/cm.^{2}/sec., if active at all, will only slightly modify the effect of daylight. Reared in groups of 20–30 at 25° C., 70–80 % of the beetles will reproduce, but rearing in isolation on foliage of optimum quality will increase this percentage to 90 or more. This suggests the importance of nutrition with regard to diapause induction.

We have shown that both the quantity and the quality of the food are significant in this respect. Interrupting the feeding period increases the percentage of diapause (de Wilde, Duintjer & Mook, 1959), and feeding with senescent potato leaves induces diapause under conditions otherwise promoting reproduction (de Wilde & Ferket, 1967).

Temperatures up to 28° C. have little effect on photoperiodic induction of diapause, but a further increase above 30° C. reverses SD induction and tends to lower the critical photoperiod (Goryshin, 1958) (Fig. 2).

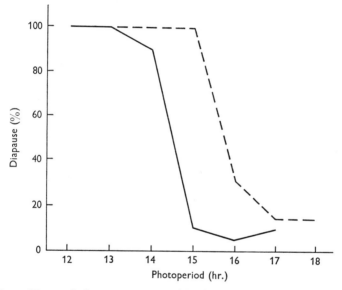

Fig. 2. Photoperiodic response curve of the Colorado beetle, influenced by temperature. (Solid line, 30° C. broken line, 26° C.) (after Goryshin, 1958).

This explains why in continental climates the beetles may produce a second generation under photoperiods which would otherwise be prohibitive to reproduction. Photoperiod, temperature and food are therefore three mutually correcting factors, governing the physiological state of diapause, which may be considered as a protective adaptation syndrome.

As mentioned above, break of diapause occurs gradually and a period of chilling is not required. Stored at 30° C., beetles will emerge from diapause within 2–3 weeks (de Wilde, 1954). At room temperature, emergence will start after 2–3 months. Chilling delays 'diapause development' and the synchronization of emergence with the favourable season is governed by soil temperature. As will be shown in the following pages, the moment of leaving the soil is a very critical one and a large part of 'diapause development' takes place during the first few hours after emergence. As competence for spontaneous locomotory activity is gradually acquired, the beetles move to superficial soil layers when the temperature exceeds 9° C. (Le Berre, 1965) and emerge when soil temperature exceeds 12–15° C. (Trouvelot, 1936; Grison, 1950; Le Berre, 1965). Competence for feeding activity is acquired 1–2 days after emergence.

Under field conditions, photoperiod cannot play any role in break of diapause, but reversal of diapause may be effected experimentally by LD treatment until at least 10 days after its initiation (Hodek & de Wilde, 1969). Sensitivity to photoperiod gradually disappears, and after post-diapause emergence the beetles are indifferent to photoperiodic treatment and oviposit under LD's as well as under SD's. This is ecologically important, as emergence may take place when SD conditions are prevailing (de Wilde *et al.* 1959).

RECEPTION AND STORAGE OF ENVIRONMENTAL INFORMATION INDUCING DIAPAUSE

Photoperiodic induction of diapause is not affected by removal of the eye function, either by diathermic cauterization or by covering the compound eyes with black lacquer (de Wilde *et al.* 1959). Light may easily penetrate the transparent head capsule and reach the brain. This suggests that, as in *Megoura viciae* (Lees, 1964) and *Anthereya pernyi* (Williams & Adkisson, 1964), the brain is the photoperiodic receptor organ.

There is some evidence that the termination of the reproductive phase on feeding senescent foliage is due to a deficiency of lecithin (Grison, 1953), but this needs confirmation as far as diapause induction is concerned.

Although prolonged photoperiodic treatment of the adult is decisive for the final state of activity, larval treatment is not without influence. LD treatment of the larvae delays the subsequent induction of diapause in the adult by SD's and results in initial oviposition. In an outdoor strain used in the beginning of our experiments this delayed effect of larval induction lasted only a few days (de Wilde *et al.* 1959), but at present in our laboratory strain, where a considerable selection for non-diapause development has taken place, its effect is apparent for 3 weeks and more.

The way in which larval LD induction is stored and transmitted through the pupal stage will be discussed later.

THE DIAPAUSE SYNDROME

The behavioural component

Diapause induced by SD (and also diapause induced by allatectomy) is preceded by a period characterized by intensive feeding and absence of oviposition. Copulation takes place during this pre-diapause period, and most of the females enter diapause with receptacula seminis filled with sperm.

At 25° C. pre-diapause lasts 10–11 days. Beetles may show ambivalent behaviour during the last days of pre-diapause and alternatively feed and burrow. In the presence of slightly moist sand (completely dry sand never elicits burrowing behaviour) the beetles finally disappear underground. By this time, geotaxis has attained a positive sign and phototaxis has lost its strong positivity characteristic for newly energed beetles (de Wilde, 1954). Burrowing behaviour comes to an end at a depth of 25–40 cm., depending on the type of soil. The above behaviour characteristics occur synchronously in males and females. The beetles now assume a resting attitude. When during diapause they are taken from the soil and placed under illumination at room temperature, they will show locomotory behaviour and soon hide under objects such as leaves and filter paper. In insectaria they will come to rest in dark corners, and when placed upon sand they will soon burrow again. It is not known which sensory criteria determine their upward movement after break in diapause.

The structural component

As has been found by de Kort (1969) in our laboratory, newly moulted adult beetles under SD treatment have immature flight muscles. The muscle fibrils are very small in diameter ($\pm 0.3\,\mu$). The sarcosomes have a protomitochondrial appearance, bearing a double membrane but almost no cristae.

After feeding has started, the flight muscles develop incompletely and attain a *status quo* at about 6 days, lasting until the beginning of burrowing behaviour. Fibril diameters of 1.6–$1.7\,\mu$ can be measured. Atrophy of the flight muscles is now started, and attains its maximum within 2–3 weeks. At this stage, lipid droplets have accumulated in the thorax, the sarcosomes have virtually disappeared and the muscle fibrillae are strongly reduced in diameter (Plate 1).

In the female reared *ab ovo* with SD the ovaries remain in a state of quiescence. Their germaria are reduced and their vitellaria small and inactive (Fig. 3). If, previous to the SD treatment, the beetles had been reproducing, oosorption is observed and the bases of the ovarioles contain orange-coloured, crystalline conglomerates.

In diapausing males, both the testes and the accessory glands are reduced in size as compared with the active state.

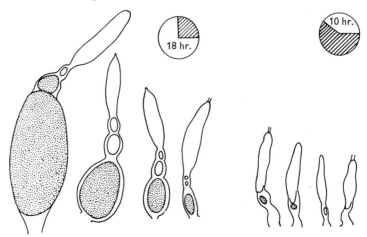

Fig. 3. Ovarioles of LD-treated beetles (left) and SD specimens (right).

Metabolic aspects of the diapause syndrome

Determinations of body composition in terms of protein, carbohydrate, lipid and water have been published by several authors (Busnel, 1939; de Wilde, 1954; de Kort, 1969). In Table 2 the data by de Kort are given,

Table 2. *Body composition of reproducing and diapausing beetles, according to de Kort (1969)*

	Active About 14 days old		Diapause			
			1–2 months		More than 3 months	
	♀	♂	♀	♂	♀	♂
Fresh body weight (mg.)	183 (43)†	129 (43)	142 (38)	117 (38)	142 (18)	122 (14)
Dry-weight (%)	30·5 (6)	35·8 (6)	44·5 (4)	44·4 (6)	37·2 (4)	36·0 (6)
Glycogen*	2·64 (18)	5·06 (21)	8·74 (7)	3·35 (8)	3·54 (3)	1·96 (5)
Soluble carbohydrate*	1·68 (14)	1·19 (17)	0·25 (8)	0·43 (8)	0·62 (3)	0·59 (5)
Triglycerides*	12·72 (14)	52·30 (16)	101·3 (7)	73·9 (8)	53·1 (3)	31·4 (4)
Total protein*	100·3 (15)	103·8 (6)	94·0 (15)	98·2 (9)	—	—
Alcohol-soluble protein*	5·28 (12)	4·97 (15)	7·25 (8)	7·15 (3)	6·21 (3)	6·16 (3)

* Expressed in mg./g. wet weight.
† The numbers in brackets indicate the number of beetles tested in each case.

obtained from beetles grown under standard conditions. With the exception of a lowered water-content and an increased content of lipids characteristic of diapausing beetles, not much information is provided by these data. It appears that at the end of diapause, lipid content is reduced and water content increased. In fact, it has been shown by Le Berre (1965) that, after leaving the soil, post-diapause beetles may loose weight extremely rapidly. This is in contrast to the course of body weight during diapause, which is remarkably constant (Fig. 4).

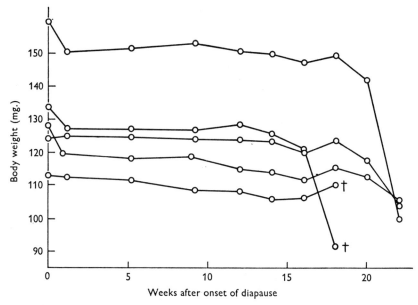

Fig. 4. Course of body weight during diapause. The beetles, stored at 25° C., show remarkably constant body weight during the diapause period (after de Wilde, 1954).

In the older literature much attention was given to water excretion as a means of gaining resistance to cold. We have shown that in *Leptinotarsa* the rate of water excretion is not increased during pre-diapause (de Wilde, 1954). Cold resistance is not strongly developed in the Colorado beetle, as during diapause the lethal minimum temperature is −7° C.

One of the most striking features of the diapause syndrome is the low rate of respiratory activity (Fig. 5). The rate of oxygen consumption is lowered to a level of 15–20 % of the rate observed in active beetles (de Wilde & Stegwee, 1958; Marzusch, 1952; Precht, 1953). This is not attained by the over-all defective state of the cytochrome system as in the case of pupal diapause in *Hyalophora cecropia* (Schneidermann & Williams, 1954), since the diapausing beetles may be set into locomotion at any moment, and are easily killed by exposure for a few minutes in hydrocyanide gas.

According to Le Berre (1965), respiratory activity is greatly increased near the end of diapause, when the beetles are present in the superficial soil layers. A significant further increase only takes place after the beetles have definitely left the soil.

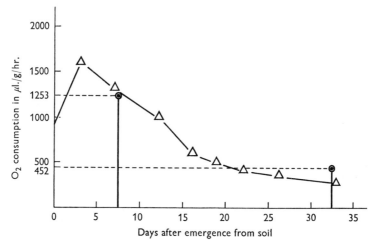

Fig. 5. Respiratory activity after allatectomy (curve) as compared with SD treatment (data after de Wilde & Stegwee, 1958). △—△, Allatectomized; ⊙—⊙, SD treatment.

CLOSER EXAMINATION OF TWO METABOLIC FEATURES OF DIAPAUSE

Energy metabolism

Contrary to the situation in the heteropterous bug *Pyrrhocoris* (Slama, 1964), the inactivity of the ovaries is not the major factor lowering the respiratory rate in *Leptinotarsa*, as has been shown in castration experiments by El Ibrashy (1965). The main causal factor is the degeneration of the flight muscles. As has been pointed out by Stegwee (1964), these muscles account for 80% of the basic metabolism in *Leptinotarsa*. A reduction to about 5% would lead to an over-all respiratory rate of less than 25% of the original value, which agrees well with the actual figures (de Wilde & Stegwee, 1958). Interest is therefore centred on the activity of sarcosomal enzymes before and during diapause. Stegwee (1962, 1964) examined some mitochondrial enzymes. He found by measuring the oxygen consumption of isolated mitochondria that in diapause the mitochondrial succinate oxidase activity was much reduced. Also other mitochondrial enzyme activities were lowered, and the cytochrome content of the degenerate sarcosomes present during diapause was greatly changed and diminished (Fig. 6).

Respiratory activity of sarcosomal preparations from beetles at different

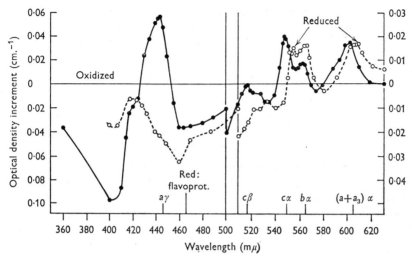

Fig. 6. Extinctions of respiratory pigments in sarcosomes of active beetles (drawn line) as compared with sarcosome fragments of diapausing beetles (broken line) (data after Stegwee, 1960, 1964).

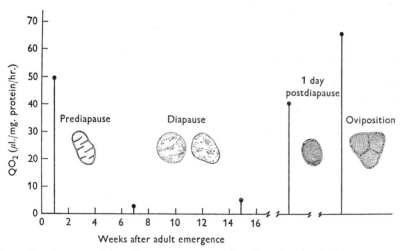

Fig. 7. Respiratory activity of sarcosome preparations in different physiological phases. The structure of the sarcosomes is given schematically. (Data after Stegwee, 1964.)

stages of activity in the presence of pyruvate and malate agreed well with the state of intactness of the sarcosomes (Fig. 7), except at the end of diapause. Here it was a remarkable fact that in beetles on the verge of terminating diapause the sarcosomes were structurally intact but metabolically inactive. Respiratory activity was restored, however, within 1 day after the insects had left the soil. This explains the discontinuous increase of O_2 consumption in beetles during post-diapause emergence as described by Le Berre (1965). De Kort (1969) determined the activity of some mitochondrial and extramitochondrial enzymes in the flight muscles, representative of three important metabolic pathways: glycolysis, fatty-acid oxidation and the tricarboxylic acid cycle. Measurements were carried out in SD beetles throughout the pre-diapause period. Sarcosome membrane-bound enzymes (succinate dehydrogenase and α-glycerophosphate dehydrogenase) increased first and reached a constant level after 4–5 days. Soluble cytoplasmic enzymes (glyceraldehyde phosphate dehydrogenase and glycerophosphate dehydrogenase) increased more gradually to a constant level several days later. In the same way, mitochondrial enzymes decreased first and cytoplasmic enzymes much more gradually at the onset of diapause. The efficiency of oxidative phosphorylation as expressed by the P:O ratio is decreased in sarcosomes obtained from diapausing beetles (Stegwee, 1964).

The soil as an environment for respiratory exchange is characterized by a relatively low and very variable partial pressure of oxygen at the depth at which *Leptinotarsa* usually hibernates. Ege (1910) found the following air composition in sandy agricultural soil at a depth of 30 cm.

Weather	O_2 (mm.)	O_2 (%)	CO_2 (mm.)	CO_2 (%)
Dry	153	20·7	1·5	0·2
6 hr. after rainfall	64	8·6	4·6	6·3

During most of the winter, when the upper soil layers are alternatively frozen and thawed, the conditions will be comparable with those during rainfall. Under these periodically rather anaerobic conditions it is not surprising that during diapause, aerobic metabolism must be at a low level (Ushatinskaya, 1956).

Protein synthesis

Diapause, as an alternative to reproduction, implies the absence of proteid yolk synthesis. As yolk proteins are taken up from the haemolymph by the ovarian follicle cells, blood-protein composition may provide important information on the biochemical mechanism underlying diapause. In our laboratory, A. de Loof has carried out electrophoretic and immunological studies of blood proteins in beetles under LD and SD treatment. These studies will shortly be published and the author has kindly allowed me to quote some results.

It appears that the protein pattern of SD beetles already in pre-diapause differs in several respects from the pattern in LD beetles. Three proteins are found by electrophoretic and immunological methods to be specific for the SD pattern. Only the most abundant of these three fractions is present at all, and then only in very small amounts in the haemolymph of LD beetles. The SD pattern of females is also characterized by the absence of at least two sex-specific proteins, which are only present in the haemolymph of LD females. One of these proteins appears to be the main component of the proteid yolk ($\pm$ 75%). There is no difference between the pattern of SD males and females. It is interesting to note that the total protein concentration in the haemolymph increases very quickly in the pre-diapause period and reaches a maximum of about 3 times the concentration in the LD haemolymph about 10 days after emergence. It thus appears that the haemolymph during pre-diapause and diapause is characterized by a high total protein concentration, which must be due mainly to the synthesis and storage of the three specific SD proteins.

THE ENDOCRINE BASIS OF DIAPAUSE IN *LEPTINOTARSA*

Preliminary observations

Already in an early stage of our research our attention was directed to the corpus allatum as a possible endocrine organ involved in adult diapause. Partly this was due to the work of Joly (1945) with *Dytiscidae*, partly to the experiments by Wigglesworth (1938) concerning the endocrine regulation of oogenesis in *Rhodnius*.

We were especially struck by the resemblance between the resting and oosorption condition of the ovarioles in the diapausing Colorado beetle and the pattern observed in allatectomized *Rhodnius*. Moreover, the size of the corpora allata during diapause was much reduced compared with the active state (de Wilde, 1954). Some morphometric data are given in Table 3.*

Table 3. *Means and standard errors of surfaces of largest cross-section of germarium, largest ovum (sample of twelve ovarioles per beetle) and corpus allatum (two per beetle) of beetles reared under two different photoperiods (after de Wilde & de Boer, 1961)*

Photoperiod (hr./day)	No. of samples	Germarium (mm.^{2}/360)	Mean largest ovum (mm.^{2}/360)	Mean number of follicles per ovariole	Cc. allata (mm.^{2}/1000)
18	14	94.7 ± 4.9	188.8	2.2	23.4 ± 1.5
10	7	53.1 ± 5.5	1.1	0.1	14.6 ± 1.8
Difference	—	41.6 ± 7.4	—	—	8.8 ± 2.3

* In the original version, some decimals were misprinted and have been corrected in the present table.

As implantation experiments with active corpora allata during prediapause met with some success (de Wilde, 1955), we felt encouraged to start an extensive series of experiments to investigate whether the diapause syndrome in *Leptinotarsa* is the result of corpus allatum deficiency. We realized, however, that activation of the corpora allata had never been found to occur spontaneously, but was generally believed to be effected by the neurosecretory cells of the brain.

The histological picture of the medial neurosecretory cells during diapause and reproduction

In the pars intercerebralis of the brain the Gomori-positive A cells in particular show characteristic changes at the onset of diapause and during 'diapause development' (Schooneveld, 1969). In LD beetles entering the reproductive period, these cells are enlarged and have small and dispersed Gomori-positive inclusions (Plate 2a). Presence of the same material throughout their axons suggests an active transport. In SD beetles on entering diapause, the inclusions accumulate to form larger clusters and after 1–2 weeks in diapause the cell volume is reduced and its interior packed with Gomori-positive material in large clusters (Plate 2b).

Nuclear volume is now considerably reduced, and the axons are devoid of Gomori-positive material or contain only a few droplets. In the course of diapause this picture gradually changes, and after 2–3 months at 25° C. the material is much more dispersed and transport restored.

Allatectomy-diapause

We managed to extirpate the corpora allata, either separately or together with the corpora cardiaca, by way of the cervical membrane and occipital cavity.

We have never observed a difference between the effects of both operations. They invariably result in a physiological condition which in nearly all respects is identical with diapause. For obvious reasons we performed most experiments with beetles reared under LD conditions. We will mention some elements of the syndrome in greater detail.

(1) When the operation is carried out at emergence a period of prediapause ensues, characterized by intensive feeding and absence of oviposition. When it is carried out during the reproductive period, oviposition is continued for some time. The length of this period is about the same as in SD-reared beetles after switching over to SD (Fig. 8) (de Wilde & de Boer, 1961, 1969).

(2) When allatectomy is performed in females after oogenesis has started, oosorption is induced (de Wilde & de Boer, 1961).

(3) In the same way as in beetles, reared under SD, the flight muscles start degenerating at the onset of behavioural diapause (Stegwee, 1964).

(4) The prevailing protein pattern in the haemolymph induced by LD persists, and is supplemented by the elements of the SD pattern (A. de Loof, 1968, personal communication).

(5) The rate of oxygen consumption, after an initial rise over approximately 5 days, is gradually lowered to a value even below that in normal diapause (de Wilde & Stegwee, 1958).

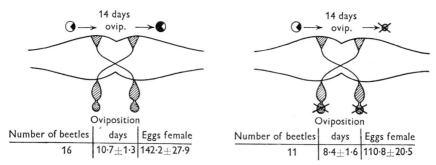

Number of beetles	days	Eggs female	Number of beetles	days	Eggs female
16	10.7±1.3	142.2±27.9	11	8.4±1.6	110.8±20.5

Fig. 8. Number of oviposition days after allatectomy and SD treatment. The beetles had previously been exposed to LD's and had oviposited during 14 days. (After de Wilde & de Boer, 1969.)

(6) When placed at 25° C. in soil of proper moisture content, beetles can survive in allatectomy-diapause for 4-5 months. The maximum period we observed was 210 days. Allatectomy diapause is broken neither spontaneously nor by any physical treatment.

Castration experiments and the induction of the behavioural component

The eventual role of the ovaries in the induction of behavioural activity was studied by combined castration and allatectomy and subsequent implantation of active corpora allata obtained from reproducing females. It appeared that castration in the male increased the percentage of burrowing responses, but in the female it was without effect on behaviour. Subsequent allatectomy induced the usual sequence of pre-diapause and diapause behaviour, and reimplantation of active glands restored the active condition. It follows that in female beetles the central nervous effect of the corpus allatum is in all probability a direct one (de Wilde & de Boer, 1969).

In conclusion it may be mentioned here that castration does not induce flight-muscle degeneration, and that its effect on basal metabolism is **not** very pronounced (El Ibrashy, 1965).

Effects of implanted corpora allata

Allatectomy diapause may be reversed by implanting 2–4 active corpora allata or postcerebral complexes and subsequent reproductive activity persists under LD. SD diapause, on the contrary, is only reversed after implantation of many more active glands, and the effect is transient, though dependent on the number of glands (de Wilde & de Boer, 1961, 1969; de Wilde, 1965).

When inactive glands, obtained from diapausing beetles, are implanted in females which had been subjected to allatectomy, the reproductive state can be induced by LD treatment, and repeated inactivation and activation is induced by successive periods of SD and LD. We may assume that, as a consequence of these treatments, the corpora allata are successively inactivated and activated (de Wilde, 1965; de Wilde & de Boer, 1969).

As a direct effect of photoperiod on the corpora allata is extremely unlikely, and implanted glands can only be regulated by the humoral pathway, we assumed that photoperiodic control was effected by way of cerebral neurosecretion.

The diapause syndrome in relation to levels of juvenile hormone (JH)

Our reasoning on different states of activity of the corpus allatum was based on indirect criteria. We have therefore attempted to obtain more direct evidence by estimating the JH content of beetles in different states of activity by means of a bioassay.

We were fortunate to find that active beetles contain a sufficient quantity of JH to allow for a reproducible estimate in ether extracts of 20–30 beetles by means of the *Galleria* wax test (Gilbert & Schneidermann, 1960; de Wilde *et al.* 1968). Although our tests with these 'total extracts' already provided interesting information (de Wilde & Oostra, 1969) we continued our efforts to measure the JH titre directly in the haemolymph, as only here is the hormone in exchange with the tissues.

Since in our laboratory continuous cultures under controlled conditions are available (de Wilde, 1957), we made synchronous cultures of large numbers of beetles *ab ovo* under 10 and 18 hr. photoperiods. Samples of 30–40 beetles were taken at regular intervals and the corresponding JH titres are given in Fig. 9. By calibrating our bioassay with JH solutions of known concentrations we were able to measure the titre in units/ml. blood.*

Generally, the course of JH titre during different physiological states of the insect corresponds well to our predictions based upon indirect evidence. The size of the corpus allatum fairly parallels the titre of its hormone. In the emerging young adult the JH is nearly at zero level, independent of pretreatment. Directly after emergence the titre starts increasing. In SD beetles it soon decreases again, reaching a minimum by the time diapause

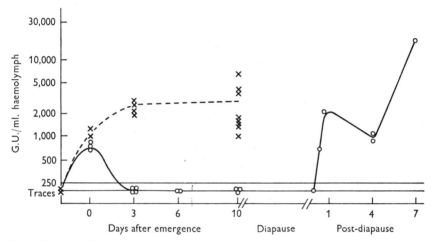

Fig. 9. Titre of JH in the haemolymph of female Colorado beetles, as a function of photoperiodic treatment. × -- ×, 18 hr.; ○—○, 10 hr. 1 *Galleria* unit is equivalent to 5×10^{-9} mg./ml. DL-JH (Röller). (After de Wilde *et al.* 1968.)

is started. At the end of 'diapause development' the titre starts increasing, and during post-diapause oviposition the level is the same as in reproducing young LD females.

Our data support the idea that reproduction, pre-diapause feeding, burrowing behaviour and diapause are induced by decreasing levels of JH.

We are now also able to give some information concerning the 'stored' effect of larval LD treatment discussed earlier. As the hormone titre at emergence is the same in beetles with SD and LD larval pretreatment, there is no question that larval LD treatment results in an increased JH titre persisting throughout adult development. The corpus allatum is merely activated more readily after adult emergence. Most probably, the information is stored in the centre activating the corpus allatum, i.e. in the cerebral neurosecretory system.

* We are grateful to Professor H. Röller for providing us with a quantity of pure juvenile hormone.

The relation between juvenile hormone and basal metabolism

In the initial stages of our experiments we supposed that JH would influence respiratory metabolism by an effect on the mitochondria. Indeed, we initially found a stimulatory effect of added, active corpora allata and JH-containing *Cecropia* extracts, on O_2-consumption of tissue homogenates obtained from diapausing beetles (de Wilde & Stegwee, 1958; de Wilde, 1960). Stegwee (1960) found that *Cecropia* extract stimulated oxidative phosphorylation by isolated sarcosomes of *Leptinotarsa*. We have not been able to reproduce the effects on O_2-consumption, but the effect of JH on P:O ratio has since been corroborated by Stegwee (1964) and Minks (1967), the latter working with sarcosomes from *Locusta migratoria*.

This direct effect of JH on metabolism is, however, negligible in comparison with the dramatic changes in the level of basal metabolism produced by its effect in the synthesis and maintenance of the flight muscles with their sarcosomes. For many years we have been investigating the mode of action of JH in this remarkable morphogenetic process. De Kort (1969) has followed the biochemical events in the degenerating flight muscle during SD diapause, as compared with the atrophic condition produced by denervation. As the effects show many similarities, the possibility remains that JH exerts its effect via the nervous system.

THE REGULATION OF CORPUS ALLATUM ACTIVITY

On the basis of foregoing evidence, we may assume that the corpus allatum is the central gland inducing the elements of the diapause syndrome by lowering the levels of its activity.

As stated before, this gland does not regulate its own function but is under the control of the neuroendocrine system, as is already suggested by its endocrine and nervous connections. Neurosecretory axons from the mid-brain taking their course in the NN corporis cardiaci and NN corporis allati penetrate into the gland, which is situated close to the corpus cardiacum, a neurohaemal end-organ of the cerebral neurosecretory cells. From the suboesophageal ganglion the gland is innervated by the nervus allato-suboesophagealis.

Removing the medial neurosecretory cells of the brain

By means of diathermic cauterization, we were able to destroy the section of the mid-brain where the medial neurosecretory cells are located (de Wilde & de Boer, 1969). Performing these operations on LD beetles, we

PLATE I

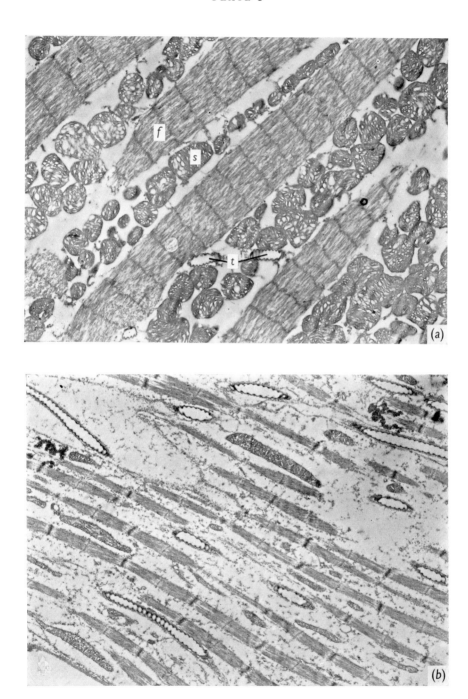

Electron micrographs of the intact flight muscle of a reproducing beetle (a) and the degenerated flight muscle of a diapausing beetle (b). (After Stegwee et al. 1963.) f, Muscle fibril; s, sarcosome; t, tracheole. Magnification: ×7,860.

PLATE 2

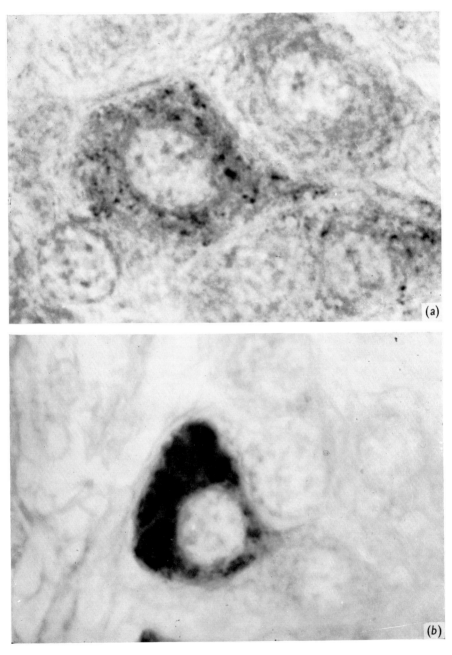

Neurosecretory A cells as stained by the Gomori technique. (*a*) Female, 5 days after adult moult, entering oviposition. Large cell volume, small and dispersed neurosecretory particles. (*b*) Female 2 weeks in diapause. Reduced cell volume, dense accumulations of colloid. (Courtesy H. Schooneveld.) Magnification: × 3,300.

could produce several elements of the diapause syndrome. Sometimes the operation inhibited feeding activity. As nutrition itself is a factor in diapause induction, such beetles were excluded from further observation. In other cases the operation interfered with diuresis, resulting in a considerable degree of swelling by retention of body water. A fair number of operations, however, merely interfered with oogenesis and in some cases induced the characteristic burrowing in moist sand. To test whether the operation interfered with oogenesis by inhibition of corpus allatum activity, the operated beetles were implanted with active postcerebral complexes obtained from reproducing beetles. In a significant number of cases oviposition took place.

These experiments, in connection with the histological observations mentioned earlier, render it probable that the activity of the medial neurosecretory cells is needed for initiation and maintenance of corpus allatum activity.

Denervation experiments

Completely denervated corpora allata or postcerebral complexes, implanted under the integument in the thorax, can be activated by varying daylength. This clearly shows that control can take place by the humoral pathway only, and that the nervous connections are not indispensable. But is the innervation non-functional in initiating diapause? We have found that transection of the N corporis allati does not influence the rate of inactivation by SD treatment.

Transecting the nervus allato-suboesophagealis resulted in prolonging the SD pre-diapause period to almost twice its length. This nerve probably contributes to speeding up the effect of SD (de Wilde & de Boer, 1969).

The effect of physiologically aged foliage

Grison (1957) supposed the inhibitory effect of feeding with aged foliage on oogenesis to be due to nutritive deficiency and found that the rate of oviposition rises again when lecithin is added to the aged leaf material. But is this effect really due to a lack of essential vitellogenic nutrients? In experiments made by Ferket in our laboratory, the corpora allata were measured after feeding with aged foliage and their size found to be reduced. Moreover, implantation of active corpora allata into beetles in which the rate of oviposition was depressed resulted in a significant increase in egg deposition. Also, after prolonged feeding with aged foliage the beetles will enter into diapause (de Wilde & Ferket, 1967).

This evidence favours the idea that the effect of aged foliage is transmitted to the neuroendocrine system regulating the activity of the corpora allata. Lecithin still may be the essential factor, but its effect is certainly not a simple nutritive one.

THE NATURE OF THE 'DIAPAUSE MECHANISM'

Many Coleoptera, e.g. *Tenebrio molitor*, do not show imaginal diapause and after allatectomy merely react by arrest of oogenesis. El Ibrashy (1965), studying the case of *Tenebrio*, did not find any change in locomotory activity nor in the structure of the thoracic musculature after allatectomy. Also, it is known that reproduction in *Tenebrio* does not respond to daylength. The fact therefore that the Colorado beetle has an adult diapause is due, first, to the reaction of its neuroendocrine system to seasonal tokens, and secondly, to the dependency of ovarian and flight-muscle growth, haemolymph protein synthesis and behavioural activity on the neuroendocrine system and more especially on the titre of JH governed by this system.

In all cases studied the genetic character 'diapause' has been shown to be dependent on several genes (Lees, 1955). In the Colorado beetle some of these may be related to photochemical receptor systems involved in photoperiodism, others to pathways of nervous conduction within the central nervous system. The fact that oogenesis and flight-muscle synthesis have different critical levels of JH already precludes the interaction between the JH and merely one 'regulator gene'. Several 'operon' systems must participate in the complex syndrome which constitutes diapause in *Leptinotarsa*.

DISCUSSION

The picture we have given is far from being complete. Much more histological evidence is needed, and the mechanisms involved in the extrasensory reception of seasonal tokens have barely been touched. Incomplete as it is, the picture is one of overwhelming complexity at one side and simplicity at the other. The complexity is in the syndrome, and simplicity in its endocrine control. Is the role of neurosecretion and reflectory innervation indeed as limited as our experiments seem to indicate? Is inhibitory innervation of the corpora allata by the brain virtually absent or non-functioning at the onset of diapause? We can only suggest that a more complicated role of neurosecretion and inhibitory innervation is probably played during the reproductive cycles. The fact that the protein pattern of the haemolymph in pre-diapause beetles is not merely the LD pattern minus something, but is characterized by its own proteins, suggests that endocrine induction of diapause may be more complex, and may involve endocrine antagonism of an unknown nature. The striking resemblance, however, between natural diapause and allatectomy-diapause characterizes the corpus allatum as the master-gland in seasonal adaptation of the adult Colorado beetle.

From a histological and histochemical study of another diapausing chrysomelid beetle, *Galeruca tanaceti* L., Siew (1965) has derived the hypothesis that pre-diapause, diapause and ovarium maturation are controlled by different levels of activity of the neurosecretory system, and thereby of the corpus allatum. Our physiological data support this idea.

More readily reversible types of dormancy, such as 'repos' or 'sommeil' (Le Berre, 1965), 'summer rest' and 'summer sleep' (Minder & Petrova, 1966), 'summer dormancy' and 'winter dormancy' (Ushatinskaya, 1966), may now perhaps be interpreted as states of incomplete activity of the cerebral neurosecretory A cells and a resulting intermediate activity of the corpus allatum. But it is evident that we can only understand the regulation of the resulting JH levels if we know more about the processes eliminating the hormone: their location, nature, and the factors determining their intensity.

REFERENCES

ADKISSON, P. L. (1961). *J. econ. Ent.* **54**, 1107.
ADKISSON, P. L., BELL, R. A. & WELLSO, S. G. (1963). *J. Insect Physiol.* **9**, 299.
ANDREWARTHA, H. G. (1952). *Biol. Rev.* **27**, 50.
BENOIT, J. (1937). *Bull. biol. Fr. Belg.* **71**, 393.
BOCZKOWSKA, M. (1944). *C. r. hebd. Séanc. Acad. Sci., Paris* **30**, 80.
BUSNEL, R. G. (1939). Études physiologiques sur le *Leptinotarsa decemlineata* Say. Thesis, Lib. le François, Paris.
DANYLIEVSKY, A. S. & KUNETSOVA, J. A. (1968). In *Photoperiodic adaptation*. Comm. Leningrad University, 1968, p. 5.
EGE, R. (1916). *Vid. Meddel. Dansk. Naturh. Foren.* **67**, 14.
EL IBRASHY, M. T. (1965). *Meded. LandbHogesch. Wageningen* **65**, 11.
FINK, D. F. (1925). *Biol. Bull. mar. biol. Lab., Woods Hole*, **49**, 381
GILBERT, L. I. & SCHNEIDERMANN, H. A. (1960). *Trans. Am. microsc. Soc.* **79**, 38.
GIRAULT, A. A. (1908). *Ann. ent. Soc. Am.* **1**, 155.
GIRAULT, A. A. & ROSENFELD, A. H. (1907). *Psyche* **14**, 45.
GIRAULT, A. A. & ZETTEK, J. (1911). *Ann. ent. Soc. Am.* **4**, 71.
GORYSHIN, N. I. (1956). *Dokl. Akad. Nauk. S. S. S. R.* **109**, 205.
GORYSHIN, N. I. (1958). In *The Colorado Beetle and its Control*, p. 136. Akademia Nayk, SSR, Moscow.
GRISON, P. (1939). *Proc. VIIth Int. Congr. Ent.*, Berlin, p. 2663.
GRISON, P. (1944). *C. r. hebd. Séanc. Acad. Sci., Paris* **218**, 342.
GRISON, P. (1950). *Proc. VIIIth Int. Congr. Ent.*, Stockholm, p. 226.
GRISON, P. (1953). *Trans IXth Int. Congr. Ent.*, Amsterdam, p. 331.
GRISON, P. (1957). *Annls Épiphyt.* **3**, 304.
HODEK, I. & WILDE, J. DE (1969). *Entomologia exp. appl.* (In the Press).
JERMY, I. & SARINGER, GY. (1955). *Acta agron. hung.* **5**, 419.

JOLY, P. (1945). *Archs Zool. exp. gén.* **84**, 47.
KORT, C. A. D. DE (1969). *Symp. Insect Endocrinology*, Brno 1966. (In the Press.)
LE BERRE, J. (1965). *C. r. Séanc Soc. Biol.* **159**, 2131.
LEES, A. D. (1955). *The Physiology of Diapause in Arthropods.* Cambridge University Press.
LEES, A. D. (1964). *J. exp. Biol.* **4**, 119.
MARZUSCH, K. (1952). *Z. vergl. Physiol.* **34**, 75.
MINDER, I. P. & PETROVA, D. W. (1966). In *Ecology and Physiology of Diapause in the Colorado beetle.* Akademia Nayk, SSR, Moscow.
MINKS, A. K. (1967). *Archs néerl. Zool.* **17**, 175.
PRECHT, H. (1953). *Z. vergl. Physiol.* **35**, 326.
SCHNEIDERMANN, H. A. & WILLIAMS, C. M. (1954). *Biol. Bull. mar. biol. Lab., Woods Hole* **106**, 238.
SCHOONEVELD, H. (1969). *Symp. Insect Endocrinology*, Brno, 1966. (In the Press.)
SIEW, Y. C. (1965). *J. Insect Physiol.* **11**, 463.
SLAMA, K. (1964). *J. Insect Physiol.* **10**, 283.
STEGWEE, D. (1960). *Proc. XIth Int. Congr. Ent.*, Vienna, vol. III, 218.
STEGWEE, D. & VAN KAMMEN-WERTHEIM, A. R. (1962). *J. Insect Physiol.* **8**, 117.
STEGWEE, D., KIMMEL, E. C., BOER, J. A. DE & HENSTRA, S. (1963). *J. Cell. Biol.* **19**, 519.
STEGWEE, D. (1964). *J. Insect Physiol.* **10**, 97.
TOWER, W. L. (1906). *Publs. Carnegie Instn* **48**, 1.
TROUVELOT, B. (1936). *Revue Zool. agric. appl.* **3**, 33.
USHATINSKAYA, R. (1956). Ber. 100– *J.Feier Deutsch. ent. Ges. Berlin*, p. 250.
USHATINSKAYA, R. (1966). In *Ecology and Physiology of Diapause in the Colorado beetle.* Akad. Nayk. SSR, Moscow.
WIGGLESWORTH, V. B. (1938). *Q. Jl microsc. Sci.* **79**, 91.
WILDE, J. DE (1949). *Bijdr. Dierk.* **28**, 543.
WILDE, J. DE (1950). *Proc. VIIIth Int. Congr. Ent.*, Stockholm, p. 310.
WILDE, J. DE (1954). *Archs néerl. Zool.* **10**, 375.
WILDE, J. DE (1955). *Meded. Lab. Ent., Wageningen* **2**, 1.
WILDE, J. DE (1957). *Z. PflKrankh. PflPath. PflSchutz* **64**, 589.
WILDE, J. DE (1960). *Proc. XIth Int. Congr. Ent.*, Vienna, vol. III, p. 215.
WILDE, J. DE (1965). *Archs Anat. microsc. Morph. exp.* **54**, 547.
WILDE, J. DE & BOER, J. A. DE (1961). *J. Insect. Physiol.* **6**, 152.
WILDE, J. DE & BOER, J. A. DE (1969). *J. Insect Physiol.* (In the Press.)
WILDE, J. DE & BONGA, H. (1958). *Entomologia exp. appl.* **1**, 301.
WILDE, J. DE, DUINTJER, J. C. S. & MOOK, J. (1959). *J. Insect Physiol.* **3**, 75.
WILDE, J. DE & FERKET, P. (1967). *Meded. Rijksfac. Landbwet. Gent.* **32**, 387.
WILDE, J. DE & OOSTRA, H. G. M. (1969). *Symp. Insect Endocrinology*, Brno, 1966. (In the Press).
WILDE, J. DE, STAAL, G. B., KORT, C. A. D. DE, LOOF, A. DE & BAARD, G. (1968). *Proc. Sect. Sci. K. ned. Akad. Wet.* (In the Press.)
WILDE, J. DE & STEGWEE, D. (1958). *Archs néerl. Zool.* **13**, 277.
WILLIAMS, C. M. (1956). *Biol. Bull. mar. biol. Lab., Woods Hole* **110**, 201.
WILLIAMS, C. M. & ADKISSON, P. L. (1964). *Biol. Bull. mar. biol. Lab., Woods Hole* **127**, 5111.

PHOTOPERIODISM AND THE ENDOCRINE ASPECTS OF INSECT DIAPAUSE

By C. M. WILLIAMS

The Biological Laboratories, Harvard University,
Cambridge, Massachusetts 02138

INTRODUCTION

The insect brain is known to be the highest centre, not only of the nervous system, but also of the endocrine system. The endocrine functions of the brain include its ability to synthesize and secrete a hormone appropriately termed the 'brain hormone'. The principal role of the brain hormone is to control other endocrine organs and, more particularly, the prothoracic glands (Williams, 1947, 1952). Under the tropic stimulation of brain hormone, the prothoracic glands secrete 'ecdysone'—a hormone prerequisite for nearly all aspects of insect growth and metamorphosis (Wigglesworth, 1964).

Ecdysone and ecdysone analogues

Ecdysone was isolated by Butenandt & Karlson (1954). In a massive research effort they extracted 30 mg. of α-ecdysone from a ton of *Bombyx mori* silkworms. A small amount of a more polar material, β-ecdysone, was also obtained and purified. Twelve years elapsed before the chemistry of α-ecdysone was clarified (Karlson *et al.* 1965). To everyone's surprise, it proved to be a sterol.

In 1966 the synthesis of α-ecdysone was independently announced by two teams of scientists (Kerb *et al.* 1966; Harrison, Siddall & Fried, 1966). Because of the ten asymmetric centres in the molecule, the synthesis was an exceedingly difficult one and for a time it appeared that only vanishingly small amounts of ecdysone would ever be available for study.

The situation changed overnight with the discovery, by Japanese and Czechoslovakian investigators, that certain plants contain amazing amounts of substances resembling or even identical with α- and β-ecdysones (for review, see Williams & Robbins, 1968). Up to the present time a total of 15 'phytoecdysones' have been isolated from certain weeds, ferns, and evergreen trees, and that number is certain to increase. With a few exceptions, all these materials possess the same substituent groups in the sterol ring system and differ among themselves only in terms of the chemistry of the side-chain. Biological tests have revealed the surprising fact that certain of the phytoecdysones are up to 20 times as active as authentic α- or β-ecdysones (Williams, 1968); this extra activity can be accounted for in terms of their resistance to inactivation within the living insect (Ohtaki & Williams, 1969).

Developmental arrest

In the absence of ecdysone, growth and metamorphosis come to an abrupt halt. Nature has exploited this state-of-affairs to provide for the overwintering of immature insects in a state of developmental standstill called 'diapause'. The immediate cause of diapause is the cessation of ecdysone secretion by the prothoracic glands. Months later, with the arrival of spring, ecdysone is again secreted and development is resumed where it had left off.

Many species of insects traverse more than one generation each year. In that case, the prothoracic glands do not 'shut down' except at some predictable stage in the metamorphosis of the final brood that must overwinter. Whether the prothoracic glands will secrete or fail to secrete ecdysone is a decision not made by the glands themselves. It is made by the brain. The situation, in short, is reminiscent of the control of the vertebrate thyroid or adrenal cortex by the tropic hormones of the anterior pituitary.

Brain hormone

This brings us back to our starting-point in the brain. The brain hormone is synthesized and secreted by neurosecretory cells located in the central region, including the so-called pars intercerebralis. In a favourite experimental animal, the silkworm *Hyalophora cecropia*, there appear to be about 40 of these cells arranged as shown in Fig. 1: a pair of medial groups containing a total of 22 cell bodies of four distinguishable types, a pair of lateral groups containing a total of 14 cell bodies of two distinct types, and a pair of posterior groups containing a total of four cells of a single type (Herman & Gilbert, 1965; Herman, 1968).

For 50 years the insect brain has been known to be an endocrine organ. But until 10 years ago the hormone resisted all efforts to obtain it apart from the living insect. Despite this handicap the function of brain hormone was worked out in considerable detail by biological manoeuvres such as the surgical removal and transplantation of living brains. For example, by removing the brain one could shut off the secretion of ecdysone and provoke the onset of diapause (Williams, 1952). The same operation performed on insects that were already in diapause caused a prolongation of the dormant condition until such time as one implanted an active brain obtained from another individual of the same or even of a different species (Williams, 1946). Moreover, at any time one could by-pass the entire brain-centred mechanism by injecting ecdysone or by implanting activated prothoracic glands.

The chemistry of brain hormone is still a matter of controversy. There is general agreement that it is a large, non-dialysable, heat-stable moiety

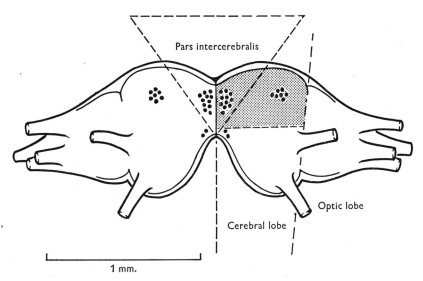

Fig. 1. A diagrammatic representation of the pupal brain of the silkworm *Hyalophora cecropia*, showing the several brain regions referred to in the text. The stippled area indicates the endocrinologically active region in one of the brain hemispheres. The solid circles show the number and location of the cell bodies of the medial (22), lateral (14), and posterior (4) neurosecretory cells as described by Herman & Gilbert (1965).

that may form aggregates of different molecular weights, all in excess of 10,000. It is generally presumed to be a peptide or protein. However, as discussed elsewhere (Williams, 1967), this conclusion is by no means a certainty: there are substantial reasons for believing it to be a mucopolysaccharide.

PHOTOPERIOD AND DIAPAUSE

In this bird's-eye view of insect endocrinology we see that the brain controls the prothoracic glands and the latter secrete ecdysone to control the tissues. This leaves us with the fresh problem of what controls the brain.

The answer to this question has gradually been worked out. The endocrine function of the brain is controlled by token signals from the external environment and, more particularly, by photoperiod and temperature.

Termination of diapause in pupae of Antheraea pernyi

At the Harvard Laboratory we have studied in special detail the critical effect of photoperiod on the termination of pupal diapause in the Chinese oak silkworm, *Antheraea pernyi*. When pupae of this species are placed at 25° C. and exposed to light for a certain number of hours each day, one obtains the results summarized in Fig. 2 (Williams & Adkisson, 1964).

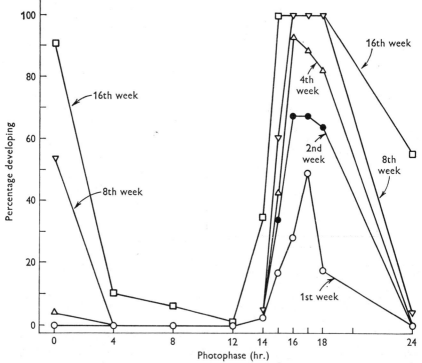

Fig. 2. The effects of photoperiod on the termination of diapause at 25° C. by previously chilled (2–3 months at 2° C.) *Antheraea pernyi* pupae (1963). Each datum is based on a homogeneous group of 50 pupae. (From Williams & Adkisson, 1964.)

Daily illumination for 15–18 hr. provokes the termination of diapause as signalled by the initiation of adult development. By contrast, daily illumination for 4–14 hr. sustains the dormant condition. The most effective 'long-day' proves to be 17 hr.; the most effective 'short-day' is 12 hr. Our calculations suggest that *A. pernyi* pupae can measure the length of the day or night with an accuracy of about 15 min.

It will be recalled that the cocoon of the silkworm *A. pernyi* is a dense and apparently opaque object. Yet, as shown in Fig. 3, pupae enclosed in cocoons are just as susceptible to photoperiod as are pupae removed from cocoons. We shall come back to that paradox.

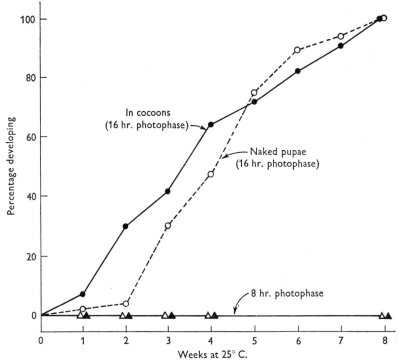

Fig. 3. The effects of long- and short-day regimens on naked *Antheraea pernyi* pupae (open symbols) and pupae in cocoons (closed symbols). Each datum is based on 50 pupae. (From Williams & Adkisson, 1964.)

Brain-centred control of diapause

In *A. pernyi* pupae there is conclusive evidence that photoperiod acts by controlling the secretion of brain hormone. For example, even a 17 hr. photophase is unable to provoke the development of brainless pupae. But if the operation is postponed until the brain has already secreted its hormone, the formation of the adult moth continues to completion without any further influence of photoperiod (Williams & Adkisson, 1964). The brain, in short, can be identified as the vehicle for the response to photoperiod.

We were able to show that only the anterior ends of *A. pernyi* pupae are sensitive to photoperiod. In the experiments in question individual pupae were removed from their cocoons, placed in an opaque diaphragm, and exposed to a long-day at one end and a short-day at the other. The results were clear cut: long-day conditions provoked the termination of diapause only when they acted on the head-end. So, also, the short-day regimen was effective in maintaining diapause only when it acted on the head-end. By contrast, exposure of the abdomen to either long- or short-day conditions was inconsequential.

The simplest assumption is that photoperiod acts directly on the brain itself. This theory seemed particularly attractive in view of the unpigmented, transparent cuticle overlying the brain of *A. pernyi* pupae (Plate 1 *a*). Even if the rest of the pupa is jet-black, the facial cuticle is always pigment-free. By moistening this region with alcohol, one can look inside and see the underlying brain. This being so, it is very tempting to think that the transparent facial cuticle has something to do with the transmission of light to the brain itself.

Of the many experiments that Adkisson and I performed on diapausing *A. pernyi* pupae, the most informative is the one illustrated in Plate 1 *b*. Here, the brain has been removed from the head of each of a series of pupae and implanted under a plastic window at the tip of the abdomen. The pupae were then placed in an opaque diaphragm and subjected to a photoperiod gradient, as described above. Here again the results were clear: by transplanting the brain to the tip of the abdomen, sensitivity to photoperiod was likewise shifted to the hind-end. This shows that photoperiod can control the activity of the neurosecretory cells of the brain even when the brain is disconnected from the rest of the nervous system. Evidently, the entire photoperiodic mechanism is brain-centred in the case of *A. pernyi* pupae. In principle, the minimal mechanism must include a pigment for the reception of light, a computer which counts the hours of daylight or darkness, and an output which controls the secretion of brain hormone.

SURGICAL MANIPULATIONS OF THE BRAIN OF *ANTHERAEA PERNYI*

In recent years I have carried out numerous surgical procedures on the brain of *A. pernyi* in search of the localization of the photoperiod mechanism. The strategy of these experiments was to do as little as possible to the brain until one destroyed its sensitivity to photoperiod. Each surgical procedure was carried out on the brains of 40 diapausing pupae. Twenty were then exposed to the stimulatory 17 hr. photophase and the other 20 to the inhibitory 12 hr. photophase.

The results of these unpublished experiments are summarized in Table 1. It may be recalled that as long as the photoperiod mechanism is fully functional, the short-day regimen will permit none of the pupae to terminate diapause; by contrast, the long-day regimen will cause 100% to do so. Therefore, it is clear that none of the seven procedures recorded at the top of Table 1 has any effect on the photoperiod response. In order to damage the mechanism it was necessary to cut away most of the lateral and ventral regions of the brain, leaving intact only the central dorsal region containing the medial and lateral neurosecretory cells (Fig. 1). Even then, the mechanism was damaged but not destroyed; this is indicated by the residual differences between the response to the short- and long-day regimens. In order to eliminate this difference, it was necessary to reduce the brain still further by cutting away the bilateral areas containing the lateral neurosecretory cells. Only then, as noted in the bottom line of Table 1, did the response to short and long days become identical.

Table 1. *Surgical procedures on the brains of diapausing pupae* of* Antheraea pernyi: *effects on subsequent response to long- and short-day photoperiods at* 25° C.

Procedure	Total no. of viable pupae	% initiating development in 2 months	
		SD†	LD‡
Cut circumoesophageal connectives	40	0	100
Cut tracheal connections to brain	40	0	100
Cut nerves to antennae and eyes	40	0	100
Cut nerves to corpora cardiaca	40	0	100
Dissociate pigmented tips of brain	40	0	100
Excise pigmented tips of brain	40	0	100
Bisect brain in mid-line	40	0	100
Dissociate optic lobes from cerebral lobes	40	45	80
Excise entire brain; reimplant cerebral lobes	40	45	85
Excise entire brain: reimplant dorsal half of cerebral lobes	40	30	75
Excise entire brain; reimplant pars intercerebralis	40	25	25

* First brood pupae (1965) stored at 2–3° C. for 8–11 weeks prior to the experiments.
† Short day: 12 hr. light:12 hr. dark.   ‡ Long day: 17 hr. light:7 hr. dark.

The significance of this result is considered in Table 2. It is not without interest that the behaviour of the operated pupae was the same as that of 'controls' maintained in continuous light. This is indeed a puzzling finding since one might have anticipated that pupae deprived of the photoperiodic mechanism would behave as if exposed to continuous darkness.

Evidently, in each hemisphere of the brain of *A. pernyi* the photoperiodic mechanism is located in a tiny mass situated just lateral to the medial neurosecretory cells. Moreover, this crucial region includes the lateral neurosecretory cells.

There is convincing evidence that this mechanism controls, not the synthesis of brain hormone, but its translocation along the axons of the medial neurosecretory cells and its release into the blood. Indeed, as pointed out previously (Williams, 1967), the synthesis of brain hormone continues under short-day conditions and brains of this type provide the richest source of extractable hormone.

Table 2. *The pars intercerebralis isolated from rest of the* Antheraea pernyi* *brain behaves as a brain exposed to continuous light*

		% initiating development in 2 months			
Procedure	Total no. of pupae	Cont. darkness	SD†	LD‡	Cont. light
Controls					
Remove from cocoons (no surgery)	80	85	5	100	25
Excise and reimplant entire brain	80	50	10	85	30
Excise entire brain and reimplant *pars intercerebralis*	40	—	25	25	—

* Diapausing first brood pupae (1965) stored at 2–3° C. for 8–11 weeks prior to experiment at 25° C.
† Short day: 12 hr. light: 12 hr. dark.   ‡ Long day: 17 hr. light: 7 hr. dark.

Light reception by the brain of Antheraea pernyi

Manifestly, the reception of photoperiod signals requires a mechanism for the absorption of light. In previously published studies (Williams, Adkisson & Walcott, 1965) we have demonstrated that the effective wavelengths extend over the lower region of the visible spectrum, including violet, blue, and blue-green light. Therefore we may confidently predict the presence of a pink pigment for the absorption of these wavelengths. On superficial examination of *A. pernyi* brains no such pigment is seen in the lateral neurosecretory cells or any other part of the protocerebrum; however, we may rest assured that it is there.

Light integration

As previously mentioned, *A. pernyi* pupae are fully sensitive to photoperiod even when they are inside apparently opaque cocoons. This paradox has been subjected to detailed study (Williams *et al.* 1965). Our measurements confirmed that the cocoon is truly opaque to the direct transmission of light. However, the geometry of the cocoon is such that it serves as an effective vehicle for the collection and integration of scattered light. The light is collected within the cavity of the cocoon and then penetrates the transparent facial cuticle to act directly on the brain itself. Our measurements suggest that the brain's photoreceptive mechanism is fully saturated by intensities not in excess of 1 ft.-candle (10·8 lux) of blue light.

The unpigmented facial cuticle proves to be four times as transparent as the pigmented cuticle found elsewhere in the pupa. The Russian investigator Shakhbazov (1961) has reported that the photoperiodic response of *A. pernyi* pupae can be eliminated by coating the facial window with opaque material. This finding was re-examined in the experiment shown in Plate 2. After the application of opaque paint to the facial region, each of a series of 40 pupae was returned to its cocoon in the head-up position, the cut-open lower end of the cocoon being pressed into plasticine. Despite the opacity of the facial window the response to photoperiod was fully preserved—that is, short days continued to inhibit and long days to stimulate development. What this implies is that the pupa, itself, acts as a light-integrating object: sufficient light is collected and internally reflected within the anterior end of the pupa to saturate the brain-centred mechanism. Consequently it appears that the transparent facial cuticle so typical of the pupae of the genus *Antheraea* is a safety device for cocoons in shady situations. Under that circumstance the extra fourfold transparency of the facial cuticle could make all the difference in synchronizing the life-history with the seasons.

INDUCTION OF DIAPAUSE

Up to this point I have considered the role of photoperiod in controlling the termination of pupal diapause in the silkworm *A. pernyi*. Manifestly, the other half of the story is the induction of diapause—a decision as to whether the insect will become dormant after pupation or proceed at once to develop into an adult moth.

Fortunately, this aspect of the matter has been studied by Tanaka (for summaries of these Japanese papers see Lees, 1955; Williams & Adkisson, 1964; Beck, 1968). Thus, when *A. pernyi* larvae are reared under day-lengths longer than 14 hr. they develop without any pupal diapause. At temperate latitudes, photoperiods of this sort are peculiar to late spring and early summer when the season is propitious for a second brood. By contrast, larvae reared under day-lengths shorter than 14 hr. (as in late summer and autumn) transform into diapausing pupae.

Tanaka's data on the induction of diapause are summarized as the hatched line in Fig. 4. For comparison, I record as the unbroken line our data for the photoperiodic termination of diapause. It is clear that those photoperiods which are effective in inducing diapause are also effective in stabilizing diapause once the latter has begun. Moreover, photoperiods which are effective in preventing the onset of diapause are precisely the same as cause the termination of pre-existing diapause. The obvious inference is that the same photoperiodic mechanism which controls the induction of diapause is retained by the pupa to control the termination of diapause.

It is important to note that a photoperiod which induces diapause does not immediately shut off the secretion of hormone by the larval brain. If it did so, pupation would be blocked and one would observe a larval rather than a pupal diapause. In the larva the action of photoperiod is to programme either the shut-down or the sustained activity of the brain after

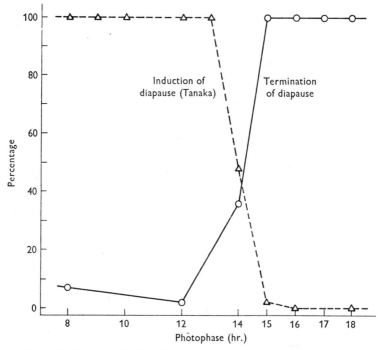

Fig. 4. The solid line records the effects of photoperiod on the termination of pupal diapause by previously chilled (2–3 months at 2° C.) *Antheraea pernyi* pupae. The hatched line shows the influence of photoperiod on the induction of diapause (data of Tanaka). (From Williams & Adkisson, 1964.)

the latter has secreted sufficient hormone to cause pupation. This state of affairs points to some unknown mechanism for the integration and latent storage of daily photoperiod signals accumulated during larval life.

Influence of temperature

In our preoccupation with photoperiod we would be ill-advised to neglect the influence of environmental temperature in conditioning the endocrine activity of the brain. Thus, in the case of diapausing *A. pernyi* pupae, one can greatly accelerate the response to long-day conditions by subjecting the pupa to several months of preliminary exposure to 2–3° C. But if the low-temperature treatment is prolonged for five or more months, the pupal brain recovers its endocrine activity and cannot be shut down by any subsequent exposure to short-day conditions.

The effects of low temperature in potentiating the termination of diapause appear to be more ubiquitous among insects than are the effects of photoperiod. A case in point is the Cynthia silkworm (*Samia cynthia*)—a species in which the termination of pupal diapause shows an absolute requirement for prolonged exposure to temperatures lower than 20° C. In my experience, unchilled pupae of *S. cynthia* cannot be aroused by exposure to long days, nor can previously chilled pupae be maintained in the dormant condition by exposure to short-day regimens.

The effects of environmental temperature have been studied in particular detail in the case of the silkworm *Hyalophora cecropia* (Williams, 1956). Here the termination of pupal diapause is potentiated by preliminary exposure to low temperatures and, especially, to temperatures in the range 10–15° C. By the excision and transplantation of *H. cecropia* brains it was possible to show that low temperature acts directly on the brain itself. As soon as the brain has regained its endocrine activity, the catalytic effects of low temperatures cease. The actual secretion of hormone and the developmental response that follows are then favoured by higher temperatures in the range 20–28° C. By this brain-centred endocrine mechanism the low temperatures of winter potentiate the initiation of adult development the following spring.

Influence of photoperiod on pupae of Hyalophora cecropia

The cocoon of the silkworm *H. cecropia* is an elongate and exceedingly opaque object consisting of a dense outer capsule, an intermediate spongy layer, and a dense inner capsule enveloping the diapausing pupa. This being so, it is not surprising to find that pupae of *H. cecropia* show no detectable response to photoperiod as long as they remain in cocoons. Under the modest light intensities of photoperiod chambers, both long-day and short-day regimens have the same effect as continuous darkness.

But what if one removes pupae of *H. cecropia* from their cocoons and repeats the study on 'naked' pupae? Under that circumstance one can demonstrate a rudimentary response to photoperiod in the sense that prolonged exposure to 16 hr. of daily illumination at 25° C. causes a larger number of pupae to initiate adult development than under the short-day regimen of 12 hr. This difference is amplified when the experiment is performed on naked *H. cecropia* pupae that have received several weeks of preliminary chilling at 5° C. But here again, if the preliminary chilling is continued for 10–15 weeks, the brain is fully activated and loses all its sensitivity to photoperiod.

Surgical procedures on brains of Hyalophora cecropia

In the case of *A. pernyi* the entire photoperiodic mechanism, commencing with the absorption of blue light and culminating with the secretion of brain hormone, is brain-centred. Thus the response to photoperiod is fully preserved when brains of *A. pernyi* are disconnected from the rest of the nervous system. The situation is different in the case of *H. cecropia*. For example, when the brains of naked pupae of *H. cecropia* are cut free from all their connections and reimplanted, exposure to short days is no longer able to inhibit the termination of diapause. Additional experiments involving the cutting of individual nerves suggest that the sites of light reception are numerous neurones located in the ganglia of the central nervous system as well as in the anlagen of the future compound eyes. Appropriate signals are then conveyed to the neurosecretory cells via the circumoesophageal connectives and the optic nerves. It is not without interest that a pink pigment is self-evident within the ganglion cells, which are presumed to be light-sensitive. It is also noteworthy that pupae of *H. cecropia* are not equipped with a transparent facial cuticle and that both anterior and posterior ends are equally sensitive to photoperiod.

Studies of six species of saturniid pupae show the following correlations: when a transparent facial cuticle is present (as in *Antheraea pernyi*, *A. polyphemus* and *A. mylitta*), no pink pigment is evident in the ganglia of the central nervous system. By contrast, in species lacking a transparent facial cuticle (*Hyalophora cecropia*, *H. gloveri* and *Samia cynthia*), a pink pigment is conspicuous in nearly all ganglia. A further novelty is that species lacking pigmented ganglia possess a pink pigment in the integument precisely overlying the ganglia. Perhaps this integumentary pigment helps 'shade' the underlying ganglia from blue or blue-green light.

EFFECTS OF TETRODOTOXIN ON THE RESPONSE TO PHOTOPERIOD

In conclusion, I direct attention once again to the *A. pernyi* silkworm and its brain-centred photoperiodic mechanism. As we have seen, the location of this mechanism was pinpointed in a pair of bilateral brain areas situated on each side of the medial neurosecretory cells. Each tiny area is characterized by the presence of the cell bodies of seven neurosecretory cells embedded in a mass of neuropile.

PLATE I

(a) A pupa of *Antheraea pernyi*, showing the transparent facial cuticle overlying the brain.

(b) Pupae of *A. pernyi* placed in an opaque partition and exposed to LD at one end and SD at the other end. In 50% of the pupae the brain has been removed from the head and implanted under a plastic window at the tip of the abdomen.

(*Facing p.* 296)

PLATE 2

Opaque paint has been applied to the facial window of certain pupae of *Antheraea pernyi* and the pupae returned to their cocoons in the head-up position.

Is it possible that the entire photoperiodic mechanism is contained within the lateral neurosecretory cells? A test of this hypothesis would, in principle, require one to dissociate the function of the neurosecretory cells from that of the surrounding neuropile. Surprisingly enough, this audacious goal is easily attained: all one has to do is to inject tetrodotoxin (TTX).

Interest in TTX ('puffer fish toxin') stems from its ability to block the 'gates' or 'channels' in electrically excitable membranes through which sodium ions must flow to generate the rising phase of the action potential. When these channels are blocked by tetrodotoxin, the nervous system is silenced and rendered incapable of conducting spike potentials (Mosher, Fuhrman, Buchwald & Fisher, 1964; Kao, 1966; Moore & Narahashi, 1967).

When administered to silkworm pupae at levels exceeding 0·05 μg./g. live weight, TTX shuts down the nervous system and thereby provokes a prompt and persistent paralysis. On superficial examination the insects appear to be dead. Clearly, this is not so because the heart continues to beat and the intersegmental muscles can usually, but not always, be caused to contract by direct electrical stimulation. Due to its incredibly selective action, the paralytic dose can be increased at least 200-fold without killing.

Experiments carried out on previously chilled pupae of *H. cecropia* and *S. cynthia* revealed the surprising fact that even the highest doses of TTX failed to interfere with metamorphosis. Thus, when placed at 25° C. paralysed pupae terminated diapause and underwent adult development to form paralysed but otherwise normal adult moths. The continued presence of TTX was documented by the ability of the moth blood to cause flaccid paralysis when injected into a series of fresh pupae. In additional experiments performed on mature larvae of three species of silkworms (*H. cecropia*, *A. polyphemus* and *A. pernyi*) the injected larvae metamorphosed into flaccid pupae. These findings give assurance that TTX fails to interfere with the secretion of brain hormone.

For present purposes we can now ask the crucial question: does TTX affect the response to photoperiod? To examine this question 1 μg. of TTX was injected into each of a series of forty diapausing *A. pernyi* pupae. Twenty were then placed at 25° C. under 12 hr. of daily illumination; the other 20 under 17 hr.

The results are summarized in Table 3. All individuals exposed to the long-day regimen terminated diapause and ultimately developed into flaccid moths. By contrast, the short-day regimen remained fully effective in maintaining diapause.

What this implies is that the conduction of nerve impulses by the neuropile of the brain of *A. pernyi* plays no role in the photoperiodic mechanism. Full assurance in this conclusion will require the completion of our current neurophysiological studies to prove whether or not TTX permanently blocks the conduction of nerve impulses in the brain of *A. pernyi*. In that event the entire photoperiodic mechanism can truly be said to reside in at least certain of the lateral neurosecretory cells.

Table 3. *Photoperiodic response at 25° C. of diapausing* Antheraea pernyi *pupae injected with tetrodotoxin**

Daily photophase (hr.)	No. of pupae	Cumulative % developing after (weeks):				
		1	2	3	4	5
12	20	0	0	0	0	0
17	20	5	35	75	95	100

* Each pupa received 1 µg. tetrodotoxin in 10 µl. H_2O.

Additional observations of the effects of tetrodotoxin on insect metamorphosis

TTX promises to be a powerful tool for studying the role of the nervous system in other aspects of insect development. In conclusion, I shall illustrate this prospect by two examples.

During the transformation of a pupa into a moth the formation of new muscles from myoblasts requires that the latter be innervated. For example, the differentiation of all the thoracic flight-muscles as well as the leg muscles does not take place if the corresponding myoblasts are denervated (Finlayson, 1956; Williams, 1958; Nüesch, 1968). TTX fails to interfere in any way with this trophic action of nerve on myoblasts. Thus, the flaccid moths derived from pupae injected with TTX show the full and complete formation of all the new muscles characteristic of the adult.

The final example has to do with the transformation of the central nervous system during the course of metamorphosis. In the brain, as elsewhere, tens of millions of new connections must be established with utmost precision to renovate the nervous system and provide for the metamorphosis of behaviour. The phenomenon as a whole constitutes one of the most baffling problems in developmental biology.

Can these new connections be established by neurones incapable of conducting nerve impulses? This, indeed, proves to be the case. Pupae receiving 50 µg. of TTX show the full and complete metamorphosis of the central nervous system.

This finding transcends the more pedestrian objective of accounting for the response of insects to photoperiod. Evidently, we should begin to think of the primordial nervous system as having the initial role of channelling specific chemicals to specific places. Then, presumably, hundreds of millions of years ago the action potential was evolved and organisms began to exploit these elongate cells for swift signalling by electrical impulses. Now, the promise of TTX is to unmask the primordial system and see what it can do. In the case of insects, the indications are that it can do some astonishing things.

The Harvard study was assisted in part by grant GB-3232 from the National Science Foundation. The photographic prints were prepared by Frank White from coloured transparencies by Muriel V. Williams. I am grateful to Professor Lynn M. Riddiford and Mr James Truman for critical reading for the manuscript.

REFERENCES

BECK, S. D. (1968). *Insect Photoperiodism.* New York: Academic Press.
BUTENANDT, A. & KARLSON, P. (1954). *Z. Naturf.* **9***b*, 389.
FINLAYSON, L. H. (1956). *Q. Jl. Microsc. Sci.* **97**, 215.
HARRISON, I. T., SIDDALL, J. B. & FRIED, J. H. (1966). *Tetrahedron Lett. Vol.* 3457.
HERMAN, W. S. (1968). In *Metamorphosis,* p. 107. Ed. W. Etkin and L. I. Gilbert. New York: Appleton-Century-Crofts.
HERMAN, W. S. & GILBERT, L. I. (1965). *Nature, Lond.* **205**, 926.
KAO, C. Y. (1966). *Pharmac. Rev.* **18**, 997.
KARLSON, P., HOFFMEISTER, H., HUMMEL, H., HOCKS, P. & SPITTELLER, G. (1965). *Chem. Ber.* **98**, 2394.
KERB, U., SCHULTZ, G., HOCKS, P., WIECHERT, R., FURLENMIER, A., FÜRST, A., LANGEMANN, A. & WALDVOGEL, G. (1966). *Helv. chim. Acta* **49**, 1601.
LEES, A. D. (1955). *The Physiology of Diapause in Arthropods.* Cambridge University Press.
MOORE, J. W. & NARAHASHI, T. (1967). *Fedn Proc. Fedn Am. Socs. exp. Biol.* **26**, 1655.
MOSHER, H. S., FUHRMAN, F. A., BUCHWALD, H. D. & FISHER, H. G. (1964). *Science, N.Y.* **144**, 1100.
NÜESCH, H. (1968). *A. Rev. Ent.* **13**, 27.
OHTAKI, T. & WILLIAMS, C. M. (1969). *Biol. Bull. mar. biol. Lab., Woods Hole.* (In the Press.)
SHAKHBAZOV, V. G. (1961). *Dokl. Akad. Nauk SSSR* **140**, no. 1 (AIBS), 914.
WIGGLESWORTH, V. B. (1964). In *Advances in Insect Physiol.* p. 247. Ed. J. W. L. Beament, J. E. Treherne and V. B. Wigglesworth. New York: Academic Press.
WILLIAMS, C. M. (1946). *Biol. Bull. mar. biol. Lab., Woods Hole* **90**, 234.
WILLIAMS, C. M. (1947). *Biol. Bull. mar. biol. Lab., Woods Hole* **93**, 89.
WILLIAMS, C. M. (1952). *Biol. Bull. mar. biol. Lab., Woods Hole* **103**, 120.
WILLIAMS, C. M. (1956). *Biol. Bull. mar. biol. Lab., Woods Hole* **110**, 201.
WILLIAMS, C. M. (1958). In *The Chemical Basis of Development,* p. 794. Ed. W. D. McElroy and B. Glass. Baltimore: Johns Hopkins Press.

WILLIAMS, C. M. (1967). In *Insects and Physiology*, p. 133. Ed. J. W. L. Beament and J. E. Treherne. London: Oliver & Boyd.
WILLIAMS, C. M. (1968). *Biol. Bull. mar. biol. Lab., Woods Hole* **134**, 244.
WILLIAMS, C. M. & ADKISSON, P. L. (1964). *Biol. Bull. mar. biol. Lab., Woods Hole* **127**, 511.
WILLIAMS, C. M., ADKISSON, P. L. & WALCOTT, C. (1965). *Biol. Bull. mar. biol. Lab., Woods Hole* **128**, 497.
WILLIAMS, C. M. & ROBBINS, W. E. (1968). *Bioscience* **18**, 791.

DIAPAUSE AND PHOTOPERIODISM IN THE PARASITIC WASP *NASONIA VITRIPENNIS*, WITH SPECIAL REFERENCE TO THE NATURE OF THE PHOTOPERIODIC CLOCK

By D. S. SAUNDERS

Department of Zoology, University of Edinburgh

INTRODUCTION

In the temperate regions of the world growth and reproduction of most insects can only take place during the favourable months of the year and winter is passed in a state of dormancy. The type of winter dormancy most frequently observed is a temporary cessation of the neuroendocrine control of growth or reproduction (diapause), which is generally under photoperiodic control (de Wilde, 1962). Thus 'long-day' insects will develop and breed during the summer months when the days are long (and nights short) but enter diapause in the autumn when day-lengths fall below a critical value. Competence for renewed development in the spring is brought about by a prolonged experience of low temperature, but the actual recovery of neuroendocrine activity is often also under the control of a photoperiodic mechanism (Williams & Adkisson, 1964).

In many respects the parasitic wasp *Nasonia* (= *Mormoniella*) *vitripennis* is a typical LD species. It is an external parasite of the pupae of Cyclorrhaphous flies, the immature stages living in the space between the pupa and the puparium. *Nasonia* produces several consecutive generations during the summer months and then enters diapause within the host puparium as a fourth instar larva just before defecation and the pupal moult. However, unlike the majority of insects with a facultative diapause of this type, the photoperiodic stimulus acts maternally and not upon the larvae directly (Saunders, 1965). Clearly a system as complex as this involves a large number of 'stages' between the reception of the photoperiod by the female parent and the production of diapause in the progeny. However, it would appear that day-length controls a switch mechanism, at long day-length directing development down one path (to pupation) and at short day-length down another (to diapause). Probably the most important stages—and those currently receiving the most attention—are the mechanisms involved in the measurement of the components of the light: dark cycle, and the nature of the endocrine systems which are presumably geared to the 'clock'. In addition there is the problem in *N. vitripennis* of the transmission of the photoperiodic information from the female parent, through the egg to the larva, and the subsequent programming of larval development to allow eclosion and three uninterrupted larval moults to occur before diapause intervenes.

The work described here is concerned with the development of *N. vitripennis* under conditions of 'normal' photoperiods and temperatures, and with the nature of the photoperiodic 'clock'.

THE INDUCTION OF DIAPAUSE IN *NASONIA VITRIPENNIS*

The effects of 'normal' photoperiods

When females of *N. vitripennis* are kept at short day-lengths (or long nights) or LD 6:8 to LD 14:10 and supplied each day with two pupae of their fleshfly hosts, *Sarcophaga barbata*, they produce most of their offspring as developing larvae for the first few days of imaginal life and then 'switch' to the production of diapause larvae after 5–11 SD cycles. For each female this changeover is very abrupt and mixed broods of developing and diapausing larvae only occur on the day of the 'switch'. For a population of females the rate of switching provides a curve as in Fig. 1. Once a female has switched to the production of diapause larvae she continues to produce them until her death.

At long day-lengths (or short nights) of LD 16:8 and over, females produce practically no diapause larvae, the switch, if occurring, doing so at the end of adult life after 20–25 light:dark cycles have been experienced.

Between 14 and 16 hr. light per day an abrupt change in the photoperiodic response occurs (Fig. 1) producing a well-defined critical day-length (Fig. 2), which in the C strain* is in the region of LD 15:9 to LD 15·25:8·75. At these intermediate day-lengths some females react as though at short day-length and switch early in life and others respond as though at long day-length and switch later. Altering the length of the light period at this point by as little as 15 min. may produce a significant difference in the photoperiodic response, thus demonstrating both the instability of the mechanism at the critical day-length and the accuracy of the photoperiodic clock.

Females kept in darkness also switch to the production of diapause larvae (Saunders, 1962), but react as though at 'intermediate' day-length; this response is typical of insects kept in 'constant darkness' (Lees, 1955).

* Isolated in Cambridge, 1961, by Dr G. Salt.

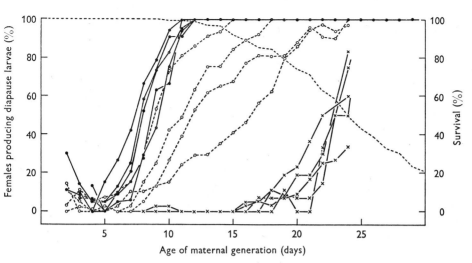

Fig. 1. The effect of photoperiod on the production of diapause larvae by females of *N. vitripennis* (C strain) at a constant temperature of 18° C. ●—●, At 'strong' short day-lengths (6, 8, 10, 12, 14 hr. per day). ○ - - ○, At intermediate day-lengths (from left to right: 14·5, 14, 15, 15 hr. per day). ×—×, At 'strong' long day-lengths (15·5, 15, 16, 18, 20 hr. per day). The dotted line shows the survival rate for the 400 females used in this experiment.

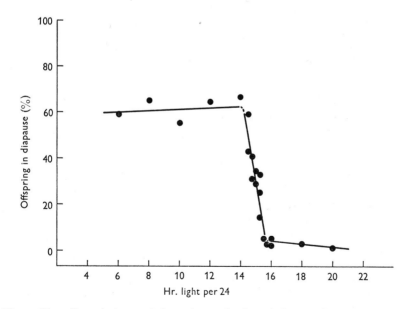

Fig. 2. The effect of photoperiod on the production of diapause larvae by females of *N. vitripennis* (C strain), showing the abrupt change at the critical day-length. The line has been fitted mathematically to subjectively selected groups of points.

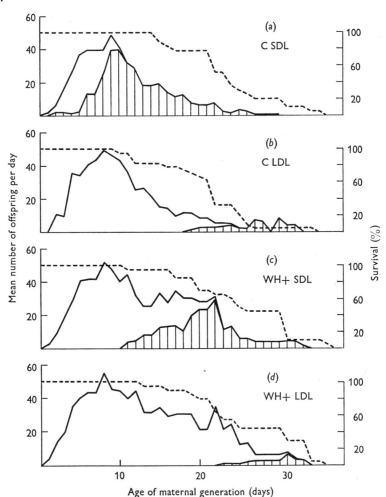

Fig. 3. The production of diapause larvae by females of *N. vitripennis* at short and long day-length (18° C.), showing its relation to egg production. (*a*) Nineteen C-strain females at short day-length (LD 14:10); (*b*) 24 C-strain females at long day-length (LD 16:8); (*c*) 20 WH+-strain females at short day-length (LD 10:14); (*d*) 19 WH+-strain females at long day-length (LD 16:8). The solid line shows the rate of egg production, with the shaded portion representing the diapause larvae. The dotted line shows the survival rate for the females in the group.

In a second strain (WH+)* photoperiod had a similar controlling effect on diapause production, except that a greater number of short-day cycles were needed to effect the switch (Fig. 3) and the critical day-length was in the region of LD 13·5:10·5.

Experiments in which the offspring of *N. vitripennis* were kept at different photoperiods and temperatures during their development showed

* Isolated at Woods Hole, Massachusetts, 1951, by Dr P. Whiting.

that the induction or inhibition of larval diapause is wholly determined by the photoperiod experienced by the adult female, and that the developmental 'future' of the egg is determined by the time it is deposited within its host (Saunders, 1966a). Sensitivity to day-length begins in the pupa; and transferring females from short to long day-length, or vice versa, will reverse the photoperiodic response (Saunders, 1965) (Fig. 4). These observations make it clear that the photoperiodic clock controlling diapause induction resides in the mother and not in the larvae she produces.

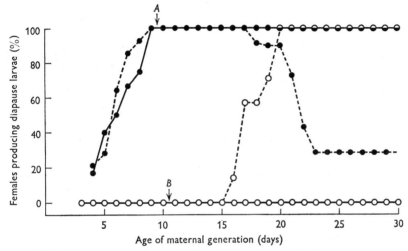

Fig. 4. The effect of transferring *N. vitripennis* (C strain) females from short to long, and long to short, photoperiods (15° C.). ●—●, Fifteen females at LD 6:18; ●--●, 15 females at LD 6:18 for 9 cycles, then transferred to natural long day-length at *A*; O—O, 15 females at natural long day-length; O--O, 7 females at natural long day-length for 10 cycles, then transferred to LD 6:18 at *B*.

A maternally operating photoperiod such as this is unusual but by no means unique. Other examples include the control of polymorphism in the aphid *Megoura viciae* (Lees, 1959) and the induction of larval diapause in the braconid *Coeloides brunneri* (Ryan, 1965) and the green blowfly *Lucilia caesar* (Ring, 1967). Similar maternal influences operate, or probably operate, in *Bombyx mori* (Kogure, 1933, Fukuda, 1951) and in the wasps *Spalangia drosophilae* and *Cryptus inornatus* (Simmonds, 1946, 1948).

The effects of temperature

The effect of temperature on the photoperiodic response is complex and involves at least two distinct aspects. On the one hand it appears to play a modifying role, egg production and longevity being controlled by normal temperature-dependent processes so that at high temperature females

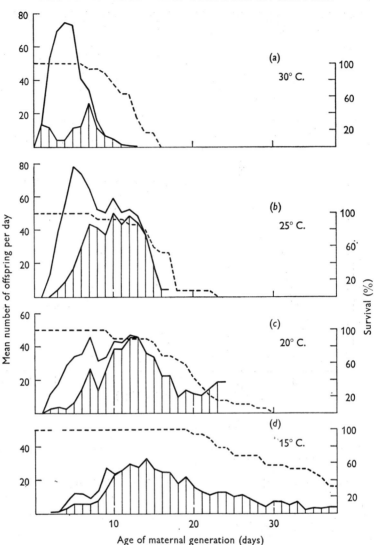

Fig. 5. The production of diapause larvae by females of *N. vitripennis* (C strain) at different temperatures and constant photoperiod (LD 12:12), showing its relation to egg production. (*a*) Seventeen females at 30° C.; (*b*) 15 females at 25° C.; (*c*) 19 females at 20° C.; (*d*) 19 females at 15° C. The solid line shows the rate of egg production, with the shaded portion representing the diapause larvae. The dotted line shows the survival rate for the females in each group.

have a short life-span and a high rate of egg production and at low temperature the reverse. On the other hand, the mechanisms associated with the photoperiodic clock are, in part at least, temperature-compensated. In particular, females kept at temperatures between 15° and 30° C in 'strong' short day-lengths (e.g. LD 12:12) require the same number of light: dark cycles to effect the switch even though the length of life and oviposition rate are markedly affected. The number of LD cycles needed to effect the switch has a Q_{10} of about 1·04.

Since temperature affects the rate of egg production but not the rate of switching to diapause it has a profound effect on the proportion of diapause larvae produced (Fig. 5). At 15° C., for instance, females lived for over 32 days and produced over 90% of their progeny as diapause larvae, whilst at 30° C. they died after about 12 days with less than 30% of their offspring in diapause. Females producing all-diapause broods are also more frequent at lower temperatures, thus exaggerating this effect. This mechanism is obviously of great ecological importance: if temperatures are high in the autumn the active period continues even though the critical day-length has been passed, and if low, diapause intervenes earlier than usual. Therefore, although the photoperiodic clock is so accurate, this mechanism provides a degree of flexibility.

A number of experiments have also been performed using a period of chilling during the early part of adult life (Schneiderman & Horwitz, 1958; Saunders, 1965). This treatment increases the proportion of diapause larvae in the offspring because egg production is delayed for a greater period of time than is the rate of switching to diapause. Chilling at very low temperatures (2° C.), however, has little effect since both egg production and the mechanism associated with the summation of LD cycles are stopped (Saunders, 1965, 1968).

The effects of host shortage

When females of *N. vitripennis* withdraw the ovipositor from the host they feed on the haemolymph which exudes from the puncture (Edwards, 1954). The host pupa thereby serves both as a place in which to deposit the eggs and as a source of protein to further egg production (Roubaud, 1917).

Shortage of hosts has two effects on the physiology of the insect. First, if newly emerged females are deprived of hosts for a few days they undergo a period of starvation in which the few eggs produced from larval reserves are resorbed (King, 1963; King & Hopkins, 1963). When the insects eventually find hosts, feeding and oviposition are resumed. Full egg production is therefore delayed during the period without hosts. Secondly, if host pupae are available but in short supply, several females may be induced to oviposit simultaneously in one host pupa. This results in superparasitism, competition for food between the resulting larvae, and hence small-sized adults in the next generation.

When these effects of host shortage were examined (Saunders, 1966*b*)

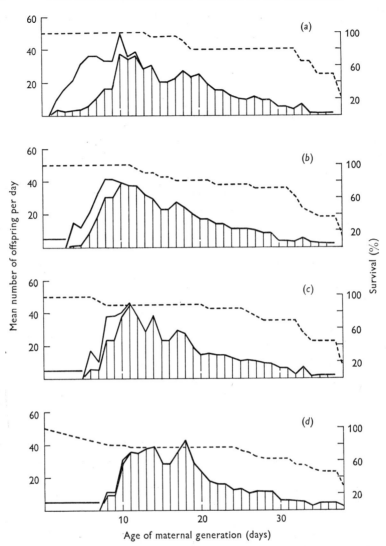

Fig. 6. The effect of host deprivation on the production of diapause larvae by females of *N. vitripennis* (C strain) at 18° C. and LD 12:12. (*a*) Nineteen females provided with two host *Sarcophaga* pupae daily; (*b*) 20 females deprived of hosts for 3 days; (*c*) 18 females deprived of hosts for 5 days; (*d*) 31 females deprived of hosts for 7 days. The solid line shows the rate of egg production, with the shaded portion representing the diapause larvae. The dotted line shows the survival rate for the females in the group.

it was found that body size had a profound effect on longevity and the number of offspring produced, but had no effect on the rate of switching to diapause. The proportion of diapause larvae produced, therefore, was independent of body size. Keeping females without host pupae for 3, 5 or 7 days at a short day-length (LD 12:12), however, had little effect on longevity or the over-all number of offspring, but had a marked effect on the proportion (and number) of diapause larvae produced because egg production was delayed whilst the mechanism summating LD cycles was not (Fig. 6). For instance, 7 days without hosts raised the proportion of diapause larvae from about 73 to 99 %.

This last result is especially interesting since it demonstrates that the mechanism associated with the summation of LD cycles is not only temperature-independent but also continues in starved females, and is therefore independent of ovarian function. At the ecological level it shows that host shortage acts in conjunction with short day-length to raise the proportion (and number) of diapause larvae produced. This effect would become most important in the autumn, when the day-length falls below the critical value and hosts become scarce due to a recession in the breeding blowfly population.

THE PHOTOPERIODIC CLOCK IN *NASONIA VITRIPENNIS*

The results described above show that the induction and inhibition of larval diapause in *N. vitripennis* are controlled by a photoperiodic mechanism and, since this effect is maternal, temperature and host shortage modify the response through alterations in the rate of egg production. Apart from this maternal aspect, which is unusual, mechanisms like this which serve to synchronize development with the seasons are well known in insects (de Wilde, 1962; Danylievsky, 1965) and are usually supposed to be governed by reference to some sort of 'physiological clock' (Bünning, 1960a).

The main evidence for the existence of such a 'clock' is provided by the photoperiodic response curve (Fig. 2). This curve shows a characteristically sharp change in response at the critical day-length, which demonstrates that the insect is capable of measuring the length of the day, the length of the night, or the length of both day and night, with a considerable degree of accuracy. It is usually assumed that this measurement is accomplished by reference to an inherited time-scale, the critical day-length.

Theories and models for the photoperiodic clock

Several theories and models to account for time measurement in photoperiodism have been proposed. These fall into two main categories: (*a*) those embodying an endogenous circadian rhythm, and (*b*) those in which the dark period (or the light) is measured by an 'hour-glass'.

As long ago as 1936 Bünning suggested that photoperiodism had a rhythmic basis. He proposed that the 24 hr. cycle was made up of two half-cycles, a 12 hr. 'photophil' ('light-requiring') and a 12 hr. 'scotophil' ('dark requiring'). The rhythm was then phased by the light and if the 'day' extended into the second or scotophil half-cycle long-day effects were produced, but short-day effects were produced when the light was restricted to the photophil. Evidence in favour of this hypothesis has been obtained in insects by interrupting a night of inductive length by short light-pulses ('night-interruption' experiments). In *Pieris brassicae* such light-pulses placed a few hours after dusk produced a strong inhibition of larval diapause (Bünning & Joerrens, 1960). Similar reversals of photoperiodic effect have been obtained in *Metriocnemus knabi* (Paris & Jenner, 1959), *Ostrinia nubilalis* (Beck, 1962), *Pieris rapae* (Barker, 1963) and a number of other insects. In plants, considerable evidence that a 24 hr. periodicity is involved in photoperiodism has been obtained by the use of abnormal LD cycles with long variable dark phases (Hamner, 1960), and by interrupting a very long night with supplementary light-pulses, e.g. the 'bidiurnal' and 'tridiurnal' cycles of Bünsow (1960), Melchers (1956) and Schwabe (1955).

Working with the pink bollworm, *Pectinophora gossypiella*, Adkisson (1964, 1966) showed that interrupting an inductive night with light-pulses produced two peaks of diapause inhibition. One of these was placed about 14 hr. after lights-on and the other about 14 hr. before lights-off, regardless of the main LD cycle employed. These observations led to the formulation of the 'coincidence model' by Pittendrigh & Minis (1964) and provide some of the most powerful evidence in favour of a circadian basis for photoperiodism.

The coincidence model to account for time measurement in photoperiodism is seen by Pittendrigh & Minis (1964) as an explicit version of Bünning's general hypothesis. It is based on Adkisson's (1964) data for *Pectinophora* together with their own for *Drosophila* eclosion rhythms and *Pectinophora* oviposition rhythms (Minis, 1965), and its strength lies in the close similarities between the behaviour of these phenomena in 'night-interruption' experiments or 'asymmetrical skeleton' photoperiods. According to Pittendrigh & Minis's (1964) original model there is a circadian rhythm of a substrate (s) which is phase-set by the light:dark cycle, the peak of the rhythm (s-max) coming to lie at about 14 hr. circadian time when the strongest (i.e. longest) short day-lengths are employed. There is also an 'enzyme' which is activated in the light but inactive in the dark. In animals no such 'enzyme' is known, but its role in plants could be filled by the pigment phytochrome (Hendricks, 1960). The coincidence model depends upon the temporal relationship between s-max and the active phase of the 'enzyme', and thus takes into account the dual role of light—as inducer and entraining agent—which Pittendrigh & Minis stress as the crux of the model.

The model operates in exactly the same way as Bünning's. For instance, at short day-length the light does not coincide with *s-max* and diapause is induced, but at long day-length coincidence does occur and diapause is inhibited.

When subjected to night-interruption experiments the moth *Pectinophora gossypiella* accepts the main light period and the pulse as one. If the pulse is placed early in the night it is accepted as dusk and if placed late in the night as dawn. Between the two is a 're-setting' zone where the pulse cannot be accepted as either. The first peak of diapause inhibition (A) is thus interpreted as the point at which the light pulse coincides with *s-max*, producing a reaction between the substrate and the activated 'enzyme'. The second peak (B) is seen as the point at which the light-pulse serves to phase-set the substrate rhythm so that *s-max* coincides with the end of the main light period and diapause is again averted. Between the two, where light-pulses have no effect, is where a 'phase-jump' occurs. This is precisely the way in which *Drosophila* eclosion and *Pectinophora* oviposition rhythms behave in similar photoperiodic regimens.

In a later paper, Pittendrigh (1966) modified this model to some extent. First, in order to be less committed to the concept of a substrate oscillation as such, *s-max* is replaced by the term 'photoperiodically inducible phase (ϕ_i)'. Secondly, he raises the point that ϕ_i could be either at peak A or peak B. The *Drosophila pseudoobscura* eclosion data (Pittendrigh, 1966) which have inspired this model show that for complete photoperiods of 12 hr. duration or more, the onset of darkness always corresponds to circadian time (Ct) 12. Therefore any extension of the main light period would be such that 'dawn' moves backwards into the previous night. Pittendrigh therefore suggests that ϕ_i lies late in the night (at B) rather than at A, as suggested in the original coincidence model. In all other respects the model remains the same.

An alternative hypothesis has been suggested by Beck (1964, 1968) for the European corn borer, *Ostrinia nubilalis*. In this insect it is claimed that there are two endogenous rhythms with an 8 hr. periodicity; a secretory rhythm in the lateral neurosecretory cells of the brain is phase-set by the lights-on stimulus of the photoperiod, and a second secretory rhythm (of the hormone proctodone) by the ileal epithelium which is phase-set by the onset of darkness. Photoperiodic induction, according to this hypothesis, then depends on the phase relationship between the two rhythms, development occurring when they are held in an in-phase relationship, and diapause resulting when they are held out of phase. A two-oscillator model for the photoperiodic clock was also proposed independently by Goryshin & Tyshchenko (1968).

Many of the early workers in photoperiodism favoured an 'hour-glass' concept to account for time measurement. In this hypothesis it was supposed that a chemical reaction was set in motion either at dawn or dusk which, after a definite length of time, produced a threshold amount of a substance which triggered the photoperiodic switch. The phytochrome system in plants, with the conversion of P_{735} to P_{660} when the light goes off, has been equated with this model (Hendricks, 1960).

Evidence in favour of an 'hour-glass' which measures the duration of the night has been produced in recent years (Lees, 1965, 1966). Furthermore, in *Megoura viciae* the data used by Lees in support of this hypothesis were provided by identical experiments (abnormal light: dark cycles and 'night interruption' experiments) to those which produce evidence for a circadian basis for photoperiodism.

Working with the aphid *M. viciae*, Lees (1965, 1966) found that a night of inductive length produced a full effect (ovipara production) even when coupled with 'days' of up to 36 hr. duration. This immediately suggests that the photoperiodic clock in this species involves a timer which measures the duration of the night. Similar results were also obtained with the mite *Panonychus ulmi* (Lees, 1953) and with the silkworm *Antheraea pernyi* (Tanaka, 1950), in which long nights were still effective when coupled with 'days' as long as 59 hr. Stronger evidence for an hour-glass in *Megoura* was obtained by the use of night-interruption experiments (Lees, 1965, 1966). In this species two peaks of diapause inhibition were obtained which were very similar to those with *Pectinophora gossypiella* (Adkisson, 1964). However, the positions of the peaks were identical after a light period of 8, 13·5 or 25·5 hr. Therefore, since the essence of any circadian model for photoperiodism requires that a rhythm of susceptibility to light is phase-set by the main light: dark cycle, and no 'higher' organism is known which can be entrained to a cycle as long as 36 hr. (Bruce, 1960), these results are regarded as evidence that the photoperiodic clock in *Megoura* is not based on a circadian rhythm but embodies an interval timer which measures the dark. One of the most important consequences of this work is that it raises an important question as to the fundamental nature of photoperiodic timing mechanisms. That is, despite the many striking similarities between different species in their reactions to both normal and abnormal photoperiodic regimens, and the considerable weight of circumstantial evidence in favour of a circadian model, are there two or more mechanisms involved?

Since the differences between the two hypotheses—hour-glass and circadian—appear to be so fundamental, an investigation of the nature of the photoperiodic clock in *N. vitripennis* was commenced, paying particular attention to its possible circadian aspects. The experimental approach included the use of abnormal light:dark cycles, night-interruption experiments and chilling.

Abnormal light:dark cycles

Abnormal light:dark cycles, in which one component of the cycle is held constant and the other varied, have provided a considerable amount of information about the nature of the photoperiodic clock. For instance, by exposing the Biloxi variety of soybean and other plants to cycles containing a short light period and a long dark period of variable length, Hamner (1960) showed that a cyclical response was produced so that maximum flowering was obtained at cycle lengths of 24, 48 and 72 hr. Pittendrigh (1966) refers to such experiments as T-experiments (T = period of the Zeitgeber) and considers that the results demonstrate that ϕ_i appears during the extended night with circadian frequency.

In the experiments with *N. vitripennis*, two types of abnormal LD cycles were investigated. In the first, the light component was held constant at 12 hr. and the length of the dark varied, and in the second, the length of the night was held constant at 12 hr. and the length of the day varied. Cycle lengths between 19·2 and 48 hr. were used, and all experiments were conducted at 18° C. A cycle of LD 12:12 was used throughout as a control.

When the light period was held constant at 12 hr. and the length of the dark varied a high degree of induction was obtained only within the quite narrow limits of 23–25·5 hr. cycle length (Fig. 7). Above and below these limits the females reacted as though at 'intermediate' or long-day conditions even though (at LD 12:18, for instance) the night-length may have been longer than the day. At cycle lengths longer than about 36 hr. (LD 12:24) females reacted as though they were in continuous darkness and produced an intermediate response. Although the results for the longer cycle lengths are rather incomplete, it is possible that this result shows a cyclical occurrence of ϕ_i similar to that demonstrated by Hamner (1960) for the Biloxi soybean.

When the dark period was held constant for 12 hr. and the light component varied a similar response was obtained in that high induction was observed only within the range of 21–26 hr. cycle length. Above and below these limits the females behaved as though at intermediate or long day-lengths. The limits for high induction are slightly wider in the case of cycles with a 12 hr. night being held constant, perhaps indicating that night-length is more important than day-length. However, whilst it is obvious that the lengths of the light and dark periods are important, it is clear that the over-all cycle length is the dominant feature. Indeed, the range of cycles which produces a high rate of diapause induction is reminiscent of the ranges of entrainment of overt circadian rhythms—such as activity rhythms—in higher plants and animals (Bruce, 1960). This result therefore could be regarded as circumstantial evidence for the involvement of a circadian rhythm in the *Nasonia* photoperiodic clock.

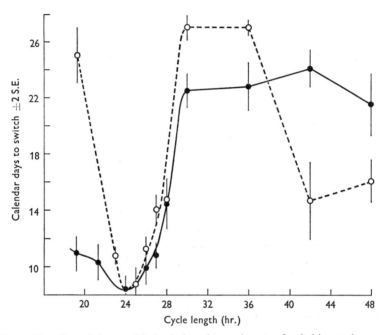

Fig. 7. The effect of abnormal light:dark cycles on the rate of switching to the production of diapause larvae in *N. vitripennis* (C strain). ○ - - ○, twelve hr. light held constant; ●—●, 12 hr. dark held constant. Vertical lines represent 95 % fiducial limits of the mean; 370 females were used in this experiment.

Night-interruption experiments

Probably the most valuable approach to the problem of time measurement in photoperiodism has been that of night interruption. The discovery that short light-pulses suitably placed in the dark period could produce significant reversals of photoperiodic response, whereas similar dark periods placed in the light did not, is one of the strongest pieces of evidence that the duration of the night is the important component of the LD cycle. Moreover, the data obtained from night interruption experiments with *Pectinophora gossypiella* led to the formulation of the most explicit version of Bünning's hypothesis so far produced (Pittendrigh & Minis, 1964;

Pittendrigh, 1966). In experiments of this nature the dark period is interrupted by a short light-pulse (1-2 hr. or as little as a few seconds), each experimental group receives a pulse in a different position, and the pulse is repeated as a daily signal.

In the present work with *N. vitripennis* three night-interruption experiments using cycles with a 24 hr. period (LD 14:10, LD 12:12 and LD 10:14) and 1 hr. pulses have been performed (Fig. 8). All of these photoperiods are strong short day-lengths when uninterrupted.

When the 10 hr. night of an LD 14:10 cycle was interrupted systematically by 1 hr. supplementary light periods two peaks of diapause inhibition were obtained, very similar to those observed in *Pectinophora gossypiella* (Adkisson, 1964), except that the peaks appear 'reversed' since 'low diapause' is at the top of the ordinate (Fig. 8). The first peak (*A*) occurred about 1-2 hr. after lights-off (Zt 15-16) and the second peak (*B*) about 2-3 hr. before lights-on (Zt 21-22). If the coincidence model is followed the pulse at peak *A* is accepted as 'dusk' and defines with the main photoperiod a long day-length of 16 hr., and peak *B* serves as an entraining agent so that the inducible phase (ϕ_i) falls at the end of the main light component, once more definining a long-day of 16 hr. Between the two, interruptions failed to inhibit diapause.

When a shorter light period was employed (i.e. LD 12:12 or LD 10:14) two peaks of diapause inhibition were again obtained. Peak *A* was at Zt 15-16 in both LD 14:10 and LD 12:12 but tended to 'drift' towards the end of the main light component (to Zt 13-14) when a 10 hr. 'day' was employed. Similarly, peak *B* occurred 15-17 hr. before lights-off in the two longer photoperiods, but drifted to 15-16 hr. before lights-off in a basic LD 10:14 regimen. Similar tendencies for the peaks to drift are discernible in the data for *Pectinophora gossypiella* (Pittendrigh & Minis, 1964).

Peak *A* was in all cases weaker than peak *B*. One reason for this could be that 1 hr. light is ample to act as an entraining agent (at *B*) but provides barely sufficient energy for the photochemical processes associated with diapause inhibition (at *A*). In addition, since some females exposed to an LD 14:1:1:8 regimen reacted as though at long day-length and some as though at intermediate conditions (i.e. switching to diapause larvae after 9-16 cycles) it is possible that a genetic diversity between females exists, so that some females are able to accept the *A* pulse as dusk and others are unable to do so.

Figure 9 illustrates how the photoperiodic clock in *N. vitripennis* probably operates. Since the position of peak *A* remains constant in relation to lights-on and that of peak *B* constant in relation to lights-off, their relative positions within the night alter as the length of the main light period is

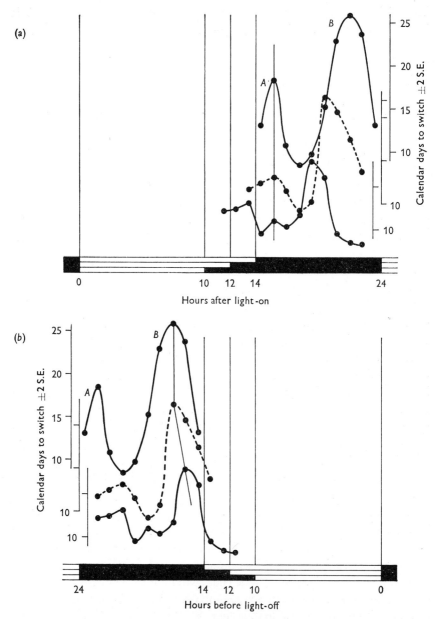

Fig. 8. The effects of 1 hr. supplementary light-pulses ('night interruptions') in the dark components of three cycles with a 24 hr. period. (a) Results presented as hours after lights-on, and (b) results presented as hours before lights-off. Top curve, results for LD 14:10; middle curve, for LD 12:12; lowest curve, for LD 10:14. Note that peak A occurs 15–16 hr. after lights-on, and peak B 15–16 hr. before lights-off, with some tendency for the peaks to 'drift' in the LD 10:14 regimen.

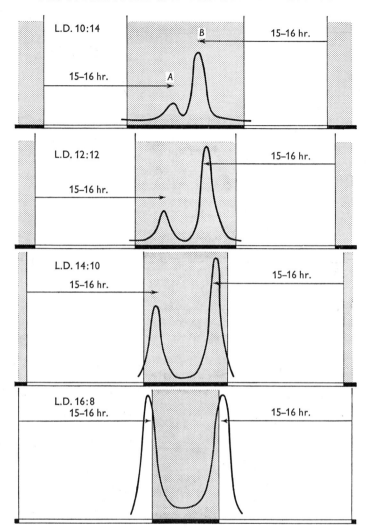

Fig. 9. The photoperiodic clock in *Nasonia vitripennis* (schematic). Two peaks of diapause inhibition (light sensitivity) are shown in the dark component. Peak A is shown as occurring 15–16 hr. after lights-on in all light:dark cycles, peak B at 15–16 hr. before lights-off. As the day-length increases the peaks move closer to either dusk or dawn and also increase in amplitude. In the lowest panel (LD 16:8) the peaks of diapause inhibition have passed from the dark into the light and long-day effects are produced.

changed. As the night becomes shorter, A moves towards dusk and B towards dawn and, as they do so, their 'amplitude' increases. Eventually, when the light component of the cycle increases to 15–16 hr. (i.e. the critical day-length) the peaks of diapause inhibition 'move' into the light so that long-day effects are produced.

Chilling

When females of *N. vitripennis* were chilled at 2° C. for 4 hr. daily, spectacular reversals of photoperiodic response were sometimes obtained (Saunders, 1967) which provide some information as to the nature of the photoperiodic clock. For instance, at LD 14:10, a photoperiod just short of the critical point (LD 15·25:8·75), chilling at the beginning or in the middle of the long night converted the response from that of a short day to a long one (Table 1). Chilling in the light period of this cycle merely had a strengthening effect on the short-day response. In the converse experiment, at LD 16:8, chilling in the dark had no effect but chilling in the light reversed the response from that of a long day to that of a short day. When this type of experiment was repeated with a cycle of LD 8:16, which is well short of the critical day-length, no reversal was obtained, although chilling in the dark strengthened the short-day effect.

Reversals of photoperiodic effect such as these are rare in the literature on photoperiodism. The closest parallel is probably in the moth *Acronycta rumicis*, in which chilling at 5° C. for 3 hr. at the beginning or the end of a 17 hr. day converted the long-day effect to a short one (Danylievsky, 1965). In this case, however, chilling in the middle of the light had no such effect; neither did chilling in the dark reverse an otherwise short day-length

Table 1. *The reversal of photoperiodic effect in* Nasonia vitripennis *by a daily period of chilling*

Temperature (° C.)	Chilling applied (Zeitgeber time)	Females (no.)	Mean days to 'switch' (±S.E.)	Delay (+) or acceleration (−) in 'switch' (days)
	LD cycle 14:10			
18	No chilling (control)	58	9·7±0·35	
2	0–4	40	8·3±0·37	−1·4
2	5–9	18	8·6±0·50	−1·1
2	14–18	39	24·0*	+14·3
2	17–21	36	23·0*	+13·3
	LD cycle 16:8			
18	No chilling (control)	36	23·0*	
2	0–4	16	10·6±0·70	−12·4
2	6–10	19	12·7±0·72	−10·3
2	16–20	20	23·0*	0·0
	LD cycle 8:16			
18	No chilling (control)	36	10·6±0·69	
2	0–4	16	12·2±1·09	+1·6
2	8–12	19	7·6±0·65	−3·0
2	14–18	18	9·5±0·80	−1·1

* None of the females had 'switched' after 23–24 cycles.

(LD 14:10). In plants, reversals of the photoperiodic control of flowering in the short-day plant, *Xanthium pensylvanicum*, by chilling have been recorded by de Zeeuw (1957) and Nitsch & Went (1959). Schwemmle (1960) also reported reversals by chilling in *Hyoscyamus niger* and *Perilla ocymoides*. Most of these authors remark that chilling in the light has a similar effect to that of darkness, and Schwemmle (1960) considered that the results indicated that the endogenous time-measuring process started with the beginning of the main light component.

The simplest way of looking at the results with *N. vitripennis* is to suppose that chilling slows down or stops the mechanism of time measurement so that at photoperiods close to the critical day-length shortening either the light or the dark component may convert the effect from short to long day or vice versa. These results therefore suggest that although it is the duration of the dark component which is measured, light and dark are both important in the photoperiodic clock. Furthermore, reversal of photoperiodic effect by chilling in the dark could be accounted for by an hour-glass mechanism measuring only the dark, but the reversal from long- to short-day effect by chilling in the light of an LD 16:8 cycle cannot. In other words, time measurement must begin with lights-on and not merely with lights-off, and involve the whole light:dark cycle.

Night interruption with a shortened cycle length or chilling

The crux of the coincidence model is the entrainment of a circadian rhythm of light sensitivity by the light:dark cycle so that at different photoperiods ϕ_i comes to lie at different points in relation to the Zeitgeber. Therefore one way of testing whether a circadian rhythm is involved in the photoperiodic clock is to vary the length of the light component whilst keeping the night-length constant and scanning the night with light-pulses to find the positions of the peaks of diapause inhibition. If a rhythm is involved one would expect it to attain a different relation to the Zeitgeber so that the positions of the peaks would move when compared with a similar night-length in a 24 hr. cycle. This was the test Lees (1965, 1966) applied to *Megoura viciae*; he found that the peaks did not move and therefore came to the conclusion that an interval timer rather than a circadian rhythm was involved in the *Megoura* clock.

Working with *N. vitripennis* the light component was shortened by two methods. In the first, an abnormal cycle length of 21·3 hr. was used, and in the second, the insects were chilled for a period (3 or 4 hr.) during the light component. Both of these treatments were compared with an LD 14:10 cycle as a control.

When the light component was shortened to 11·3 hr. and the 10 hr. night scanned with 1 hr. supplementary light-pulses two peaks of diapause inhibition were obtained as in the control regime, using an LD 14:10 cycle. Peak *A*, however, had moved by about 2 hr. further into the night (Fig. 10), so that it fell 3–4 hr. after lights-off instead of 1–2 hr. after lights-off as in LD 14:10. This result confirms that time measurement begins with 'dawn' rather than with 'dusk' and shows that shortening the

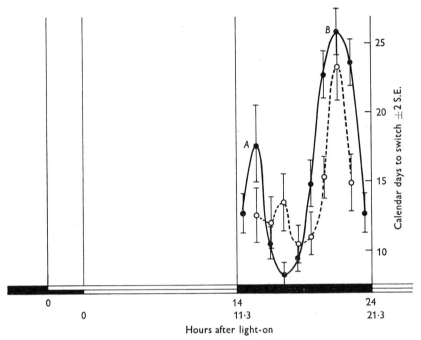

Fig. 10. The effect of interrupting the dark component of an LD 11·3:10 cycle with 1 hr. light pulses, showing the shift of peak *A* by about 2 hr. further into the night when compared with a similar experiment at LD 14:10. ●—●, At 14:10; ○--○, at LD 11·3:10. Vertical lines represent 95% fiducial limits of the mean; 252 females were used in this experiment.

light component alters the position of ϕ_i in the night. This result is predictable from the coincidence model if it is assumed that the rhythm is entrained to an LD 11·3:10 cycle. However, it would also be expected that peak *B* would move to the left. There was little evidence of this occurring but peak *B* occurred at about the same time before lights-off (13·3–14·3 hr.) as peak *A* did after lights-on (14·3–15·3), a situation which is to be expected from the model.

This result is very different to that obtained by Lees (1966) and is strong evidence that an interval timer as envisaged by Lees for *Megoura* does not operate in *Nasonia*.

PHOTOPERIODISM AND DIAPAUSE IN NASONIA 321

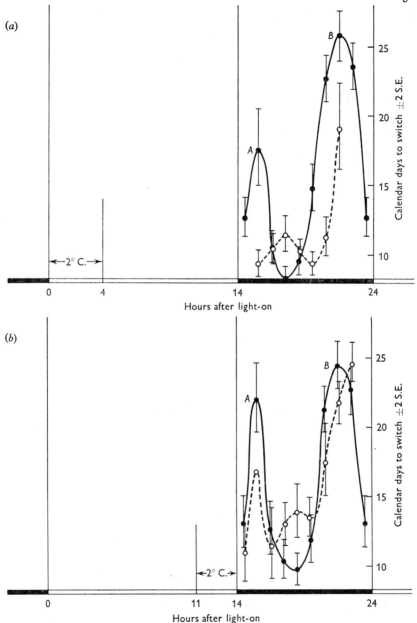

Fig. 11. The effects of chilling (2° C.) in the light component of LD 14:10 on the positions and amplitudes of the peaks of diapause inhibition as revealed by night-interruption experiments. ●—●, LD 14:10, unchilled; ○ – – ○, LD 14:10 chilled. Vertical lines represent the 95% fiducial limits of the geometrical means. (a) Chilling for 4 hr. at Zt 0–4. Peak A moves about 2 hr. farther into the night (to Zt 17–18) and is reduced in amplitude; 251 females were used in this experiment. (b) Chilling for 3 hr. at Zt 11–14. Peak at Zt 15–16 reduced in amplitude; evidence of a second peak at about Zt 18–19; 277 females were used in this experiment.

21 WDA 23

A similar shift in the position of peak A was obtained after chilling for 4 hr. (Zt 0–4) at the beginning of the light component (Fig. 11a). This again shows that time measurement begins with lights-on, and involves the whole of the light:dark cycle. The period at low temperature is seen as slowing down or stopping the mechanism of the clock, thus reducing the time available for time measurement from 24 to 20 hr. each day. If a circadian rhythm is involved this would mean that the rhythm is now forced to oscillate within the 20 hr. available at suitable temperature. If it is then assumed that the rhythm is able to entrain to a period as short as 20 hr. and remains symmetrical, it is to be expected that peak A would move farther into the night. This is precisely what the results have shown.

When females of *N. vitripennis* were chilled at the end of a 14 hr. light period (Zt 11–14) a similar shift of A was recorded (Fig. 11b), although a peak was still obvious at its 'normal' position. One explanation for this result could be that when chilling was applied at the beginning of the light component it prevented lights-on being accepted as a Zeitgeber ('dawn') so that the effective cycle was reduced to LD 11:10. However, when chilling was applied at the end of the main light component, lights-on and lights-off (the end of the pulse) were unaffected. This therefore suggests that some of the individual rhythms in the population were able to continue as in unchilled females because the oscillation was able to proceed at a temperature as low as 2° C. This explanation, however, is conjectural and has no experimental support without the concurrent analysis of a suitable rhythm.

If the explanation for the shift of peak A to the right by chilling in the light is accepted, then, for similar reasons, chilling in the dark should move peak A to the left (Saunders, 1968). This possibility has been investigated by chilling females for 3 hr. periods during the dark component of an LD 12:12 cycle, whilst scanning the night with 1 hr. supplementary light pulses.

When the females were chilled early in the night (Zt 14–17)—at a point where the first peak occurred in the controls—peak A appeared to move to the left (to Zt 13–14) although its 'amplitude' was lowered and the differences between the experimental group and the control were not significant (Fig. 12a). Light-pulses applied during the period at 2° C. were apparently 'not seen' by the females since the rate of switching to diapause was no different from the control groups chilled from Zt 14–17 without light-pulses. The most significant effect of chilling in this experiment was the considerable increase in diapause inhibition in the latter half of the night.

In a second experiment of this type females were chilled in a position

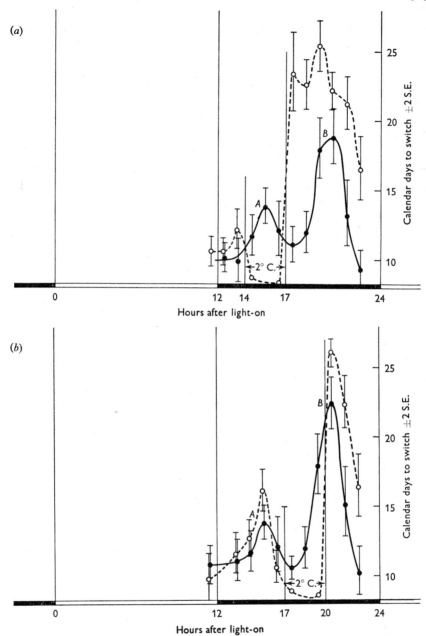

Fig. 12. The effects of chilling (2° C.) in the dark component of LD 12:12 on the positions and amplitudes of the peaks of diapause inhibition as revealed by night interruption experiments. ●—●, LD 12:12 unchilled; ○– –○, LD 12:12 chilled. Vertical lines represent the 95% fiducial limits or the geometrical means. (*a*) Chilling for 3 hr. at Zt 14–17. Diapause reversal is suppressed during the period at 2° C. Peak *A* appears to move to the left with its amplitude reduced. The amplitude of peak *B* is significantly increased; 313 females were used in this experiment. (*b*) Chilling for 3 hr. at Zt 17–20. Diapause reversal is suppressed during the period at 2° C. Both peaks show significantly increased amplitude; there is also evidence that peak *A* moved to the left and peak *B* to the right; 333 females were used in this experiment.

(Zt 17–20) between the two peaks of diapause inhibition. In this case both peaks increased their 'amplitude' (peak A at Zt 15–16, $t = 2\cdot32$, $P < 0\cdot05$) and there was evidence that peak A shifted somewhat to the left and peak B to the right, although only the latter was significant (Fig. 12b). Both of these shifts are predictable from the coincidence model if it is assumed that the circadian oscillation is forced to operate in the shorter time available at suitable temperature (21 hr.), although the shifts were less dramatic than expected.

These results are interesting for two reasons. First, the results are strong —although still circumstantial—evidence for the involvement of a circadian rhythm in the *Nasonia* photoperiodic clock. Particularly important in this respect are the findings that chilling in the light shifts peak A farther into the night, and that chilling in the dark shifts peak A towards dusk. The first effect would not have been obtained had an interval timer (as envisaged by Lees for *Megoura*) been operating, and the latter demonstrates a 'carry-over' effect from one cycle to another which is most consistent with a rhythmic hypothesis.

Secondly, these results provide an explanation for the reversals of photoperiodic effect obtained when females are chilled in the light or the dark in cycles close to the critical point. For instance, in females chilled for 4 hr. at Zt 0–4 in an LD 14:10 cycle the position of peak A (ϕ_i?) is shifted from Zt 15–16 to Zt 17–18. If a similar shift had occurred in a light:dark cycle of LD 16:8, ϕ_i would have moved from the light to the dark and produced a short-day effect. Similar arguments can be put forward for the conversion of LD 14:10 from short to long day by the shifting of peak A to the left when the females are chilled in the dark. However, it appears that the shift in the peaks may be less important in this respect than the changes in *amplitude*. Chilling in the light reduced the amplitude of the peaks, thereby encouraging the appearance of short-day effects, and chilling in the dark increased the amplitude of the peaks, thereby encouraging the appearance of long-day effects. How an identical treatment (chilling for 3 or 4 hr.) has different effects on the rate of switching to diapause in different positions of the light:dark cycle is not explained.

DISCUSSION

The reversal of photoperiodic response obtained when females of *N. vitripennis* were chilled during the light component of a long-day cycle (LD 16:8), and the narrow range of abnormal light:dark cycles (21–26 hr.) which gave a high rate of induction, show that time measurement (1) begins with the lights-on signal rather than with lights-off, (2) involves the whole light:dark cycle and not just the dark, and (3) is probably based on a rhythmic or circadian mechanism. This inference is strengthened by the results from night-interruption experiments. Those with a cycle length of 24 hr. were very similar to those for *Pectinophora gossypiella* (Adkisson, 1964) and can therefore be interpreted in terms of the coincidence model. Those in which the main light component was shortened by using an abnormal LD cycle, or by chilling, confirmed that time measurement begins with lights-on, and show that the positions of the 'peaks' of diapause inhibition (light sensitivity) are related to the duration of the main light component. These results are most adequately explained by the coincidence model (Pittendrigh & Minis, 1964) and are strong, although circumstantial, evidence for a circadian component in the *Nasonia* photoperiodic clock.

In the context of photoperiodism in general the important questions seem to be: how universal is a circadian basis to time measurement; are there one or two 'peaks' of diapause inhibition in response to night interruption experiments; and is the inducible phase (ϕ_i) early or late in the subjective night? These questions will be dealt with below.

Circadian rhythm or interval timer?

There are several difficulties with the hypothesis that photoperiodic time measurement involves a circadian rhythm, both in terms of the fundamental nature of the mechanism and in the specific features of the coincidence model. There is, for instance, a considerable body of evidence to suggest that the seasonal photoperiodic clock is quite different in different organisms, and is not always based on the circadian system despite the apparent universality of these rhythms. The majority of plant and animal physiologists appear satisfied that the clock is essentially circadian, but Lees (1965, 1966) has produced seemingly irrefutable evidence which indicates that the clock in *Megoura* is of the hour-glass type, measuring only the dark component of the cycle. The evidence in favour of this conclusion centres essentially on the facts that a high rate of induction is obtained with a night of inductive length coupled with a very wide range of 'days', and that altering the duration of the main light component fails to alter the positions of the peaks of diapause inhibition in night interruption experiments. Both of these results are in contrast to those observed for *N. vitripennis*.

Pittendrigh (1966) has attempted to explain Lees's data in order to preserve the concept of the universality of circadian rhythms in photoperiodism. He points out that the clock in *Megoura* could be based on a circadian rhythm, but if this rhythm is 'damped out' by light periods greater than 12 hr. and the onset of darkness always corresponds to Ct 12—as with the eclosion rhythm in *Drosophila pseudoobscura*—the oscillation would 'restart' again at dusk (Ct 12) and 'measure' the duration of the dark as an interval timer. He also suggests that Lees's experiment using a night of inductive length coupled with a very long light period would be expected to produce a high rate of induction for the same reason: the oscillation would stop during the extended 'day', restart at dusk and measure the night as an interval timer. However, this explantion does not account for Lees's failure to find a rhythm of sensitivity to light in a very long night scanned by short light-pulses (Lees, 1966) or in cycles which contain a short day and long variable night lengths (Lees, 1965). In addition, although Pittendrigh's explanation is sufficient for Lees's failure to shift the peaks of diapause inhibition in an LD 25·5:10·5 cycle (Lees, 1965) it is not for a similar result with a shortened light component (LD 8:10·5) (Lees, 1966). At the moment therefore it must be accepted that the photoperiodic clock in *Megoura* is not based on a circadian rhythm. This suggests that despite the similarities between different organisms in their reactions to identical photoperiodic regimens, more than one type of clock probably exists, and the similarities between them are probably the result of all organisms having evolved in an environment with a 24 hr. periodicity.

One peak or two?

Although considerable evidence in favour of a circadian base to photoperiodism has been obtained in many plants, birds and insects, the individual details of the clock in different species appear to differ. For instance, two peaks of diapause inhibition have been found in night interruption experiments with *Pectinophora gossypiella* (Adkinsson, 1964), *Pieris rapae* (Barker, Cohen & Mayer, 1964), *P. brassicae* (Goryshin & Tyshchenko, 1968), *N. vitripennis* (Saunders, 1968) and *Megoura viciae* (Lees, 1965), but in plants and birds, and many insect species, a single mid-point reversal of photoperiodic effect has been recorded. A single peak of diapause inhibition was observed in the codling moth, *Carpocapsa pomonella* (Peterson & Hamner, 1968), the depression occurring at Zt 14–15 in a cycle of LD 6:18 but moving to Zt 16–17 in a cycle of LD 13:11. A similar mid-point depression was recorded in *Ostrinia nubilalis* (Beck, 1968), but in this species night interruption experiments have not been systematically explored.

Pittendrigh & Minis (1964) suggest that time measurement is accomplished by a temperature-compensated system and quote as an example the 'temperature-independence' of the position of the critical day-length in *Pieris brassicae* (Bünning & Joerrens, 1960). However, in several other species the critical day-length does change with temperature, and in the moth *Acronycta rumicis* it shifts to shorter values by about 1·5 hr. for every 5° C. rise in temperature (Danylievsky, 1965). From these examples it would appear that the coincidence model is inadequate for some species.

Is the 'inducible phase' early or late in the subjective night?

For those species of insect which show two peaks of diapause inhibition in response to night-interruption experiments, Pittendrigh (1966) has pointed out that although both peaks represent places where light acts as an entraining agent, only one represents the position of the inducible phase (ϕ_i). Furthermore, from data of this type it is not possible to say whether ϕ_i falls on A or B. Since the pupal eclosion data for *D. pseudoobscura*—upon which the coincidence model is partly based—show that the oscillation is damped out by light periods longer than about 12 hr. and that the light:dark transition always corresponds to Ct 12, Pittendrigh has suggested that ϕ_i falls at B. As the day-length increases beyond 12 hr. 'dawn' moves 'backwards' into the previous night until it coincides with B and long-day effects are produced.

The work of Barker (1963) and Barker, Cohen & Mayer (1964) suggest that this is the case with *Pieris rapae*. Two peaks of diapause inhibition are found, one at Zt 13–14 and one at Zt 20–21. When pulses of light as short as 5 min. are used, diapause inhibition at A is greater than at B, and when a photoflash with a discharge half-life of 0·8 msec. is used, peak B disappears altogether. Since short flashes of light are known to phase-set and initiate rhythms, this result seems to indicate that sufficient light was present at A to act as an entraining agent, but insufficient at B for the photochemical processes associated with diapause inhibition. In other words, ϕ_i is at B in *P. rapae*. In *N. vitripennis* (and *P. gossypiella*), however, peak A is always smaller than B, and this suggests that ϕ_i is at A as in the original coincidence model (Pittendrigh & Minis, 1964). In the house sparrow the inducible phase also seems to lie early in the subjective night (Menaker & Eskin, 1967). These examples seem to indicate tht ϕ_i may lie at either A or B in different species, depending on the phase-relationship between the rhythm and the Zeitgeber.

I would like to acknowledge the assistance of Mrs D. Sutton and Mrs M. H. Downie. This work was financed by a grant from the Science Research Council.

REFERENCES

ADKISSON, P. L. (1964). *Am. Nat.* **98**, 357.
ADKISSON, P. L. (1966). *Science, N.Y.* **154**, 234.
BARKER, R. J. (1963). *Experientia* **19**, 185.
BARKER, R. J., COHEN, C. F. & MAYER, A. (1964). *Science, N.Y.* **145**, 1195.
BECK, S. D. (1962). *Biol. Bull. mar. Biol. Lab., Woods Hole* **122**, 1.
BECK, S. D. (1964). *Am. Nat.* **98**, 329.
BECK, S. D. (1968). *Insect Photoperiodism.* New York and London: Academic Press.
BRUCE, V. G. (1960). *Cold Spring Harb. Symp. quant. Biol.* **25**, 29.
BÜNNING, E. (1960a). *Cold Spring Harb. Symp. quant. Biol.* **25**, 1.
BÜNNING, E. & JOERRENS, G. (1960). *Z. Naturf.* **15b**, 205.
BÜNSOW, R. C. (1960). *Cold Spring Harb. Symp. quant. Biol.* **25**, 257.
DANYLIEVSKY, A. S. (1965). *Photoperiodism and Seasonal Development of Insects.* London: Oliver and Boyd.
EDWARDS, R. L. (1954). *Q. Jl microsc. Sci.* **95**, 459.
FUKUDA, S. (1951). *Proc. imp. Acad. Japan* **27**, 672.
GORYSHIN, N. I. & TYSHCHENKO, V. P. (1968). In *Photoperiodic Adaptations in Insects and Acari.* Ed. A. S. Danylievsky. Leningrad University Press.
HAMNER, K. C. (1960). *Cold Spring Harb. Symp. quant. Biol.* **25**, 269.
HENDRICKS, S. B. (1960). *Cold Spring Harb. Symp. quant. Biol.* **25**, 245.
KING, P. E. (1963). *Proc. R. ent. Soc. Lond.* A **38**, 98.
KING, P. E. & HOPKINS, C. R. (1963). *J. exp. Biol.* **40**, 751.
KOGURE, M. (1933). *J. Dep. Agric. Kyushu imp. Univ.* **4**, 1.
LEES, A. D. (1953). *Ann. appl. Biol.* **40**, 487.
LEES, A. D. (1955). *The Physiology of Diapause in Arthropods.* Cambridge University Press.
LEES, A. D. (1959). *J. Insect Physiol.* **3**, 92.
LEES, A. D. (1965). In *Circadian Clocks*, p. 351. Ed. J. Aschoff. Amsterdam: North-Holland.
LEES, A. D. (1966). *Nature, Lond.* **210**, 986.
MELCHERS, G. (1956). *Z. Naturf.* **11b**, 544.
MENAKER, M. & ESKIN, A. (1967). *Science, N.Y.* **157**, 1182.
MINIS, D. H. (1965). In *Circadian Clocks*, p. 333. Ed. Aschoff. Amsterdam: North-Holland.
NITSCH, I. P. & WENT, F. W. (1959). In *Photoperiodism and Related Phenomena in Plants and Animals*, p. 311. Ed. R. B. Withrow. Washington: American Association for the Advancement of Science.
PARIS, O. H. & JENNER, C. E. (1959). In *Photoperiodism and Related Phenomena in Plants and Animals*, p. 601. Ed. R. B. Withrow. Washington: American Association for the Advancement of Science.
PETERSON, D. M. & HAMNER, W. M. (1968). *J. Insect Physiol.* **14**, 519.
PITTENDRIGH, C. S. (1966). *Z. Pflanzenphysiol.* **54**, 275.
PITTENBRIGH, C. S. & MINIS, D. H. (1964). *Am. Nat.* **98**, 261.
RING, R. A. (1967). *J. exp. Biol.* **46**, 123.
ROUBAUD, E. (1917). *Bull. scient. Fr. Belg.* **1**, 425.
RYAN, R. B. (1965). *J. Insect Physiol.* **11**, 1331.
SAUNDERS, D. S. (1962). *J. Insect Physiol.* **8**, 309.
SAUNDERS, D. S. (1965). *J. exp. Biol.* **42**, 495.
SAUNDERS, D. S. (1966a). *J. Insect Physiol.* **12**, 569.
SAUNDERS, D. S. (1966b). *J. Insect Physiol.* **12**, 899.
SAUNDERS, D. S. (1967). *Science, N.Y.* **156**, 1126.

SAUNDERS, D. S. (1968). *J. Insect. Physiol.* **14**, 433.
SCHNEIDERMAN, H. A. & HORWITZ, J. (1958). *J. exp. Biol.* **35**, 520.
SCHWABE, W. W. (1955). *Physiologia Pl.* **8**, 263.
SCHWEMMLE, B. (1960). *Cold Spring Harb. Symp. quant. Biol.* **25**, 239.
SIMMONDS, F. J. (1946). *Bull. ent. Res.* **37**, 95.
SIMMONDS, F. J. (1948). *Phil. Trans. R. Soc.* B **233**, 385.
TANAKA, Y. (1950). *J. Seric. Sci. Japan.* **19**, 580. (Quoted from Lees, 1955.)
DE WILDE, J. (1962). *A. Rev. Ent.* **7**, 1.
WILLIAMS, C. M. & ADKISSON, P. L. (1964). *Biol. Bull. mar. Biol. Lab., Woods Hole* **127**, 511.
DE ZEEUW, D. (1957). *Nature, Lond.* **180**, 558.

THE SURVIVAL OF INSECTS AT LOW TEMPERATURES

By R. W. SALT

Research Station, Canada Department of Agriculture,
Lethbridge, Alberta

INTRODUCTION

In the limited time at my disposal I cannot review the entire field of insect cold-hardiness, nor even a major part of it. Two fairly recent reviews exist, covering different aspects of the subject: that of Asahina (1966) stressing frost injury and protective mechanisms, and that of Salt (1961) stressing the factors affecting supercooling, nucleation, and freezing. I shall deal here with a few selected topics that have not heretofore received much attention. In addition, I think it wise to bring up to date the topic of nucleative freezing because of its fundamental importance to many aspects of insect cold-hardiness. First of all, however, I would like to correct a widespread misconception about supercooling.

Everyone is familiar with the common phenomenon of water freezing to ice, and is aware also that in many situations water supercools before freezing. Accepting that water freezes at 0° C., one logically concludes that supercooling must involve a kind of force or situation that *prevents* freezing. In other words, ice formation is looked upon as the normal course of events at temperatures of 0° C. and lower, whereas supercooling is regarded as something that interferes with the normal process. In fact, however, the very opposite is true, for supercooling, not freezing, is the norm. Freezing is *only* possible, indeed, when certain conditions have been met, these leading to the birth of a submicroscopic ice crystal known as a *nucleus*. The birth of the ice crystal is called *nucleation* and its subsequent *growth* is the process we call *freezing*.

It is possible for water molecules *alone* to form an ice crystal nucleus (this is homogeneous nucleation), but the conditions are very stringent; the water must be virtually pure. Not only the water itself but also its containing surfaces must be free of matter that bears favourable nucleation sites. And, even if these conditions were met, the water would not freeze until its temperature approached −40° C. The actual temperature of such freezing varies because it is determined by the statistical probability of molecular aggregations forming a nucleus; a probability which, in turn, is a function of water volume or mass.

Conditions leading to homogeneous nucleation are highly unlikely to exist in nature. Water invariably contains or touches other matter, which is certain to provide suitable sites for the formation of ice nuclei. The more effective the sites in this respect, and the more numerous they are, the more readily will an ice nucleus form (i.e. at a higher temperature). Nucleation will be discussed in more detail later, whereas here I wish only to stress that it is necessary for non-aqueous matter to be present in or bounding water in order to obtain freezing at temperatures much above $-40°$ C. Such matter is, of course, always present in living organisms and tissues.

In essence, then, supercooling is the primary circumstance in water cooled below its freezing point, whereas freezing is a secondary feature that must be induced. We are easily misled by most of our common outdoor observations of freezing; for example, puddles on roads, the surfaces of ponds, and automobile surfaces covered with condensed moisture are all densely contaminated with effective nucleators that reduce supercooling almost to nil.

NUCLEATIVE FREEZING IN INSECTS

Birth of an ice nucleus

Freezing has two stages: *birth* and *growth*. Birth comes first, naturally, and the process is oversimplified as follows. Water molecules aggregate into small clusters called embryos, which are very transient in nature. They quickly grow or decay according to the forces attracting or repelling their constituent molecules. If an embryo succeeds in attaining a certain critical size, which is temperature-dependent, its further growth is assured. This is the instant of birth—nucleation. Subsequent events fall under the heading of ice growth.

When a surface-dry intact insect is cooled, an ice-crystal nucleus is eventually formed somewhere in its body from water molecules arranged on the surface of a nucleator. The reason for spontaneous growth of the nucleus is the existence of supercooling—a metastable state having an energy level higher than that of the solid state. Reactions move spontaneously from higher to lower levels once they have been started, and in this case the nucleus is the starting impulse. The drop in energy level means that heat is produced—the familiar latent heat of crystallization. It is liberated when and where ice is formed; that is, at the moving ice front. The front is therefore warmed towards the initial freezing point of the body fluids, but will not quite attain it for two reasons: (1) heat is dissipated outwardly because there it is colder, the temperature of the insect having been approximately the same as ambient temperature at the instant of nucleation, and the heat is absorbed by all of the tissue substance, not by the water alone; (2) ice formation concentrates solutes, thereby lowering the freezing point of adjacent body liquids.

In addition to being temperature-dependent, nucleation is time-dependent as a result of the element of chance that is involved in the production of an ice nucleus. I have recently elaborated (Salt, 1966b, c) on the subject of the time course of the incidence of freezing of supercooled

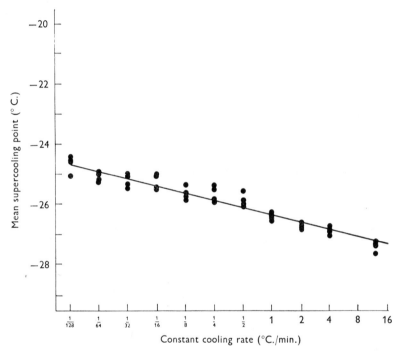

Fig. 1. Relation between mean freezing temperatures (supercooling points) of *Cephus cinctus* larvae and rate of cooling. Doubling the cooling rate lowers the mean freezing temperature by 0·24 degrees.

insects, both from the standpoint of rate of cooling, which is of particular concern in the laboratory, and also in regard to the long exposures to subzero temperatures endured by many hibernating insects. The extent to which each factor affected hibernating larvae of the wheat-stem sawfly, *Cephus cinctus* Nort., was determined. Doubling the rate of cooling lowered the freezing temperature by 0·24 degrees throughout a 1472-fold range of constant cooling rates (Fig. 1). All of these rates were sufficiently slow that they provided *uniform cooling* of the body, an essential requirement for accurate data. Figure 2 illustrates the progressive incidence of freezing within samples of larvae held at various constant subzero temperatures.

334 SURVIVAL OF INSECTS AT LOW TEMPERATURES

Theoretically, the rate of freezing remains constant, as shown by the five straight lines calculated from data obtained at −24° to −28° C. Actual data obtained at 25·5° C. are shown by the dots. Some irregularity of freezing rate is to be expected because of variability of the insects, but in

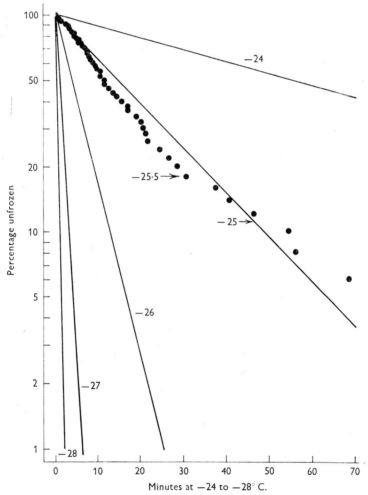

Fig. 2. Progression of freezing of *Cephus cinctus* larvae at constant temperatures. The straight lines represent ideal results of constant-rate freezing in homogeneous material. The point curve was obtained experimentally at −25·5° C.

addition there is opportunity for changes in the insects when the rate of freezing is so slow as to involve long periods of time.

Plotting the means of a family of freezing-rate curves produces a straight-line curve illustrated in Fig. 3. At constant subzero temperatures the time required for 50% of the larvae to freeze doubled with each 0·53 degree rise in temperature, ranging from 1·2 sec. at −30° C. to more than a year at −17° C. From this curve it would be possible to predict the progressive average mortality occurring in a hibernating population exposed under natural conditions to a lengthy regime of monitored fluctuating temperatures (Salt, 1966c).

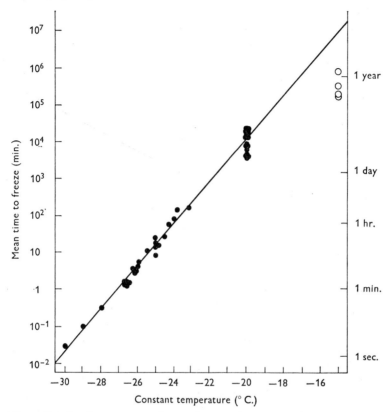

Fig. 3. Relation between mean time to freeze and constant subzero temperature exposure for hibernating larvae of *Cephus cinctus*.

The chemical and structural identity of insect nucleators is still vague, though their usual location is known. I have shown (Salt, 1966a, 1968) that the most efficient insect nucleators, the ones that actually get a chance to function, are contained in the lumen of the gut. Nucleation has never been observed elsewhere, to my knowledge, except in refrozen or injured insects. Although gut contents nucleate more readily than do tissues, the nucleators of sequestered appendages are effective at temperatures only a few degrees lower than those for the intact insects.

The efficiency of nucleator action, in terms of probability of successfully co-operating in the formation of an ice-crystal nucleus, appears to have both qualitative and quantitative aspects. Various attributes have been suggested as responsible for nucleator quality, but the problem is not yet resolved. Nevertheless, various substances when subjected to testing differ in effectiveness as nucleators and this is usually taken to mean that real qualitative differences exist, e.g. AgI is better than kaolinite. Basically, however, many of these differences are quantitative in nature at the ultra-micro level. On the other hand, quantitative effects in terms of numbers or concentration of nucleators per unit volume of liquid have often been

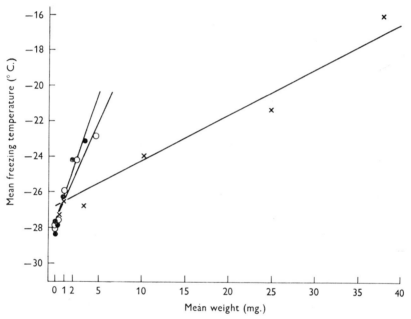

Fig. 4. Relation between mean freezing temperatures and weight of excised legs of the grasshopper *Melanoplus sanguinipes*. ●—●, Forelegs; ○—○, midlegs; ×—×, hind legs.

demonstrated experimentally, particularly with water drops of varying sizes (Mason, 1956). I have recently succeeded in demonstrating the same relation in insect material (Salt, 1968). I found the supercooling of sequestered grasshopper legs to be a decreasing function of leg weight (Fig. 4). Had the investigation stopped there it would have conformed to what might be expected on the basis of nucleation theory and observations on freezing water drops of various sizes. Fortunately, however, it was also possible to detect the site of nucleation in the legs, and it was seen that seven out of every eight legs nucleated in a particular place—the femorotibial joint. Thus, most of the substance of the legs, including most of the water, took no active part in the nucleation process. We are left with the clear indication that some attribute of the femorotibial joint, possibly a structural one, is usually superior in quality or quantity of nucleators. It is readily shown that quantity is dominant over quality. The original experimental observation was that freezing temperature was an increasing function of leg weight, a quantitative relation. Because of the preponderence of joint-located nucleation, 'leg weight' can be amended to read 'joint weight or size', for joint size is obviously a function of leg size. It is of interest, too, that although the quantitative aspects of the nucleators (numbers, area, concentration) dominated in seven out of eight legs, the eighth leg presumably possessed nucleators superior to those in the femorotibial joint, but these were located elsewhere. Such a situation, if I interpret it correctly, indicates dominance of quality in one out of eight cases and of quantity in the remaining seven.

In the present incomplete state of our knowledge of nucleators it is very easy to get into difficulties of interpretation of what represents quality. For example, unless quality is an extremely transient attribute, one seems bound to conclude that the single nucleator that is superior to all others present will produce the first ice nucleus, no matter how many other nucleators are present, and so numbers or quantity would not matter. Yet we know that quantity *does* matter, for its effects can be demonstrated experimentally.

Ice growth

As soon as an ice crystal nucleus is formed, freezing of the remainder of the water begins. The nucleus grows three-dimensionally and very rapidly attains visible size unless the system is very viscous, when growth is slow. Most biological systems freeze as dilute solutions of low viscosity, and it is usually convenient to think of freezing in these terms, adding complications such as membranes, permeability rates, solute concentrations, and viscosity to the model as required.

In the earliest stage of ice growth, immediately after formation of the nucleus, the first and most important consideration is the amount of supercooling that existed at the instant of nucleation. At this time supercooling can be regarded as a heat deficit, and obviously a good deal of the heat of crystallization that first results from freezing will go to make up that deficit. This is manifested by the well-known rebound or temperature rise that signals the beginning of freezing in a supercooled system. I propose now to examine this early stage of freezing, the portion represented by the temperature rise or rebound. In order to do this, we must consider where the heat of crystallization is liberated, where it then goes, how much is produced and why, and the results in terms of the body temperature of a small organism such as an insect.

It should be apparent that the site of liberation of heat of crystallization is at the ice front, and also that the front moves. The heat is dissipated in all directions away from the front, which is the warmest spot, to the colder surroundings. Rate of ice growth, therefore, or the speed of the ice-front advance, must depend on the extraction of heat from the front, and this will be influenced by the heat capacities and conductivities of the tissues and of their nearby environment. The general pattern of warming will be radially outwards from the site of nucleation or inoculation, like an expanding sphere. The ice front tends to expand in the same form, but in biological tissues a multiplicity of structures divert the ice and irregular growth results. Temperature gradients will be quite pronounced at this time but, in general, the heat flow will be outward.

A gram of water at 0° C., in freezing to 1 g. of ice at 0° C., liberates 80 cal. of heat. Conversely, 80 cal. are required to melt 1 g. of ice at 0° C. to water at 0° C. This appears to mean that, at 0° C., just enough heat would be produced by a given mass of forming ice to melt itself. And this is precisely true! Of course, the fact of the matter is that *freezing does not occur* under these conditions, for these are equilibrium conditions at 0° C. For ice to grow in the first place, there must be removal of heat from the system—that is, the ambient temperature must be below 0° C. Hence, the extent, direction, and rate of ice formation is determined rigidly by the conditions of cooling. I cannot emphasize too strongly the importance of this fact in the freezing of biological materials, for it means that ice growth is a highly variable process in such a complex system. Uniformity of freezing is likely to be an elusive experimental goal.

From the above, it follows that a small amount of supercooling is necessary to support *continuing* ice formation. However, the amount of supercooling may be insignificant in most circumstances, and indeed may be better regarded as the cooler part of a temperature gradient. When a biologist talks of supercooling, he is usually referring to the supercooling that exists *before* any freezing has occurred; that is, before nucleation or inoculation. Such supercooling is usually appreciable in amount, say a few degrees at least, and *can* be quite large; 20–35 degrees is common in insects. Let us now see what influence supercooling of a few degrees or more has on the production, distribution, and dissipation of heat of crystallization, the ensuing thermal pattern, and its effects in terms of ice distribution and ice form.

It is obvious, I think, that the initial portion of the heat of crystallization will be absorbed by *all* components of the supercooled system for the simple reason that they are initially colder than the ice front, where the heat is being liberated. This phase of freezing will continue until the temperature of the tissues of the organism, assuming that it is small or moderate in size, has risen almost to the initial freezing point of the tissues. Actually, the freezing point falls as ice is formed, since ice formation is concentrating the body liquids, and so the original freezing point is never quite reached.

Calculation of a compensated freezing point for any set of conditions can be readily made, but this is of no particular interest at the moment. Of more significance is the calculation of the portion of ice formed, and heat of crystallization liberated, in compensation for supercooling. This is a very simple calculation, given by Getman & Daniels (1931, 5th ed., p. 178) as follows:

If the overcooling of the solution in degrees is represented by u, the heat of fusion of 1 gram of solvent at the freezing point by l_f, and the specific heat of the solvent by c, then the fraction, f, of the solvent, which will solidify may be calculated by the formula $f = cu/l_f$. When water is used as the solvent, $c = 1$, and $l_f = 80$. Therefore, for every degree of overcooling, the fraction of the solvent separating as ice will be 1/80, and the concentration of the original solution will be increased by just so much.

The amount of ice that forms solely as a result of the original supercooling also determines the amount of heat of crystallization liberated at this stage of freezing. Ideally this heat should be absorbed only by supercooled water, but in fact it is absorbed by all components of the organism and will in turn be lost to the surroundings, including highly conductive metallic thermocouples. Since heat loss to the environment begins almost immediately after ice formation begins, a larger fraction of ice actually forms than the $u/80$ predicted by the formula, which is therefore a minimum figure.

To use a simple example, an insect that is supercooled 20 degrees and then nucleated or inoculated to produce freezing, quickly freezes to the extent of somewhat more than 20/80 or 1/4 of its water content. By the time it has done so, and this time is a small fraction of a second in most species, the body temperature, though not strictly uniform throughout, will have risen to its equilibrium freezing point. The fact that the insect was originally supercooled 20 degrees, however, implies that the surroundings are about $-20°$ C. and are, therefore, drawing heat from the freezing insect. In accord with principles that we have already established, subsequent ice formation in this insect will be determined by the rate at which heat is withdrawn from it by the surroundings.

Now I wish to proceed to a complication of the simple example just given, to speculate about what happens when ice growth is impeded, inhibited, or even stopped by barriers. Cell walls and membranes within an organism are, generally speaking, permeable to water, but the rate of water transfer may vary greatly, so that a membrane that allows only a slow transfer of water may be regarded as inhibitory. Suppose, then, that most of the body water of a particular insect is contained within cells; say 90% intracellular and 10% extracellular water, almost all of the latter being haemolymph plasma. When such an insect nucleates and freezes after 20 degrees supercooling, a quarter of the body water should freeze very quickly to compensate for supercooling. Assuming that nucleation and initial ice-growth are extracellular—and this can be substantiated—there will be a shortage of extracellular water to fulfil the requirements of the first phase of freezing (1/10 available, 1/4 needed). Of course, intracellular water will immediately start to leave the cells, driven by osmotic forces, and will freeze extracellularly. The critical matter here is the rate of water outflow from the cells as compared to the 'demand' for ice formation supplied by continued supercooling. Supercooling will be reduced, of course, but it will not be eliminated until a quarter of the body water is frozen. In case the calculation is of interest, when the 10% extracellular water is frozen and before water has begun to leave the cells, the amount of supercooling will have been reduced by 8 degrees from 20° to 12° C. Further extracellular ice formation will then depend on the ease of water transfer out of the cells; the persistence of supercooling, and its magnitude, will be likewise determined. In addition, supercooling will be augmented and prolonged by heat loss from the insect to its environment, especially when the rate of cooling is great. It is quite conceivable, therefore, when the proportion of extracellular water is low, supercooling great, and cell permeability low (or slow, inhibited, or delayed), that the body temperature will not reach the freezing point and supercooling will persist. The critical factor is the ease with which water can leave the cells.

SINGLE VERSUS MULTIPLE NUCLEATION IN INSECTS

One consequence of the persistence of supercooling *after* nucleation has occurred seems to be that further nucleations are possible. I shall argue that multiple nucleation is highly improbable in an organism such as an insect, whereas in plants it is apt to occur under certain circumstances.

Multiple *temperature rebounds* (not *nucleations*) have often been observed in freezing plants but rarely in insects (Salt, 1933). The observations were always made in the laboratory and thus might be artifacts. However, the observation of more than one temperature rebound, even if authentic, does not necessarily mean that multiple nucleation has taken place. For example, my first experience with multiple rebounds occurred while freezing hibernating adults of the Colorado beetle, *Leptinotarsa decemlineata* (Say), in 1933. They supercooled only 4–8 degrees before freezing and gave 1–5 temperature rebounds. The recording thermocouple was situated between the elytra and the dorsal abdominal surface. Upon dissection the beetles were found to contain no free-running haemolymph, but only a paste-like coating of the haemocoele, a condition suggesting an explanation for the multiple rebounds. The dryness and viscosity of the tissues and extracellular liquids probably slowed ice growth considerably, but more in some places than in others, making growth-rate irregular. The consequently fluctuating supply of heat of crystallization *reaching the thermocouple* would be recorded by it as warming, cooling, or constant according as it exceeded, was less than, or equal to the heat being lost by the thermocouple to its environment. Therefore, cessation of freezing between rebounds is not necessary for their observation, and indeed is unlikely. The paradox is accentuated by the high conductivity of the metallic thermocouple, which makes a slow irregular-rate process appear to be an intermittent one. There is no need to invoke multiple nucleation, which is improbable, though less improbable than in a fully hydrated insect.

A similar situation exists in plants. In spruce needles two rebounds were recorded when ice growth was retarded by the endodermis after slight dehydration (Salt & Kaku, 1967). It was not stopped, however, but reached and permeated the mesophyll after a period of time proportional to the amount of drying. There was no evidence of two nucleations. Multiple *nucleation* does, however, seem possible and even probable in large plants where temperature differences exist between widely separated parts, or between aerial and subterranean parts, or where tissues or parts are separated by dry barriers. Very rapid cooling can also produce conditions favouring multiple nucleation when steep temperature gradients are created. However, this is a highly artificial situation.

Under natural conditions, the objections to a second nucleation in an insect are twofold. Firstly, supercooling is necessarily less than it was at the instant of first nucleation, and therefore a nucleator that *could* function with this lesser amount of supercooling would probably have already done so. The second objection is based on the circumstances existing when the extracellular water is freezing or has been frozen and supercooling persists. If a second nucleation could occur at this time, it would have to occur inside a cell, whose contents would therefore freeze. On reaching the cell boundary, the ice would either stop growing or grow through the cell membrane. If it stopped, only one cell would have been frozen and the event would remain unknown. Alternatively, if ice grows through the membrane, it meets either existing extracellular ice or else an adjoining unfrozen cell. It cannot inoculate the adjoining cell, however, for if this were possible, the extracellular ice would already have done so. In summary, then, further nucleation in a partially frozen and still somewhat supercooled insect is highly improbable.

INOCULATIVE FREEZING

Inoculative freezing is distinctly different from the more common nucleative freezing (Salt, 1963). It arises from the growth or propagation of *external* ice into the body of the insect by any route, resulting in the freezing of body fluids. It may occur in almost any insect that cools below its freezing point when in contact with water. Hibernation in a moist or wet environment that cools below 0° C. is therefore a precarious existence.

Several factors determine the ease with which inoculation takes place. The permeability of the cuticle to water is crucial when ice must grow through it. Also, the area of cuticle wetted affects the probability of inoculative freezing, or the ease of such freezing. Entry of ice may also occur via openings of glands or of the reproductive, digestive, or respiratory systems, specific sites which must be covered by freezing water to allow the entry of ice. The speed of ice propagation and ensuing inoculation of the interior is increased as the temperature falls, other things being equal. Consequently, time is involved, and when ice propagation is slow the temperature may become critical. For example, if the temperature drops only slightly below the freezing point of the insect and the contact water freezes, the ambient temperature may rise above the freezing point of the insect before inoculation can occur. Heat of crystallization assists warming and may even be totally responsible for a short period of time for the prevention of freezing. The freezing point of the body fluids of the insect will usually be somewhat lower than that of the external water.

Inoculative freezing of an insect occurs after the water that is in contact with the insect freezes. It rarely occurs when ice that is already formed touches the insect's exterior, possibly because the contact is apt to be poor in both extent and adhesion, for only the high spots of ice and insect will come into contact.

In the laboratory, inoculative freezing is a time-dependent process in some species, the insect freezing as much as a day or two after the contact water has frozen, while in other species there is no significant delay (Salt, 1963). The delay is thought to represent the time required for ice to grow through the cuticle via minute pores or channels whose hydrophobic air-filled outer ends allow ice penetration only by accretion from the vapour phase (Fig. 5). In the laboratory, pretreatment—with a detergent, for example—that makes the cuticle more permeable reduces or eliminates the delay. In nature, the pre-freezing conditions are often such as to favour little or no delay of inoculative freezing, for the insect will probably have been in contact with water for a relatively long time before the water freezes. The delay aspect may therefore be largely a laboratory artifact. Nevertheless, the phenomenon of inoculative freezing is an important factor in the overwintering of insects in moist hibernacula. Here, survival depends either on escape from freezing through some type of water-proofing (of the insect or of an encasing cell, cocoon, etc.) or on tolerance to some degree of freezing.

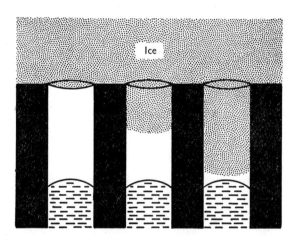

Fig. 5. Schematic representation of the progressive accretion of ice, left to right, into the hydrophobic termini of three pores containing water proximally, as in hypothetical insect cuticle. The ice grows into the pore by freezing of vapour from the liquid which does not recede, being replenished by the body fluids connected to it proximally.

In the laboratory most insects can tolerate a little ice formation in the body for a very short time. More ice, more time, or both, lead to injury in freezing-susceptible forms. In nature, inoculative freezing upon minimal exposure will likewise be expected to produce no injury, but since exposures are usually long in nature, injury is to be expected eventually unless the temperature rises. In this connection it is not known if injury is accumulative after multiple mild doses of inoculative freezing, or if restorative processes occur. Cold-hardening of some type before freezing occurs may improve the insect's chances of survival. Increasing the concentration of the solute reduces the amount of ice formed at any given temperature and solutes such as glycerol may reduce injury in this way.

When cooling is both rapid and extensive, ice nucleation may occur before inoculation. Both types of freezing have a negative temperature coefficient, so the lower the temperature the more probable each type of freezing becomes. I found that larvae of *Cephus cinctus* (Salt, 1963) were inoculated more readily at $-10°$ than at $-5°$ C. but that there was no further enhancement at $-15°$ C. I explained this by showing that the difference in vapour pressure between ice and supercooled water reached a maximum at about $-12°$ C., and surmised that this energy difference, the chemical potential, provided the driving force to move water vapour across an air barrier and to give ice growth from the vapour phase. In a private communication Dr Peter Mazur kindly pointed out that chemical potential is proportional to the ratio of the two vapour pressures, not their difference. I therefore repeated the experiments on inoculative freezing of *C. cinctus* larvae at $-5°$, $-10°$, and $-15°$ C. with an improved technique which standardized the area of larva that was wetted. In the original method the S-shaped larvae lay on their sides on a wet blotter, with a variable amount of wetting. When moisture coverage was uniform the larvae froze more readily at $-15°$ than at $-10°$ C., showing that the temperature coefficient was negative down to the limit of supercooling, as it should be.

INTRACELLULAR FREEZING BY INOCULATION

Earlier, a situation was described where the amount of extracellular water was insufficient to provide enough heat of crystallization to warm the organism to its freezing point, so that some supercooling would persist unless or until intracellular water could leave the cells and freeze outside them. The persistence of even a small amount of supercooling for even a short time is conducive to the inoculative freezing of cell contents, as I shall now try to describe.

Water in very fine pores or capillary spaces has a lower freezing point than water in bulk. The freezing-point depression is negligible, however, until pore diameter is reduced to a fraction of a micron (Jackson & Chalmers, 1958). If a pore is so small that water in it freezes only below, say, $-3°$ C., then, of course, ice alongside the pore cannot grow into it until the temperature falls to $-3°$ C. Should this happen, and the pore is in a cell membrane, intracellular freezing will result. Note that such freezing is inoculative, not nucleative. The possibility that ice will grow into the interior of a cell is thus fixed by *temperature* and the *size* of the largest pores or water-filled pathways traversing the enveloping membranes of the cell.

Now, also, it becomes evident why the persistence of supercooling after nucleation, even in small amount and of short duration, is vital to intracellular freezing, and thus to survival itself since protoplasmic freezing is usually fatal. The situation resolves itself into a contest between forces inducing water to leave the cells and forces inducing ice to grow into them. The balance of power is influenced by a number of factors, but basically by water permeability.

Insect muscle and nerve cells may not freeze under natural conditions because of their small size, whereas large cells are likely to freeze. The very large fat cells and salivary gland cells of the goldenrod gall fly, *Eurosta solidaginis* (Fitch), freeze internally (Salt, 1959), probably by inoculation through large pores. Any cell can, of course, be frozen internally simply by cooling it fast enough. The influence of rate of cooling or intracellular freezing has been investigated by Mazur (1966), who calculated the rates required for intracellular freezing in cells of extreme sizes and permeabilities. They increased from 0·05 degrees/min. for very large impermeable cells to 3000 degrees/min. for small, highly permeable cells like human erythrocytes. Fast cooling produces very steep temperature gradients and thus allows multiple nucleation, at the same time reducing the time during which water can leave the cells. *Intracellular nucleation* is the norm under these conditions. On the other hand, at very slow rates of cooling, freezing must be inoculative unless the cells are completely impermeable to water, which is unlikely. At some intermediate range of cooling rates, of course, either type of freezing is possible; the determining factor will be the rate at which water leaves the cells.

Intracellular freezing is sometimes observed in the laboratory on microscope-stage freezers, fairly commonly with plant tissues but also with very large animal cells such as the fat cells of insects. Individual cells or groups of cells here and there are observed to darken suddenly, a process known to botanists as flashing, and considered to represent intracellular freezing. Its suddenness, as well as its sporadic distribution, seems to indicate that intracellular nucleation has taken place. This would indeed be possible if steep temperature gradients were present, but not otherwise. Flashing is very probably inoculative freezing, its suddenness merely indicating that penetration of the enveloping cell membrane has been somewhat hindered or delayed, but that once inside, ice growth is rapid. Although supercooling is not essential to this type of freezing, it does enhance the process.

It seems, then, that water leaves the cells initially, freezing extracellularly, and at some time later the cell membrane may, as a result of shrinkage from within or pressure from without, suddenly allow the penetration of ice inwardly as though rupture had occurred. Alternatively, penetration may be delayed by dehydrated barriers such as the endodermis of dry spruce needles (Salt & Kaku, 1967) or by the attainment of a lower temperature necessary for ice penetration.

PATHWAYS OF ICE GROWTH IN INSECTS

I have shown (Salt, 1966a, 1968) that nucleation of a supercooled insect usually occurs in the gut contents. This raises the interesting and important question of how ice reaches the haemocoele, which it appears to do without delay. The first wave of freezing takes only a small fraction of a second to sweep throughout the insect. Either the cells of the digestive tract walls readily allow ice penetration or else there are passages of some sort between the cells that provide good freezing pathways. If such passages are present, they are probably widespread, for in all cases I have observed, ice has grown as readily transversely through the gut walls as it has along the gut lumen.

Up to this point I have supported the opinion that intracellular freezing of insect cells is unlikely except in large cells with large pores, or when supercooling is considerable. Now it is necessary to reconcile this opinion with the hard fact that ice is observed to grow through the gut walls without hindrance. This can best be done, I think, by drawing an analogy with plant cells. It is now considered likely that freezing *within* plant cell walls is fairly common (Idle, 1968; Salt & Kaku, 1967). Botanists do not regard this as intracellular freezing, which they restrict to that part of the cell inside the plasma membrane. Animal cells have no such rigid, thick walls, of course, but they do have flexible membranous walls that usually appear in electron micrographs to be double and about 20–50 mμ thick. Adjoining walls are sometimes separated by intercellular material. Basement membranes and other connective tissues often form a lining around tissues, groups of cells, and even individual cells. They are fibrous, contain collagen, and are about 0·05–5 μ in thickness (Ashhurst, 1968). Connective tissue and intracellular liquids are thus sufficiently thick to support internal ice growth, and, though not present between all adjoining cells, they are ubiquitous enough to support the observation that ice readily penetrates the gut walls. At the same time, the cell membranes themselves may be impervious to ice growth if their pore size is very small ($<$ 20 mμ).

THE ICE MASS IN FROZEN INSECTS

Initial form

One of the consequences of single ice nucleus formation in insects is that the resulting ice mass is a single crystal, though much ramified and irregular in structure. Its form is dictated largely by the innumerable barriers that deflect its growth in body tissues, and partly by temperature. The typical dendritic growth along the four axes of hexagonal symmetry, so evident when bulk water freezes, is greatly modified by solutes and non-aqueous matter, which becomes concentrated as water turns to ice and is enmeshed by ice branches. There are many consequences of such action—mechanical contraction, chemical concentration, increased viscosity, decreased mobility and diffusion, for example—and among these lies the responsibility for freezing injury.

Recrystallization

Ice usually changes its structure after its initial formation, especially if supercooling has been appreciable. Its branched form is changed to a massive form during a process of recrystallization that is due to slight differences in surface energy existing between crystal faces, edges, and corners. Vapour pressure and chemical potential are greater at edges and corners, where convexities are greater, than they are on flat surfaces, thereby providing a driving force that leads to growth of larger surfaces at the expense of smaller ones. Adjustment is faster near the freezing point than at lower temperatures because reaction rates are higher, energy differences greater, and the unfrozen portion of the solution more fluid.

Recrystallization is undoubtedly a normal feature of frozen insects, for even if temperatures fail to approach 0° C. after nucleation, long exposures to lower temperatures compensate for the lower rate of recrystallization. Ultimately, too, it will occur before melting. Luyet, Sayer & Gehenio (1967) have intensively studied recrystallization of frozen gels and solutions and I have observed this phenomenon *in vitro* while freezing small droplets of insect haemolymph. Many droplets under 50 μ in diameter apparently failed to freeze when cooled to $-45°$ C., but upon warming they assumed a frozen appearance at a fairly characteristic higher temperature. Upon further investigation it was seen that the process required time and that this was inversely proportional to temperature. Therefore, when warming was fairly rapid the droplets recrystallized—they appeared to be freezing—within a characteristic temperature range only 2 or 3 degrees wide. When warming was very slow, recrystallization could be discerned to start at lower temperatures and could be completed there if given time. In view of Luyet's work it may be assumed that recrystallization had been taking place from the time of first freezing.

CHILLING AND CHILL-HARDENING

In temperate and colder climates native insects must spend the winter in a chilled state, perhaps cold enough to be supercooled much of the time. During this time they must be or become chill-tolerant to temperatures as low as their freezing temperatures for as long as these may occur. Within their normal range of winter temperatures native species are not in danger but more prolonged chilling may begin to produce injury, especially if the temperature is at an unfavourable level. Examination of the temperature–time interaction is thus of interest.

When a hibernating species is experimentally exposed to a series of constant low temperatures, two types of chill-injury are seen to exist

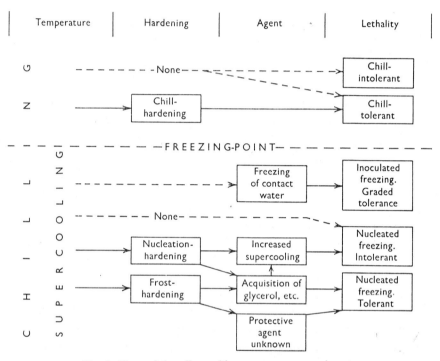

Fig. 6. Chart of the effects of low temperatures on insects.

(Fig. 6). One type is readily recognized simply as interference with or prevention of development by cold temperatures not far below what is often called the threshold of development. Typically some development takes place, but is eventually blocked at a crucial point, such as pupation, that requires a slightly higher temperature for successful completion. Such interference with development by cold is common in the laboratory after long storage but may be rare in nature; a mild winter followed by a late and cold spring would seem to favour it.

The second type of chill-injury occurs at temperatures well below the developmental range at a rate that is inversely related to temperature. Two examples are given in Fig. 7 which trace the beginning of injury in larvae of the wheat-stem sawfly, *Cephus cinctus*, and eggs of the grasshopper *Melanoplus bivittatus* (Say), on exposure to constant subzero temperatures. Only non-frozen insects were considered in this experiment. Injury first appeared in *Cephus cinctus* larvae upon incubation after > 16 weeks at −10°, > 8 weeks at −15°, and > 2 weeks at −20° C. At −20° C. all larvae exposed for 4 weeks and still unfrozen were affected to some extent, and after 8 weeks the injury was irreversible. Eggs of *Melanoplus bivittatus* showed injury starting after > 4 weeks at −10°, > 2 weeks at −15°, and

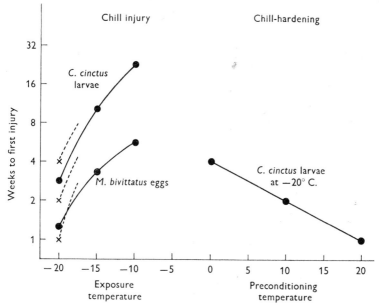

Fig. 7. First appearance of injury in two hibernating species exposed to constant subzero temperatures, and the effect of preconditioning temperatures. The levels of cold-hardiness after chill-hardening at 20°, 10° and 0° C. are indicated by the dotted lines.

> 1 week at −20° C. However, the values obtained for each species are not fixed but depend on the pretreatment history of the insects, i.e. on their chill-hardening. The right-hand curve of the figure shows the result of chill-hardening *Cephus cinctus* larvae at 20°, 10° and 0° C. for 3–6 months before exposure to −20° C. Injury began in the non-frozen larvae after about 1 week in those that had been stored at 20° C., 2 weeks in those from 10° C., and 4 weeks in those from 0° C.

The chill-injury to pupae of *Platysamia cecropia* L. and other saturniid species reported by Williams (1956), and which he called winter sickness, is of the same type that occurs in *Cephus cinctus* larvae at subzero temperatures. The rate of sickening was inversely related to temperature, being greater at 2·5° than at 6° C. In *Cephus* there was a temperature range from about 0 to $-10°$ C. where neither type of chill-injury was appreciable within the time limits used, whereas in *P. cecropia* the injury extended up to 6° C. along with a slight amount of development. *C. cinctus* larvae are therefore more chill-tolerant than are *P. cecropia* pupae, and moreover there is no overlapping of the two types of chill-injury in *C. cinctus* as there is in *P. cecropia*.

Chilling can also produce effects that are beneficial to the insect. The stimulus inducing neurosecretory cells in the brain to terminate diapause is most often a period of chilling. Another beneficial result of chilling may be chill-hardening, where mild doses of chilling produce an increase in the ability to withstand more intense chilling. An additional advantage accrues when chilling also increases the supercooling potential (nucleation-hardening), providing a greater probability of escape from freezing.

REFERENCES

ASAHINA, E. (1966). In *Cryobiology*, ch. 9. Ed. H. T. Meryman. London: Academic Press.
ASHHURST, D. E. (1968). *A. Rev. Ent.* **13**, 45.
GETMAN, F. H. & DANIELS, F. (1931). *Outlines of Theoretical Chemistry*, 5th ed. New York: Wiley.
IDLE, D. B. (1968). *Sci. J.* **4**, 59.
JACKSON, K. A. & CHALMERS, B. (1958). *J. appl. Phys.* **29**, 1178.
LUYET, B. J., SAYER, D. & GEHENIO, P. M. (1967). *Biodynamica* **10**, 123.
MASON, B. J. (1956). *Sci. Progr., Lond.* **44**, 479.
MAZUR, P. (1966). In *Cryobiology*, ch. 6. Ed. H. T. Meryman. London: Academic Press.
SALT, R. W. (1933). M.S. thesis, Montana State College.
SALT, R. W. (1959). *Nature, Lond.* **184**, 1426.
SALT, R. W. (1961). *A. Rev. Ent.* **6**, 55.
SALT, R. W. (1963). *Can. Ent.* **95**, 1190.
SALT, R. W. (1966a). *Can. J. Zool.* **44**, 117.
SALT, R. W. (1966b). *Can. J. Zool.* **44**, 655.
SALT, R. W. (1966c). *Can. J. Zool.* **44**, 947.
SALT, R. W. & KAKU, S. (1967). *Can. J. Bot.* **45**, 1335.
SALT, R. W. (1968). *Can. J. Zool.* **45**, 329.
WILLIAMS, C. M. (1956). *Biol. Bull. mar. biol. Lab., Woods Hole* **110**, 201.

SOME MECHANISMS OF MAMMALIAN TOLERANCE TO LOW BODY TEMPERATURES

By R. K. ANDJUS

Department of Physiology, Faculty of Science and
Cryobiology Unit, Institute for Biological Research,
University of Belgrade, Belgrade, Yugoslavia

INTRODUCTION

Attempts to define the physiological basis of tolerance to low body temperature in mammals meet with difficulties arising not only from the diversity of mechanisms limiting tolerance at different levels of hypothermia, but also from the diversity of criteria of tolerance. The difficulties may be largely overcome by considering separately (a) limits compatible with spontaneous or unassisted recovery and (b) limits concerning assisted recovery, including revival by specific reanimation or resuscitation procedures.

Spontaneous recovery from induced hypothermia, achieved solely by unassisted physiological activity after the elimination of the cooling agent, is undoubtedly of major interest from the point of view of tolerance to cold in free nature. Experimental results indicate that many mammalian species, especially those of small body mass, may easily be forced into hypothermia by conditions occurring almost regularly in their natural habitats. External cold, especially when combined with wetting or partial immobilization, not to mention inadequate food supply, may induce relatively rapid and deep body cooling in such mammals. The remarkable ability to rewarm spontaneously after a drop of some 20° C. below normal body temperature has been described, under laboratory conditions, for a number of mammalian species among non-hibernators, and it is considerably greater in hibernators, even when artificially cooled. This ability may have an unsuspected survival value for their natural populations.

Spontaneous rewarming may be slow in re-establishing the normal level of body temperature, but appears nevertheless as strikingly efficient. In our own experience a cold-acclimated rat, rewarming from a body temperature of 19° C. at an environmental air temperature of the same magnitude, will restore its thermal homeostasis in less than 2 hr., while a cat will achieve the same in 10–12 hr., according to Britton (1922). One of our rhesus monkeys, cooled not lower than 16° C. body temperature, needed more than 6 hr. to re-establish its normal body temperature at room temperature (26° C.). A frostbitten human patient, accidentally cooled below a rectal temperature of 20° C. and left to rewarm spontaneously at a room temperature of 21° C., rewarmed to normal body temperature in a matter of 19 hr. (Laufman, 1951). In agreement with Britton (1922), but opposing Simpson & Herring (1905) who first used the rather unfortunate expression 'artificial hibernation' in describing deep hypothermia, Hamilton insisted that in artificially cooled mammals there is no neutral point similar to hibernation: 'the animal, unless in the later stages of respiratory and cardiac failure, is homoiothermic', meaning that it shows the tendency of increasing its temperature above that of the environment (Hamilton, 1937).

Much of this old statement, stressing the closeness between temperature limits for spontaneous rewarming and for the major physiological functions, as well as the dissimilarity between induced hypothermia and natural hibernation in respect to the tendency of rewarming, remains valid. Our own experiments aimed at defining with precision the limits compatible with unassisted recovery, showed indeed that, in some species at least, spontaneous rewarming always occurs unless the animal is cooled to body temperatures which stop breathing and heart beating by their direct action, or unless hypothermia has been maintained beyond certain time limits (Andjus, 1956; Andjus & Petrović, 1959; Andjus, Matić, Petrović & Rajevski, 1964). In contrast to induced hypothermia, hibernation may be characterized by the absence of this tendency until arousal mechanisms are activated.

Thus, physiological mechanisms of spontaneous rewarming, as well as mechanisms setting time and temperature limits to hypothermia compatible with such rewarming, remain of supreme interest when the study of tolerance mechanisms is approached by taking into account only unassisted recovery or recovery under natural conditions.

When *assisted recovery* is considered as well, the study of tolerance broadens considerably. After reactivating successfully an isolated rabbit heart cooled to −2° C., Kodis started wondering, at the beginning of the present century, whether the revival of a whole similarly cooled rabbit might be just a matter of devising adequate techniques for reactivating the animal's heart *in situ* (Kodis, 1902). At the same time, impressed by his own achievements in reviving insects and later bats cooled to temperatures somewhat lower than zero, Bachmetiev became convinced that any mammal might be brought into a state of reversible anabiosis of unlimited duration at a sufficiently low temperature (Bachmetiev, 1902). 'Hypothermia as such', wrote Heymans some fifty years ago, 'is not lethal to any organ'; the organs, however, 'stop functioning, although they are not killed' (Heymans, 1919). In other words, cold-induced arrest of physiological activities does not necessarily mean death and leaves the possibility of recovery.

A great many of these old statements, beliefs and predictions proved to be valid. We know now that mammals, including non-hibernators, can tolerate body temperatures considerably lower than the physiological zero for the pumping activity of their hearts, provided that adequate techniques of reanimation are applied in due time. Temperature and time limits have been described: no adult mammal has ever been revived in the laboratory after its central body temperature has dropped below about $-7°$ C. or after 7 hr. have elapsed following a cold-induced cardiac arrest (Andjus, 1964). It must be stressed, however, that our definitions of tolerance limits are almost entirely based upon results obtained by applying a given reanimation procedure to experimental animals cooled by a chosen technique. In devising these techniques we learned that the limits of tolerance, determined by testing the possibility of recovery, greatly depend on the adequacy of the procedure adopted. The lack of success may be due to irreversible changes taking place not only at low temperature, but also during cooling or rewarming by inadequate techniques. Tolerance limits defined at present may thus reflect our abilities, rather than real possibilities, as was the case so many times in the past. It should not be forgotten that ways and means have already been found of safely cooling isolated mammalian cells, tissues and even organs to temperature levels approaching absolute zero and ensuring the preservation of viability for practically unlimited periods. Theoretically, therefore, indefinite bioconservation should not be regarded as inaccessible for the whole mammalian organism before this is unequivocally disproved. Meanwhile, the tolerance limits determined to date should be considered as temporary and valid for a given set of experimental conditions. A better understanding of mechanisms responsible for the present limitations to revival becomes ever more necessary if our knowledge of further possibilities and our achievements are to be substantially improved.

When considering the various mechanisms involved, two major consequences of extreme cooling become of utmost importance: (*a*) anaerobic conditions created at the tissue level by cooling below the physiological zero for the breathing and cardiac activity, and (*b*) tissue freezing after cooling below the freezing point of body fluids.

Let us first consider the effects of cold-induced anaerobic conditions. It is well known that hypothermia, by virtue of its retarding effect on metabolic processes, affords protection against anoxia of various origins (see Andjus, 1960, for a detailed discussion of the protective effects of internal cold). One is justified in assuming that low body temperature will protect, by means of the same mechanism, against anoxic conditions created by cooling. To a considerable extent therefore tolerance to extreme hypothermia may be regarded as cold-enhanced tolerance to cold-induced anaerobic conditions. However, our knowledge of the *in vivo* temperature-dependence of the effects of anoxia leaves much to be desired. The relationship between temperature and metabolic events taking place under anaerobic conditions is far from being defined for the whole region of temperatures compatible with revival and for all processes that may interfere with recovery. Other side effects of low body temperature, alone or in combination with anaerobic conditions, may have a tolerance-limiting value even before the challenge of freezing sets in. *In vitro* studies have shown, for instance, that qualitative changes in some enzyme activities, of vital importance for the functional integrity at the cellular level, may occur at temperatures well above zero (Bowler & Duncan, 1968). On the other hand it has been claimed that in the range of extreme hypothermia, time limits of tolerance do not show a clear relationship to temperature and that blood oxygen levels do not seem to be critical (Adolph, 1956). Metabolic correlates of anaerobic conditions and their tolerance-limiting role may thus be greatly modified under conditions of hypothermia. They may differ qualitatively from those studied at higher body temperatures or under *in vitro* conditions. Thus, a detailed analysis of the effects of anoxia under conditions of hypothermia imposes itself as a major component of the study of tolerance-limiting mechanisms.

Freezing seems to set a particularly rigid barrier to further safe cooling of the whole mammalian organism. Efforts to overcome this barrier are made at present along two main lines. One concerns the avoidance of freezing by supercooling techniques. It includes the study of mechanisms responsible for the temperature limits of supercooling in mammals and of the possibly specific mechanisms limiting endurance to the supercooled state. The other approach concerns the possibility of revival from the frozen state. It was greatly encouraged by achievements in definitely reviving partially frozen hibernators (Smith, Lovelock & Parkes, 1954; Lovelock & Smith 1956), or in reanimating temporarily similarly frozen non-hibernators (Andjus, 1951*a* and *b*, 1955). The future of this approach may, however, reside in the search for means of chemical protection against freezing damage. It finds its basic support in the very active field of cryobiological studies concerned with mechanisms of freezing injury to isolated cells, tissues and organs, as well as with mechanisms of protection afforded to them by a number of very efficient chemical protectors (see Smith, 1961). Efforts to identify cryobiological peculiarities of different components of the mammalian body appear as the most relevant if the limits of tolerance of the whole mammalian organism are to be extended to temperatures sufficiently low for long-term preservation.

In reviewing our own studies concerning tolerance mechanisms, and without hesitating to discuss work still in progress, we shall select results concerning the major problems just outlined: mechanisms ensuring and mechanisms limiting recovery by spontaneous rewarming; effects of anoxia as modified by low body temperature, especially from the point of view of the tolerance-limiting role played by metabolic correlates of anaerobic conditions created in or by hypothermia; and further possibilities of revival from subzero body temperatures. Mechanisms limiting survival of animals maintained for longer periods at levels of hypothermia compatible with spontaneous breathing and cardiac activity, and mechanisms responsible for the failure of specific functions during cooling, will be left out of the present discussion.

From the point of view of general methodology, our work has been governed by two main approaches. One concerns the comparative analysis of phylogenetically and ontogenetically determined differences in tolerance among animals. The other approach is represented by studies of artificially induced tolerance.

The striking variability in tolerance to internal cold among mammals is best evidenced by a comparison between three types of mammalian organisms: the hibernators, the non-hibernators and the neonates with immature thermo-regulatory responsiveness. Comparative studies of mechanisms responsible for the outstanding differences between them may greatly improve our knowledge of the tolerance-limiting mechanisms. In comparison to non-hibernators, the hibernators and the neonates may be regarded as ideal natural models for studying mechanisms by which tolerance to internal cold may be improved. The ground squirrel, as a typical hibernator, the adult and the newborn rat were used as such models in our studies.

On the other hand, the study of induced tolerance, orientated towards the identification of means by which tolerance can be artificially improved, usually adds to our understanding of tolerance mechanisms themselves. One of the classical ways of exploring the possibility of improving tolerance is to test its responsiveness to adaptive treatments, including those of the cross-adaptation type. This way is being followed systematically in our studies.

UNASSISTED RECOVERY: LIMITS AND MECHANISMS OF SPONTANEOUS REWARMING

Critical temperatures of warming

The critical temperature of warming (an expression borrowed from Adolph & Richmond, 1955) has been defined by us as the lowest body temperature from which 50% of cooled animals will rewarm completely if left at a constant air temperature equal to the body temperature to which they were cooled. Thus defined, the critical temperature of warming (CTW) proved to be a very sensitive parameter for defining temperature limits of spontaneous recovery unassisted even by external heat. In addition to interspecific CTW variability, considerable differences due to temperature-independent seasonal changes, to changes brought about by natural hibernation, to thermal acclimation and to endocrine status, have also been found (Andjus & Petrović, 1959; Andjus et al. 1964). Thus, ground squirrels (*Citellus citellus*), cooled artificially on the next day following induced arousal from natural hibernation, showed CTW values close to the physiological zero for their heart (about 3° C.). Animals prevented from hibernating by an artificial warm environment showed at the same period considerably higher critical temperatures of warming (9° C.). Although living constantly at thermal neutrality, they showed lower CTW values in autumn and winter than in summer. Acclimation to different environmental temperatures (0° and 30° C.) had little influence on their CTW values, at least in autumn, in comparison to the effect of acclimation found in rats. In the latter, temperatures allowing for spontaneous rewarming are much higher than in the hibernator. Despite this, in cold acclimated rats, the CTW value was between 14 and 15° C., a temperature very close to the one incompatible in this species with prolonged breathing and heart activity. Acclimation to a thermo-neutral or hot environment (30 and 38° C.) reduced greatly the capacity to rewarm spontaneously. A CTW value of 16·5° C. was found in rats acclimated to thermal neutrality.

It was found further that adrenalectomy, to a greater extent than thyroidectomy, resulted in increased CTW values. While thyroidectomy affects mainly cold-acclimated rats by making their CTW values similar to those found both in normal and thyroidectomized rats acclimated to thermal neutrality, adrenalectomy increases considerably the CTW values in animals from thermal neutrality as well as in cold-acclimated rats, without abolishing differences between them. Single doses of cortical steroids administered prior to cooling enhance considerably the rewarming capacity of adrenalectomized rats, acclimated to cold before their adrenals were

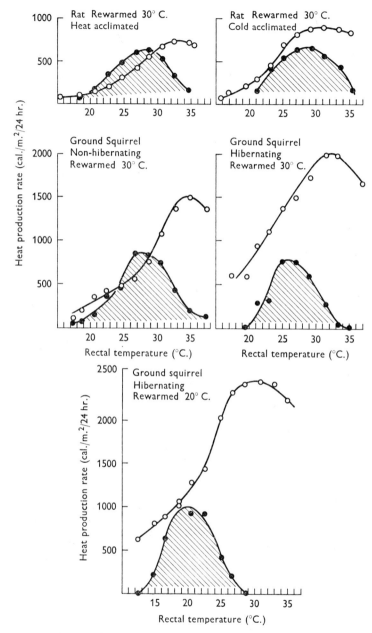

Fig. 1. Rates of 'shivering' and 'non-shivering' heat production (closed and open circles respectively) in rats and non-hibernating ground squirrels rewarming from induced hypothermia at 30° C. ambient temperature and in hibernating ground squirrels rewarming from natural hibernation at ambient temperatures of 30 or 20° C. (see text). Rats were previously acclimated for a month to 28–30° C. ('heat') or to 10–14° C. ('cold'). Means from 6–9 animals.

removed. Given, however, to normal rats acclimated to thermal neutrality, the same doses do not provoke any change of CTW values in the direction of cold acclimation.

Rewarming thermogenesis

Differences are also found between hibernators and non-hibernators in heat-production rates recorded during rewarming. The heat-production rates also vary in differently acclimated animals, in hibernating and non-hibernating hibernators and in animals rewarming at different temperatures (Fig. 1). The two curves found in each of the graphs are representative of the so-called shivering thermogenesis (lower curves limiting the shaded regions) and non-shivering thermogenesis (upper curves). They were calculated from oxygen-uptake data obtained in comparative groups of normal and curarized animals, the latter being considered as deprived of heat production originating from muscular shivering activity (for details of the technique see El Hilali & Andjus, 1966a). It is evident that higher rates of heat production are associated with the same levels of body temperature after cold acclimation in the rat. Higher rates are found in ground squirrels than rats, including the cold-acclimated ones. However, rewarming heat production rates appear as considerably higher in animals arousing from hibernation than in those rewarming from induced hypothermia at the same ambient temperature of 30° C. Finally, a lower ambient temperature (20° C.) is shown in the last graph to be associated with higher rates of heat production at identical levels of body temperature during rewarming from hibernation, thus illustrating the stimulating effect of external cold in the presence of internal cold. Figure 1 stresses, however, the fact that the majority of these differences are mainly due to changes in those heat-generating processes which appear as independent from muscular shivering activity. The contribution of shivering thermogenesis is greatest at relatively low levels of body temperature during rewarming (25° C.), when it may be greater than the contribution of other heat-generating mechanisms (heat-acclimated rat, non-hibernating ground squirrel, rewarming at 30° C.). In rats rewarming at an ambient temperature of 20° C. (not shown in the figure) the contribution of shivering thermogenesis appears critical: their body temperature will not increase substantially beyond that level if shivering is prevented by curarization (El Hilali & Andjus, 1966a). Rewarming does occur in the curarized ground squirrel under identical environmental conditions, but this animal is characterized by greater rates of non-shivering thermogenesis at low body temperatures than the rat (Fig. 1). In contrast to the large contribution of shivering thermogenesis to heat production between 20° and 30° C. body

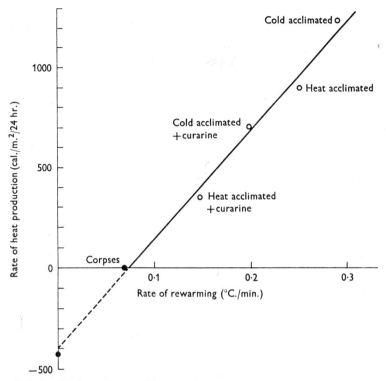

Fig. 2. Relationship between rate of heat production and rate of rewarming in rats during rewarming from 23 to 28° C. rectal temperature at 30° C. ambient. Same groups as in Fig. 1.

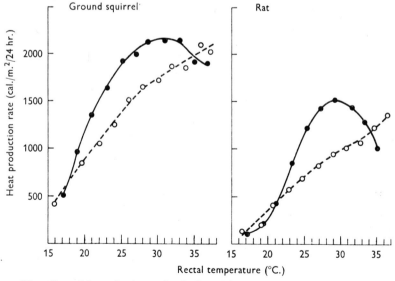

Fig. 3. The effect of thyroxine (2 mg./kg. body weight on alternate days for two weeks) on heat production during rewarming from induced hypothermia in ground squirrels and rats during rewarming at ambient temperatures of 20 and 30° C. respectively (means from 6–8 animals per group). —, Control; – – –, thyroxine.

temperature, above 30° C. in both species the final re-establishment of body temperature appears to be accomplished solely by heat-generating mechanisms of a different nature, as evidenced both by the results presented in Fig. 1 and by accompanying electromyographic studies.

The region between 23° and 28° C., which is critical for successful rewarming to less-dangerous levels of hypothermia, is characterized by a steady increase of body temperature and heat production in successfully recovering animals. In this region the rate of rewarming, as influenced by cold acclimation and by curarization in the same species (the rat), appears to be linearly related to the rate of heat production (Fig. 2). It is worth mentioning that the point of intersection with the y axis, showing the extrapolated rate of heat removal needed to prevent warming in dead rats at an ambient temperature of 30° C., is close to the value (closed circle) predicted by using, in the corresponding linear equation, the figure of 0·82 for specific heat, although this figure has been obtained in cooled dead mice by a different procedure (Hart, 1951).

As noted earlier, cold-acclimated rats, when deprived of their thyroids, fail to show an increased capacity to rewarm spontaneously, as revealed by the CTW test. Figure 3 shows, however, that excess thyroxine does not seem to favour the production of heat during rewarming, either in rats or in ground squirrels. On the contrary, animals showing increased basal metabolic rates after thyroxine treatment display considerably lower rates of rewarming thermogenesis, as calculated from oxygen uptake data. Peak rates, associated with body temperatures approaching 30° C., seem to be particularly affected, and the curves become almost linear. At near-normal temperatures, however, the rate of heat production again becomes greater in thyroxine-treated animals than in the controls. Without discussing in detail the significance of our heat-production figures, calculated from oxygen-uptake data (El Hilali & Andjus, 1966b), we may emphasize that a decreased oxygen-consumption rate was found in thyroxine-treated animals within the range of body temperatures characterized during rewarming by an intense shivering activity in untreated animals (20–30° C.), while rates of oxygen uptake were higher in relation to controls at temperatures at which shivering did not occur. Thus, it may well be that thyroxine has different effects on processes linked to shivering and non-shivering thermogenesis during rewarming.

It must be stressed that lower rates of rewarming, accompanying lower rates of oxygen consumption, were also found in thyroxine-treated animals, especially in ground squirrels (El Hilali & Andjus, 1966b). Thyroxine-treated rats showed substantially lower rates of rewarming than cold-acclimated ones. The difference was greatest when rewarming occurred at a lower ambient temperature (20° C). In a warmer environment, however, the rates of rewarming in the thyroxine-treated animals seemed to be less affected.

ASSISTED RECOVERY: MECHANISMS OF TOLERANCE TO EXTREME HYPOTHERMIA ABOVE THE FREEZING POINT

Metabolic correlates of anaerobic conditions in extreme hypothermia

Soon after cooling below 15° C. body temperature, breathing and heart beats are arrested in the rat. Nevertheless, recovery remains possible even from a central body temperature of 0° C. if adequate re-animation procedures are applied in less than 2 hr. (Andjus, 1951a, 1955; Andjus & Smith, 1955; Andjus & Lovelock, 1955).

In order to establish whether, and to what extent, metabolic events taking place in the anoxic brain are responsible for the time limits of extreme hypothermia, rats cooled to and maintained at 0° C. have been compared to animals in which anaerobic conditions have been created by tracheal clamping at higher levels of hypothermia (for technical details see Andjus *et al.* 1964). Figure 4 stresses the basic similarity between metabolic events induced in the brain simply by cooling from 15° C. down to zero (right-hand graph) and those taking place at 15° C. after tracheal occlusion (curves on the left). Except for a considerably lower rate (note the difference in time scales), metabolic changes in rats cooled to zero appear as equivalent to those induced by severe asphyxia.

Another point deserves to be stressed, however. A closer inspection of our curves reveals that in rats rapidly cooled from 15° C. downwards by covering them completely with crushed ice, anaerobic conditions created by cold-induced respiratory and cardiac arrest provoke nevertheless considerable metabolic changes in the brain, even before the level of 0° C. is actually reached by the central body temperature. These changes are comparable to those recorded beyond the respiratory survival time in animals asphyxiated at higher temperatures, as shown by results outlined in the next section. This stresses the necessity of taking into account, when studying time limits of tolerance to a chosen level of extreme hypothermia, changes occurring *during* cooling to this level. Especially in those mammals in which a considerable body mass limits the rate of cooling, irreversible changes may take place during cooling, before the desired level of body temperature is actually reached and without this level being particularly intolerable in itself.

Both levels of body temperature explored so far represent critically deep hypothermia: 15° C. is close to the temperature limit of unassisted survival, while 0° C. is close to the challenge of the freezing point. Thus, before attempting to evaluate the tolerance-limiting role of metabolic changes just described, and in order to characterize them better, it would be useful to know whether these changes differ substantially from those observed under anaerobic conditions at normal body temperature or in shallower hypothermia. A close inspection of the families of curves illustrated in Figs. 5

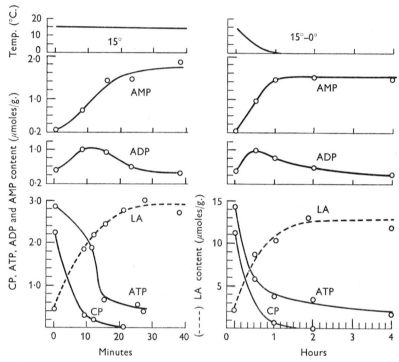

Fig. 4. Metabolic changes in brains of rats cooled from 15° down to 0° C. (right-hand curves) and in rats asphyxiated at 15° C. (left-hand curves). ATP, Adenosine triphosphate; ADP, adenosine diphosphate; AMP, adenylic acid; LA, lactic acid; CP, creatine phosphate. Means from 3–8 animals per point.

and 6, obtained at four different levels of body temperature under conditions of induced asphyxia, reveals that some characteristics other than the rate of changes seem to be specific to the respective levels of hypothermia. We shall briefly discuss only those which might be particularly important from the viewpoint of tolerance. In this respect the beginnings and the terminal parts of our curves deserve special attention.

Our initial values were obtained in animals which were breathing normally before their oxygen supply had been interrupted. They are representative, therefore, of the situation determined solely by the respective levels of body temperature. It may be seen that, in general, energy-rich compounds (ATP, CP) which are rapidly used by the anoxic brain tend to show higher initial levels at lower body temperature. The opposite is true of metabolites known to accumulate under anaerobic conditions: brain lactate (LA) and AMP tend to show lower initial values in deeper

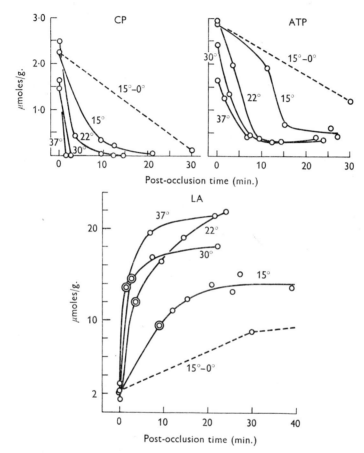

Fig. 5. Changes in the concentration of brain creatine phosphate (CP) ATP and lactate (LA) in rats after tracheal occlusion at different levels of hypothermia (plain curves) or after the onset of cooling from 15 to 0° C. body temperature (dashed curves). Double circles: values at the end of the respiratory survival time. Means from 3–6 animals per point.

than in shallower hypothermia or at normal body temperature. There is no evidence of changes induced by hypothermia itself, which might be interpreted as due to an inadequate oxygen supply of the brain: while cooling below 15° C. induces changes characteristic of anaerobic conditions, cooling to 15° C is not associated with hypoxic changes, as testified by metabolic parameters exceedingly sensitive to lack of oxygen. On the contrary, in relation to survival, the initial situation at the level of brain

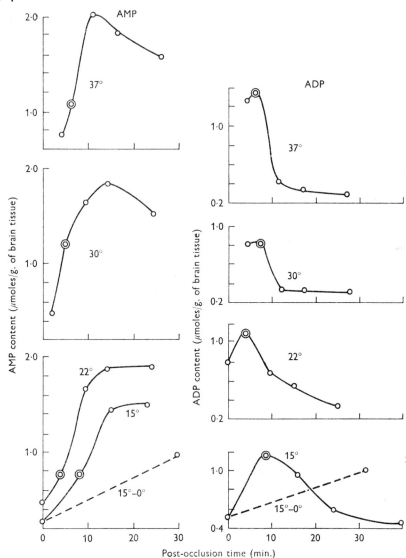

Fig. 6. Changes in the concentration of AMP and ADP in the brain of rats asphyxiated at different levels of hypothermia. Same symbols as in Fig. 5.

energy stores might be regarded as more favourable, if anything, in deeper hypothermia than at higher body temperature.

It is tempting to interpret the observed temperature-dependent differences in initial values of metabolites, as reflecting different dynamic equilibria between energy-using and energy-storing processes determined by different rates of metabolism, lower rates at lower temperatures being responsible for equilibria characterized by higher levels of primary energy stores. Alternatively, the recorded differences might be due, at least to some extent, to the limitations of the sampling technique itself. It was stressed by other authors (Lowry, Passonneau, Hasselberger & Schulz, 1964) that the procedure of *in vivo* freezing of the head in liquid nitrogen, although extremely rapid, may include a period, however short, of anoxia with possible metabolic consequences. The importance of this period in causing errors should be expected to vary with the initial level of body temperature in such a way as to engender precisely the same type of apparent differences as those actually found. Therefore, initial values obtained in deeper hypothermia might be closer to reality than those obtained at higher levels of body temperature. Some of our curves, illustrating the course of changes in the asphyxiated brain at relatively high levels of body temperature, appear indeed as deprived of their natural beginnings when compared to curves obtained in deeper hypothermia (see ATP and ADP curves in Figs. 5 and 6, for instance). Nevertheless the fact which is essential for our argument remains: in animals cooled to the very limit of unassisted survival, and before respiratory and cardiac arrest are induced by further cooling, the levels of the most sensitive brain-energy stores appear as undisturbed.

As to the terminal parts of our curves, illustrating the extent to which changes can proceed, two peculiarities deserve attention. First, the tendency of AMP levels to decrease after reaching a maximum is less marked at 15° and 0° C. than at higher temperatures. We shall return to this point when discussing the efficiency of AMP-eliminating mechanisms in deep hypothermia. Secondly, the terminal lactate levels are decidedly lower in deep hypothermia. It is important to note that a comparatively low terminal level of lactate is found not only at zero, in animals with arrested hearts, but also at 15° C. where the possibility of a contribution of lactate from other sources to the brain region, by way of circulation, is not abolished during a considerable period of asphyxia.

Tolerance correlates of anaerobic brain metabolism

As explained in the introductory section, the basic similarity of metabolic events taking place under anaerobic conditions at different temperatures does not permit the conclusion that they are necessarily of the same consequence to recovery throughout the region of hypothermia explored. Since our primary interest in the metabolic changes taking place in the anoxic and hypothermic brain concerns their roles in the tolerance-limiting mechanism, we tried to correlate them with selected tolerance parameters. Asphyxia induced by tracheal occlusion at 15° C., and cold-induced respiratory and cardiac arrest at 0° C., remained the two main comparative situations, while the respiratory survival time (RST, duration of the post-occlusion respiratory movements) and the 50 % revival time (time allowing for half of the animals tested to be revived) were chosen as parameters of tolerance. Since marked differences in tolerance to extreme hypothermia between hibernators and non-hibernators were already found to be paralleled by similar, but essentially temperature-independent differences in tolerance to anoxia (Andjus *et al.* 1964), the ground squirrel and the adult rat were chosen as comparative models for studying metabolic correlates of tolerance to anoxia and extreme hypothermia. Studies on newborn rats were also initiated because the well-known tolerance of neonates to low body temperature as well as to anoxia (Adolph, 1948) has often been compared to that of adult hibernators and suggested a possible similarity of the underlying mechanisms (Kayser, 1959; Burlington & Wiebers, 1965). Detailed data defining differences in tolerance limits between the three mammals studied—albino rats, hooded rats and ground squirrels—are shown in Fig. 7, while Figs. 8 and 9 illustrate the relationship of these tolerance parameters to metabolic events taking place in the brain.

Comparative results obtained at 15° C. in asphyxiated animals are shown in Fig. 8. Obviously, in spite of the same level of temperature and in spite of the essentially identical initial levels of brain metabolites, the rates of changes induced by asphyxia are very different in the three mammals. This is stressed by the differences in the time scales, 2·5 and 6 times reduced, for the ground squirrel and the newborn rat respectively, in comparison to the time-scale used for the adult rat. Otherwise, the changes appear as basically similar. Important differences are revealed however by a closer inspection of the curves. In the newborn rat, right-hand graph, except for the exceedingly slow rate of metabolic changes, stressed by the 6 times-reduced time-scale, the events remain basically similar to those recorded in the adult animals, left-hand graph, both in respect to the depletion (CP, ATP) and the accumulation processes (AMP, LA). In contrast, events recorded in the ground squirrel, middle graph, are greatly different. While the depletion processes, considerably slower than in the adult rat but not to the same extent as in the newborn, seem again to differ mainly by their rate, the changes concerning the accumulating metabolites differ significantly by their amplitude from those observed in rats: brain lactate reached exceedingly high values, while the accumulation of AMP appears as markedly restricted.

The horizontal, partly shaded, bars in the same figure, symbolize the tolerance parameters. The first shaded portion indicates the duration of the first, so-called aerobic, series of post-occlusion respiratory movements. In the adult rat this period appears as characterized by a rapid depletion

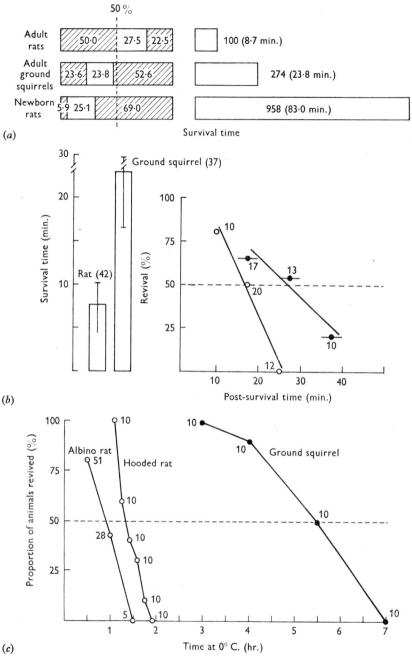

Fig. 7. (a) Respiratory survival times in adult rats, in adult ground squirrels and in newborn rats asphyxiated at 15° C. body temperature. Hatched areas indicate the proportion of the total respiratory survival time taken by the first and the second series of post-occlusion respiratory movements separated by the period of primary apnoea. (b) Survival times (left) and post-survival periods compatible with revival by artificial respiration (right) in rats and ground squirrels asphyxiated at 15° C. body temperature. The numbers of animals used in each experiment are indicated on the figures. (c) Proportion of animals revived from 0° C. body temperature after varying lengths of time. Albino rats (Wistar) and ground squirrels (*C. citellus*) represent animals used in the presently reported metabolic studies. Data on hooded rats from Andjus (1955). The numbers of animals in each experiment are indicated on the curves.

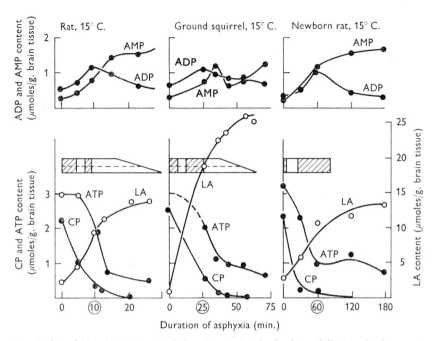

Fig. 8. Correlation between metabolic changes in the brain and limits of tolerance to asphyxia at 15° C. body temperature in adult and newborn rats, and ground squirrels (see text). At least 3 animals per point.

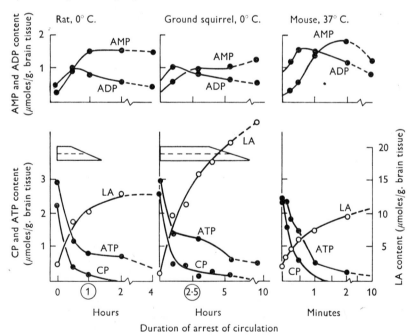

Fig. 9. Correlation between metabolic changes in the brain and time limits of arrested circulation compatible with recovery in rats and ground squirrels maintained at a body temperature of 0° C. At least 3 animals per point. Data replotted from Lowry's results in the mouse at 37° C. are shown for comparison. (See text.)

of CP, while ATP stores remain very little disturbed, if at all. The second and last or anaerobic series of gasps, and the preceding period of primary apnoea, are symbolized, respectively, by the second shaded portion of the bar and the preceding unshaded region. It is during this 'anaerobic period' that CP almost disappears, while the depletion of ATP accelerates, the rate of lactate accumulation reaches its maximum and the concentration of ADP attains its peak value before entering its decreasing phase. The remaining portion of the bar symbolizes the period of secondary apnoea

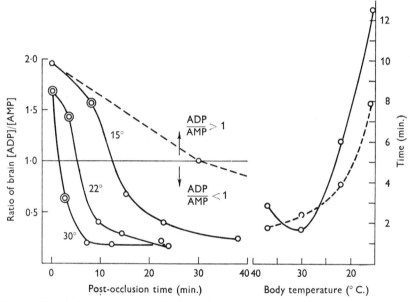

Fig. 10. Correlation between the ADP/AMP ratio and the respiratory survival time in rats asphyxiated at different levels of body temperature. Survival times are indicated by double circles in the left-hand curves and by the interrupted curve in the right-hand graph. The solid line in the right-hand graph shows the time taken for the ADP/AMP ratio to reach unity at different body temperatures.

compatible with revival by artificial respiration. Its height indicates the revival rate. Thus, the 50 % revival time is reached when the bar narrows by one half, the dashed horizontal line indicating the 50 % level. The beginning of this period compatible with revival (end of the respiratory survival time) correlates well with the drop of the ADP/AMP ratio below unity, since ADP enters its decreasing phase, while AMP continues to accumulate. This correlation seems to be valid for the rat throughout the region of hypothermia compatible with survival, as shown in detail by Fig. 10. As to our ATP figures at the end of the survival period, they were close, at 37° and 30° C. (1·2 and 1·3 μM/g.; see Fig. 5) to the value of 1 μM/ATP/g. reported by other authors as limiting for survival of the rat at normal body temperature (Samson, Balfour & Dahl, 1959). At 15° C., however, as well as at 22° C., our figures for ATP level commensurate with survival were considerably higher. A higher ATP concentration associated with survival at lower body temperatures has also been found by Samson *et al.*, although their data concern rats pretreated with iodoacetate. Finally, our data indicate it is by the end of the revival period that the metabolic changes still in progress approach their terminal values.

In the ground squirrel (middle graph, Fig. 8) the second series of gasping movements occupies a comparatively greater portion of the whole survival period than in the rat. This is in agreement with our earlier results, indicating that in this animal the respiratory survival time appears as significantly extended in comparison to the rat, mainly because of a 6–7 times longer period of anaerobic gasps (Andjus *et al.* 1964; see also Figs. 11 and 12). It is evident, on the other hand, that the survival time as a whole represents in the ground squirrel a proportionately greater part of the revival period than in the rat. The correlation between metabolic and tolerance parameters, at the end of the survival times, does not seem to be significantly different from the one just described for the rat. Lactate levels, however, at the end of both tolerance periods (the survival and revival time) are marked by considerably higher absolute values than in the rat, while the opposite is true of AMP.

As for the newborn rat, our data concerning tolerance parameters are still incomplete (data on revival time are lacking). They do not seem, however, to correlate with metabolic parameters in the same way as in adult rats or the ground squirrel. The exceedingly long respiratory survival time (83 min. on the average), though characterized by occasional respiratory movements, spread by several minutes, seems to be compatible with more advanced metabolic changes in the brain.

Figure 9 describes the situation at 0° C., this time associated with a cold-induced arrest of circulation. The comparison between adult rats and ground squirrels reveals, however, the same type of differences: the amplitude of lactate accumulation in particular is significantly greater in the ground squirrel, while other changes differ from those in the rat mainly by their lower rates. The graph at the far right deserves special comment. It concerns the mouse, another non-hibernator, a differently induced arrest of brain circulation (decapitation) in a different thermal situation (normal body temperature), and data obtained by a different research team (Lowry *et al.* 1964). Yet, when plotted in our way, against a time-scale with minute intervals made equal to hours in the other two graphs, these data appear as strikingly similar to those concerning our ice-cold rats, and significantly different from the ground squirrel.

COLD TOLERANCE IN MAMMALS 371

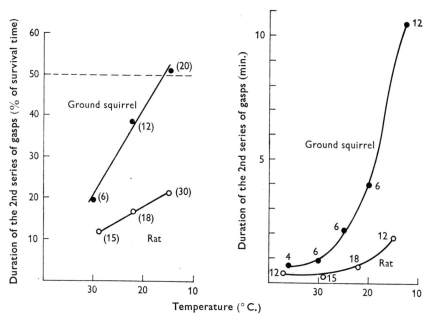

Fig. 11. On the left the duration of the second series of 'anaerobic gasps' is shown as the percentage of the total respiratory survival time at different levels of body temperature in rats (○) and ground squirrels (●). The right-hand graph shows the change in the absolute values of the duration of the second series of gasps in these species. Number of animals per group shown against the points.

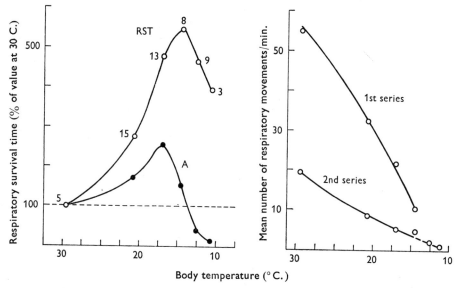

Fig. 12. Changes in the respiratory survival time (RST) and in the total number of respiratory movements in the same periods (A) in rats asphyxiated at different body temperatures (values at 30° C. taken as 100%; left-hand graph). The mean rates of respiratory movements during the first and second series of gasps are plotted against body temperature in the right-hand graph. Number of animals given against the RST curve.

24-2

372   COLD TOLERANCE IN MAMMALS

The difference in revival times between rat and ground squirrel is again decidedly in favour of the hibernator, even more so than at 15° C. under conditions of asphyxia. Although revival from 0° C. necessitates a more

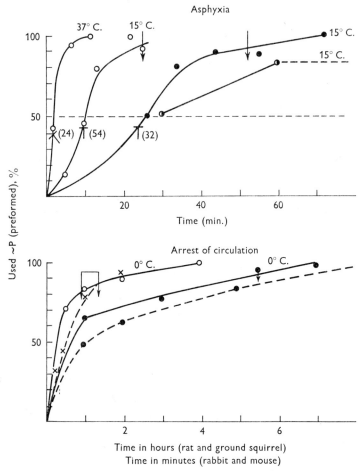

Fig. 13. Changes in the combined concentrations of initially available CP, 2ATP and ADP ('preformed $\sim$P') in different animals during asphyxia and arrest of circulation. Means from 5–13 animals per point. Body temperature levels indicated on the curves (see text). T, Respiratory survival time (number of animals in parentheses); ↓, 50% revival time (based on results from Fig. 7c); ○, rat; ◑, newborn rat; ●—●, ground squirrel; ●--●, rabbit; ×--×, mouse.

complex reanimation procedure, including the reactivation of the heart by microwave diathermy (Andjus & Lovelock, 1955), revival times seem again to be correlated with the period when the most lasting among the recorded metabolic events in the brain approach their end.

Figure 13 may help in further defining the correlation between metabolic and tolerance parameters. Changes in the sum of the molar concentrations of CP, 2ATP and ADP, expressed as a percentage of their total change during the whole exposure period and considered as changes of the initially available energy-rich phosphates ('preformed $\sim$P'), are plotted against exposure time. In spite of considerably different rates of change due to different body temperature in the adult rat (37° and 15° C.), or to interspecific differences between rats and ground squirrels studied at the same body temperature (15° C.), the end of the respiratory survival time (T-shaped indicators, upper curves) always coincides with a strikingly similar metabolic stage: a 40–45 % depletion of the 'initially available' sources of energy-rich phosphate bonds. Data on asphyxiated newborn rats, although incomplete, suggest the slowest rate of $\sim$P use. At the 50 % level, however, the difference in respect to the ground squirrel does not seem to be outstanding. From this point of view, therefore, one is justified in speaking of a similarity. The respiratory survival time (not shown in the figure) coincides with a much more advanced stage of $\sim$P use in the newborn rat (outside the limits covered by the graph). The 50 % revival times (arrows), as determined for the asphyxiated adult rat and ground squirrel at 15° C., differ greatly in their absolute values, but coincide in both cases, with a stage when about 90 % of the 'initially available $\sim$P' seems to be used up. A strikingly similar metabolic stage seems to set the limit to revival from 0° C. (lower curves), especially in the ground squirrel. In the adult rat, data on revival appear to coincide with a somewhat earlier stage of $\sim$P depletion. The additional dashed curves were constructed by borrowing data from other authors. They concern the brain of the decapitated mouse (Lowry *et al.* 1964) and the ischaemic brain of the rabbit (Thorn, Pfleiderer, Frowein & Ross, 1955), both studied at normal body temperature. And yet, by their shape they fall into the same category as our curves concerning cold-induced circulatory arrest.

Our results suggest therefore a close correlation between tolerance to extreme hypothermia and metabolic parameters concerning brain energy stores. Other authors described the revival rate of deeply cooled rats as correlated to the degree of cerebral oedema attributable to a direct temperature-induced inhibition of active cation transport (Reulen *et al.* 1966). These authors admit however that below 15° C,. when hypothermia becomes associated with circulatory disturbances, tissue hypoxia plays a major role in the mechanism of brain swelling, since the hypoxic exhaustion of cerebral ATP and CP stores leaves active transport short of energy supply. *In vitro* studies suggesting that cold may reduce energy expending activities directly and critically, and that the mechanism of cation transport may be more sensitive than the supporting respiration (Willis, 1968), seem to be more

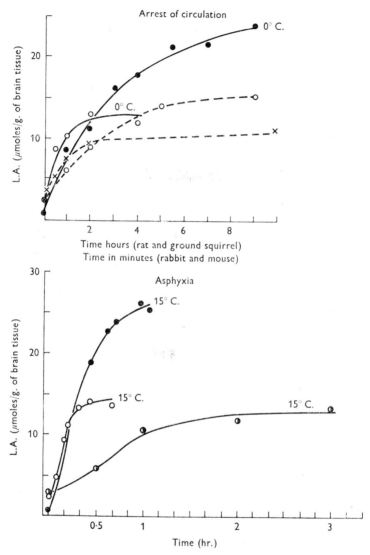

Fig. 14. The accumulation of lactate (LA) in the brain in different mammals during asphyxia at 15°, or during cardiac arrest at 0° in rats and ground squirrels (5–13 animals per point). Figures for the mouse at normal body temperatures were plotted from the data of Lowry *et al.* (1964) and for the rabbit at normal body temperature from the data of Thorn *et al.* (1961). ○, Rat; ◐, newborn rat; ●—●, ground squirrel; ○ - - ○, rabbit; ×--×, mouse.

relevant to the interpretation of mechanisms leading to the failure of specific functions at reduced temperature in spite of adequate oxygenation. Under conditions of interrupted oxygen supply, the failure of energy-yielding processes may become the predominant tolerance-limiting factor.

Figure 14 shows the differences in lactate accumulation mentioned earlier. The newborn rat differs from the adult in having a much lower rate of lactate accumulation, the terminal levels attained are, however, almost identical (Fig. 14, lower graph). In contrast, the rate at which lactate accumulates in the brain of the asphyxiated ground squirrel seems very close to that recorded initially in the adult rat. However, at a time when the lactate concentration stops increasing in the brain of the rat, it continues to accumulate in the ground squirrel brain at a high rate and reaches far higher terminal values. By referring to Fig. 5 one sees that these terminal values are actually higher than the highest lactate levels reached in the rat at any temperature level, including normal body temperature (see also Fig. 15). In comparison to the newborn rat, the situation in the ground squirrel appears this time very different indeed: the accumulation of lactate not only proceeds at a much higher rate, but also attains far higher terminal levels. This certainly speaks strongly against the supposed similarity between mechanisms responsible for the striking tolerance to anoxia in neonates and hibernators.

At $0°$ C. body temperature, under conditions of cardiac arrest induced by cooling (Fig. 14, top graph), the difference between rat and ground squirrel closely resembles the one characterizing asphyxiated animals. Borrowed data (dashed curves), concerning two other non-hibernators (rabbit and mouse), although obtained under very different conditions of circulatory arrest and body temperature, further suggest that the extremely high lactate accumulation in the ground squirrel is specific to the hibernator.

Figure 15 brings together the two metabolic parameters already discussed, macroergic phosphates (the sum of 2ATP, ADP and CP) and lactate, attempting to define the mutual relationship between changes in their concentrations. In the ground squirrel, a considerably greater accumulation of lactate becomes associated with the decrease of the initial level of $\sim$ P-containing compounds than in the rat studied at the same temperature ($15°$ C.). The drop of the initial $\sim$ P concentration which coincides with the respiratory survival time (RST) is of very similar magnitude in the asphyxiated rats and ground squirrels, but it is associated with a considerably greater accumulation of lactate in the hibernators. The same is true of the still greater drop of the initially available $\sim$ P sources which appears to limit revival (50% RT): similar in both species, this drop is also accompanied by a considerably higher lactate concentration in the ground squirrel. The great difference between the ground squirrel and the rat remains even when the rat is represented by results obtained at higher body temperatures ($22°$ and $37°$ C., middle curve) associated with higher levels of lactate accumulation than at $15°$ C. On the other hand, data obtained from new-

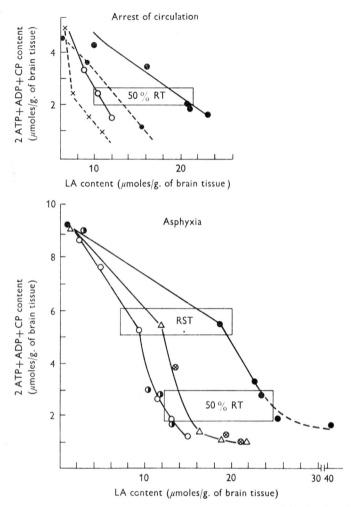

Fig. 15. Correlation between the decrease in the initially available levels of combined energy-rich phosphates and the increase in lactate during asphyxia (lower graph) and after the arrest of circulation (upper graph). ◐, newborn rats at 15° C. Adult rats: ○, at 15° C.; ⊗, 22° C.; △, 37° C.; ●, Ground squirrels at 15° C. in the lower graph. Rats and ground squirrels at 0° C. are represented in the upper graph by open and solid circles respectively. The figures for mouse and rabbit at normal temperature (dotted lines) are plotted from the publications quoted in Fig. 14. RST, Respiratory survival time; RT, revival time.

born rats are shown in the figure (half-occluded circles) to fit quite well into the region occupied by results belonging to adult rats studied at the same body temperature. In this respect the newborn rat seems closer to the adult rat than to the hibernator. Data obtained after the arrest of circulation at 0° C. indicate basically the same difference between the ground squirrel and the rat, at the level of the 50% revival time in particular. Again, data published by other authors and concerning other non-hibernators fit with our results obtained in the rat, while those recorded in our hibernator remain very different.

Thus in summarizing results concerning brain lactate, it should be stressed that (a) equal degrees of lactate accumulation appear as linked to a greater reduction of $\sim$P levels in rats than in the ground squirrel, (b) in the hibernator the accumulation of lactate continues until much higher terminal levels are reached, and (c) in the hibernator the rate of accumulation of lactate is at least as high as in the adult rat within its survival period and far greater than in the newborn rat.

In addition to the decrease of energy-rich phosphate levels, lactate accumulation may be considered as a measure of energy utilization, provided that metabolic sources of lactate are known (Lowry et al. 1964). The apparently 'sparing' effect of lactate accumulation in respect to P depletion seems to be greater in the ground squirrel. This could be due to a greater efficiency of energy-releasing processes responsible for the generation of lactate. It is well known for instance that breakdown of glycogen to lactate yields more energy than the breakdown of glucose to lactate. In this respect, as well as in respect to mechanisms responsible for the amplitude of lactate accumulation, our preliminary results concerning brain glycogen are worth mentioning (Andjus & Genci, unpublished). In the ground squirrel the brain is provided with a significantly greater amount of available glycogen than in the rat, as judged by the amount of total glycogen disappearing during a 30 min. period after asphyxiation at normal body temperature, as well as by the decrease of glycogen concentration within the respective revival periods after tracheal occlusion at 15° C. Determined in this way, the available glycogen stores, at normal temperature, amounted to 2560 μM/g. in the rat (cf. 2250 μM/g. for the mouse (Lowry et al. 1964), while 4970 μM/g. were found in the ground squirrel (means from seven and five individuals respectively). The absolute levels of total brain glycogen were 1·6 times greater in the ground squirrel than in the rat, and similar glycogen values were obtained for animals at 15° C. and at normal body temperature. In animals asphyxiated at 15° C. the amounts of glycogen within the periods formerly defined as the 50% revival times (25 and 51 min. in the rat and ground squirrel respectively), were 1400 μM/g. in the rat and 2700 μM/g. in the ground squirrel. Although these figures suggest a similar mean rate of glycogen utilization for the entire revival periods, results obtained so far do not exclude the possibility of greater rates in the ground squirrel at particular stages of asphyxia. The possibility of different relationships, in rats and ground squirrels, between the utilization of glycogen and the breakdown of other different lactate-generating sources, also remains. In addition, results from other laboratories (Burlington & Klain, 1967) testify to an increased glyconeogenic capacity induced by hypoxic exposure in tissues of the ground squirrel, in great contrast to a decrease of this capacity in the identically treated rat. This should also be taken into account when evaluating the significance of the rate at which glycogen appears to be used when determined solely by measuring changes in its level. Additional factors determining the low rate of $\sim$P depletion in the hibernator should also be taken into account, such as possible peculiarities of anabolic $\sim$P-consuming processes, particularly under anaerobic conditions. In this respect the finding that in two species of ground squirrels no labelling occurred in a variety of tissues, after the administration of tritiated thymidine as a precursor for DNA synthesis (Adelstein, Lymen & O'Brien, 1964), may become of particular interest.

By contrast to the ground squirrel, in the asphyxiated hypothermic newborn rat a slow depletion of the initially available $\sim$P sources is accompanied by an exceedingly slow accumulation of lactate. This undoubtedly indicates a low rate of $\sim$P use by the immature brain in the described situation. Moreover, the tolerance, at least in terms of respiratory survival time, is less affected by the amplitude of these metabolic changes in the newborn rat under these conditions.

In this context results of *in vitro* studies should be recalled. For example, Burlington & Wiebers (1965) have shown that rates of anaerobic glycolysis in the cerebrum from hypoxic infant rats were not increased significantly in comparison to rates found in control brains, while a considerable increase was recorded in comparative experiments with brain tissue from hypoxic ground squirrels. Actual glycolytic rates were greater, however, in infant rats than in either adult rats or ground squirrels, when excised brain tissue was studied under anaerobic conditions at 37° C. (Burlington & Wiebers, 1965). However, these studies were carried out in the presence of relatively high glucose concentrations, far greater than those available to the anoxic and/or ischaemic brain *in vivo*. Brain glucose levels as well as maximal glucose fluxes are significantly lower in the immature brain, as shown in the mouse (Lowry *et al.* 1964). These authors also found significantly lower rates of glucose and glycogen utilization, as well as of the calculated total $\sim$P consumption, in 10-day-old than in adult mice by following changes taking place in the brain after decapitation at normal body temperature. This is in full agreement with our results concerning the newborn and adult rat, asphyxiated at 15° C. We may add that our preliminary measurements of brain glucose at 15° C. body temperature also indicate significantly lower initial levels in the newborn than in the adult rat, and particularly low in comparison to figures obtained in the ground squirrel under identical conditions (the relationship being newborn rat 1, adult rat 1·5, ground squirrel 3, approximately 1:1·5:3). In this respect it is worth mentioning that with isolated rat-brain preparations studied at reduced temperatures (23–25° C.) we found greater rates of glucose utilization and lactate production in the presence of higher levels of glucose in the perfusion fluid (Andjus, Suhara & Sloviter, 1967). This indicates the importance of the level of glucose as a factor determining the rate of its utilization by the hypothermic brain of one and the same animal. Comparative *in vitro* studies have shown that the brain tissues of hibernators use more glucose and produce more lactic acid than in the rat; the rate of ATP production by isolated brain-cell mitochondria is also greater in hibernators than in the rat (Kayser & Malan, 1963). This was interpreted as being related to the greater resistance to hypoxia and hypothermia in hibernators and is in agreement with the results of our *in vivo* studies concerning the brain of ground squirrels exposed to the combined effects of hypothermia and lack of oxygen.

In addition to metabolic peculiarities of hibernators which appear as essentially temperature-independent because they are present even at normal levels of body temperature, it should not be forgotten that the metabolic differences between rats and ground squirrels may be accentuated in extreme hypothermia because of a different relationship of their inherent biochemical characteristics to falling temperature. Let us recall (fig. 11) that the duration of the 2nd series of gasps, supported solely by anaerobic processes (see Gänshirt, Severiss & Zylka, 1952; Opitz & Saathoff, 1952), and its share in the total respiratory survival time, increase proportionately more with falling temperature in the ground squirrel than in the rat. In comparative *in vitro* experiments with excised heart tissue, exceedingly high values of the energy of activation were found in the rat, but not in the ground squirrel at low temperatures equivalent to extreme hypothermia. This was taken as reflecting a lowered enzyme activity in the rat, possibly due to an alteration of the molecular configuration of the enzymes (Burlington & Wiebers, 1966).

Figure 16 allows a further insight into the previously mentioned differences at the level of the other accumulating metabolite, the adenylic acid (AMP). Deprived of energy-rich phosphate bonds, but potentially toxic, AMP is treated here as a component of the total brain adenylate. Molar concentrations of AMP, recorded in the course of its accumulation under conditions of asphyxia or arrested circulation, are plotted against the decreasing values of the sum of the molar concentrations of $\sim$P-containing adenylates (ATP + ADP). The left-hand straight lines in each of the three graphs define the ideal relationship of mutual equality between the increase of AMP and the decrease of ATP+ADP, all points along these lines testifying to an unchanged total adenylate. Consequently, any departure from this slope to the right indicates a fall of total adenylate, which is known to occur whenever AMP accumulates significantly (Lowry et al. 1964). Inspection of the graph from above downwards shows how this departure becomes greater as the body temperature falls. At 37° and 30° C., in asphyxiated rats, the accumulation of AMP in the brain is equivalent to the decrease of the sum of ATP and ADP up to a point when the combined concentrations of the energy-rich components of the adenylate system become markedly reduced. When the level of total adenylates falls below 2 μM/g. there is inversion in the course of AMP changes and the further fall in combined concentration of ATP and ADP is then followed by a decrease of AMP. Data from the work of Lowry et al. (1964) on mice (crosses) also fit well into the scheme, although they concern decapitated animals. At 22° and 15° C. (Fig. 16(a), middle graph), under conditions of asphyxia, the departure from the initial slope begins earlier when the concentration of ADP+ATP falls below 3 μM/g. The data for ground squirrels (solid circles) at 15° C. show a sharp drop in the level of AMP when the ADP+ATP level falls to 2 μM/g., no such change was found in rats at this body temperature. At the same time this concentration of ADP+ATP in the ground squirrel was considerably higher than the concentration (1 μM/g.) associated with a similar AMP inversion phenomenon at normal and near normal temperature in the rat and mouse (Fig. 16(a), top graph). In contrast, data concerning the newborn rats followed closely those of the adults. The situation at 0° C., under conditions of cold-induced circulatory arrest, is of particular interest from the point of view of tolerance limits to cold. The data show (Fig. 16(a), bottom graph) that at this temperature there is the earliest departure from the ideal, total adenylate-conserving relationship which is found in both the adult rat and in the ground squirrel. The difference between them, suggested by our first set of ground-squirrel data, became less probable after another group of these animals provided additional results (small closed circles).

It appears, then, that at lower temperatures there is an earlier departure from the relationship of equality between the accumulation of AMP and the loss of the other adenylates. Without entering into a detailed discussion of the possible biochemical mechanisms, it may be pointed out that the departure from this relationship might be a measure of the efficiency with which AMP is being broken down in the presence of AMP-generating processes (use of ATP and ADP). This efficiency may depend on the rate at which AMP is generated. This in turn would explain an earlier departure from the mentioned relationship at lower temperatures characterized by

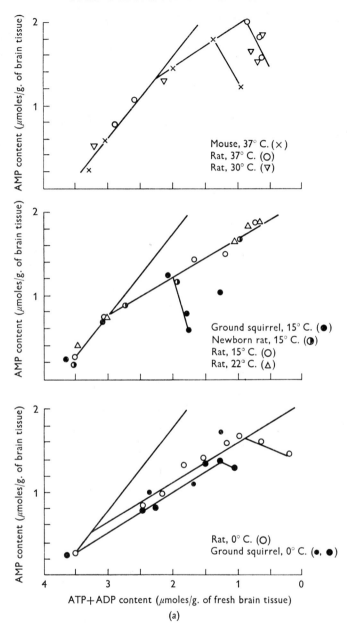

Fig. 16. (a) Increase in the concentration of AMP in the brain shown as a function of the decrease in the combined concentrations of ATP and ADP in rats and ground squirrels at different body temperatures. In the bottom graph circulation was arrested. Mouse data replotted from the data of Lowry et al. (1964). (b) Data of Lowry et al. (1964) for adult mice, immature mice and immature anaesthetized mice decapitated at normal body temperatures are replotted to show a similar relationship.

lower rates of the AMP-generating processes if the AMP-eliminating mechanisms were not equally limited by temperature. Although highly speculative, this reasoning may find support in calculations based on the data of Lowry et al. (1964); these are shown in Fig. 16(b), where they are plotted in such a way as to be comparable with Fig. 16(a). The data concern three groups of decapitated mice (adult, immature and anaesthetized), characterized by different metabolic rates in spite of normal body temperatures. The differences between them are reminiscent of those existing between our group of rats studied at different levels of body temperature. The group of anaesthetized immature mice, which is the group with the lowest metabolic rate, also shows the earliest and greatest loss of total

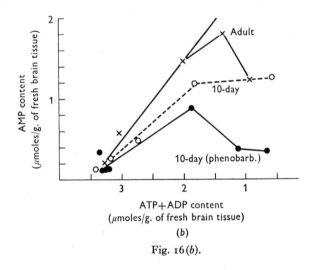

Fig. 16(b).

adenylate. A similar difference, however, between newborn and adult rats asphyxiated at one and the same level of hypothermia (15° C.) has not been noted in our experiments. Other temperature-dependent or species-specific factors, influencing the efficiency of AMP-eliminating mechanisms, may of course be envisaged in interpreting the difference illustrated by Fig. 16(b). The role of a critical concentration of AMP or ADP might be one of these.

Authors studying brain metabolism after respiratory arrest in dogs bled at different levels of hypothermia (Nossova, 1960; Gaevskaya, 1963) concluded that the elimination of AMP, presumably by way of its deamination, must be less efficient in deep hypothermia, since the tendency of AMP to decrease after reaching a peak concentration in the brain becomes less pronounced with falling temperature (a fact observed in our experiments as well). In their experiments, moreover, the fall in brain ATP and ADP content would not account entirely for the accumulation of AMP, so that the contribution of other AMP-generating sources has been postulated.

Anyhow, the question remains whether and to what extent the accumulation of brain AMP influences tolerance to extreme hypothermia and whether it contributes to the differences in tolerance between our hibernators and non-hibernators.

INDUCED TOLERANCE TO EXTREME HYPOTHERMIA

Three experimental designs have been explored so far in order to obtain induced tolerance by adaptation (Andjus, 1955; Andjus & Hozić, 1965 a and b; South, Andjus & Gumma, 1966; Andjus et al. 1967). Results from the quoted studies, some of them published only partially as yet, will be summarized here. In the first of the three designs rats were cooled repeatedly to 0° C. and maintained there for periods of 60 min. The intervals between successive coolings were dictated by the time taken for the reanimated animals to regain their initial body weights. This time was gradually decreased from 15 to 2 days.

For the second procedure the inter-trial interval was constant, held actually to 7 days, but the duration of suspended animation at 0° C. was gradually increased from 30 to 120 min.

The third design was a combination of the preceding two: the 0° C. period was increased from 40 min. to 120 min., while the inter-trial intervals were shortened from 7 to 4 days.

Figure 17 shows body-weight curves characteristic of the three types of design used. The first type of treatment (*a*) resulted in the maintenance of body weights in the vicinity of their initial value; the second type (*b*) was characterized by the resumption of growth *above* the initial level; while during the third type of treatment (*c*), characterized by the greatest initial weight loss, the body weight remained reduced throughout the treatment period, although it increased steadily towards its initial level.

If we consider the number of repeated coolings during a given period of time, say 4 weeks, we see that it is greatest in the case of the third type of treatment and smallest in the case of the second one (seven and five respectively). In other words, the third type of treatment can be considered as the most efficient one as far as the number of coolings per month is concerned.

Obviously it can be inferred from Fig. 17 that rats may tolerate a considerable number of repeated coolings to 0° C. with at least 60 min. periods of cardiac arrest at each cooling. Actually, the greatest number of repeated coolings tested so far in our experience has been 13.

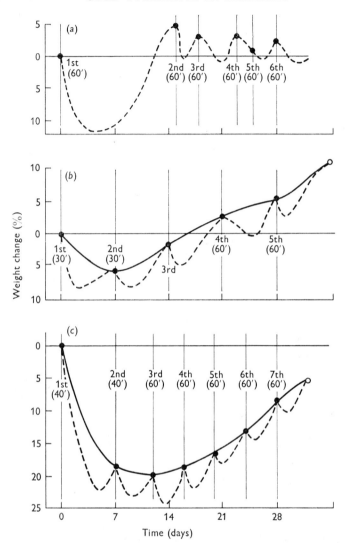

Fig. 17. Body weight changes in rats repeatedly cooled to 0° C. according to the three different experimental designs described in the text. The time (minutes) at 0° C. with arrested circulation at each trial is shown in parentheses. Points: weights at each trial. Dashed curves based on daily measurements. In (b) and (c) means from 8 and 6 male rats respectively; in (a) a single animal.

The weight curves indicate that changes of an adaptive nature do occur in the course of the treatment. In the case of our first experimental design, these adaptive changes are indicated by the fact that the time to regain initial weight after each cooling steadily decreases as the number of repeated coolings becomes greater. In the case of the other two types of treatment, the initial period of decreasing body weight (during the first three coolings or so) has been followed by a period of steadily increasing weight in spite of a continuing and even more severe treatment.

Adaptation to repeated hypothermia may also be indicated by the ability of the recently reanimated animals to thermoregulate against the cold. As outlined in Fig. 18(a), immediately following artificial rewarming to a body temperature of 35° C., the reanimated animals were tested by successively placing them in controlled environmental chambers of 33° C. (1 hr.), 0° C. (30 min.), again 33° C. (1 hr.) and finally at 23° C. for the balance of the 24 hr. Figure 18(b) shows the temperatures measured 5 times, immediately following each of these periods. The temperature fluctuations recorded on the graphs are means for the same group of six animals cooled 7 times by our type-3 treatment procedure. Results obtained after the first, second, fourth and seventh reanimation are shown. The lowest points are always those obtained after the 30 min. period of cold exposure. It is quite obvious that as the treatment proceeds the ability of the animals to withstand wide fluctuations in ambient temperature increases markedly. They actually become less affected by the preceding period of cardiac arrest at extreme hypothermia.

Figure 19(a) and (b) summarizes the data on induced tolerance obtained from the second adaptation procedure, that characterized by constant inter-trial intervals. From Fig. 19(b) it is clear that repeated exposures to either 30 or 60 min. at 0° C. hypothermia enhance the probability of success of the survivors during subsequent trials. The lower interrupted line shows that, while no animals survived an initial experience of suspended animation for 1·5–2·0 hr., a long-term survivorship somewhat greater than 50% has been obtained after the 7th trial. Figure 19(a) strongly suggests that on 50% revival time basis (observe the dashed verticals) the tolerance to cardiac arrest at 0° C. was increased in the adapted animals by a factor greater than 2. The longest period survived by individual rats was actually 2·5 hr. The means by which we have achieved this—that is, by repeated coolings—should not be considered as being either the only one or the most efficient. It may prove, however, a model procedure with which to gain insight into those physiological changes that must take place within a mammalian organism if its tolerance to extreme hypothermia is to be increased substantially.

Two preliminary experiments of the cross-adaptation type have been completed so far. In the first we found that animals that had been repeatedly cooled failed to show the increased basal metabolic rate so characteristic of cold-adapted animals (Andjus, 1955). In the second we did obtain some indication that the repeatedly cooled animals might be more resistant to oxygen deficiency at normal body temperature. A group of 12 rats cooled

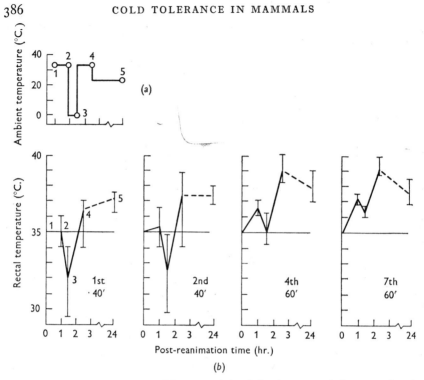

Fig. 18. Rectal temperatures at different periods of the thermoregulation test after the first, second, fourth and seventh reanimation from 0° C. The design of the test is shown in graph (a) and explained in the text. Results are taken from six repeatedly cooled male rats. (range of individual variations indicated by thin verticals).

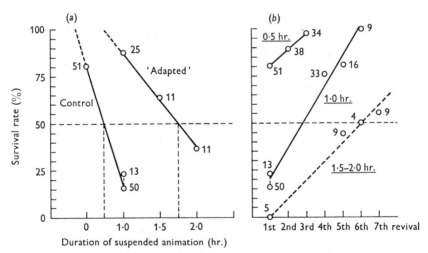

Fig. 19. The proportion of long-term survivors after varying periods of suspended animation at 0° in the control and repeatedly (5–7 times) cooled rats (adapted) is shown in (a). In (b) data on survival are plotted against the numbers of repeated coolings in rats maintained 0.5, 1.0 or 1.5–2.0 hr. at 0° C.

to, and revived from, 0° C. body temperature 5 times at constant inter-trial intervals of 7 days were tested for tolerance to hypoxia at a simulated altitude of 12,000 m. In a large group of 160 control rats and in a group of 12 simultaneous controls, ST 50 values of 6·0 and 6·7 min. respectively were observed. In the 13 rats repeatedly exposed to circulatory arrest at 0° C. body temperature the ST 50 was 9·4 min. At 6·7 min., when only 50% and 48% of animals remained alive in the group of simultaneous controls and in the large control group respectively, 83% of the 'cross-adapted' rats were still alive (10 out of 12). Up to the tenth minute of exposure, survival figures remained higher in the 12 'cross-adapted' rats than in any of the 13 subgroups of 10–16 rats each, constituting the large group of 160 control animals.

Further experiments, testing the possibility of increasing tolerance to extreme hypothermia by previous adaptation to hypoxia or to hypothermia not associated with disturbances in oxygen supply, have been initiated.

REVIVAL FROM SUBZERO TEMPERATURES

Total supercooling remains the only efficient means of reaching body temperatures substantially lower than 0° C. in mammals without losing the possibility of recovery, and little can be added in the sense of new achievements to the data outlined by A. U. Smith in her extensive study published seven years ago (Smith, 1961). The first comparative studies, based on equivalent techniques, indicated that temperature limits of supercooling are similar in the hibernator (golden hamster; Smith *et al.* 1954) and the non-hibernator (hooded rat; Andjus 1955). Figure 20 shows that another hibernator (the ground squirrel, *Citellus citellus*) and another non-hibernator (the mouse) may also be supercooled to temperatures 5° C. lower than the freezing point and revived. As shown in Fig. 21, the newborn rat can also be successfully reanimated after supercooling to temperatures between -7 and $-8°$ C. (Andjus, 1958; Andjus & Hozić, 1965*b*). These temperatures appear at the same time as the lowest compatible with supercooling in mammals, including bats—the first mammals shown to tolerate supercooling (Kalabukhov, 1933, 1960)—and newborn ground squirrels, expected to combine tolerance properties of neonates and hibernators (Popović & Popović, 1963). Lower temperatures always induce freezing, signalled by a spontaneous rise of body temperature towards the freezing point due to the liberation of the latent heat of crystallization of water. In this respect mammals do not appear to differ substantially from poikilothermic vertebrates, as indicated by our comparative experiment with the small frog, compared in Fig. 21 to the newborn rat. Other data on frogs

(Matsko, Zhmeido & Selivanova, 1950) lizards and fish point in the same direction, and similar data have even been published for invertebrates such as snails (see the monograph by Schmidt (1955) and the comparative diagram published by Andjus, 1964).

Time limits compatible with recovery from the supercooled state have not been determined with precision as yet. The longest periods reported so far do not seem in any case to surpass those obtained at levels of body

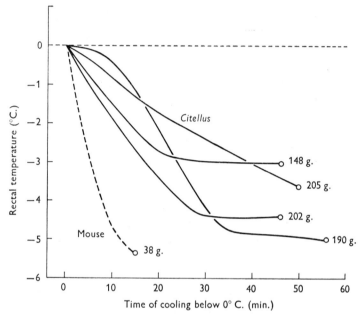

Fig. 20. Rectal temperatures of a mouse and ground squirrels during cooling below 0° C. in supercooled animals. Numbers show the weights in grams of the individual animals. The curves end at the moment reanimation with microwave diathermy was started. All these animals revived.

temperature just above the freezing point. The longest period of supercooling after which we have revived rats remains 40 min. (Andjus, 1955). This particular animal, reanimated within time-limits compatible with revival from 0° C., showed, however, an exceptionally long and difficult period of convalescence. Much longer periods of tolerance were reported for supercooled hamsters (Smith, 1956a, b), but these are still within the range of the revival times determined by us in another hibernator, the ground squirrel, cooled not lower than 0° C. (conclusive comparative data for the hamster are still lacking). The exceedingly long periods (several days) of tolerated supercooling found in bats by Kalabukhov (1960) are quite exceptional. Since, however, heart beats were recorded in the supercooled

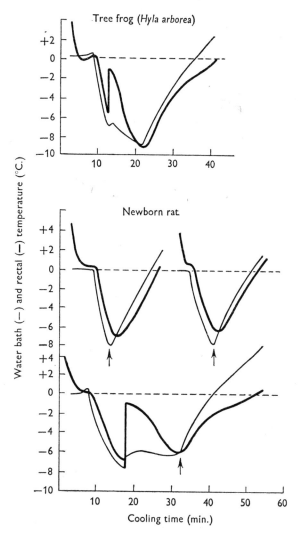

Fig. 21. Supercooling and rewarming in two newborn rats (middle and bottom graphs) immersed in 50% glycerol is shown in the upper curves (heavy lines represent the rectal temperature, the lighter lines the bath temperature; arrows indicate the onset of rewarming). In the bottom graph, supercooling was followed by freezing of body fluids when the bath temperature approached $-8°$ C. This animal failed to revive in contrast to the two in the upper curves. The top graph shows similar temperature curves in a tree frog (*Hyla arborea*) which supercooled and then froze, and failed to revive after rewarming.

state, the reported time limits of tolerance to supercooling cannot be identified with revival periods as defined in other mammals at 0° C. or below. In newborn ground squirrels, periods up to 11 hours at -3 to $-4°$ C. body temperature were compatible with recovery in spite of cardiac standstill, but the revival time decreased to 5 hours at still lower temperatures

(−6 to −8° C.), possibly because of delayed freezing (Popović & Popović, 1963). Our experiments with supercooled adult ground squirrels did not indicate longer revival times than at 0° C. On the contrary, all periods beyond those shown in Fig. 22 for *Citellus*, and substantially shorter than those compatible with recovery from 0° C., resulted in lack of recovery. Although differences between time-limits of tolerance at 0° C. and at lower temperatures have not been systematically explored as yet, the first results indicate therefore that the differences may be in favour of the higher level,

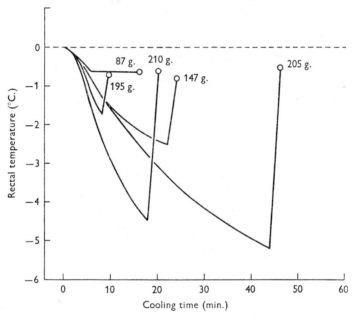

Fig. 22. Supercooling followed by spontaneous freezing in ground squirrels. Numbers against the curves show the weights in grams of the individual animals.

above the range of supercooling. In this respect *in vitro* studies may be mentioned, showing that in some isolated mammalian tissues viability periods increase with falling temperature until levels approaching 0° C. are reached, but decrease significantly at subzero temperatures even when freezing is avoided by supercooling (Wolfensone et al. 1963). It may be that in the supercooling range new changes become of increasing importance among tolerance-limiting factors.

Only temporary recovery has been obtained in our ground squirrels in which total supercooling was followed by freezing, although reanimation was attempted immediately after the onset of freezing. Animals that revived did not live longer than 2 days. Out of the ground squirrels shown in Fig. 22, (Andjus & Hozić, 1965 b) only the one which froze progressively after partial supercooling in peripheral tissues without its central temperature decreasing below the freezing plateau, survived a week after reanimation. The one which froze suddenly after the deepest total supercooling of some 5° C. did not revive at all, while the others, in which freezing was initiated at higher levels of supercooling, were reanimated but died within 2 days. Basically then, these results are in agreement with data concerning conditions and possibilities of revival of hamsters after partial crystallization, studied extensively by

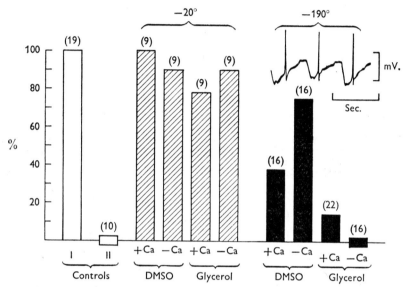

Fig. 23. Numbers of hearts from 16-day-old rat foetuses which have recovered their pulsatile activity after intraocular transplantation into adult recipients. Control series: I, non-frozen hearts (direct transplantation); II, hearts frozen in Tyrode solution at −20° C. without chemical protection. Hatched columns: frozen at −20° C. in the presence of either DMSO or glycerol and in the presence or absence of Ca ions, as shown. Black columns show similarly treated hearts which had been frozen at −190° C. The number of hearts in each series is shown above the columns. The electrocardiogram shown above was recorded 40 days after homo-transplantation of a heart which had been frozen in liquid air in the presence of DMSO and in a medium free from calcium. (From Andjus, Rajevski & Pavlović-Hournac, 1962.)

Smith (1956 a, b), although our results obtained with another hibernator are far less encouraging. Even in newborn ground squirrels, only periods of partial freezing not exceeding 2 min. proved compatible with revival in the majority of animals (Popović & Popvić, 1963).

The temperature of liquid nitrogen, known to afford definite preservation of viability in frozen mammalian tissues protected with chemical 'cryophylactic' agents, remains accessible to a very limited number of complex organs as yet. Let us underline only that great differences as to the mechanisms of freezing damage and cryoprotection seem to exist between these organs. Our comparative experiments with isolated hearts

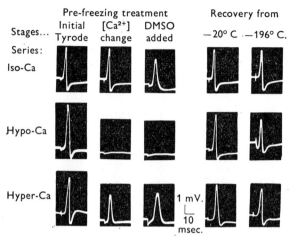

Fig. 24a. Transmitted action potentials (isolated rat superior cervical ganglion). Effect of the different pretreatment phases (the first three records in a horizontal row belong to one and the same preparation) and recovery after thawing (each record belongs to a different preparation, the one with the highest degree of recovery in the series). 'Iso-Ca': standard Tyrode solution; 'hyper-Ca': a threefold increased concentration of $CaCl_2$ in the standard Tyrode medium; 'hypo-Ca': calcium-free Tyrode. Concentration of DMSO in all series: 10 %.

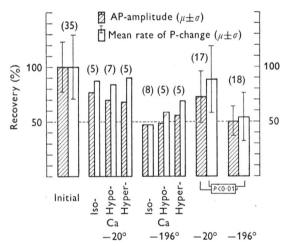

Fig. 24b. Recovery after thawing. Per cent fractions (arithmetical means and standard deviations) of the initial values obtained in the same preparations prior to cooling. The first two columns refer however to the total number of initial values (means taken as 100). Per cent values of the rate of potential change are calculated on the basis of the ratio of the potential hight (negative peak) to its duration (mV. $msec^{-1}$). Number of preparations in brackets. No indication of a marked influence of different concentration of Ca on recovery. Better recovery from -20 than from $-196°$ C, regardless of the treatment.

and ganglia, concerning the role of calcium ions in the mechanism of cryo-injury, might be an example. The protective value of calcium-free media was first demonstrated by Smith (1959) in her experiments with isolated uteri. It was amply confirmed by our experiments with isolated hearts from frogs and foetal rats (Andjus et al. 1961, 1962). As shown in Fig. 23, a high percentage of rat foetal hearts, frozen in liquid nitrogen, recovered their pulsatile activity upon thawing and transplantation into the anterior chamber of the eye of adult recipients, only if treated with calcium-free dimethyl-sulphoxide-containing media before and during freezing. In contrast, our recent experiments with isolated rat ganglia, treated in the same way, showed that the recovery of their ability to transmit nerve impulses upon thawing from the temperature of liquid nitrogen, was not affected by either the absence of calcium or its presence in increased concentrations (Fig. 24). Thus, the still unexplained role of calcium ions in the mechanism of cryo-injury might be specific to contractile organs, and not of the same importance to nervous structures.

I wish to thank my colleagues and collaborators, particularly T. Ćirković, Dr M. El Hilali, R. Genci, Dr N. Hozić, D. Pavlović, Dr V. Petrović and Dr F. E. South for allowing me to use in this review article partially published or unpublished data from our joint studies. I am greatly indebted to Dr A. U. Smith for helping me in revising the manucsript.

REFERENCES

ADELSTEIN, S. J., LYMAN, C. P. & O'BRIEN, R. C. (1964). *Comp. Biochem. Physiol.* **12**, 223.
ADOLPH, E. F. (1948). *Am. J. Physiol.* **155**, 378.
ADOLPH, E. F. (1956). In *Physiology of Induced Hypothermia*, p. 44. Ed. R. P. Dripps. *Publs natn. Res. Coun., Wash.* no. 451.
ADOLPH, E. F. & RICHMOND, J. (1955). *J. Appl. Physiol.* **8**, 48.
ANDJUS, R. K. (1951 a). *C. r. hebd. Séanc. Acad. Sci., Paris* **232**, 1591.
ANDJUS, R. K. (1951 b). *Glas, Serbian Acad. Sci., Belgrade*, **200**, 249.
ANDJUS, R. K. (1955). *J. Physiol., Lond.* **128**, 547.
ANDJUS, R. K. (1956). In *Physiology of Induced Hypothermia*, p. 214. Ed. R. P. Dripps. *Publs natn. Res. Coun., Wash.* no. 451.
ANDJUS, R. J. (1958). *J. Physiol., Paris* **50**, 111.
ANDJUS, R. K. (1960). In *Progress in Refrigeration Science and Technology*, vol. 1, p. 488. Ed. M. Jul and A. M. Singer Jul. Oxford: Pergamon Press (unabridged version in *Arh. biol. Nauka* (1961). **13**, 85).
ANDJUS, R. K. (1964). In *Researches and Development in Freeze-Drying*, p. 451. Ed. L. Rey. Paris: Herman.
ANDJUS, R. K. & HOZIĆ, N. (1965 a). *Bull. exp. Biol. U.S.S.R.* **9**, 38.
ANDJUS, R. K. & HOZIĆ, N. (1965 b). *Bull. exp. Biol. U.S.S.R.* **12**, 82.
ANDJUS, R. K. & LOVELOCK, J. E. (1955). *J. Physiol., Lond.* **128**, 541.
ANDJUS, R. K., MATIĆ, O., PETROVIĆ, V. & RAJEVSKI, V. (1964). *Ann. Acad. Sci. Fenn., Helsinki*, A **71**, 27.
ANDJUS, R. K. & PETROVIĆ, V. (1959). *J. Physiol., Paris* **51**, 388.
ANDJUS, R. K., RAJEVSKI, V. & PAVLOVIĆ, M. (1961). *Bull. Inst. Refrig.*, Annex 1961-4, p. 43.
ANDJUS, R. K., RAJEVSKI, V. & PAVLOVIĆ-HOURNAC, M. (1962). *J. Physiol., Paris* **54**, 272.

ANDJUS, R. K. & SMITH, A. U. (1955). *J. Physiol., Lond.* **128**, 446.
ANDJUS, R. K., SOUTH, F. E., HOZIĆ, N., HOUSE, W. & ĆIRKOVIĆ, T. (1967). *18th Internatn. Astronautical Congress* (Belgrade, Sept. 1967) *Book of Abstracts*, p. 314. Belgrade: Internatn. Astron. Fed.
ANDJUS, R. K., SUHARA, K. & SLOVITER, H. A. (1967). *J. Appl. Physiol.* **22**, 1033.
BACHMETIEV, P. I. (1902). *Bull. Imperial Acad. Sci., St Petersbourg* **17**, 161.
BOWLER, K. & DUNCAN, C. J. (1968). *Comp. Biochem. Physiol.* **24**, 1043.
BRITTON, S. W. (1922). *Q. Jl exp. Physiol.* **13**, 55.
BURLINGTON, R. F. & KLAIN, G. J. (1967). *Comp. Biochem. Physiol.* **20**, 275.
BURLINGTON, R. F. & WIEBERS, J. E. (1965). *Comp. Biochem. Physiol.* **14**, 201.
BURLINGTON, R. F. & WIEBERS, J. E. (1966). *Comp. Biochem. Physiol.* **17**, 183.
EL HILALI, M. & ANDJUS, R. K. (1966a). *Arh. biol. Nauka* **18**, 133.
EL HILALI, M. & ANDJUS, R. K. (1966b). *Arh. biol. Nauka* **18**, 143.
GÄNSHIRT, H., SEVERISS, G. & ZYLKA, W. (1952). *Pflügers Arch. ges. Physiol* **255**, 283.
GAEVSKAYA, M. S. (1963). *Biokhimiya mozga pri umiraniyi i ozhivlenii organizma.* Moscow: Medgiz.
HAMILTON, J. B. (1937). *Yale J. Biol. Med.* **9**, 327.
HART, J. S. (1951). *Can. J. Zool.* **29**, 224.
HEYMANS, J. F. (1919). *Archs int. Pharmacodyn. Thér.* **25**, 1.
KALABUKHOV, N. I. (1933). *Bull. Moscow Soc. Nature Investigators* **12**, 243.
KALABUKHOV, N. I. (1960). In *Recent Research in Freezing and Drying*, pp. 101–118. Ed. A. S. Parkes and A. U. Smith. Oxford: Blackwell.
KAYSER, C. (1959). *C. r. hebd. Séanc. Acad. Sci., Paris* **248**, 1219.
KAYSER, C. & MALAN, A. (1963). *Experientia* **19**, 441.
KODIS, F. K. (1902). *Bull. Imperial Acad. Sci., St Petersbourg* **17**, 129.
LAUFMAN, H. (1951). *J. Am. med. Ass.* **147**, 1201.
LOVELOCK, J. E. & SMITH, A. U. (1956). *Proc. R. Soc.* B **145**, 427.
LOWRY, O. H., PASSONNEAU, J. V., HASSELBERGER, F. X. & SCHULTZ, D. W. (1964). *J. biol. Chem.* **239**, 18.
MATSKO, S. N., ZHMEIDO, A. T. & SELIVANOVA, V. M. (1950). *Dokl. Akad. Nauk. SSSR* **75**, 883.
NOSSOVA, E. A. (1960). *Vop. med. Khim.* **6**, 264.
OPITZ, E. & SAATHOFF, J. (1952). *Pflügers Arch. ges. Physiol.* **225**, 485.
POPOVIĆ, P. & POPOVIĆ, V. (1963). *Am. J. Physiol.* **204**, 949.
REULEN, H. J., AIGNER, P., BRENDEL, W. & MESSMER, K. (1966). *Pflügers Arch. ges. Physiol.* **288**, 197.
SAMSON, F. E. Jr., BALFOUR, W. M. & DAHL, N. A. (1959). *Am. J. Physiol.* **196**, 325.
SCHMIDT, P. YU (1955). *Anabioz.* Moscow: Akad. nauk SSSR.
SIMPSON, S. & HERRING, P. T. (1905). *J. Physiol., Lond.* **32**, 305.
SMITH, A. U. (1956a). *Proc. R. Soc.* B **145**, 391.
SMITH, A. U. (1956b). *Proc. Roy. Soc.* B **145**, 407.
SMITH, A. U. (1959). In *Symposium on Hypothermia*, pp. 480–492. Ed. Andjus, R. K. Belgrade: 15th Int. Congress Milit. Med. Pharm., Belgrade 1957.
SMITH, A. U. (1961). *Biological Effects of Freezing and Super-cooling.* London: Edward Arnold.
SMITH, A. U., LOVELOCK, J. E. & PARKES, A. S. (1954). *Nature, Lond.* **173**, 1136.
SOUTH, F. E., ANDJUS, R. K. & GUMMA, M. R. (1966). *Fedn Proc.* **25**, 406.
THORN, W., PFLEIDERER, G., FROWEIN, R. A. & ROSS, I. (1955). *Pflügers Arch. ges. Physiol.* **261**, 334.
WILLIS, J. S. (1968). *Am. J. Physiol.* **214**, 923.
WOLFENSONE, L. G., KESSMANLY, N. W., FILOV, V. A., ANDREEVA, E. V. & LOZINA-LOZINSKAYA, F. L. (1963). *Cytology, Moscow* **5**, 212.

GROWTH AND SURVIVAL OF PLANTS AT EXTREMES OF TEMPERATURE—A UNIFIED CONCEPT*

By J. LEVITT

University of Missouri, Columbia, Missouri

GROWTH ADAPTATIONS AND SURVIVAL ADAPTATIONS

Both the ability to grow and the ability to survive at extremes of temperature may vary markedly from plant to plant. Since most investigators have been concerned with only the one or the other of these adaptations, a terminology has been developed to distinguish them from each other (Precht, 1967; Christophersen, 1967). The ability to metabolize, grow, and develop at extremes of temperature is called *capacity* adaptation; the ability to survive at extremes of temperature *resistance* adaptation. The resistance adaptation of an organism must at least equal, and usually exceeds, its capacity adaptation. Yet the two are not necessarily correlated. Many animals, for instance, have as well developed a capacity adaptation as do plants; but their resistance adaptation is nearly always far less pronounced. From an evolutionary point of view, this difference has presumably arisen because animals can maintain their temperatures above the low extremes or below the high extremes of the environmental temperature by (*a*) control of body temperature (homoiotherms), or (*b*) seeking shelter. They are, therefore, able to avoid the extremes. Plants, on the other hand, are poikilotherms and are immobile. They are, therefore, unable to avoid the extremes of temperature and must develop both kinds of adaptation in order to survive. Capacity adaptation, however, has not been investigated in plants as much as resistance adaptation, perhaps because of the more dramatic effects of the latter which cannot be duplicated by the vast majority of animals: for instance, the ability of adapted plants to freeze solid without suffering any injury. In some plants, the upper and lower extremes of temperature for growth and survival are essentially stable, regardless of the environment. In other plants, however, the extremes depend on the environmental conditions to which the plants were previously exposed. If these conditions lead to an extension of one or other of the extremes, the plant is said to be 'hardened' to the cold or heat respectively.

* This work is based on research performed with the aid of a grant from the National Science Foundation of the U.S.A. (NSFGB 6828).

The purpose of the following analysis is to make use of the recent explosion of information to establish the physiological bases of these adaptations.

GROWTH AT EXTREMES OF TEMPERATURE

Relation of growth to temperature

All plants show minimum, optimum, and maximum temperatures for growth (Fig. 1). For any specific plant therefore the minimum and maximum temperatures obtained in this way are the extremes of temperature

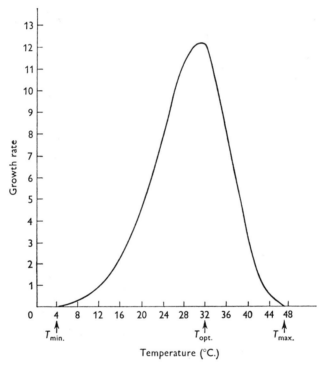

Fig. 1. Cardinal temperatures for growth of maize seedlings. T_{min} = minimum temperature for growth. T_{opt} = optimum temperature for growth. T_{max} = maximum temperature for growth. From Lundegårdh (1957).

for growth. In the case of bacteria, recent results have led to the conclusion that some thermophiles can grow even at 100° C., and it was therefore impossible to determine their maximum temperature for growth (Brock, 1967). A little below this is the maximum temperature for the growth of blue-green algae which may reach values of 80–90° C. (Biebl, 1962; Brock, 1967). It is much more difficult to obtain similar information for higher plants, partly because they are not found growing in a constant-temperature environment as in the case of the above micro-organisms growing in hot springs. The highest temperatures recorded for higher plants in the vegetative (i.e. growing) state are 60–65° C. (Biebl, 1962). But these values were maintained for very short periods of time, as in mid-afternoon, during which growth probably had temporarily ceased. We can therefore only conjecture that the upper temperature limit for growth of higher plants is somewhat below 60–65° C. For the vast majority of higher plants it is far below this range (Table 1). It may, in fact, be impossible to determine a single maximum temperature for continued growth of higher plants because this varies with the environmental conditions. Many plants, for instance, show a thermoperiodic response and are unable to grow normally in the absence of two different day and night temperatures.

Table 1. *Extremes of temperature for growth of plants*
(From Walter, 1950)

Plant	Cardinal temperatures (° C.)		
	Min.	Opt.	Max.
Plants from temperate regions			
Rye	1–2	25	30
Wheat	3–4·5	25	30–32
Barley	3–4	20	30
Oats	4–5	25	30
Plants from tropical or subtropical regions			
Maize	8–10	32–35	40–44
Rice	10–12	30–32	36–38
Beans	10	32	37
Tobacco	13–14	28	35
Melons	12–15	35	40

The minimum temperature for growth is difficult to determine for a different reason—the long time required for the very slow growth to be detectable. Snow algae must be capable of growth at 0° C., and some higher plants are also able to grow through the snow (Biebl, 1962). In the frozen state no growth can occur, since it depends on the hydrostatic pressure produced in the cells by their liquid water. It is conceivable, however, that unfrozen (e.g. supercooled) plants may be able to grow at temperatures below 0° C.

Evidence of relation to enzyme activity

The relation of growth to temperature is most instructive when graphed as an Arrhenius plot. It can then be seen that for a short temperature range (15–30° C.) a straight line is obtained, just as in the case of simple chemical reactions (Fig. 2). But unlike the simple chemical reactions the straight line for growth curves downward precipitously at both ends, due to the existence of the minimum and maximum temperatures. When, however, the chemical reactions are enzymically controlled, they give precisely the same kind of Arrhenius plot as for growth (Fig. 3). Since all the metabolic reactions required for growth are enzymically controlled, the existence of extremes of temperature for enzyme activity must lead to

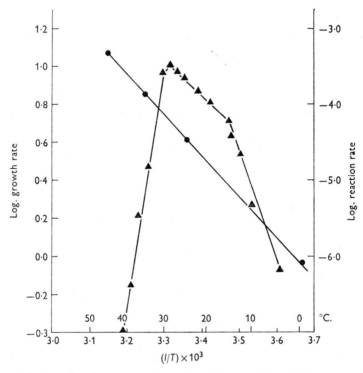

Fig. 2. Arrhenius plots for a chemical reaction (decomposition of nitrogen pentoxide, ●—●) and for growth (of roots of *Lepidium sativum*, ▲—▲). From Getman & Daniels (1937) and Talma (1918) respectively.

similar extremes of temperature for growth. An understanding of enzyme activity at extremes of temperature must therefore be a prerequisite to an understanding of growth at extremes of temperature. What, then, are the reasons for the upper and lower temperature limits for enzymically controlled reactions?

The existence of an upper limit is easy to understand, since high temperatures denature enzymes, and this leads to unfolding and therefore inactivation of the active sites on the protein molecule. Recent evidence (Brandts, 1967) has shown that low temperatures also denature proteins.

These relations can be understood by a consideration of the energy changes involved in protein denaturation (Brandts, 1967).

The native, folded protein is in the ordered state and the denatured, unfolded protein is in the disordered state. Consequently, in the absence of any other factors, denaturation would be a spontaneous process, occurring with a large decrease in free energy, or increase in entropy. This

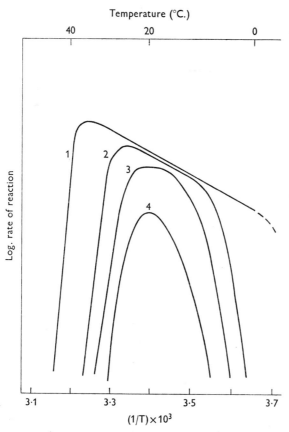

Fig. 3. Arrhenius plots for four enzymically controlled reactions. From Brandts (1967).

tendency to unfold is measured by determining the conformational entropy for conversion of the protein. Counteracting the conformational entropy, and therefore holding the protein in the native state, are two kinds of bonds: (1) the hydrophobic bonds due to the force of attraction between the hydrophobic side chains of amino-acid residues, and (2) hydrogen bonds formed mainly between parallel peptide links in the folded protein molecule. Although other bonds such as electrostatic and disulphide (SS) bonds are important in controlling protein stability in relation to pH and oxidation-reduction potential, only the hydrophobic and hydrogen bonds control protein stability in relation to temperature change. Though each of the three major forces determining protein stability involves an energy change of the order of 100–200 kcal., the sum of the two bonding components, the hydrogen and hydrophobic bonds, so nearly counterbalances the con-

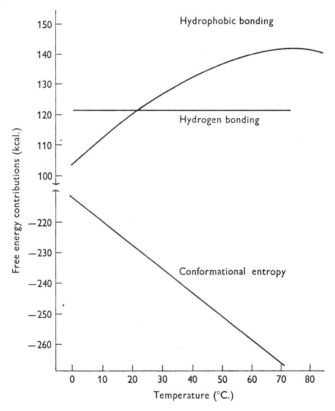

Fig. 4. The relation of hydrophobic bonding, hydrogen bonding, and conformational entropy to temperature. From Brandts (1967).

formational entropy, that the net free energy change on transition from the native to the denatured state is very small—below 10 kcal. The transition must therefore be readily reversible (unless fixed by a secondary change): $N \rightleftharpoons D$, where N is the native, D the denatured protein.

A change in temperature has markedly different effects on the above three factors (Fig. 4). The stability of the protein decreases with a rise in temperature due to the increase in negative energy value of the conformational entropy. On the other hand, the stability increases with a rise in temperature due to the increased strength of the hydrophobic bonds up to about 75° C., where it levels off and decreases at higher temperatures. The hydrogen-bond strength remains constant. At high temperatures the increase in conformational entropy exceeds the increase in hydrophobic bond strength and the protein unfolds. At low temperatures, the hydrophobic bonds become so weak that unfolding is favoured in spite of the relatively low conformational entropy. The net result of these three interacting factors is a maximum stability at temperatures below and above these ranges, for example:

$$N \underset{\text{below } 40°C.}{\overset{\text{above } 40°C.}{\rightleftharpoons}} D \quad \text{and} \quad N \underset{\text{above } 10°C.}{\overset{\text{below } 10°C.}{\rightleftharpoons}} D.$$

High temperature denaturation has been investigated for many enzymes, and may occur at any temperature from about 45° C. or lower to above 100° C., depending on the specific enzyme and the conditions such as the exposure time and the pH. Low-temperature denaturation is more difficult to determine, but methods have been developed recently (Brandts, 1967), and a relatively small number of enzymes have been investigated. The high-temperature denaturation is reversible only for a very small temperature range and even then probably for only a short period of time. Although this fact has only recently been recognized by biochemists, plant physiologists discovered it long ago. Thus, Sachs (1864) showed that when plant cells are heated, a point is reached at which cytoplasmic streaming ceases and the protoplasm solidifies. If the cells are quickly cooled, they recover; if left a little longer at this temperature, they are killed. Although only a few enzymes have been investigated at low temperatures, study of the thermodynamics of the denaturation process leads to the conclusion that it occurs to a greater or lesser degree in all proteins, though not necessarily at the same low temperature. The degree of unfolding (i.e. denaturation) and the temperature at which it occurs depends on the number and strength of the bonds responsible for the secondary and tertiary structure (i.e. the conformation) of the proteins. It is the hydrophobic bonds which are weakened with the drop in temperature; therefore, the greater the role of these hydrophobic bonds in maintaining the tertiary structure, the more effective is the low temperature in producing the unfolding.

Even if all proteins show some degree of unfolding at the minimum temperature for growth, this does not mean that all enzyme activity must stop. Denatured proteins may still retain enzymic activity to a greater or lesser degree, depending on whether or not the active sites are affected by the unfolding. Thus, the evidence indicates that RNA-ase has a thermolabile region that is reversibly disrupted by heat, and a thermostable region that is not (Ginsburg & Carroll, 1965). Some enzymes, in fact, retain activity even in the completely unfolded state. When denatured in 8 M urea, RNA-ase is induced to refold to an approximately native and active form, simply by the presence of substrate (Bello, 1966). Metabolism therefore continues in plants at temperatures below the minimum for growth, though at a slower rate. In the case of trees in temperate climates, for instance, in the autumn after growth has long ceased, proteins and nucleic acids continue to be synthesized (Siminovitch, Gfeller & Rhéaume, 1967). Similarly, potatoes in the non-growing (dormant) state at 0° C. show a marked increase in their protein content (Cotrufo & Levitt, 1958). In the case of the meristematic cells of the roots of several species of plants (*Crepis capillaris, Trillium erectum*, etc.) the minimum temperature for mitosis is above 0° C. (Grif, 1966). Yet at temperatures below the minimum, though mitosis is completely arrested, nucleic acid and protein synthesis continue, though at a slower rate.

Since metabolism does not come to a stop, there are, presumably, three possible reasons for the minimum temperature for growth: (1) a certain minimum rate of metabolism is essential, in order to produce an energy minimum needed to support even the slowest rate of growth; (2) the metabolic process is uncoupled from growth; (3) one (or more) metabolic process essential for growth does come to a stop, though the others continue at a sufficient rate to support growth. There is some evidence for the operation of all of these possible mechanisms. An example of the first is the inability of mesophiles to transport solutes across their membranes at temperatures below the minimum for growth (Rose, 1968). This is presumably due to an insufficient release of metabolic energy to support the active transport. In the case of the psychrophilic fungus *Sclerotinia borealis*, the second mechanism appears to operate at the maximum temperature, for growth is apparently uncoupled from respiration (Ward, 1968). Thus, the uncoupling inhibitors (2,4-DNP and dicoumarol) stimulated respiration relatively more at the optimum temperature for growth (0° C.) than at a temperature above the maximum for growth (25° C.), though the latter was optimum for respiration. The third mechanism seems to operate in an obligate, psychrophilic, marine organism (*Pseudomonas* sp.) since RNA synthesis could not be detected at supramaximal temperatures (Harder & Veldkamp, 1968). In all three cases, however, the limiting factor is metabolism, due presumably to denaturation and inactivation of specific enzymes.

Theories of enzyme adaptation

How, then, do some plants succeed in growing at higher or lower temperatures than other plants? On the basis of the above concept, the answer is to be sought in the enzymes. If the cardinal temperatures for the growth of plants are compared, it is usually found that plants with low maxima also have low minima; and those with high maxima have high minima (Table 1). Alexandrov (1967) explains this as follows, on the basis of the three-dimensional structure (or conformation) of the proteins. It is now known that in order to initiate or catalyse a reaction, an enzyme must undergo a certain degree of conformational change as it combines with the substrate (Koshland & Kirtley, 1966). This requires a certain degree of conformational flexibility. But each enzyme is able to maintain an adequate degree of flexibility only within a limited temperature range, which must include the optimum temperature for growth of the plant. If the habitat temperature rises above this range, flexibility can become too high (i.e. denaturation begins). If the temperature drops below this range, the structure becomes too rigid for enzymic activity. This concept of Alexandrov's would explain the fact that a low minimum temperature for growth tends to be accompanied by a low maximum temperature for growth, a high minimum with a high maximum.

Alexandrov points out that these postulated changes in conformational flexibility of proteins must be due to changes in the strength of the bonds responsible for the conformation—hydrogen, hydrophobic, electrostatic, SS, etc. But his concept would be understandable only if the strength of the bonds mainly responsible for the protein conformation were inversely proportional to the temperature. Recent results, however, lead to the conclusion that the hydrophobic bonds contribute more to the conformation of the protein molecule than do the hydrogen bonds (Koshland & Kirtley, 1966; Brandts, 1967). But hydrophobic bond strength is directly proportional to temperature (Fig. 4). It is, in fact, because of this relation that proteins denature reversibly at low temperatures (Brandts, 1967). Thus, instead of becoming more rigid at the low temperature, as predicted by Alexandrov, the protein actually becomes more flexible and unfolds.

The minimum and maximum temperatures for activity of a particular enzyme must, therefore, depend on the relative proportions of hydrophobic and hydrogen bonds. In agreement with this conclusion, Koffler (1957) found a higher proportion of hydrophobic groups in the proteins of a thermophilic bacterium than in those of a mesophile. At the higher temperature, these would become stronger. The psychrophile, on the other hand, would be expected to have proteins with the largest proportion of hydrogen bonds, since these do not become weaker with decreased temperature. Similarly, among the proteins investigated by Brandts, RNA-ase had the smallest proportion of hydrophobic groups and the lowest temperature for maximum stability of the native form.

The minimum and maximum temperatures for the growth of plants can, therefore, be readily explained on the basis of these newer concepts of conformational structure of proteins. Plants presumably have high maximum temperatures for growth because their proteins have a high proportion of hydrophobic to hydrophilic groups. Due to the increased strength of these hydrophobic groups at higher temperatures, the unfolding of the proteins would occur at a higher temperature than in the case of non-adapted plants. However, this would also lead to a high minimum temperature for growth, due to the high proportion of hydrophobic bonds, which are weakened by low temperature. Conversely, plants adapted to grow at low temperatures would have low minima for growth because their proteins have a low proportion of hydrophobic to hydrogen bonds. It should be emphasized that this relation may not necessarily hold true for *total* protein bonds, since the most important ones are those near the active sites of the protein molecule.

The proportion of hydrophobic:hydrophilic bonds in a protein is not simply a matter of conformation. It depends on the primary structure of the protein, i.e. on the proportion of amino acids with hydrophobic groups and on their arrangement in the protein molecule. It is therefore a genetically fixed characteristic of a specific protein. In agreement with this fact, Ushakov (1967) has concluded that the adaptation of animals to high temperatures is a fixed hereditary characteristic and cannot be altered by exposing an animal to moderately high temperatures. No such intensive investigations have been made on plants. It is therefore not known to what extent adapted plants automatically possess the ability to grow at extremes of low or high temperature, and to what extent they can 'harden', i.e. become adapted, to grow at extremes of temperature by exposing them to moderately low or high temperatures. On the basis of the above concept, any such hardening would require gene repression and derepression, in order to halt the synthesis of readily denatured proteins, and to replace these with a kind of protein not so readily denatured at the temperature extreme.

What evidence is there for the above concepts? Less work has been done, unfortunately, on the growth of higher plants at extremes of temperature in relation to metabolism than in the case of micro-organisms. The alga *Chlorella pyrenoidosa*, for instance, photosynthesizes at essentially the same rate at all light intensities whether the temperature is 20° or 7° C. (Steemann-Nielsen & Jørgensen, 1968). This is attributed to an increase of all the enzymes at the low temperature, for the protein content per cell at 7° C. is double the amount at 20° C. (Jørgensen, 1968). The growth-rate, however, did decrease with the temperature. The accumulation of organic matter in general, and the proteins in particular, on the other hand, increased with the decrease in temperature. This reciprocal interrelation between growth-rate and accumulation of organic matter also occurs in higher plants, even though unlike *Chlorella*, the rate of photosynthesis

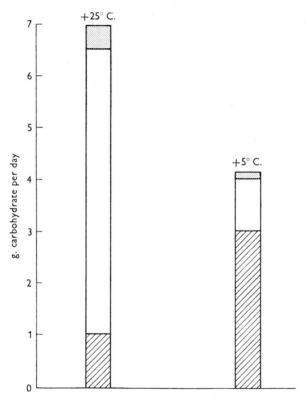

Fig. 5. Relation of growth, metabolism, and carbohydrate accumulation to temperature. From Levitt (1967b). ▦, Actual photosynthesis, respired to CO_2; □, used for growth; ▨, carbohydrate stored.

does decrease somewhat with the drop in temperature (Fig. 5). The complete interrelation is apparently one of homeostasis: (1) low temperature decreases growth more than it decreases metabolism; (2) the energy saved from the decreased growth is converted into extra synthesis of organic substances; (3) the increase in organic substances consists of both an increase in substrate and in enzymes; (4) the accumulation of substrates and enzymes leads to increased growth.

No evidence is available as to whether the newly synthesized proteins are indeed different in kind, and if so, whether the difference involves the predicted decrease in proportion of hydrophobic to hydrogen bonds. The mere increase in quantity of protein may, conceivably, lead to an increased growth at the extreme of temperature due to the increased concentration of enzymes. Even if the reversible denaturation is sufficient to cause partial but not complete inactivation, an increase in concentration of the enzyme would lead to an increased metabolism and growth of the plant. This kind of adaptation, due to increased quantity of protein without a change in quality, would presumably be limited in its effectiveness.

Dormancy and extremes of temperature for growth

This simple interaction between reduced growth and increased accumulation of metabolite may, perhaps, be sufficient to explain the changes in growth-rate of some simple algal cells, or even of some higher plants such as some winter annuals during autumn. In the case of many plants (e.g. overwintering plants and non-growing plant parts such as bulbs and tubers) rather than adapting to growth at extremes of temperature, they de-adapt. In other words, the temperature range for growth becomes narrower, either due to a rise in the minimum or a drop in the maximum temperature for growth, or both. This occurs when the plant enters dormancy or the rest period (Fig. 6). At the ultimate full dormancy or deep rest, the two extremes become equivalent, and no growth occurs at any temperature.

Obviously, when the plant emerges from dormancy, the reverse process occurs and it gradually readapts to growth at increasing extremes of temperature, until dormancy is completely broken. The control mechanism is apparently the balance of growth regulators, inhibitors such as abscisic acid and stimulators such as gibberellin (van Overbeek, 1966), but other factors such as seed-coat permeability and O_2 availability may also be involved (Vegis, 1964). This control by growth regulators, however, is only a switch mechanism which turns the metabolic processes fundamental to growth on or off. Activation or deactivation of enzymes are presumably produced by derepression and repression of the corresponding genes. These changes in extremes of temperature associated with dormancy occur within the temperature range that does not induce protein denaturation.

In summary, then, it is proposed that the temperature extremes at which a plant can grow are controlled by two distinct mechanisms:

(1) In the absence of dormancy: (*a*) The growth of the plant is limited by denaturation (and accompanying inactivation) of its enzymes. (*b*) A plant with a high extreme of temperature for growth has proteins with

a high proportion of hydrophobic:hydrogen bonds; a plant with a low extreme of temperature for growth has proteins with a low proportion of hydrophobic: hydrogen bonds. (c) Since this proportion depends on the primary structure of the proteins, it is a hereditary characteristic. Whether or not it can be changed appreciably by 'hardening' is not known. A small degree of hardening may result from an increase in the quantity of the same enzymes, as long as the denaturation is not sufficiently extreme to inactivate the enzymes completely.

(2) In the presence of dormancy, one or both of the extremes for growth becomes less extreme, and the plant de-adapts to growth at extremes of temperature. On emerging from dormancy the reverse occurs and the plant

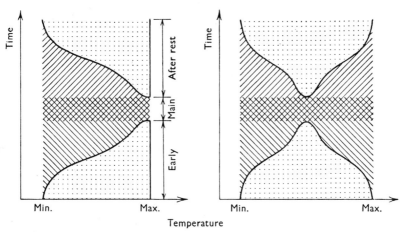

Fig. 6. Change in minimum and maximum temperatures for growth during entrance into and emergence from dormancy. ▦ = Growth. ▨ ▧ = No growth. From Vegis (1964).

readapts. These changes are due to repression and derepression of genes and therefore of enzyme synthesis, respectively. Denaturation of proteins is not involved, since the changes in the extremes of temperature occur above (for low temperatures) or below (for high temperatures) the temperatures at which denaturation occurs.

SURVIVAL AT LOW EXTREMES OF TEMPERATURE

Chilling injury

Many tropical or semi-tropical plants show injury after periods of one to several days, at 'chilling' (cool but above freezing) temperatures (usually 0–10° C.). The injury has been associated with a number of metabolic changes (Levitt, 1956). Since the metabolism of the plant is controlled by its enzymes, the above described reversible denaturation of proteins at low temperatures must play some role in the chilling injury. The simplest explanation is represented diagrammatically in Fig. 7. Denatured proteins are more readily hydrolysed than native proteins, because the unfolded parts can be more easily attacked by enzymes. Consequently, in spite of the drop in temperature, hydrolysis to amino acids would be accelerated. The proteolytic enzymes responsible for this hydrolysis may be expected to retain their activity at low temperatures due to their much greater stability. They are SS proteins, and at least some of their folds are held in place by these covalent SS bonds which, unlike the hydrophobic bonds, are not broken or weakened by low temperature. On the other hand, the enzymes involved in protein synthesis are not SS proteins and therefore suffer

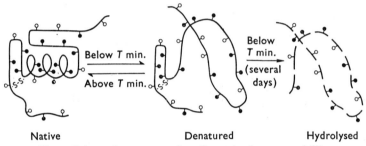

Native   Denatured   Hydrolysed

Fig. 7. Schematic representation of protein changes at chilling temperatures. Diagram of protein from Brandts (1967).

reversible denaturation. Once hydrolysis occurs, the proteins can be replaced only by resynthesis, which is now impossible due to loss of the necessary enzymes.

In favour of this explanation, it has long been known that chilling injury is accompanied by an increase in amino acids at the expense of the proteins —for instance, in the case of beans and tomatoes, which suffer chilling injury at $+5°$ (Wilhelm, 1935). There was a close relation between the injury and the rate of protein decomposition. Seible (1939) points out that the same phenomenon occurs if the plants are kept in the dark at room temperature (Table 2). This is no doubt due to the steady-state turnover of proteins in leaves. They are being constantly broken down and resynthesized, with a half-life of 7 days in tobacco leaves (Holmsen & Koch, 1964). Due to the link between protein synthesis and photosynthesis, essentially only the breakdown reaction will occur in the dark. Therefore, half of the proteins should be hydrolysed in the dark in a week's time. At the low temperature in the light on the other hand, even if the synthesis of the protein drops to zero, the breakdown rate would also be expected to decrease to one-quarter to one-ninth, due to the 20° C. decrease in temperature and a Q_{10} of 2–3. Consequently, only one-quarter to one-ninth as rapid an accumulation of amino acids would be expected. But the actual rates at $+5°$ C. and room temperature were approximately the same, at least for the first 2 days (Table 2). Consequently, the hydrolysis of proteins at the chilling temperature must have been increased 4-9 times the normal turnover rate. Since proteins are much more readily hydrolysed when in the denatured state, these results can be explained by the reversible denaturation of the proteins at the chilling temperature.

Table 2. *Protein breakdown in tomato plants exposed to low temperatures (from Seible, 1939)*

	Ratio of protein N to soluble N		
Days treated	Light controls	Cold room plants	Dark controls
2	15·9	9·7	8·2
4	19·8	9·2	6·7
7	16·9	8·2	4·6
11	—	13·35	1·95

More recent investigations, however, have failed to detect a hydrolysis of proteins at chilling temperatures. In the case of sweet potatoes chilled at $0°$ C., the respiratory activity declined after about 10 days, yet no hydrolysis of the mitochondrial proteins was detected even after 18 days (Minamikawa, Akazawa & Uritani, 1961). In such cases, perhaps the reversible denaturation is itself sufficient to disrupt the metabolism. Not all enzymes will be unfolded and inactivated to the same degree; therefore, the rates of individual reactions will be inhibited to different degrees and metabolism will eventually be disrupted.

Chilling resistance and cold shock

Most plants which suffer chilling injury of the above type probably cannot be hardened to resist low temperatures, since the injury occurs at the very temperatures normally employed for hardening. But there is a second type of chilling injury which is more correctly referred to as 'cold shock'—injury due to a brief sudden exposure to severe chilling or even freezing temperatures but without ice formation in the tissues. Hardening to this kind of chilling injury has been demonstrated by Kuraishi, Arai, Ushyima & Tazaki (1968). They exposed pea plants to $5°$ C. for 3 hr. each day. If this was done for three or more days in the light, the plants survived 3 hr. at $-3°$ C. without injury, though the control, unhardened plants were killed without ice formation. Since this cold-shock injury occurred in one-eighth to one-sixteenth of the shortest time for chilling injury, and since the temperature was $8°$ C. lower, the protein hydrolysis must have been negligible. The hardening treatment, however, did produce a biochemical change, leading to an increase in the ratio of NADPH to NADP. This would seem to indicate that the chilling treatment tends to lower the normally high reduction potential of the living cells and the hardening treatment counteracts this by developing an even greater reduction potential. Previous workers also produced evidence of a disruption in the oxidation-reduction system. In the case of cucumber, chilling injury occurred at 41° F., accompanied by an increase in respiration (Eaks & Morris, 1956). Cold-shocked cotyledons of cacao seed (8–10 min. at 4–6° C.) also showed a higher respiration rate than normal on return to normal temperatures (Ibanez, 1964).

The cold-shock effect can be explained on the basis of modern concepts of protein conformation. The hydrophobicity of amino acid side-chains ranges from 0·45 to 3·00 kcal./residue (Bigelow, 1967). The hydrophobic bonds holding the native proteins in the folded state must therefore have a corresponding range of strengths. If a protein is cooled gradually, all the hydrophobic bonds will slowly become weaker and weaker until the originally weakest bonds break. The stronger hydrophobic bonds will break at successively lower and lower temperatures. In this way the protein will unfold gradually and gently. If, on the other hand, the protein is cooled rapidly to a temperature low enough to break the strongest hydrophobic bond, both the weak and the strong bonds will break simultaneously, and the protein molecule will unfold all at once, like a spring suddenly uncoiling. The impact of two such molecules may supply the energy of activation needed for intermolecular covalent bond formation. Kuraishi *et al.* (1968) suggest that the SH → SS mechanism proposed by Levitt (1962) for freezing injury may apply here. In the presence of the high oxidation potentials resulting in the low NADPH:NADP ratios of the chilled plants, and the sudden contact between unfolding chains, the SH groups of adjacent proteins would be oxidized to intermolecular SS bonds, resulting in irreversible aggregation. Cold-shock injury would therefore occur due to an immediate aggregation on denaturation:

$$N \rightleftharpoons D \to A.$$

In favour of this interpretation, a direct correlation has been found between SH group content and NAD reduction in mitochondria (Vinogradov, Nikolaeva, Ozrina & Kondrashova, 1966). Similarly, when barley shoots are cooled to $+3°$ C. for 1–6 days, the activity of dehydrogenases almost doubles, that of cytochrome oxidase falls sharply (Godnev & Shabelskaya, 1964). Since the dehydrogenases are SH enzymes, the increase in NADPH:NADP found by Kuraishi *et al.* would be expected to enhance their activity and to inhibit the activity of cytochrome oxidase which is an SS enzyme. It is also interesting, in this connection, that chilling resistance of maize was increased by treating the seeds with an SS substance (tetramethylthiuram disulphide—Nezgovorov & Sokolov, 1965).

Freezing injury

The direct injury produced by freezing temperatures is commonly rapid, dramatic, and total. It has therefore long been assumed that the cause must be physical. It is easy to conceive of instant death by a physical process such as cell rupture on freezing. It is not so easy to explain it chemically. Chemical reactions are slowed down by lowering the temperature; so one might expect the chemical injury to become more gradual the more rapid the temperature drop and the lower the freezing temperature attained. Yet the opposite is the case—the more rapidly the temperature drops (up to

Table 3. *Comparative rates of non-enzymic reactions in frozen and unfrozen aqueous systems* (from Fennema, 1966)*

Type of reaction	Substrate	Catalyst	Rate of reaction Unfrozen	Rate of reaction Frozen
Spontaneous hydrolysis	Acetic anhydride	None	30X (+5° C.)	X (−10° C.)
	β-Propiolactone	None	Considerable	O (5 hr. at −10°)
Acid-catalysed hydrolysis	Acetic anhydride	HCl	X† (5°)	3–27X (−10°)
Base-catalysed hydrolysis	Acetic anhydride	Acetate	X (5°)	2·7X (−10°)
Imidazole-catalysed hydrolysis	β-Propiolactone	Imidazole	X (5°)	7X (−10°)
	Penicillin G (pH 7·7)	Imidazole	X (0°)	18X (−8°); 16X (−18°); 5X (−28°); 1X (−78°)
Acid-catalysed chemical dehydration	5-Hydro-6-hydroxydeoxyuridine	HCl	X (30°) O‡ (22°)	12X (−10°) Rapid (−20°)
Oxidation	Ascorbic acid (pH 5·5)	None	X (1°)	3X (−11°)
	Ascorbic acid (pH 5·5)	CuCl$_2$	X (1°)	3·5X (−11°)
Reduction	Potassium ferricyanide	KCN	Stable above freezing	Complete conversion to ferrocyanide in: 7 hr. (−12°); 2 hr. (−25°); 107 sec. (−78°)
Catalysed decomposition of peroxide	Hydrogen peroxide (pH 7·2)	FeCl$_3$	X (1°)	13–28X (−11°)
	Hydrogen peroxide (pH 7·2)	CuCl$_2$	Stable (1°)	Quite rapid (−11°)
Hydroxyaminolysis	Amino acid methyl esters (pH 7·2–7·7)	None	X (1°)	1·7–5·5X (−18°)
	Amides (pH 7·0)	Buffer	X (0°)	1·3–7X (−18°)

* Substrate concentration ranged from 0·0001 M to 0·02 M.
† X = the lower rate of the two.   ‡ O = undetectable.

a point) the greater the injury; and the lower the temperature reached (down to perhaps -30 to $-50°$ C.) the greater the injury. Nevertheless, the early physical hypothesis of cell rupture proved false; and more recent experiments have shown that chemical reactions may actually be accelerated by freezing temperatures (Table 3). According to some interpretations (Grant, 1966) this is due to the ice crystals acting as catalysts. On the other hand, the second-order kinetics, with up to 1000 times more rapid rates than in supercooled liquid solutions led Pincock & Kiovsky (1966) to conclude that it is a concentration effect. In the case of the plant frozen under normal conditions, only the latter explanation can apply, since ice forms extracellularly, and the dehydrated cell contents have no contact with the ice.

It is therefore necessary to consider the possible effects of such an acceleration of chemical reactions on freezing injury. As in the case of cell growth, in order to produce the physiological effect, the chemical reaction must involve an essential chemical constituent of protoplasm such as nucleic acids, proteins, or lipids. Although nucleic acids are essential for protein synthesis, once the proteins are formed, a chemical change in the nucleic acids would not affect them. Therefore, a sudden injury does not seem possible unless the proteins are directly affected. Sudden death due to the loss of cell semipermeability is, of course, possible. Since the cell membranes are apparently lipid-protein complexes, this could occur due to a change in either of these groups of substances. But the protein is hydrated and lipid is not. It is therefore reasonable to expect a direct effect of freezing on the water-containing protein but not on the water-free lipid. It is, in fact, possible to demonstrate a chemical change in a model protein, as a result of freezing. Thiogel (thiolated gelatin) will slowly undergo an oxidation of its SH groups to SS, in the form of intermolecular bonds which therefore aggregate the proteins. As a result, the gel structure becomes more and more rigid, and the melting point rises as more and more intermolecular SS bonds are formed. This process can be markedly accelerated by slow freezing, allowing the gel to become dehydrated by the growth of an ice layer on the gel surface (Fig. 8).

Since ice formation normally occurs extracellularly, i.e. in the intercellular spaces, freezing dehydrates the cell proteins just as in the model system thiogel was dehydrated in the freezing experiments. Therefore, the same kind of protein aggregation may be expected. The extent of this aggregation of proteins is known to increase with protein concentration (Brandts, 1967). Therefore, it must increase with the degree of freezing dehydration. A secondary result of this dehydration is cell collapse, producing a tension on the rigid cell wall. The dehydrated protoplasm also increases in rigidity, and due to adhesion to the cell wall it will also be under tension as well (Iljin, 1933). It is now apparent that this physical tension may lead to a chemical change; for when keratin is under tension, it stretches, and this leads to an increase in SS ⇌ SS interchange between the molecules (Feughelmann, 1966). When not stressed, the interchange between the keratin molecules occurred only at high temperatures (above 100° C.); when under stress, it occurred even at 20° C.

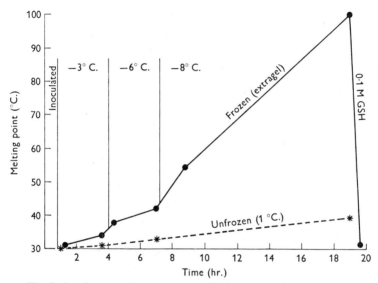

Fig. 8. Acceleration of intermolecular SS formation (shown by rise in melting point) in thiogel on freezing. From Levitt (1965).

Thus, in spite of the kinetic retardation due to the low temperature, both the dehydration of protoplasm induced by the freezing, and the physical stresses resulting from it, are capable of initiating or accelerating chemical reactions between protein molecules, leading to aggregation. But the above model proteins (thiogel and keratin) are denatured. If the protoplasmic proteins behave in the same way, the reversible denaturation of enzymes, which is found to occur at low temperatures, must precede the initiation of the above kind of chemical reaction (Fig. 9):

$$N \underset{\text{above 5° C.}}{\overset{0-5° C.}{\rightleftharpoons}} D \xrightarrow{\text{freezing}} A.$$

According to this concept, the reversible denaturation which occurs at 0–5° C., and is not of itself injurious, is converted on freezing to an irreversible aggregation of the protoplasmic proteins, which is fatal to the cell. The aggregation is due to intermolecular reactions which are induced or accelerated by (1) dehydration of the proteins, bringing them close together, and (2) tension on the proteins due to cell collapse on dehydration by freezing. On the basis of modern concepts of protein conformation the series of events may be visualized as follows. At the low, non-freezing temperature, the hydrophobic bonds become weaker and the proteins unfold. This unfolded state is stabilized by water molecules which simultaneously form a clathrate structure around these newly exposed hydro-

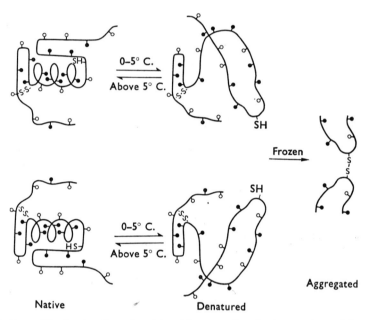

Fig. 9. Reversible denaturation of proteins at low, non-freezing temperatures, followed by extracellular freezing, dehydration, close approach of molecules, and intermolecular SS formation, which aggregates the proteins irreversibly. Diagram of protein from Brandts (1967).

phobic groups (Brandts, 1967). This structured water is not bound by the hydrophobic groups, and is therefore as free to diffuse as the non-structured water. Consequently, when ice forms outside the cells, all this water diffuses readily out of the cell to these ice crystals and the hydrophobic groups lose their protective coats of structured water. The naked hydrophobic groups of adjacent protein molecules may therefore form weakly bonded reversible aggregates. This brings together other groups capable of forming stronger irreversible bonds. When embedded in a hydrophobic environment, electrostatic and hydrogen bonds are stronger (Koshland & Kirtley, 1966), and therefore would help to strengthen these weak, hydrophobically bonded, reversible aggregates. Finally, adjacent

SH groups would form intermolecular covalent SS bonds, conferring an irreversible rigidity on the otherwise reversible aggregates.

But why should protein aggregation result in injury? The aggregates which form on freezing are apparently not large enough to precipitate, for the amount of protein remaining in the supernatant after centrifuging down all the organelles is not decreased by freezing (Morton, 1968). Furthermore, in the one case in which an enzyme was inactivated by denaturing, freezing, and intermolecular SS formation (Massey, Hoffmann & Palmer, 1962), no more than dimer formation apparently occurred. This kind of aggregation is not in itself injurious, since many enzymes are active as aggregates, in the dimer, trimer, or tetramer form. Others are active as part of an aggregated, multi-enzyme system. In both of these cases, however, it is the native protein which forms the aggregate, and therefore the active sites on the enzymes are in their normal, folded state. When denatured proteins are aggregated, this prevents a refolding of the unfolded protein, and if this includes the active sites, the reversibly denatured enzyme loses its activity irreversibly. However, the reversibly denatured proteins may still be active in the denatured form; for a large part of the protein molecule is not essential to its enzymic activity (Anfinsen, 1967). If, then, the unfolding occurred only in the inactive portion of the protein, aggregation would not, of itself, inactivate the protein. If, however, it is an SH enzyme (which owes its activity to SH groups), then intermolecular SS formation would inactivate the enzyme. But even if it does not owe its activity to SH groups, there may be masked SH groups which form hydrophobic bonds within the structure of the protein molecule (Heitmann, 1968). In such cases, the first step in freezing injury would be the breaking of such weak hydrophobic bonds, releasing the previously protected SH groups for intermolecular SS formation. Similarly, if it is an SS enzyme, SS interchange between adjacent proteins, such as occurs in stretched keratin (see above), would break intramolecular SS bonds, leading to irreversible unfolding. Again, if this SS group is part of the active site, the enzyme would be inactivated. But even if neither the SH nor the SS group were part of the active site, intermolecular SS formation would lead to further unfolding on thawing (Levitt, 1962), and this would almost certainly include the active site. This postulated secondary unfolding following intermolecular SS formation gives a reasonable explanation for the long-known thawing injury.

Direct evidence of this order of events has been produced in the case of lipoyl dehydrogenase (Massey *et al.* 1962). The oxidized (SS) enzyme is not inactivated by either freezing alone or by 6·5 M urea at 0° C. However, when these two treatments were combined—the enzyme was unfolded in

6·5 M urea and then frozen—the enzyme lost all its activity after freezing at −20° C. That aggregation did occur was shown by an increase in molecular weight. The reduced (SH) enzyme, on the other hand, was inactivated by freezing without unfolding in urea; presumably because in the absence of the covalent SS bonds, the unfolding occurs spontaneously at the low temperature.

In the case of plants, evidence of the proposed changes in protein SH has been obtained by measuring protein SH and SS content before and

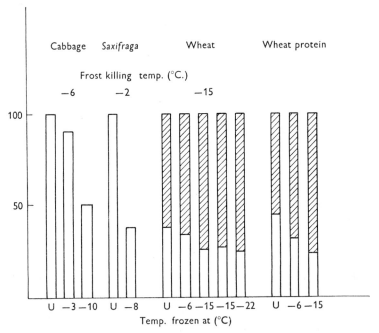

Fig. 10. Increase in protein SS as a result of freezing injury. From Levitt (1967a). U, Unfrozen; ▨ = rel. SS; ☐ = rel. SH.

after freezing. An increase in protein SS was detected when the freezing led to injury. None occurred if the freezing failed to injure (Fig. 10). More recent measurements have revealed that these early SS values are greatly exaggerated since they include the masked SH groups. The relative relationship has, however, been confirmed by labelling the proteins with [^{14}C]-p-chloromercuribenzoate and separating them by disk electrophoresis (Morton, 1968).

This evidence (Morton, 1968) indicates that only a small fraction of the SH groups are converted to intermolecular SS bonds on freezing. Frost killing of cabbage leaves produced a significant decrease in SH/unit protein, but it also resulted in a larger fraction of the remaining SH groups in the largest and slowest moving protein band of the 24 separated by disk electrophoresis. This seems to indicate that all or nearly all the SH-proteins are aggregated by forming one or a very few intermolecular SS bonds. This would (1) decrease the SH content/total protein, and (2) concentrate the SH-proteins in the top, aggregated layer. As indicated above, only one intermolecular SS bond may be sufficient to inactivate an enzyme if it is formed in a strategic position either near to or a part of the active site.

It might be suggested that a conversion of SH groups to *intra*molecular SS groups would increase the resistance of the proteins to *inter*molecular SS formation. In view of Feughelmann's (1966) results, however, this would not remove the danger, since when the protoplasm is under tension due to extramolecular ice formation, *inter*molecular exchange between *intra*molecular SS bonds would be favoured.

Freezing-injury is frequently instantaneous, and therefore not explainable by a loss of enzymic activity. In such cases, the 'active sites' may be those portions of the structural proteins combined with the lipids to form the protoplasmic membranes. Irreversible unfolding of such 'active sites' would result in loss of semipermeability, and therefore instantaneous death of the cell.

In favour of this interpretation, a greater proportion of hydrophobic amino acids has been found in membrane lipoproteins than in soluble lipoproteins, which in turn have a higher proportion than soluble non-lipoproteins (Hatch & Bruce, 1968). This suggests that the unfolding at low temperature would be more severe in these membrane proteins than in any of the others. Since these membrane lipoproteins have a marked tendency to self-aggregate after removal of the lipids (Hatch & Bruce, 1968), the unfolding induced by the low temperature could lead readily to irreversible aggregation. Freezing could, in this way, produce a loss of semipermeability and death of the cell. The denaturation of other proteins at the low temperature could markedly enhance this effect, for it has been shown that denatured haemoglobin (Heinz bodies) can attach to red-cell membranes through mixed S–S linkages of globin and membrane SH groups. This critically alters the membrane of the red cell, producing hyperpermeability and osmotic damage (Jacob et al. 1968).

Freezing resistance

In the case of hardy (i.e. resistant) higher plants, freezing resistance is a seasonal characteristic. Even though they are able to survive the lowest winter temperatures in the frozen state without injury, they are nevertheless killed by the mildest freezes (e.g. $-5°$ C.) when actively growing in spring. Under artificial conditions these plants can be induced to harden by exposure to chilling temperatures ($0-5°$ C.), provided that they are at the right stage of growth. What, then, are the changes that give rise to this hardening?

On the basis of the above concept, the hardy plant must prevent either (a) $N \rightleftharpoons D$ or (b) $D \rightarrow A$.

(a) Prevention or decrease of $N \rightleftharpoons D$

In order to harden at chilling temperatures, a plant must possess proteins with a lower temperature of denaturation than in the case of a plant which suffers chilling injury. But chilling injury would be prevented only if this is an innate, stable, hereditary characteristic which does not have to develop during hardening. In agreement with this concept, plants from temperate regions which are capable of hardening have lower minimum and maximum temperatures for growth than plants from tropical and subtropical regions which are incapable of hardening (Table 1). On the basis of the above explanations of these cardinal temperatures, the temperate plants must have proteins with a lower proportion of hydrophobic:hydrogen bonds than the tropical plants. The reversible denaturation would, therefore, require a lower temperature, and chilling injury would not occur at hardening temperatures.

But these differences would only be quantitative. There would still be enough hydrophobic bonds in the proteins to produce some $N \rightleftharpoons D$ as the plant is cooled below the chilling temperature to freezing temperatures. Freezing would then convert $D \rightarrow A$, and would injure the plant. The stable, hereditary, low ratio of hydrophobic to hydrogen bonds is therefore not enough to prevent freezing injury, and the plant must also be hardened. Can this hardening process operate by producing a further decrease in $N \rightleftharpoons D$? One method would be by a further decrease in the ratio of hydrophobic:hydrogen bonds. Since the number of hydrophobic groups present depends on the amino acid content, or primary structure of the protein, any change in this value can occur only if one kind of protein is replaced by another, due to a combination of repression and derepression of genes. That this does not occur has been shown by the many attempts to discover new proteins in hardened plants. The older electrophoresis method (Siminovitch & Briggs, 1949) and the newer disk electrophoresis method (Morton, 1968) have consistently failed to discover either the appearance of new proteins or the disappearance of old proteins.

But there is one method available to the plant, which may have some effect on the degree of denaturation. The actual unfolding of the protein requires the presence of water and it occurs more readily in dilute solutions than in concentrated solutions. By increasing the sugar content of its vacuole during hardening, the cell succeeds in partially dehydrating its protoplasm. This dehydration increases the concentration of the proteins, and therefore presumably decreases the degree of unfolding at the low temperature. A possible example of this effect is the renaturation of denatured α-amylase (Yutani, Yutani & Isemura, 1967). When the enzyme is highly diluted, the renaturation is slow; when other proteins are added, it is speeded up. A similar effect may also explain a long-known dramatic protection which up to now has defied explanation. Maximov (1912) discovered that even cells with no freezing resistance can be frozen at temperatures as low as $-30°$ C. without injury, provided that they are immersed in solutions of non-toxic and non-penetrating solutes (e.g. sugars). These results have been confirmed by many other investigators, and appear to occur as a result of plasmolysis (Åkerman, 1927). As the solution outside the cells freezes, it becomes more and more concentrated, and plasmolyses the cells more and more strongly. The protoplasmic proteins therefore become more and more dehydrated, and this leads to a prevention of further unfolding or even to a refolding of partially unfolded proteins. No such protection occurs during the normal dehydration in the absence of plasmolysis, e.g. by freezing dehydration or by desiccation in the absence of solutes outside the cells. The reason for this has been given above. Under these conditions, the whole cell collapses and a tension is exerted on the proteins, due to adhesion of the protoplasm to the contracting, rigid cell wall. This tension would favour unfolding of the proteins and prevent refolding. In agreement with this interpretation, Santarius & Heber (1967) obtained a sharp increase in SH groups (from 47 to 107) when isolated chloroplasts were dehydrated, losing 98–99 % of their water content. They interpreted this as evidence of denaturation of the proteins which would unmask previously masked SH groups. No such increase in SH groups was detected when the chloroplasts were dried in the presence of small amounts of sucrose (20 μmoles/1·5 mg. chlorophyll). The sucrose solution therefore presumably prevented the unfolding during desiccation.

The degree of denaturation may also explain the long-known inverse relation between hardiness and growth (Levitt, 1956). In the spring when growth-rate is maximal, hardiness is minimal; in autumn when growth stops, hardiness develops steadily. Artificial effects on growth-rate produce similar changes. The reason for this relation has never been clear, though the accumulation of solutes in the non-growing plants appears to be one factor.

On the basis of the present concept, the relation can be readily explained.

A rapidly growing plant is synthesizing proteins rapidly for the newly produced protoplasm. A non-growing plant simply synthesizes enough to compensate for the slow breakdown in the mature, non-growing cells. Newly formed proteins are denatured, i.e. they are merely long peptide chains as they are released from the ribosomes which synthesize them. After release, they gradually fold, forming their secondary and tertiary structure. How long this naturation process takes is not known. Under artificial conditions, unfolded RNA-ase refolds spontaneously to the native form in 10–20 hr. (Anfinsen, 1967). An enzyme present in microsomes can catalyse this reaction in a much shorter time. RNA-ase, of course, is a very small protein, and may be expected to fold in a much shorter time than the larger proteins.

In any case, it appears obvious that a rapidly growing plant will have a considerable fraction of its proteins in the newly formed, unfolded or incompletely folded state. The more rapid the growth, the larger this fraction will be. When such a plant is frozen, these unfolded proteins will be in an ideal state for conversion to irreversible aggregates and the protoplasm will therefore be injured. Non-growing plants, on the other hand, will have essentially all their proteins in the folded, native state. They will suffer injury only to the extent that their proteins are unfolded at chilling temperatures before freezing.

(b) Prevention of $D \to A$

It is impossible, of course, for the plant to develop a protein which would not unfold at any low temperature, for all proteins depend on hydrophobic bonds to a greater or lesser extent for their conformation. Therefore plants capable of hardening presumably have proteins which denature at lower temperatures than proteins of plants incapable of hardening. Consequently, though chilling temperatures do not produce a sufficient degree of unfolding in hardy plants to cause chilling injury, some protein unfolding must, nevertheless, occur before the temperature drops to the freezing point of the plant. These unfolded proteins must therefore be in danger of irreversible aggregation during dehydration by freezing. How then can the hardy plant prevent this irreversible aggregation?

It was suggested above that an accumulation of solutes (e.g. sugars) in the cell vacuole during hardening might decrease the reversible denaturation. If, however, the solutes also accumulated in the protoplasm, they would tend to prevent aggregation by osmotic retention of water, which would otherwise move out of the cell to solidify in the intercellular spaces. This water, and the solutes dissolved in it, would keep the protein molecules separated, preventing them from forming aggregates. This mechanism has been demonstrated in the model system thiogel. By including solutes in the gel, the intermolecular SS formation between the proteins, which normally occurs on freezing, was markedly inhibited (Fig. 11). Since thiogel is already in the denatured form, only the aggregation can be affected. The effect appears to be purely osmotic, since sodium chloride is just as effective as dimethyl sulphoxide (DMSO) or sucrose when present in iso-osmotic concentrations. In the case of the living cell, the superiority of some

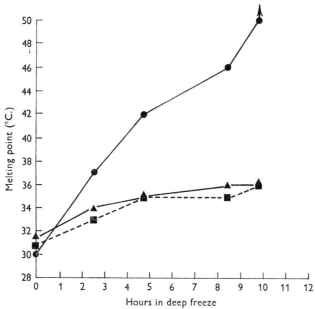

Fig. 11. Protection of thiogel by sucrose and glycerol against intermolecular SS formation (measured by rise in melting point) on freezing. From Andrews & Levitt (1967). ●—●, Control frozen; ▲—▲, sucrose frozen; ■—■, glycerol frozen.

cryoprotective agents over others may be due to their speed of penetration, e.g. glycerol penetrates more rapidly than sugars, DMSO more rapidly than glycerol. Other specific properties may also be involved. Thus glycol and glycerol are good preservatives of urease, and this is apparently related to the reversible dissociation of the enzyme without loss of activity in the presence of these substances (Blattler, Contaxis & Reithel, 1967). In other words, these protectants specifically protect urease against aggregation. In the case of other enzymes, cryoprotectants may operate by preventing or reversing $N \rightleftharpoons D$. Some that are as much as 70% denatured may be restored almost completely to the native state at 0° C. by 10% methanol, propylene glycol, or DMSO (Graves et al.—see Brandts, 1967). In the case of living cells, however, as in the case of thiogel, it is obvious that only the D → A process can be prevented by cryoprotective agents within the protoplasm, since they are effective only if they penetrate the living cells in high concentrations (e.g. 10%). On freezing, the above substances would reach such high concentrations that rather than preventing denaturation they would actually favour it (Brandts, 1967), by stabilizing the clathrate structure arising at low temperature.

It is conceivable that specific protective substances may be present in the hardy plant. It is interesting to note that such a substance has been found in a non-hardy animal system. Pullman & Monroy (1963) have shown that both the ATPase and the coupling factor F_1 in the mitochondria are cold-labile, losing their activities on freezing at $-55°$ C., or even after 3 hr. at 4° C. There is also a mitochondrial protein of low molecular weight (M.W.) which inhibits the activity of the ATPase but does not uncouple phosphorylation. When all three are present as a complex in submitochondrial particles, they survive 4 days at $-55°$ C. with the coupling activity of the particles unaltered. It is difficult to conceive of an effect on the reversible denaturation of the protein, since this is an innate characteristic of the protein. A prevention of aggregation, however, could conceivably result from a protection of the active chemical groups.

Any attempt to postulate that hardiness is due to prevention of D → A requires a knowledge of the bonds formed between proteins when they aggregate. A large body of indirect evidence and a smaller body of direct evidence point to the formation of intermolecular SS bonds (Levitt, 1962, 1967c). In the case of some proteins the aggregation reaction is so favoured that it occurs even at low concentrations, and quickly destroys the reversibility of the denaturation (Brandts, 1967). One of these is ovalbumin,

Table 4. *SH-containing enzymes inactivated by freezing (from Levitt, 1966a)*

Enzyme	Molecular weight	SH (groups/molecule)	SS (groups/molecule)	Protectant against cryoinjury
1. Lactic dehydrogenase	170,000	14		Glutathione, mercaptoethanol
2. Triosephosphate dehydrogenase	100,000	11–12	0	Mercaptoethanol
3. Glutamic dehydrogenase	1,000,000	90–120	0	Mercaptoethanol
4. Lipoyl dehydrogenase	100,000	8–12	2	Oxidation: SH → SS
5. Catalase	248,000	(Altered by SH reagents)		
6. Myosin	594,000	45		
7. 17β-Hydroxy steroid dehydrogenase		(Inactivated by SH reagents)		
8. Succinate dehydrogenase	200,000	(Inactivated by SH reagents)		
9. Phosphoglucomutase	74,000	7·5		

which has 3–4 SH groups/mol. (M.W. 46,000). Similarly, those with many SH groups are readily inactivated by freezing; but if the SH groups are prevented from forming intermolecular SS bonds, either by converting them into *intra*molecular SS bonds or by protecting them with thiols during freezing, inactivation is prevented (Table 4). Aldolase isolated from spinach chloroplasts is another SH enzyme which is unstable during storage or freezing (Jacobi, 1967). The effect of freezing on isolated proteins, however, depends markedly on the freezing method used. One of the above sensitive enzymes—lactic dehydrogenase—may retain most of its activity if frozen at a controlled, rapid rate ($1°$/min. to $-70°$ C.) and thawed rapidly (Greiff & Rightsel, 1967). But results with isolated proteins frozen in solutions are not directly relevant to the freezing of plants in nature; for the plant cells are frozen extracellularly, and it is the resulting dehydration which accelerates the intermolecular SS formation. Therefore, aggregation of proteins during freezing is far more likely to occur when they are inside the cell than when isolated in solution.

In the case of leaves, the protein fraction present in the largest quantity, accounting for 25–70% of the total leaf protein (Dorner, Kahn & Wildman, 1957), is the fraction I protein which occurs only in the chloroplast. It is dependent on SH groups for activity (Sugiyama, Akazawa & Nakayama, 1967). Ten SH groups had to be titrated before loss of any of its carboxylase activity. When 30 were titrated, all the activity was lost. The total per molecule (M.W. 550,000) was 96. None of the SH groups were buried within the molecule; so they were all available for intermolecular SS formation. It does not necessarily follow, however, that fraction I protein is the most important for freezing survival. It is quite conceivable that a protein present in smaller quantity (e.g. a membrane protein) may be decisive for survival. But not as much is known about these other proteins.

Even the protective effect of solutes on freezing injury may be due to a protection of the SH groups of proteins. Sucrose, for instance, is sometimes superior to the other substances. It has proved more effective than other sugars in increasing the hardiness of wheat and rye plants (Tumanov & Trunova, 1957), and of leaves of *Hedera helix* (Steponkus & Lanphear, 1967) and as a protector of ATPase activity during freezing at $-20°$ C. (Heber & Santarius, 1964). Although it is no more effective than other solutes in preventing intermolecular SS formation in thiogel during freezing, it is far more effective in preventing the much slower process in unfrozen thiogel (Fig. 12). Another similarity between the effects of sucrose on freezing inactivation of enzymes and on the reactivity of their SH groups is provided by two unrelated investigations. Heber (1967) found that salts interfere with the protection afforded to ATPase by sucrose; Warren &

Cheatum (1966) found that salts increase the rate of reaction of SH groups in β-amylase.

What about the above-mentioned inverse relation between hardiness and growth? The nearly complete lack of resistance in rapidly growing plants was explained above by the high content of newly formed proteins still in the incompletely folded state. But since unfolding must occur to some extent at chilling temperatures, this does not completely explain the high degree of resistance that can be developed by non-growing plants.

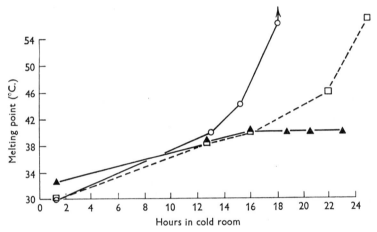

Fig. 12. Protection of thiogel, markedly by sucrose and slightly by glycerol, against intermolecular SS formation at low temperature in the absence of freezing. From Andrews & Levitt (1967). ○—○, Buffer; ▲—▲, sucrose; □—□, glycerol.

It has long been known that growth is associated with high SH content. If this applies to the proteins specifically, then they would form aggregates more readily by intermolecular SS formation. Many recent investigations have confirmed the older work relating SH groups to growth. Szalai (1959) found that reduced glutathione (GSH) rapidly increases during sprouting of potato. There is also a rapid and quantitative conversion of oxidized glutathione GSSG → GSH in the early stages of germination of seeds, and it has been suggested that this precedes protein reduction (Spragg, Lievesley & Wilson, 1962). Pilet & Dubois (1968) observed a typical gradient of SH groups in carrot-root fragments, from a high value in the growing callus tissue in contact with the air, to a low value in the non-growing base of the fragment in contact with the medium. With increasing age, the concentration of endogenous SH decreased. GSH has actually been found to promote intense callus formation (Koblitz, Grutzmann & Hagen, 1967). The addition of GSH has also yielded negative results (Bennett, Oserkowsky & Jacobsen, 1940), but this may be due to a high GSH oxidizing activity of the tissues (Waisel, Kohn & Levitt, 1962). The phytochrome-mediated photomorphogenesis may be replaced by thiols, e.g. β-mercaptoethylamine and GSH (Klein & Edsall, 1966). These substances induced expansion in darkness of disks from dark-grown bean leaves. Oxidizing agents, including GSSG, depressed the expansion induced by red light. High SH values have also been found associated with cancerous tissues (Wiman, 1964).

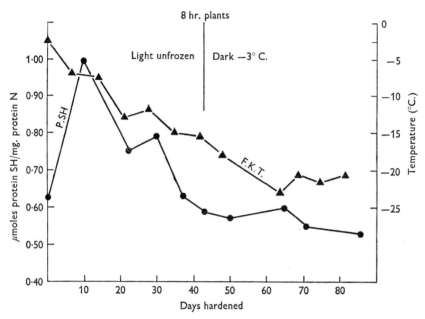

Fig. 13. Initial rise, followed by steady drop in protein SH during hardening. F.K.T., frost killing temperature. From Kohn & Levitt (1966).

A further extension of the inverse relation between growth and hardiness is the direct relation between hardiness and dormancy, which is frequently found (see Levitt, 1956). Recent results have confirmed this relation by showing an enhanced hardiness due to the application of the dormancy-inducing growth regulator, abscisic acid (as well as synthetic growth inhibitors) and a decreased hardiness due to application of the growth-stimulating regulator, gibberellin (Irving & Lanphear, 1968). It has also long been known that GSH sometimes helps to break dormancy.

The direct evidence on the relation of SH to aggregation and freezing injury in plants is, unfortunately, not too clear. The hardening of cabbage and other leaves leads to a rise in protein SH during the first week or two followed by a steady drop which parallels the hardiness of the leaf (Fig. 13). The first rise in SH may be interpreted in more than one way. Some of the evidence indicates that it may be an artifact due to more rapid oxidation of protein SH in the non-hardened plants, but further information is needed. The drop in SH which parallels hardiness agrees with the above theory. But more recent (unpublished) results have demonstrated a similar drop in non-hardened leaves simply due to ageing; and this is accompanied by only a very small increase in hardiness. As mentioned above, Kuraishi *et al.* (1968) obtained evidence of a rise during hardening in the ratio of NADPH to NADP, confirming the prediction (Levitt, 1967b) that this could protect the protein SH groups against intermolecular SS formation. According to this concept, photosynthesis at low temperature could lead to protection by an accumulation of NADPH due to a decreased synthesis of carbohydrate.

In contrast to the above results, Heber & Santarius (1964) were unable to detect either a change in SH content of ATPase (from spinach chloroplasts) during inactivation by freezing, or an effect of thiols on this inactivation. But in these tests, vesicle membranes isolated from the chloroplasts were used. Heber (1967) later suggested that freezing specifically damages the membranes which are essential for ATPase activity. In support of this explanation, the damage was prevented by low concentrations of sucrose. When suspended in water, the membranes occupied at least twice the volume they possessed when suspended in 0·19 M sucrose. He was therefore probably dealing simply with an osmotic rupture of the membrane by freezing and thawing in water, and an absence of rupture when freezing in sucrose. Salts oppose this protection because the membranes are permeable to them, and they would cause swelling and rupture even in the presence of sucrose. Consequently, negative SH results in these experiments have no bearing on the problem of freezing injury and protein aggregation.

In agreement with this conclusion, a relation between ATPase inactivation on freezing and SH content has been found in the case of myofibrillar proteins of chicken pectoralis (Khan, Davidkova & Van den Berg, 1968). When frozen at $-20°$ C., the proteins showed a loss of SH content, solubility, ATPase activity, and water-holding capacity. Blocking the SH groups with PCMB also inactivated the ATPase, but failed to alter its solubility or water-holding capacity. Oxidation of the SH groups to SS by H_2O_2 had the same effect as freezing. The logical explanation is that freezing induced an aggregation of the proteins by intermolecular SS formation; for ATPase is an SH enzyme (Koukol, Dugger & Palmer, 1967) and Heber (1967) emphasized that it does require thiol for activity. It has, of course, long been known that the extraction of active enzymes is often dependent on the addition of thiols to the extracting medium (Anderson &

Rowan, 1967). Obviously, any tying up of the SH group associated with the active site (e.g. by intermolecular SS formation) would certainly inactivate the ATPase.

The major problem in testing this concept is that one cannot generalize as to the effect of thiols on enzymes, since thiols may also inactivate others (Nagatsu, Kuzuya & Hidaka, 1967), or prevent their reactivation (Kim & Paik, 1968). There are also cases where intermolecular SS bond formation plays a definite role, e.g. in the sexual agglutination reaction of yeast (Taylor, Orton & Babcock, 1968). Similarly, fructose diphosphatase is actually activated by disulphides: oxidized mercaptoethanol (ME), ethyl disulphide, and an aromatic disulphide (Pontremoli et al. 1967). It is not even possible to generalize for any one specific case as to the effect of SH reagents; since some may be effective and others not. Thus, the germination of lettuce seed is stimulated by arsenate, and this is enhanced by BAL but not by any other thiols (McDonough, 1967). Similarly, an acetaldehyde dehydrogenase from germinating seeds is stabilized by GSH but is inhibited by GSH and ME (Oppenheim & Castelfranco, 1967).

It must therefore be concluded that SH reagents are double-edged weapons which may act to kill or cure. It is therefore difficult to predict the effect in the case of any one reagent acting on any one enzyme. It must be even more difficult to predict in the case of living cells. Thus some applied thiols may be immediately oxidized to disulphides on entering the cell, and others may not; some applied disulphides may be reduced to thiols on entering the cell, others may not (Eldjarn, 1965). First attempts failed to find any effect of GSH on the freezing resistance of sections of red cabbage cells (Levitt & Hasman, 1964). Previous results had shown that GSH is readily oxidized by the plant, so it was later replaced by artificial thiols not normally oxidized by the plant. In the case of tissue sections, mercaptoethanol, instead of protecting the cells, actually removed the protection against freezing injury produced by sugar solution (Krull, 1966). But if the above concept is correct, sugar solutions protect by preventing or reversing $N \rightleftharpoons D$, whereas thiols can protect only by preventing $D \rightarrow A$. The mercaptoethanol could conceivably remove resistance by increasing this unfolding, e.g. by breaking SS bonds holding the folds in place, or by combining with SH groups, or by inducing SS interchange. It has recently been shown that the internal, masked SH groups probably act as hydrophobic groups, forming bonds with other hydrophobic groups. Conversion to SS mercaptoethanol would break these hydrophobic bonds and lead to more pronounced unfolding.

Further attempts were made to protect the SH groups of the proteins by tying them up to other thiol reagents. Both IA and PCMB actually decreased the freezing resistance of unhardened or partially hardened cells (Levitt, 1969). Again, this is the opposite of what might be expected from the above theory, although as we have seen the thiol reagents are capable of either protection or injury. It has been shown, for instance, that when PCMB combines with protein SH groups, it also changes the conformation of the molecule.

(c) Increased protein content

Although the electrophoretic evidence indicates that there is no synthesis of any new kinds of proteins during hardening (Gerloff, Stahmann & Smith, 1967; Morton, 1968), there is considerable evidence of an increase in total protein, and of certain specific fractions. A parallel between soluble-protein content and hardiness was first found by Siminovitch and his co-workers for trees and this was later confirmed for winter annuals and herbaceous perennials (see Levitt, 1966b). The ability to incorporate glycine into soluble proteins also increases (Siminovitch & Chater, 1958) during hardening of trees in autumn. At the same time as the soluble protein rises in trees during autumn, RNA (but not DNA) shows a parallel increase (Siminovitch, Rhéaume & Sahar, 1967; Li & Weiser, 1967). A similar rise in all three components occurs in crowns and roots of alfalfa during autumn, and the quantities are larger in the hardy than in the non-hardy variety (Jung, Shih & Shelton, 1967). The higher content in the hardy variety of alfalfa is apparently due to either highly charged or low M.W. proteins, or both, judging from electrophoretic separation (Coleman, Bula & Davis, 1966). Yet no large shifts in the electrophoretic pattern of the proteins were detected with polyacrylamide gels, and no appreciable difference between hardy and non-hardy varieties (Gerloff et al. 1967). It has not been possible, in general, to show an increase in a specific protein, though two new isoenzymes with peroxidase activities were found in fully hardened alfalfa (Gerloff et al. 1967). Nitrate reductase activity in crown tissue of wheat plants did increase slightly with maximum hardening, but it increased more rapidly with dehardening (Harper & Paulsen, 1967). However, in this investigation even the water-soluble protein decreased with hardening. There have been other discrepancies (Levitt, 1966b), probably due to differences in methods of extraction, but the increase in protein on hardening has been confirmed by so many investigators that it cannot be doubted.

How can a mere increase in quantity of protein of the same kind affect the cell's resistance to freezing? If this is due to an increase in protein concentration within the protoplasm, it could perhaps affect the $N \rightleftharpoons D$, since the less water present the smaller the degree of unfolding. It is not, however, certain to what extent the increase in protein per cell is an increase in concentration within the protoplasm. As Siminovitch, Gfeller & Rhéaume (1967) point out, there is a visible increase in the amount of protoplasm per cell; and this may account for the increase in protein per cell without any increase in concentration of proteins within the protoplasm. But even if a small increase in protein concentration does occur, it would not affect the degree of unfolding sufficiently to account for a marked increase in resistance. The $D \to A$ change, on the other hand, involves a chemical reaction, if due to intermolecular SS formation. Since this is an oxidation reaction, a doubling in enzyme content per cell may significantly speed up metabolic processes at the low hardening temperature, resulting, for instance, in a marked increase in reduction potential within the cell. Such an increase may be sufficient to prevent the $SH \to SS$ oxidation.

There are reasons, however, for concluding that the protective effect of the newly synthesized proteins is due to their quality, and not to their quantity. (1) The correlation between hardiness and protein quantity does not always occur. The high protein content may remain during early spring (before bud growth) in spite of a large loss in resistance (see Levitt, 1966b). Conversely, Siminovitch showed that if sections of tree branches are subjected to hardening conditions, resistance may develop without an increase in protein content. Finally, the increase in proteins occurs in the fall before any appreciable increase in hardiness, then remains constant as hardiness increases (Li & Weiser, 1967). (2) There is an over-all increase in organic reserves of all kinds, and the increase in protein content may simply be due to a continued synthesis of reserves without the drain on them that would occur in the presence of growth.

A very recent line of evidence points to a possible explanation of the role of the newly formed proteins. As was pointed out earlier, a rapidly growing plant is a plant with minimum freezing resistance. This was explained by the rapid synthesis of proteins, and therefore the large fraction in the unfolded or incompletely folded state. In contrast to this relation, if cabbage leaves are exposed to a hardening treatment (5° C.) for as little as 24 hr., they develop a significant degree of freezing resistance, and the leaves with the maximum absolute growth-rates show the maximum degree of hardening during this 24 hr. period. At first sight this seems to contradict the well-known fact that even the potentially hardiest plants are unable to harden when actively growing, e.g. trees in the spring. The cabbage plants do, in fact, show a similar relationship. The youngest leaves in the stage of cell division or early cell enlargement do not harden. It is only the leaves that are near their full size and therefore growing solely due to cell enlargement that show this rapid hardening. It should be emphasized that the above described correlations between SH content and growth were obtained with plants growing due to cell division.

The following explanation is proposed for this rapid (24 hr.) hardening of cabbage leaves in the stage of rapid cell enlargement. Although the primary structure of a protein is fixed by the DNA and RNA templates which control its synthesis, its three-dimensional structure depends, in part, on the environment during the folding (Koshland & Kirtley, 1966). If synthesized at normal growing temperatures, then cooled to hardening temperatures, the conformation of the protein molecule will change. But it cannot change to the same conformation as it would have attained if it had been fully synthesized at the hardening temperature, for the following reasons. When a plant is cooled to the hardening temperature, the hydrophobic bonds of its proteins become weaker (Fig. 4) and some unfolding will occur. But the conformation of a protein molecule cannot attain the state of maximum stability at the hardening temperature unless all the bonds are broken and it is allowed to refold freely. This complete unfolding cannot happen by itself (or at best will happen very slowly) since the hydrogen bonds are not weakened by the hardening temperature. Therefore, the only way for this molecule to achieve a conformation with maximum stability at the hardening temperature is for it to be broken down, resynthesized, and refolded at that temperature. This breakdown and resynthesis will normally take a long time, for the half-life of leaf proteins is about 7 days even at normal temperatures. Rapidly growing leaves, on the other hand, are synthesizing proteins rapidly, and therefore have a large fraction of their proteins in the unfolded or incompletely folded state. On cooling these leaves to a hardening temperature, these proteins will now fold to the conformation of maximum stability for that temperature. This new conformation will not necessarily affect the active sites, or it may even enhance their activity. All the groups with the greatest ability to bond with other proteins at freezing temperatures will now be protected inside the newly folded molecule, and only the most weakly bonding groups will be in the unfolded portions outside the molecule. Therefore, the ability of the proteins to form aggregates by means of intermolecular bonds will be greatly decreased.

Investigations of pure enzymes indicate that the reactivity of the SH or SS groups may be specifically controlled by differences in conformation of the above type. Thus, Na borohydride reduces selectively two SS bonds in trypsinogen and trypsin which are not essential for activity (Light & Sinha, 1967). Chymotrypsinogen and chymotrypsin are not reduced under these conditions. Yet the SS groups which are modified in trypsinogen are located in a homologous position in chymotrypsinogen, which suggests that the two proteins differ only in conformation in this region. One conformation protects the SS groups, the other does not.

How does this explain the role of the increase in protein content during hardening? Siminovitch, Gfeller & Rhéaume (1967) showed that this is due to an increase in RNA content, leading to a marked increase in rate of protein synthesis. In other words, during the hardening period there must be a nearly complete turnover of proteins. This means that each newly formed protein chain must fold at a temperature which prevents the formation of hydrophobic bonds. They must all therefore fold with the hydrophilic, reactive groups within the folds and the hydrophobic, weakly reactive groups on the outside. All the newly formed proteins must therefore be resistant to aggregation and the plant can therefore undergo freezing dehydration without injury. This would also explain why frost hardening is increased by exposure to lower and lower temperature (e.g. $5°$, $0°$, and $-3°$ C.). The lower the temperature, the weaker the hydrophobic bonds, and the more completely the protein molecules will be turned inside out. It also explains why, under natural conditions, though the major hardening is accomplished in 2 weeks in the field, a slower and slower increase in hardening continues until at least midwinter. For if the half-life of protein turnover is 7 days, nearly 100% turnover would require 70 days.

At first sight the evidence seems to oppose this concept of a changed conformation from a hydrophobic interior to a hydrophobic exterior, since the major increase during hardening is in the water-soluble proteins, which would therefore presumably have a hydrophilic exterior. However, the hydrophobic groups which would now be external to the protein molecule would be stabilized by the water forming a clathrate structure around them (Brandts, 1967), and the protein molecule would remain soluble. Furthermore, the increase in the water-soluble fraction is apparently an increase in low M.W. proteins. This interpretation is corroborated by observation of the 24 bands separated by disk electrophoresis (Morton, 1968). Though all the bands were in the same place in hardened and unhardened plants, there was an increase in the proportion of the rapidly moving (and therefore presumably low M.W.) proteins. This can be explained by the common formation of dimers, trimers, etc., under normal conditions, by aggregation of the *native* monomer protein. If, however, the protein formed at the low temperature has mainly clathrated hydrophobic groups external to it, this polymerization could not occur, and the low M.W. monomers would accumulate.

Results with the model protein thiogel indicate how these low M.W. proteins might protect the plant from freezing injury. Although the high

M.W. thiogel (M.W. 100,000) showed a markedly accelerated rise in melting point on freezing, the low M.W. thiogel (M.W. 10,000) did not (Levitt, 1965). Since it is already in the denatured state, the only possible conclusion is that aggregation during freezing by intermolecular SS formation is favoured by high M.W., opposed by low M.W. Presumably the low M.W. proteins which accumulate during hardening would be similarly protected against aggregation during freezing.

Although, as mentioned above, the changed conformation of the proteins at the hardening temperature may not necessarily affect the active sites, or may even increase their activity in some cases, in others the activity of the enzymes would be expected to decrease or even to disappear. Several investigators (e.g. Bamberg, Schwarz & Tranquillini, 1967) have found that the most resistant conifers show a steady decrease in photosynthetic rate (measured under standard, optimum conditions) during autumn and winter, which parallels the increase in hardiness, dropping to a rate of practically zero in the state of maximum hardiness. Even in the case of species which do not enter a true dormancy during autumn or winter, the most frost-resistant varieties commonly have the slowest rate of growth at hardening temperatures, though they grow just as rapidly as the less-resistant varieties at normal growing temperatures (Levitt, 1956).

This also explains the role of the high accumulation of proteins in the autumn *before* appreciable hardening. In order to achieve maximum hardening, the proteins must be hydrolysed and resynthesized. Highly active proteinases and protein synthesizing components must therefore be available to produce a sufficiently rapid breakdown and resynthesis, especially if these enzymes themselves become less active. On the basis of this concept, the proteins accumulated during or preceding hardening should be primarily enzymes involved in protein synthesis and hydrolysis. In agreement with this prediction is the accumulation of RNA on hardening (see above).

Though this explanation is largely speculative, it is supported by the hydrophobic and hydrogen-bonded forms in which polylysine can occur (Davidson & Fasman, 1967). The conversion of the β (hydrophobic) to the α (hydrogen-bonded) form is very slow. This explains why the plant must resynthesize the proteins at low temperatures in order to permit rapid formation of the hydrogen-bonded form. That this proposed change from mainly hydrophobic to mainly hydrogen bonds would produce the postulated increase in stability has also been shown for polylysine. The α-helical poly-L-lysine is stabilized largely by intra-amide hydrogen bonds and is the most stable form at $4°$ C.; the β-form appears to owe a large part of its stability to hydrophobic interactions between the lysyl residues and is the most stable form at $50°$ C. (Davidson & Fasman, 1967).

From the above considerations it may be concluded that although the net increase in protein content per cell is not likely to affect the $N \rightleftharpoons D$ reaction, the resynthesis of new proteins at the low temperature will have the effect of producing a mainly hydrophilically bonded instead of a mainly hydrophobically bonded molecule. It will therefore be only partially unfolded, and it will resist aggregation.

What then are the possibilities for artificial changes in freezing resistance? By spraying plants with purines and pyrimidines, Jung et al. (1967) have in some cases increased their proteins, their nucleic acids, and their hardiness. In general, however, it would presumably be necessary to by-pass all the protein changes, and to attempt to protect the non-resistant proteins by the addition of specific chemicals. The cryoprotective reagents in use so far are primarily, if not solely, physical protectants, i.e. they protect by maintaining a large fraction of the water in the unfrozen state at moderate freezing temperatures. They therefore keep the proteins far enough apart to prevent intermolecular bond formation and therefore aggregation. If the freezing temperature is low enough, even this water will freeze, but only at a temperature too low for the reaction to occur or for the intracellular freezing to cause damage.

These physical cryoprotectants must be used in such high concentrations that they sometimes produce injury instead of protection. A more elegant method would be to use specific chemical protectants for the chemical groups on the protein molecule which are in danger of forming intermolecular bonds, in this way preventing aggregation. If the most important groups are the sulphhydryls, then it should be possible to protect these. As indicated above, thiol reagents are double-edged weapons and many will have to be tried under a number of different conditions. So far they have succeeded only in decreasing freezing resistance. The most elegant method would be by prevention of $N \rightarrow D$, since the second step $(D \rightarrow A)$ could not then occur. But it is not clear, at present, how this could be done.

SURVIVAL AT HIGH EXTREMES OF TEMPERATURE

Heat injury

It has long been held that high-temperature injury is due to protein denaturation (Belehradek, 1935); for the denaturation at high temperatures was known long before the denaturation at low temperatures. Three main theories were proposed some time ago (see Belehradek, 1935; Levitt, 1956) to explain the growth and survival of thermophiles at high temperatures. These are still current (Campbell & Pace, 1968).

(a) Essential cell components are more heat-stable in thermophiles than in mesophiles.

(b) Resynthesis is rapid enough in thermophiles to compensate for the heat destruction, but is not rapid enough in mesophiles.

(c) Injury is due to the melting of lipids, and the lipids of thermophiles have higher melting points than those of mesophiles. Many attempts have been made to support this theory, but without success (Belehradek, 1935; Campbell & Pace, 1968).

The first theory has long been interpreted as due to heat-stability of the proteins. It is not so commonly recognized that the second theory depends on this same protein stability, at least in the case of those enzymes which control the resynthesis. More recently, however, attempts have been made to implicate the nucleic acids. No relation exists between the heat stability of DNA and thermophily, since both the base composition and the melting temperature are identical in thermophiles and mesophiles (Campbell & Pace, 1968). Similarly, the thermal denaturation of the sRNA's from a mesophile (*Escherichia coli*) and a thermophile (*Bacillus stearothermophilus*) are virtually identical. Yet the ribosomes of the thermophiles are much more heat-stable than those of the mesophiles. With few exceptions, the guanine and cytosine of the rRNA tended to increase, the adenine and uracil to decrease with increasing growth temperature. As in the case of DNA and sRNA, however, the rRNA is not significantly different in mesophiles and thermophiles, either as to its thermal denaturation of its gross base composition. Similar results have been obtained with a psychrophile (*Micrococcus cryophilus*). It is unable to grow above 25° C. due to an inhibition of protein synthesis (Malcolm, 1968). This was not due either to an inability to synthesize mRNA or to a degradation of existing RNA. On the other hand, three enzymes (tRNA synthetases) were found to be temperature-sensitive.

The evidence to date therefore indicates that neither the lipids nor the nucleic acids can explain thermophily. Furthermore, in neither case do the melting points approach the heat-killing temperatures of most cells. On the other hand, the range of temperatures for the heat denaturation of proteins does overlap the range for heat-killing of cells.

Brock (1967) has recently suggested that inactivation of heat-sensitive enzymes cannot be the answer, because heat injury follows first-order kinetics. But first-order kinetics are characteristic of monomolecular reactions, and protein denaturation is simply the unfolding of a protein molecule. Consequently, Brock's objection is actually a point in favour of protein denaturation as the cause of heat injury. Unfortunately, the kinetics of the denaturation process are difficult to determine for the following reason.

Unlike the denaturation at low temperatures, protein denaturation at high temperatures was assumed, in the past, to be irreversible. This is undoubtedly due to the rapid rate of aggregation at the high temperatures. Irreversible aggregation therefore quickly follows the reversible denaturation and leads to difficulties in determining the kinetics of the denaturation process. In the case of at least a number of proteins, it has now been shown that the loss of the native conformation due to high temperature is completely reversible (Brandts, 1967). The order of events is therefore as follows.

As the temperature rises, the conformational entropy favouring the denatured state increases more rapidly than the increase in strength of the hydrophobic bonds (Fig. 4) and a temperature is finally reached at which unfolding begins. The first effect of the high temperature on the proteins is therefore denaturation. In the absence of desiccation this denaturation cannot be followed by dehydration, in the usual sense, as happens on freezing. But as the temperature rises the free energy of the water molecules increases, decreasing the solvation of the protein molecules. This decrease in solvation is equivalent to a dehydration, for the uncovered chemical groups of adjacent protein molecules become more and more free to react with each other, leading to aggregation. Heat injury can therefore be explained in the same way as freezing injury:

$$N \underset{\text{below } 40°\text{C.}}{\overset{\text{above } 40°\text{C.}}{\rightleftharpoons}} D \xrightarrow{\text{above } 40°\text{C.}} A.$$

Heat resistance

Marked differences in heat resistance have been found between plant species, and even between varieties of a species (Levitt, 1956). The relation of this resistance to the plant's proteins has only recently received attention. Malate dehydrogenase is much more resistant to high temperature inactivation when obtained from an ecotype of *Typha latifolia* native to a hot climate than when obtained from an ecotype native to a cool climate (McNaughton, 1966). This difference was obtained even though the two ecotypes were grown under the same conditions in a growth chamber, and it is therefore presumably genetically determined. On the other hand, glutamate-oxaloacetate transaminase was quite resistant regardless of origin, and aldolase was rapidly inactivated regardless of origin. This may be interpreted to mean that only certain specific enzymes are important in the heat resistance of plants, or that some enzymes are not resistant unless the plant is first hardened (see below). On the other hand, the heat resistance of an enzyme in a living cell may be quite different from that in extracts of the cell, due to pH differences, presence of substrate, etc. Since, in the above experiments, the leaves were blended with 30 volumes of water and no buffer was used, only the stablest enzymes would be unaltered. Aldolase is an unstable enzyme (see above) and any differences between plants would therefore quickly disappear.

Earlier evidence (Levitt, 1956) indicated that exposing plants to moderately high temperatures actually decreased heat resistance due to exhaustion of reserves. More recent experiments, however, have shown that very brief exposures to moderately high temperatures definitely can increase the heat resistance of the plant (Alexandrov, 1964; Yarwood, 1967). This hardening treatment has now provided evidence for the long-held belief that heat injury is due to protein denaturation. Malic dehydrogenase extracted from heat-hardened bean plants is significantly more heat-stable than when extracted from unhardened plants (Kinbacher, Sullivan & Knull, 1967). Similar results have been obtained by Feldman (1966, 1968) for ATPase, acid phosphatase, and urease. On the basis of the above concept, both the fixed heat-hardiness of thermophiles and the increase which occurs during hardening may conceivably be due to prevention of (*a*) $N \to D$ or (*b*) $D \to A$.

(a) Prevention of $N \to D$

This could be produced by strengthening the protein bonds. Thus Koffler's (1957) analyses of the amino acid content of the proteins led him to conclude that the heat resistance of the proteins from thermophiles is due to their higher hydrophobicities. Similar results were obtained by Ohta, Ogura & Wada (1966). Since the strength of hydrophobic bonds increases with rise in temperature until about 75° C. (Brandts, 1967), these proteins would remain in the folded, native state at temperatures high enough to denature proteins with a lower proportion of hydrophobic bonds. In agreement with this concept, the β form of polylysine is the most stable form at 50° C. and appears to owe a large part of its stability to hydrophobic interactions between the lysyl residues. The α-helical polylysine is stabilized largely by interamide hydrogen bonds and is the most stable form at 4° C. (Davidson & Fasman, 1967).

Bigelow (1967) has calculated the hydrophobicities of the amino acid side-chains and has used these values to obtain a quantitative comparison between those proteins of thermophilic and mesophilic organisms for which amino acid analyses are available. The phycocyanins of five mesophilic species yielded average hydrophobicities of 980–1110 cal./residue, compared to 1190 cal./residue for the one thermophilic species. The α-amylases of four mesophilic species had hydrophobicities of 1020–1130 cal./residue, compared to 1210 for the one thermophile. Friedman (1968), on the other hand, has compared the amino acid analyses for the ribosome protein of a mesophilic and a thermophilic bacterium and has concluded that there is no obvious difference between the two. But this conclusion was based solely on visual comparison. Calculation of the hydrophobicities, with the use of Bigelow's values, supports his conclusion, yielding average values of 1095 cal./residue and 1062 cal./residue respectively (Table 5). However, some of the hydrophobicity values for the side-chains are so low (0·45–1·70 kcal./residue) that they may not be able to form bonds strong enough to hold the protein folds in place. Furthermore, these differences in strength are exaggerated at high temperatures (Brandts, 1967). The strength of the weaker hydrophobic bonds rises only slightly with temperature to a maximum at 50° C., only 200 cal. above the value at 0° C. The strength of the stronger hydrophobic bonds rises much more steeply with temperature to a maximum at 80° C., which is 750 cal. above the value at 0° C. If only those side-chains with values of 2 kcal./residue and above are compared, in all five cases the amino acid residues are found to be present in larger quantity in the thermophilic than in the mesophilic protein (Table 5). On the basis of these five amino acids, the average hydrophobicity of the protein of the thermophile is 666 cal./residue compared to 566 cal./residue in the mesophile. In terms of numbers of strong hydrophobic bonds, there would be a maximum of 1 bond/8 residues in the thermophile, only 1/9·3 residues in the mesophile. It is even conceivable that some of the weaker hydrophobic side-chains may be effective in the proteins of the thermophile because of proximity to the one strong SS bond which apparently occurs in the protein of the thermophile but not in the

Table 5. *Relative hydrophobicities of proteins from a thermophile* (Bacillus stearothermophilus) *and a mesophile* (Escherichia coli) *(data from Bigelow, 1967, and Friedman, 1968).* (*Numbers in italics indicate averages.*)

Amino acids with hydrophobic side chains	Hydrophobicity (kcal./res.)	B. stearothermophilus			E. coli		
		Mole (%)	kcal.	Av. cal./res.	Mole (%)	kcal.	Av. cal./res.
Trp	3·00	—	—	—	—	—	—
Ile	2·95	6·64	19·55	—	5·51	16·25	—
Tyr	2·85	2·16	6·15	—	1·78	5·07	—
Phe	2·65	3·55	9·40	—	3·03	8·03	—
Pro	2·60	4·44	11·52	—	3·67	9·55	—
Leu	2·40	8·34	20·00	—	7·40	17·75	—
Val	1·70	9·07	15·40	*666*	9·63	16·35	*566*
Lys	1·50	6·30	9·45	—	9·01	13·50	—
Met	1·30	2·33	3·02	—	2·40	3·12	—
Cys/2	1·00	0·82	0·82	—	0·53	0·53	—
Ala	0·74	10·51	7·90	—	10·98	8·23	—
Arg	0·75	5·01	3·76	—	7·30	5·46	—
Thr	0·45	5·53	2·49	—	5·22	2·35	—
Totals		64·70	109·46	*1095*	66·46	106·19	*1062*

protein of the mesophile. A similar analysis of Matsubara's (1967) data for the heat-resistant protein thermolysin again reveals no higher value for total hydrophobicity (1010 cal./residue) but a very high value for the six groups having high hydrophobicities (706 cal./res.). This mechanism could conceivably also arise due to hardening. If the proteins are broken down and resynthesized during this exposure to moderately high temperature, the newly synthesized proteins, though having the same amino acid sequence, would fold in a somewhat different manner due to the increased affinity of the hydrophobic groups for each other at the higher temperature. They would therefore have a higher proportion of hydrophobic to hydrogen bonds and would be more stable at the high temperatures. Similarly, even if the proteins are not broken down and resynthesized, if some of the hydrogen bonds are broken, this may permit the formation of new and stronger hydrophobic bonds between groups previously separated sterically, and which now have a stronger affinity for each other because of the high temperature.

As mentioned earlier, the change in conformation may or may not alter the activity of the enzyme, depending on whether or not the active site is altered. Thus, heat-hardening of leaves, which leads to increased resistance of their cells also increases the heat resistance of their enzymes by $1 \cdot 5$–$7°$ C. in the case of ATPase, acid phosphatase, and urease (Feldman, 1968). In the case of the first two, this was accompanied by a decrease in activity; but the activity of urease was not affected by its increase in heat resistance.

Evidence of this conversion during hardening has been produced by Jacobson (1968) in the case of *Drosophila*. An increase in heat-stability accompanies the conversion of the electrophoretically slowest-moving isoenzyme of alcohol dehydrogenase to the fastest-moving form. The increased stability indicates that a conformational change has occurred. An increase in hydrophobic bonds would certainly lead to such an increase in electrophoretic mobility, without any decrease in the number of hydrophilic groups, due to the larger ratio of hydrophobic to hydrophilic groups *within* the folded molecule, and therefore a larger charge external to the molecule.

It does not necessarily follow, however, that all heat-resistant proteins must have a high proportion of hydrophobic side-chains. The strongest of all intramolecular bonds responsible for the conformation of the proteins is the covalent SS bond. The importance of this bond in the heat resistance of proteins can be seen by comparing heat-stable with heat-labile enzymes. The heat-stable are characterized by absence of SH groups and presence of SS bonds (Table 6). Similarly, when different lysozymes are compared, hen's-egg lysozyme has four SS bonds and is the most heat-stable, human lysozyme has three SS bonds and is less heat-stable, goose-egg lysozyme

Table 6. *Properties of heat-stable enzymes*
(*from* Levitt, 1966a)

Protein	Inactivation temperature (° C.)	Time (min.)	Inactivation (%)	Molecular weight	SH (groups/molecule)	SS (groups/molecule)
1. Pepsin	65 (in acid)	15	50	35,000	0	3
2. α-Amylase (*Bacillus subtilis*)	65	30		48,700	0	0
3. Arginase	70	177	50	140,000	0	0
4. α-Amylase (thermophile)	90	60	10	15,000	0	2
5. Inorganic pyrophosphatase	90–100	—	—	63,000	Trace	
6. Cytochrome *c*	> 100	—	—	13,000	0	0
7. Muramidase (lysozyme)	> 100	—	—	15,000	0	4
8. Ribonuclease	> 100 (in acid)	—	—	12,700	0	4
9. Trypsin	> 100	—	—	24,000	0	6
10. Myokinase (adenylate kinase)	> 100	—	—	21,000	2	0

has two SS bonds and is the least stable (Jollès, 1967). Similar results have been obtained with bovine-lens proteins (Mehta & Maisel, 1966).

Although analyses of pure plant proteins have not yet been made, indirect evidence points to a possible role of SS bonds in the heat resistance of plants. Fraction I protein was found to be more heat-resistant when isolated from the leaves of heat-hardened bean plants than from unhardened plants (Sullivan & Kinbacher, 1967). Blocking the SH groups with PCMB did not change the heat-stability of the protein from hardened leaves. On the other hand, cleavage of the SS bonds with ME and sodium sulphite decreased the heat-stability of the protein from hardened plants to that of the protein from unhardened plants. No significant difference was found, however, between the number of SS bonds in the hardened and unhardened plants. It was therefore suggested that the hardening process led to a repositioning of the SS bonds in such a way as to increase protein stability.

(b) Prevention of $D \to A$

This could conceivably be produced by eliminating the chemical groups which interact with each other to form intermolecular bonds. As in the case of intramolecular bonds, the strongest *inter*molecular bond that can be formed is the covalent SS bond. In agreement with this, SH-containing enzymes are heat-labile (Table 7), presumably because the SH groups of adjacent molecules can combine to form intermolecular SS bonds. The heat-stable enzymes, on the other hand, are mostly free of SH groups (Table 6). The most striking difference is between the same enzyme (α-amylase) from a heat-resistant and a non-resistant species of bacterium.

Table 7. *Properties of heat-labile enzymes*
(*from Levitt, 1966a*)

Enzyme	Inactivation temperature (° C.)	Time at temperature (min.)	Molecular weight	SH (groups/ molecule)	SS (groups molecule)
1. β-Galactosidase	55	1	750,000	12	2
2. L-Glutamate dehydrogenase	55	—	1,000,000	90–120	
3. α-Glycerophosphate dehydrogenase	55	1 (53% activity lost)	78,000	15–16	
4. Glyceraldehyde-3-phosphate dehydrogenase	Room temperature (stabilized by ethylenediaminetetraacetate at 39)	—	120,000	11±2	0
5. Succinic dehydrogenase	Unstable even at 25	—	200,000	SH enzyme	
6. Xanthine oxidase	56	—	290,000	SH enzyme	
7. Glucose oxidase	Above 40	—	154,000	SH enzyme	

Similarly, the proteins of some thermophilic bacteria are free of any SH or SS groups (Koffler, 1957; Ohta *et al.* 1966; Matsubara, 1967).

In the absence of SH groups (e.g. in the case of proteins from thermophiles), Awad & Deranleau (1968) suggest that intermolecular hydrophobic bonding is possible. But nearly all the hydrophobic groups are internal to the native molecule and therefore unavailable for intermolecular bond formation as long as the molecule remains native. When it is denatured, unfolding occurs because the hydrophobic bonds are too weak to hold the folds in place. Consequently they would also presumably be too weak to form intermolecular bonds, and if any such weak bonds did form, the aggregation would be reversible. Therefore it must be concluded that in the absence of SH groups high-temperature aggregation of proteins must involve the formation of covalent bonds other than SS. What these are has not as yet been determined. They must, however, be formed only at higher temperatures than the readily formed SS bonds.

An apparently opposite relation has been suggested by Berns & Scott (1966). They conclude that the protein from a thermophilic Cyanophyta (*Synechococcus lividus*) has a greater number of charged and polar groups than that of mesophiles. According to the above concept, this should lead to an increase in both $N \rightleftharpoons D$ and $D \rightarrow A$ at high temperatures and therefore to enzyme inactivation. At moderate temperatures, of course, the strength of the polar bonds would prevent $N \rightleftharpoons D$. But if the polar groups are sufficiently numerous, $N \rightleftharpoons A$ might occur without denaturation. According to their interpretation, this aggregation reversibly inactivates the enzymes at 25° C., and only by raising the temperature to 50° C. would the smaller, active form of the enzyme occur. This explanation would seem to be valid only if Alexandrov's concept (see above) were also involved. Since at 50° C. there would be some $N \rightleftharpoons D$, we would have to conclude that this degree of unfolding would be just sufficient to activate the active sites which are too tightly folded at 25° C.

On the basis of the above results, heat-stability of proteins depends on their possession of one or more of the following characteristics: (1) a high proportion of strong hydrophobic bonds, (2) intramolecular SS bonds, (3) absence of SH groups. Heat resistance of plants is presumably dependent on these same characteristics.

A COMPARISON OF HEAT AND FREEZING RESISTANCE

A paradoxical relation between freezing-resistance and heat-resistance has long puzzled investigators. When a plant hardens in the autumn it develops resistance to freezing injury. At the same time its heat-resistance also rises, parallel to its freezing-resistance (Levitt, 1958; Alexandrov, 1964). During the summer, on the other hand, the heat-resistance of a plant rises without any concomitant rise in freezing-resistance (Levitt, 1958). The nature of these two different kinds of heat-resistance now becomes clear from the above concept, for the following reasons. Denaturation at low temperature is due to the sharp drop in hydrophobic bond strength, leading to unfolding. Consequently, when a plant hardens at low temperatures, its resistance to $N \to D$ can only be due to an increase in intramolecular hydrogen bonds. Conversely, denaturation at high temperature occurs at temperatures of maximum hydrophobic bond strength, and unfolding must be initiated by rupture of the weaker hydrogen bonds. Therefore, when the plant hardens at high temperatures, its resistance to $N \to D$ can only be due to an increase in intramolecular hydrophobic bonds.

These two kinds of hardening would be mutually exclusive. On the other hand, when a plant develops resistance to $D \to A$, the resistance will remain at both freezing and high temperatures if it is due to the absence or protection of chemical groups capable of interacting with those of adjacent protein molecules. The proteins of the frost-hardy plant would therefore denature at high temperatures but the denaturation would remain reversible. On this basis, we must conclude that hardening at low temperatures primarily involves an increase in resistance to aggregation, and therefore carries with it an increased resistance to heat injury. Hardening at high temperatures primarily involves an increase in resistance to denaturation at high temperatures, due to an increase in intramolecular hydrophobic bonds. It therefore does not carry with it an increase in resistance to freezing injury, since resistance to denaturation at low temperature would be decreased.

Further evidence that hardening against freezing injury is due primarily to an increased resistance to D → A is the correlation that has frequently been found between freezing-resistance and drought-resistance. Since drought injury occurs at ordinary growing temperatures there is no reversible N ⇌ D. A resistance to drought injury would therefore presumably be a resistance against aggregation, i.e. against N → A. This aggregation could result, for instance, by intermolecular SS bond formation between the stressed protein molecules in the collapsed cells, just as was suggested for freezing injury. Consequently, the only kind of resistance developed during low-temperature hardening which could also induce drought hardiness would be resistance to aggregation.

This concept also explains the tremendous difference between the length of time required for frost-hardening (about 2 weeks for the major hardening) and heat-hardening (seconds or minutes). In the case of frost-hardening, resynthesis of the proteins is required for maximum hardening, and this may require weeks or even months because of the slow reaction rates at the low hardening temperatures. In the case of heat-hardening, all that may be necessary is a refolding of the proteins to take advantage of the sharp increase with temperature in the hydrophobic bond strength. This may explain why some hardening has been detected as a result of even a single second at the heat-hardening temperature (Alexandrov, 1964).

It is still a question as to which chemical groups are responsible for the D → A. Considerable evidence has been cited above, pointing to the conversion of SH groups to intermolecular SS bonds during freezing. Since, as mentioned above, an increase in freezing-resistance is accompanied by an increase in heat-resistance, the same chemical groups must be involved in protein aggregation at both temperatures. In favour of this conclusion, a decrease in protein SH and an apparent increase in SS groups has been found during heat-killing (Levitt, 1962). Similarly, the results of Levy & Ryan (1967) indicate that heat-inactivation of the relaxing site of actomyosin occurs because certain labile SH groups are oxidized to the SS form. Dithiothreitol led to both prevention and reversal of this effect. Mishiro & Ochi (1966) found that human serum albumin becomes turbid at 60–95° C. This turbidity was completely prevented in 0·05 M solution in the presence of 10^{-3} M-Na dipicolinate, which decreased the number of SH groups.

However, even the SH-free proteins of thermophilic bacteria are inactivated if the temperature is high enough. Even gelatin, which is denatured and free of SH groups, forms covalent cross-links between adjacent molecules when the water content falls below 0·2 g./100 g. protein (Yannas & Tobolsky, 1967). It therefore seems reasonable to conclude that as the temperature rises, and more and more bound water is converted to free water, previously protected chemical groups become available for intermolecular covalent bond formation. But the SH groups do not bind water appreciably, and they are furthermore readily oxidized to SS under suitable conditions (e.g. the presence of the necessary enzymes, cofactors, and oxidation potential). Consequently it can be reasonably concluded that at moderately high temperatures aggregation is primarily due to intermolecular SS bonds; at still higher temperatures it is due to other covalent bonds.

This may also explain the nearly invariable fatality of intracellular freezing injury. Ice formation within the protoplasm cannot usually be fully uniform. Water will crystallize between some proteins, but not between others. As a result, water moves quickly to these suddenly formed ice nuclei from between proteins which are not separated by ice crystals. These proteins therefore are suddenly naked of protective water coats, and may also be pressed together by the expanding ice crystals on the other sides. As a result, intermolecular covalent bonds are quickly formed, resulting in irreversible aggregation. Only if the intracellular freezing is absolutely uniform—e.g. when the crystals are so small and numerous that none of the protein molecules are displaced towards each other—is it non-injurious. This kind of harmless freezing occurs when the crystals are so small that they are detectable only by X-ray analysis, i.e. only when the crystals are within the size range of molecules.

SUMMARY

On the basis of the available experimental evidence with plants, and of modern concepts of protein conformation in relation to temperature, the following unified concept is proposed:

(1) Both growth and survival at extremes of temperature are due to protein denaturation at these temperatures.

(2) Growth extremes are due to a reversible denaturation of the proteins $\left(N \underset{T_{\text{normal}}}{\overset{T_{\text{extreme}}}{\rightleftharpoons}} D\right)$, leading to inactivation of at least some of the enzymes. As a result, metabolism is insufficient for, or uncoupled from growth. A high maximum temperature for growth is due to a high ratio of hydrophobic to hydrogen bonds in the proteins, leading to a higher temperature of denaturation. A low minimum temperature for growth is due to a low ratio

of hydrophobic to hydrogen bonds in the proteins leading to a lower temperature of denaturation.

(3) Chilling injury, which occurs after one or more days at cool but above freezing temperatures, is due to denaturation of the proteins which may or may not be followed by their hydrolysis to amino acids:

$$N \underset{T_{normal}}{\overset{T_{chilling}}{\rightleftharpoons}} D \overset{T_{chilling}}{\longrightarrow} \text{amino acids.}$$

Cold shock, which occurs on sudden, short (minutes or hours) exposure to freezing temperatures without ice formation, is due to denaturation of proteins accompanied by aggregations: $N \overset{T_{freezing}}{\longrightarrow} D \rightarrow A$. The plant cannot be hardened against slow chilling injury, since the hardening temperatures are those producing the injury. It can be hardened against cold shock, and the hardening prevents the aggregation ($D \rightarrow A$) by an increase in reduction potential. Presumably the high ratio of NADPH to NADP prevents intermolecular SS formation.

(4) Freezing injury is due to protein denaturation at chilling or freezing temperatures followed by aggregation during freezing:

$$N \underset{T_{normal}}{\overset{T_{chilling}}{\rightleftharpoons}} D \overset{freezing}{\longrightarrow} A.$$

The denaturation is caused by a weakening of the hydrophobic bonds at low temperature, leading to unfolding. The most important reactive groups that lead to irreversible aggregation and injury are the SH groups, which can combine to form intermolecular SS bonds, and may even lead to further denaturation during thawing. Frost-hardening is possible only in plants with a specific stable, hereditary character which prevents chilling injury—their proteins must have a low enough ratio of hydrophobic to hydrogen bonds to prevent injurious denaturation at chilling temperatures. Such plants are capable of hardening during exposures (for days or weeks) to chilling temperatures. This hardening leads to a prevention of protein aggregation ($D \rightarrow A$) at freezing temperatures, and is produced by a resynthesis of the proteins at hardening temperatures. As the temperature is lowered more and more, within (and beyond) the chilling range, the resynthesized proteins form fewer and fewer hydrophobic bonds and more and more hydrophilic bonds. This results in a conformation with the reactive groups protected within the protein molecules, and therefore unavailable for aggregate (i.e. intermolecular bond) formation. It also prevents the newly formed proteins from polymerizing to the normal dimers, trimers, etc., and therefore leads to the increase in low-M.W., water-soluble proteins normally found during hardening.

(5) Heat injury is due to protein denaturation at high temperatures followed by aggregation: $N \underset{T_{normal}}{\overset{T_{high}}{\rightleftharpoons}} D \overset{T_{high}}{\longrightarrow} A$. Since hydrophobic bond strength increases with temperature, the denaturation must be due to breaking of hydrogen bonds, unlike the unfolding at low temperature. Thermophilic organisms are able to survive high temperatures because of the high ratio of hydrophobic to hydrophilic groups in their proteins. Heat-hardening occurs during short (seconds or minutes) exposures to high temperatures and is due to the change in protein conformation at the heat-hardening temperature. Heat-hardening is possible only in plants with proteins that have a high ratio of hydrophobic to hydrophilic groups. Since the higher the temperature the stronger the hydrophobic bonds, the proteins refolded at the high temperatures have a higher proportion of hydrophobic bonds, and this results in a higher temperature of denaturation. Heat-hardened plants therefore resist heat injury due to a resistance to protein denaturation $\left(N \underset{T_{normal}}{\overset{T_{high}}{\rightleftharpoons}} D \right)$. Extreme heat-hardiness may be due to an additional factor—a decrease in the number of groups (e.g. SH) capable of forming covalent intermolecular bonds.

(6) Hardening against freezing-injury leads to both freezing-resistance and heat-resistance, because it is primarily due to protection against $D \rightarrow A$. Hardening against heat injury does not lead to freezing-resistance because it is primarily due to protection against $N \overset{T_{high}}{\longrightarrow} D$ and this leads to increased $N \overset{T_{low}}{\longrightarrow} D$.

ABBREVIATIONS

N, native. D, denatured. A, aggregated. SH, sulphhydryl. SS, disulphide. PCMB, parachloromercuribenzoate. DMSO, dimethylsulphoxide. GSH, glutathione. GSSG, oxidized glutathione. IA, iodoacetate.

REFERENCES

ÅKERMAN, Å. (1927). *Studien über der Kältetod und die Kälteresistenz der Pflanzen.* Lund, Sweden: Berlingska Boktryckeriet.
ALEXANDROV, V. Y. (1964). *Q. Rev. Biol.* **39**, 35.
ALEXANDROV, V. Y. (1967). In *Molecular Mechanisms of Temperature Adaptation*, p. 53. Ed. C. L. Prosser. Washington, D.C. American Association for the Advancement of Science.
ANDERSON, J. W. & ROWAN, K. S. (1967). *Phytochemistry* **6**, 1047.
ANDREWS, S. & LEVITT, J. (1967). *Cryobiology* **4**, 85.
ANFINSEN, C. B. (1967). In *Molecular Organization and Biological Function*, p. 1. Ed. J. M. Allen. New York: Harper and Row.
AWAD, E. S. & DERANLEAU, D. A. (1968). *Biochemistry* **7**, 1791.
BAMBERG, S., SCHWARZ, W. & TRANQUILLINI, W. (1967). *Ecology* **48**, 264.
BELEHRADEK, J. (1935). *Temperature and Living Matter.* Protoplasma Monogr. no. 8. Berlin: Borntraeger.
BELLO, J. (1966). *Cryobiology* **3**, 27.

BENNETT, J. P., OSERKOWSKY, J. & JACOBSEN, L. (1940). *Am. J. Bot.* **27**, 883.
BERNS, D. S. & SCOTT, E. (1966). *Biochem.* **5**, 1528.
BIEBL, R. (1962). *Protoplasmatologia* **12** (1), 344.
BIGELOW, C. C. (1967). *J. theoret. Biol.* **16**, 187.
BLATTLER, D. P., CONTAXIS, C. C. & REITHEL, F. J. (1967). *Nature, Lond.* **216**, 274.
BRANDTS, J. F. (1967). In *Thermobiology*, p. 25. Ed. Anthony H. Rose. New York: Academic Press.
BROCK, T. D. (1967). *Science, N.Y.* **158**, 1012.
CAMPBELL, L. L. & PACE, B. (1968). *J. appl. Bact.* **31**, 24.
CHRISTOPHERSEN, J. (1967). In *Molecular Mechanisms of Temperature Adaptation*, Ed. C. L. Prosser. Washington, D.C. American Association for the Advancement of Science.
COLEMAN, E. A., BULA, R. J. & DAVIS, R. L. (1966). *Pl. Physiol., Lancaster* **41**, 1681.
COTRUFO, C. & LEVITT, J. (1958). *Physiologia Pl.* **11**, 240.
DAVIDSON, B. & FASMAN, G. D. (1967). *Biochem.* **6**, 1616.
DORNER, R. W., KAHN, A. & WILDMAN, S. G. (1957). *J. biol. Chem.* **229**, 945.
EAKS, I. L. & MORRIS, L. (1956). *Pl. Physiol., Lancaster* **31**, 308.
ELDJARN, L. (1965). *Int. Symp. Radiosensitizers, Radioprotective Drugs*, Milan **1**, 173.
FELDMAN, N. L. (1966). *Dokl. Akad. Nauk SSSR* **167**, 946.
FELDMAN, N. L. (1968). *Planta* **78**, 213.
FENNEMA, O. (1966). *Cryobiology* **3**, 197.
FEUGHELMANN, M. (1966). *Nature, Lond.* **211**, 1259.
FRIEDMAN, S. M. (1968). *Bact. Rev.* **32**, 27.
GERLOFF, E. D., STAHMANN, M. A. & SMITH, D. (1967). *Pl. Physiol., Lancaster* **42**, 895.
GETMAN, F. H. & DANIELS, F. (1937). *Outlines of Theoretical Chemistry*. London: John Wiley & Sons.
GINSBURG, A. & CARROLL, W. R. (1965). *Biochemistry* **4**, 2159.
GODNEV, T. N. & SHABELSKAYA, E. F. (1964). *Fiziologiya Rast.* **11**, 818.
GRANT, N. H. (1966). *Discovery* **27** (8), 26.
GREIFF, D. & RIGHTSEL, W. A. (1967). *Cryobiology* **3**, 432.
GRIF, V. G. (1966). *Tsitologiya* **8**, 659.
HARDER, W. & VELDKAMP, H. (1968). *J. appl. Bact.* **31**, 12.
HARPER, J. E. & PAULSEN, G. M. (1967). *Crop Science* **7**, 205.
HATCH, F. T. & BRUCE, A. L. (1968). *Nature, Lond.* **218**, 1166.
HEBER, U. (1967). *Pl. Physiol., Lancaster* **42**, 1343.
HEBER, U. & SANTARIUS, K. A. (1964). *Pl. Physiol., Lancaster* **39**, 712.
HEITMANN, P. (1968). *Europ. J. Biochem.* **3**, 346.
HOLMSEN, T. W. & KOCH, A. L. (1964). *Phytochemistry* **3**, 165.
IBANEZ, M. L. (1964). *Nature, Lond.* **201**, 414.
ILJIN, W. S. (1933). *Protoplasma* **20**, 105.
IRVING, R. M. & LANPHEAR, F. O. (1968). *Pl. Physiol., Lancaster* **43**, 9.
JACOB, H. S., BRAIN, M. C., DACIL, J. V., CARRELL, R. W. & LEHMANN, H. (1968). *Nature, Lond.* **218**, 1214.
JACOBI, G. (1967). *Z. Pflanzenphysiol.* **56**, 262 (D 68).
JACOBSON, K. B. (1968). *Science, N.Y.* **159**, 324.
JOLLÈS, P. (1967). *Bull. Soc. Chim. Biol.* **49**, 1001.
JØRGENSEN, E. G. (1968). *Physiologia Pl.* **21**, 423.
JUNG, G. A., SHIH, S. C. & SHELTON, D. C. (1967). *Cryobiology* **4**, 11.
KHAN, A. W., DAVIDKOVA, E. & VAN DEN BERG, L. (1968). *Cryobiology* **7**, 184.
KIM, S. & PAIK, W. K. (1968). *Biochem. J.* **106**, 707.
KINBACHER, E. J., SULLIVAN, C. Y. & KNULL, H. R. (1967). *Crop Science* **7**, 148.
KLEIN, R. M. & EDSALL, P. C. (1966). *Pl. Physiol., Lancaster* **41**, 949.

Koblitz, H., Grutzmann, K. & Hagen, I. (1967). *Z. Pflanzenphysiol.* **56**, 27.
Koffler, H. (1957). *Bact. Rev.* **21**, 227.
Kohn, H. & Levitt, J. (1966). *Pl. Physiol., Lancaster* **41**, 792.
Koshland, D. E. & Kirtley, M. E. (1966). In *Major Problems in Developmental Biology*, p. 217. Ed. M. Locke. New York: Academic Press.
Koukol, J., Dugger, W. M., Jr. & Palmer, R. L. (1967). *Pl. Physiol., Lancaster* **42**, 1419.
Krull, E. (1966). Investigations of the frost hardiness of cabbage in relation to the sulfhydryl hypothesis. Ph.D. dissertation, University of Missouri, Columbia, Missouri.
Kuraishi, S., Arai, N., Ushyima, T. & Tazaki, T. (1968). *Pl. Physiol., Lancaster* **43**, 238.
Levitt, J. (1956). *The Hardiness of Plants*. New York: Academic Press.
Levitt, J. (1958). *Frost, Drought and Heat Resistance*. Protoplasmatologia **8** (6), 87. Vienna: Springer.
Levitt, J. (1962). *J. theoret. Biol.* **3**, 355.
Levitt, J. (1965). *Cryobiology* **1**, 312.
Levitt, J. (1966a). *Cryobiology* **3**, 243.
Levitt, J. (1966b). In *Cryobiology*, p. 395. Ed. H. T. Meryman. New York: Academic Press.
Levitt, J. (1967a). In *The Cell and Environmental Temperature*, p. 260. Ed. A. S. Troshin. Oxford: Pergamon Press.
Levitt, J. (1967b). In *Cellular Injury and Resistance in Freezing Organisms*, p. 51. Ed. E. Asahina. Inst. of Low Temp. Sci., Hokkaido University, Japan.
Levitt, J. (1967c). In *Molecular Mechanisms of Temperature Adaptation*, p. 41. Ed. C. L. Prosser. Washington, D.C.: American Association for the Advancement of Science.
Levitt, J. (1969). *Cryobiol.* **6**. (In Press).
Levitt, J. & Hasman, M. (1964). *Pl. Physiol., Lancaster* **39**, 409.
Levy, H. M. & Ryan, E. M. (1967). *Science, N.Y.* **156**, 73.
Li, P. H. & Weiser, C. J. (1967). *Proc. Am. Soc. hort. Sci.* **91**, 716.
Light, A. & Sinha, N. K. (1967). *J. biol. Chem.* **242**, 1358.
Lundegårdh, H. (1957). *Klima und Boden*. Jena: Gustav Fischer.
McDonough, W. T. (1967). *Physiologia Pl.* **20**, 455.
McNaughton, S. J. (1966). *Pl. Physiol., Lancaster* **41**, 1736.
Malcolm, N. L. (1968). *Biochim. biophys. Acta* **157**, 493.
Massey, V., Hofmann, T. & Palmer, G. (1962). *J. biol. Chem.* **237**, 3820.
Matsubara, H. (1967). In *Molecular Mechanisms of Temperature Adaptation*, p. 283. Ed. C. L. Prosser. Washington, D.C. American Association for the Advancement of Science.
Maximov, N. A. (1912). *Ber. dt. bot. Ges.* **30**, 52.
Mehta, P. D. & Maisel, H. (1966). *Experientia* **22**, 818.
Minamikawa, T., Akazawa, T. & Uritani, I. (1961). *Pl. Cell Physiol. Tokyo* **12**, 301.
Mishiro, Y. & Ochi, M. (1966). *Nature, Lond.* **211**, 1190.
Morton, Wm. (1968). Effects of hardening and freezing on the sulfhydryl content of protein fractions of cabbage leaves. Ph.D. dissertation, University of Missouri, Columbia, Missouri.
Nagatsu, T., Kuzuya, H. & Hidaka, H. (1967). *Biochem. biophys. Acta* **139**, 319.
Nezgovorov, L. A. & Sokolov, A. K. (1965). *Fiziologiya Rast.* **12**, 1093.
Ohta, Y., Ogura, Y. & Wada, A. (1966). *J. biol. Chem.* **241**, 5919.
Oppenheim, A. & Castelfranco, P. A. (1967). *Pl. Physiol., Lancaster* **42**, 125.
Pilet, P. E. & Dubois, J. (1968). *Physiologia Pl.* **21**, 445.
Pincock, R. E. & Kiovsky, T. E. (1966). *J. Am. chem. Soc.* **88**, 4455.

PONTREMOLI, S., TRANIELLO, S., ENSER, M., SHAPIRO, S. & HORECKER, B. L. (1967). *Proc. natn. Acad. Sci., U.S.A.* **58**, 286.
PRECHT, H. (1967). In *The Cell and Environmental Temperature*, p. 307. Ed. A. S. Troshin. New York: Pergamon Press.
PULLMAN, M. E. & MONROY, G. C. (1963). *J. biol. Chem.* **238**, 3762.
ROSE, A. H. (1968). *J. appl. Microbiol.* **31**, 1.
SACHS, J. (1864). *Flora* **47**, 5.
SANTARIUS, K. & HEBER, U. (1967). *Planta* **73**, 109.
SEIBLE, D. (1939). *Beitr. Biol. Pfl.* **26**, 289.
SIMINOVITCH, D. & BRIGGS, D. R. (1949). *Archs Biochem.* **23**, 8.
SIMINOVITCH, D. & CHATER, A. P. J. (1958). In *The Physiology of Forest Trees*, p. 219. Ed. K. V. Thimann. New York: Ronald Press.
SIMINOVITCH, D., GFELLER, F. & RHÉAUME, B. (1967). In *Cellular Injury and Resistance in Freezing Organisms*, p. 93. Ed. Eizo Asahina. Inst. Low. Temp. Sci., Hokkaido Univ. Japan.
SIMINOVITCH, D., RHÉAUME, B. & SAHAR, R. (1967). In *Molecular Mechanisms of Temperature Adaptation*, p. 3. Ed. C. L. Prosser. Washington: American Association for the Advancement of Science.
SPRAGG, S. P., LIEVESLEY, P. M. & WILSON, K. M. (1962). *Biochem. J.* **83**, 314.
STEEMANN-NIELSEN, E. & JØRGENSEN, E. K. (1968). *Physiologia Pl.* **21**, 401.
STEPONKUS, P. L. & LANPHEAR, F. O. (1967). *Pl. Physiol., Lancaster* **42**, 1673.
SUGIYAMA, T., AKAZAWA, T. & NAKAYAMA, N. (1967). *Archs Biochem. Biophys.* **121**, 522.
SULLIVAN, C. Y. & KINBACHER, E. J. (1967). *Crop Science* **7**, 241.
SUTHERLAND, R. M. & PIHL, A. (1967). *Biochem. biophys. Acta* **135**, 568.
SZALAI, I. (1959). *Acta Biol. hung.* **9**, 253.
TALMA, E. G. C. (1918). *Recl Trav. bot. néerl.* **15**, 366.
TAYLOR, N. W., ORTON, N. L. & BABCOCK, G. E. (1968). *Archs Biochem. Biophys.* **123**, 265.
TUMANOV, I. I. & TRUNOVA, T. I. (1957). *Fiziologiya Rast.* **4**, 379.
USHAKOV, B. P. (1967). In *The Cell and Environmental Temperature*, p. 332. Ed. A. S. Troshin. New York: Pergamon Press.
VAN OVERBEEK, J. (1966). *Science, N.Y.* **152**, 721.
VEGIS, A. (1964). *Pl. Physiol., Lancaster* **15**, 185.
VINOGRADOV, A. D., NIKOLAEVA, L. V., OZRINA, R. D. & KONDRASHOVA, M. N. (1966). *Biokhimiya* **31**, 501.
WAISEL, Y., KOHN, H. & LEVITT, J. (1962). *Pl. Physiol., Lancaster* **37**, 272.
WALTER, H. (1950). *Grundlagen des Pflanzenlebens*. Stuttgart: Eugen Ulmer.
WARD, E. W. B. (1968). *Can. J. Bot.* **46**, 385.
WARREN, J. C. & CHEATUM, G. G. (1966). *Biochemistry* **5**, 1702.
WILHELM, A. F. (1935). *Phytopath. Z.* **8**, 111.
WIMAN, L. G. (1964). *Protein-Bound Sulfhydryl Groups in Pulmonary Cytodiagnosis and Their Significance in Papanicolaou Technique and Acridine-Orange Fluorescence Microscopy*. Stockholm: Almquist and Wiksell.
YANNAS, I. V. & TOBOLSKY, A. V. (1967). *Nature, Lond.* **215**, 509.
YARWOOD, C. E. (1967). In *Molecular Mechanisms of Temperature Adaptation*, p. 75. Ed. C. L. Prosser. Washington, D.C. American Association for the Advancement of Science.
YUTANI, K., YUTANI, A. & ISEMURA, T. (1967). *J. Biochem.* **62**, 576.

THE PHYSIOLOGY OF DORMANCY AND SURVIVAL OF PLANTS IN DESERT ENVIRONMENTS

By D. KOLLER

Department of Botany, The Hebrew University of Jerusalem, Israel

THE RELATIONSHIP BETWEEN DORMANCY AND SURVIVAL

Dormancy in plants may be defined as a process by which physiological activities become capable of ceasing entirely, in a reversible manner. This applies primarily to activities concerned with growth, and therefore with meristems, but may spread to other metabolic activities, such as photosynthesis, protein turnover, water exchange, and even respiration. The dormant state thus results in a reduced dependence on the environment and consequently in an enhanced tolerance of adverse environments. A variety of tissues and organs are capable of exhibiting symptoms of dormancy. However, dormancy in seeds and buds is associated with the formation of specialized organs, tissues and structures, which have specific and well-defined roles in increasing tolerance of adverse environments, as well as in the actual control of dormancy itself. The following discussion will be confined to dormancy phenomena in seeds and regeneration buds, which are specialized plant adaptations to the dormant state.

Dormancy is of survival value where it spans unfavourable periods in the natural environment. Onset and relaxation of dormancy must therefore be under environmental control. Evidence is accumulating that even where dormancy itself is not under environmental control, the autonomous control is only apparent (Vegis, 1964). In such cases, dormancy follows as an 'inevitable' part of the developmental sequence, set in motion by an environmentally controlled component process, which may have no direct bearing on dormancy itself. A mechanism of gene control for such sequences has been suggested by Heslop-Harrison (1963) to account for phenomena associated with sex-expression. This is apparently the case in onset of seed dormancy, which is entrained by environmental control as far back as the flower-initiation stage of development, or soon after. There still exists the possibility that dormancy may be controlled by an endogenous annual periodicity. However, the biological clock has also to be environmentally entrained to maintain synchrony with astronomical periodicity (Aschoff, 1960). This synchrony is essential for survival in habitats incorporating a season unsuitable for growth.

DORMANCY AND SURVIVAL IN DESERTS

The desert environment presents specific problems to plant survival by means of dormancy. Deserts differ, but in most desert environments the single dominant factor which is critical for plant survival is the negative balance between precipitation and potential evapo-transpiration. If total annual rainfall were to be evenly distributed throughout the year, it would all evaporate at the soil surface and become ineffective for plant existence (with the possible exception of lichens). Plant life in the desert becomes possible only because rainfall is unevenly distributed in time and space, and because it generally becomes even more so by differences in soil surface topography and permeability. A transient supply of soil moisture is thus sometimes able to accumulate and serve as the unique source supporting plant life. In deserts, therefore, stress is synonymous with lack of soil moisture. Growth is limited to periods which are characteristically short, in absolute terms, as well as in relation to the time required for completion of the intricate processes and metamorphoses involved in the formation of new dormant seeds (Amen, 1968), or regeneration buds (Smith & Kefford, 1964), in preparation for the oncoming adverse interval. The requirement for dormancy to be initiated may therefore follow quite closely behind the onset of growth. This places an additional responsibility on the timing of bud sprouting and of seed germination, which now have to take into account and provide for conditions which are not only favourable for growth, but also ensure the time and environment required for completion of the dormant structures. Furthermore, dependence of the plant on the environment is increasing during relaxation of dormancy and decreasing during its onset. Susceptibility to adverse environments is similarly affected. The environmental control of bud sprouting and of seed germination thus assumes a more vital role in species survival under desert conditions than the environmental control of onset of dormancy (flowering, formation of regeneration buds). Nevertheless, it is not inconceivable that onset of stress might initiate dormancy. Supply of cytokinins from the root may be curtailed by decrease in water potential of the root medium (Itai, Richmond & Vaadia, 1968). Cytokinins are involved in RNA synthesis, probably by incorporating into tRNA (Letham, 1967), and in the relaxation of certain types of dormancy (Khan, 1966; Sachs & Thimann, 1967). Furthermore, cytokinins counteract dormancy imposed by abscisic acid (dormin) (van Overbeek, 1968). Onset of soil-moisture stress might therefore induce bud dormancy by affecting the balance between inhibitors and promoters (Wareing, 1965; Khan, 1966). It is noteworthy that the ocotillo (*Fouquieria splendens*) of the North and Central American deserts, reputedly sprouts lateral clusters of ephemeral leaves along its elongated basal branches after every rainfall. There are no indications that onset of soil moisture stress may be involved in the onset of seed dormancy, e.g. by promoting flower initiation. However, desiccation accelerates the onset of dormancy in seeds of certain Papilionaceae, possessing water-impermeable seed coats. As the seed dries, water evaporates from the interior of the seed and diffuses through a fissure in the hilar region of the seed coat. This fissure is capable of acting as an hygroscopic valve which regulates the entry of liquid water into the seed, but is prevented from doing so as long as the volume of the imbibed colloidal contents of the seed remains large. As these colloids dry their volume becomes smaller and the distended seed coat shrinks, bringing into operation the valve responsible for the dormancy of such seeds (Hyde, 1954; Evenari, Koller & Gutterman, 1966).

Seeds are deprived of the large amounts of reserve food which are available to the regeneration buds in the storage tissues of the 'mother' plant, to which they are attached. Regeneration buds also have at their disposal a much more efficient system for the supply of soil moisture, by using the persistent roots, or by capacity of the perennial part of the plant for rapid *de novo* production of roots. Conceivably, they may also be able to obtain some metabolic water through respiration of their more abundant food. Consequently, activated regeneration buds are better equipped than germinated seeds to sustain survival over longer unfavourable periods. For this reason, under desert conditions, dependence of germination on environmental control must be greater than that of sprouting. Bud dormancy is of immediate survival value for the individual, whereas seed dormancy is of sole value for survival of the species. It is not surprising, therefore, that evolution has led to an almost universal imposition of regulating mechanisms on seed germination, which will prevent it from using up the entire reproductive capacity of the species in any given habitat, except where and when the probability of survival is maximal. Some seeds, but by no means the majority, apparently require only rehydration in order to germinate over a wide range of environmental conditions. Prompt germination as soon as water becomes available would appear to have a high survival value for plants under desert conditions, allowing them to make the longest possible use of this water. However, because of their small size, seeds are able to sample the water potential in a relatively minute volume of soil. Furthermore, seeds must generally germinate close to the soil surface, to become autotrophically established sooner. Soil moisture conditions are least stable at the surface. Consequently, the estimates of soil moisture availability which the seed can sense may be quite misleading. Prompt germination upon rehydration would thus appear to be a property detrimental for survival, particularly under desert conditions.

Survival under desert conditions faces the additional threat of the 'uncertainty factor' inherent in the pattern of rainfall. Rain is not merely scarce, but extremely irregular in timing, in localization and in the amounts it eventually contributes to soil moisture. As a result, water in amounts necessary to complete the developmental cycle may become available in the wrong time or place. The more predominant is this 'uncertainty factor', the more important it becomes for germination to be controlled by precise perception of the environment, in the interests of survival. We may therefore start by reviewing some of the response mechanisms which regulate germination and speculate on their possible contribution to the survival of the species under desert conditions.

SELECTIVITY AND HETEROGENEITY IN THE CONTROL OF SEED GERMINATION

Broadly speaking, regulation of germination is based on selectivity, on heterogeneity, or on their combination. Selectivity creates response mechanisms to specific environmental signals, with the result that reproductive capacity of the seed population is selectively used up in specific environments and is conserved in others. Germination in the desert is frequently of an epidemic nature: species whose seedlings are practically non-existent in certain years or locations may germinate profusely in others. All this because of selective germination. Heterogeneity, on the other hand, creates quantitative differences in the germination requirements within the seed population produced by single individuals. In this fashion the reproductive capacity of any individual plant is made incapable of being used up in a single attempt at establishment.

The role of selectivity

Temperature, light and humidity are the environmental variables most commonly used by the seed for identifying its environment. These variables may be used in combination, or some may predominate over others in their control capacity.

In its simplest form, control by temperature serves as a seasonal indicator.

As such, it may act in determining the seasonal distribution of summer and winter annuals in deserts where rainfall is non-seasonal, yet may equally contribute to soil moisture in all seasons (Went, 1949; Juhren, Went & Phillips, 1956), for example, because the soil is permeable, or because it is able to trap water in rock fissures. It may act similarly by preventing germination in seasons when rainfall is generally ineffective—for example, because soil is impermeable and evaporation potential is high. The temperature range within which germination can occur may serve as an indicator of the specificity of this control mechanism. Under laboratory conditions this range may appear too wide. Under field conditions, however, it might be a great deal narrower. Additional stresses, such as reduced aeration, reduced water potential, and salinity, tend to narrow the range considerably. Moreover, the micro-environment in the field changes rapidly—for instance, by drying out of the soil, by crust-formation, by increase in salinity, or by onset of competition. The lag-phase in onset of germination, as well as the rate of germination, now become critical for establishment and survival (Fig. 1). The optimal temperature-range for effective germination is thus narrowed down considerably.

A different form of control by temperature exists in seeds whose germination is triggered by an alternation in temperature. In *Oryzopsis miliacea* (Gramineae) dark germination at the optimal temperature (20° C.) does not exceed 31 %, but exposure of the imbibed seed for as little as 1 hr. to 30° C. increases germination to 64 %. There is practically no germination at 30° C. and diurnal fluctuations between these two temperatures increases germination even further. In the desert, diurnal fluctuations in temperature are an almost inevitable feature of the soil surface, because of the daily change in direction of energy flux. These fluctuations are increasingly damped with depth of soil, and the degree of damping depends on soil properties and soil moisture (Van Wijk & De Vries, 1963). Soil depth is of critical importance for a germinating seed, which has to strike a balance between two conflicting requirements that are a function of soil depth. One is the requirement for a dependable supply of soil moisture, the other is a need for rapidly reaching the soil surface and becoming autotrophic. The latter requirement is even more critical in soils which tend to form a crust. The mechanism underlying the requirement for temperature alternation is far from clear (Koller, Mayer, Poljakoff-Mayber & Klein, 1962). On the functional level, however, such a response-mechanism serves as an indicator of soil depth. A situation such as this has been described in the bulbils of *Allium ampeloprasum* (Liliaceae). These remain totally dormant at their normal depth (about 15 cm.) but sprout readily when planted more shallowly (3 cm.), in response to the greater amplitude in temperature fluctuations. This geophyte inhabits soils which are frequently disturbed by cultivation. The capacity for controlling dormancy by gauging soil depth replaces uprooted individuals and thus increases the potential for survival (Galil, 1965).

Competition for light does not play a major role in plant survival under desert conditions. On the other hand, the light regime in the soil is a function of soil depth and structure (Wells, 1959), as well as of its topography and stability (i.e. proneness to disturbance). Photocontrol of germination

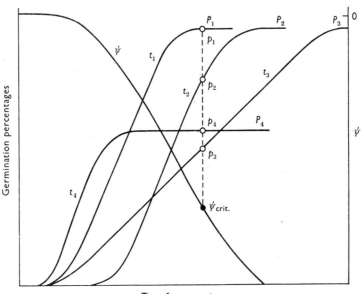

Fig. 1. Combined effect of temperature-dependent time-course of germination and of decreasing water potential (assumed to be independent of temperature) on germination percentages (schematic). At $\psi = 0$ (e.g. Petri dish), germination at temperatures t_1, t_2, t_3 and t_4 will be $P_1 = P_2 = P_3 > P_4$. In the field, ψ will decrease to its critical value for germination, ψ_{crit}, causing germination to stop at $P_3 < P_4 < P_2 < P_1$, because of differences in kinetics at the various temperatures (it is assumed that these kinetics are not appreciably affected above ψ_{crit}).

could thus lend selectiveness to germination with respect to depth of germination and physical characteristics of the soil habitat. It is conceivable, for instance, that requirement for light might be of value in soils which form crusts readily, or in soils which are frequently disturbed and the seeds may become buried too deeply for emergence. Similarly, inhibition by light might be of value in soil where surface conditions rapidly deteriorate and become unfavourable for growth. For instance, surface salinity may increase by capillary rise and evaporation of soil moisture, or moisture content in the surface layer may decrease by the combined action of infiltration and evaporation. This may explain the strong inhibition by light in seeds of *Calligonum comosum* (Koller, 1956) and *Citrullus colocynthis* (Koller, Poljakoff-Mayber, Diskin & Berg, 1963), which inhabit coarse sandy soils in the Negev desert. It may be of significance that in both instances the site of light sensitivity is located in the radicular tip of the dispersal unit.

Kinetic studies of germination have shown that though the receptor-pigment system may become fully operative soon after imbibition, build-up of its substrate(s) may be a lengthier process, thereby delaying the achievement of full responsiveness to light. An additional delay may be caused if full promotion requires prolonged irradiation. As this requirement is usually satisfied by several short irradiations repeated at intervals, it presumably arises because active phytochrome is rapidly inactivated thermally and requires repeated re-promotion (Borthwick & Cathey, 1962), or because phytochrome (Isikawa & Yokohama, 1962) or its substrate (Koller, Sachs & Negbi, 1964) are in dynamic equilibrium with a non-reactive precursor. On the functional level, a requirement is thus created for prolonged maintenance of the imbibed state, which could serve as a gauge forecasting availability of soil moisture.

Some of these aspects are illustrated by the germination responses of *Artemisia monosperma* (Compositae), a common evergreen shrub inhabiting bound and shifting sands in arid and semi-arid regions (Koller, Sachs & Negbi, 1964). The seeds have an absolute requirement for light but are extremely responsive to it—as little as 0·1 m.-candle-sec. sufficed to increase germination from 7% to 45%. Germination was promoted by light transmitted through a sand filter 2 mm. thick, when the source (mixed incandescent and fluorescent) supplied about 1·5% of full sunlight at filter level. Germination is thereby prevented in seeds buried too deeply by the shifting sands. The responsiveness to light increased with duration of imbibition, apparently by build-up of the substrate(s) for light action (Table 1). Even when full responsiveness to light had been reached, maximal germination could not be obtained by a single saturating irradiation, only by several such irradiations applied at intervals. The build-up of the substrate for light action was not affected by intervention of desicca-

Table 1. *Build-up of responsiveness to light in germination of* Artemisia monosperma (*from Koller, Sachs & Negbi*, 1964)

Irradiation		Final germination (%)		
Timing	Duration	10° C.	15° C.	20° C.
2nd day	1 min.	1	25	26
2nd day	100 min.	1	25	20
12th day	1 min.	24	60	58
12th day	100 min.	18	57	54

tion, and the response depended solely on the total duration in the imbibed state prior to irradiation (Table 2). Thus, in the field, the seeds will become increasingly responsive to light and increasingly germinable the longer the sand remains moistened. As rainfall is the sole natural source for soil moisture, this response-mechanism will act as an integrating rain-gauge. The fate of the seed will be determined by the duration of the imbibed state and by perception of light. Since soil moisture regime improves with depth, optimal soil depth for germination would increase with aridity of the climate, until the limits of light perception are exceeded. Beyond that point, it seems that germination must depend on shifting of the sands to expose fully responsive seeds to the promotive action of light.

Table 2. *Effects of desiccation on build-up of responsiveness to light in germination of* Artemisia monosperma *(from Koller, Sachs & Negbi, 1964)*

Days imbibed* before irradiation†	Final germination at 15° C. (%)	
	Dried	Undried
4	41	38
6	58	60
10	67	59

* Not counting 48 hr. desiccation 2 days before irradiation.
† Ten min.

The photocontrol of germination by light of spectral-energy distribution similar to that of the sun ranges from promotive, through neutral, to inhibitory. Light responses may be strongly modified by temperature (Koller *et al.* 1962), as well as by water potential (Kahn, 1960). In some cases white light has a dual action: promotive to begin with, and becoming inhibitory as irradiation is prolonged (Negbi & Koller, 1964). While there is scarcely any doubt that the promotive action of light is due to phytochrome activation, there is still controversy as to whether the inhibitory action is due to phytochrome, to a different pigment system, or to a modified form of phytochrome (Mohr & Appuhn, 1963; Hartmann, 1966; Rollin & Maignan, 1966; Negbi, Black & Bewley, 1968). On the functional level, this imposes a photoperiodic control over germination, as in *Atriplex dimorphostegia* (Chenopodiaceae) (Fig. 2). The photoperiodic response may be strongly modified by temperature, as in *Hyoscyamus desertorum* (Solanaceae) (Fig. 3). A temperature-sensitive photoperiodic control mechanism may easily be envisaged as capable of exerting delicate seasonal control over germination, particularly if one bears in mind that within the soil, photoperiod may become strongly attenuated with depth, as optical path length in the soil and reflection from its surface change through the day

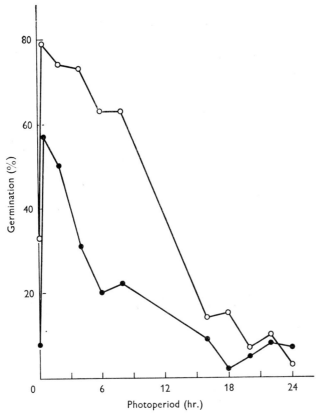

Fig. 2. Germination of 'flat' (○) and 'humped' (●) fruits of *Atriplex dimorphostegia* at 15° C., as influenced by photoperiod.

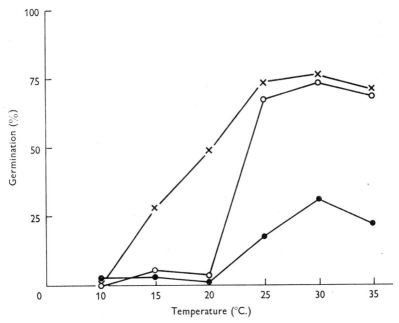

Fig. 3. Effects of temperature on germination of *Hyoscyamus desertorum* seeds in darkness (●), continuous irradiation (○) and in intermittent irradiation of 10 min. light at 24 hr. intervals (×). (From Roth-Bejerano, 1967.)

with angle of incidence of sunlight. The effective photoperiod will therefore become shorter with depth.

The transient and uncertain nature of soil moisture supply is by far the predominant stress facing plant survival in the desert. The role which moisture relationships might play in the control of seed germination is therefore of particular interest.

Control of germination by hydration is of survival value only if the mechanism is able to provide information on the environment, other than mere indications of the availability of sufficient water for hydration of the seed itself. The most clear-cut case is that of seeds that germinate to a greater extent in moist soil which had been previously exposed to rain, than in similar moist soil which had not (Went, 1949). Such seeds apparently contain water-soluble, water-leachable chemicals which inhibit germination. In wet soil they will diffuse out of the seed, but their concentration gradient will be steeper and their diffusion therefore more rapid under advective conditions. This most commonly happens when water moves through the soil by gravity, after rainfall. Clearly, both velocity and duration of flow will play decisive roles in determining the rate at which concentration of the inhibitor will diminish, and consequently the time when the concentration will cease to be inhibitory. This mechanism therefore meters the amount of water moving into the soil, as well as its flow rate. It thus serves as a gauge for the availability of soil moisture for future use by the seedling.

A different type of mechanism, which controls the potential capacity of certain papilionaceous seeds to germinate according to preceding moisture conditions, has been described by Hyde (1954). In such seeds, the pathway for entry of water (the hilar fissure) is capable of changing its width by hygroscopic movement. It is also apparently coated with a water repellent. Its hygroscopically controlled width therefore determines whether or not liquid water enters the seed, but allows exchange of water vapour. In atmospheres of high, or gradually increasing, humidity the incoming water vapour imbibes the colloidal contents of the seed and thus leads to distension of the coat and passive opening of the pore. When the dimensions of the pore reach a critical value, liquid water can enter the seed (Fig. 4). This condition is, however, reversible by drying. Redried seeds will be as impermeable to liquid water as mature, unhumidified seeds. In other species, such as *Lathyrus*, there does not appear to be a critical threshold beyond which liquid water can enter the seed. Atmospheric humidity does not affect the percentage of imbibable seeds, only the timing of the onset of imbibition and its rate (Table 3). The different behaviour of the two *Lathyrus* spp. from that of *Medicago laciniata* may be due to their lack of water-repellent and possibly lack of hygroscopic control over the pathway(s) for water entry. On the functional level, the two mechanisms may be of survival value in habitats where effective rainfall is usually preceded by an

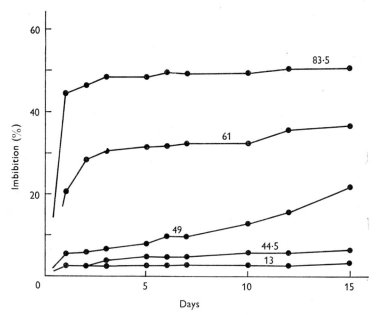

Fig. 4. Time-course of imbibition (percentage of seeds swelling) of *Medicago laciniata* seeds after 2 months storage at various relative humidities (percentages indicated in drawing).

Table 3. *Kinetics of imbibition in* Lathyrus *seeds as affected by humidity during storage*

Relative humidity during 2 months storage at 25° C. (%)	Imbibition*			
	L. hierosolymitanus		*L. aphaca*	
	Onset (days)	Rate (%/day)	Onset (days)	Rate (%/day)
13·0	4·2	11	7·9	14
44·5	1·9	13	5·3	11
49·0	1·3	13	4·5	15
61·0	0·6	20	2·4	14
74·0	0·4	25	1·9	13
80·5	0·3	60	0·8	20
83·5	0·2	94	0·2	93

* Final percentages 100 in both spp.

increase in humidity of the soil atmosphere. However, the *Medicago* type will allow prompt and uniform germination of a certain proportion of the seed population, which is determined solely by the preconditioning humidity and not at all by how long the soil remains moistened by the rain. In the *Lathyrus* type, on the other hand, the preconditioning humidity determines the degree of dependence of germination on how long the soil remains moistened by rain.

Panicum turgidum (Gramineae) inhabits extremely arid deserts. Its grains may be capable of obtaining an estimate of availability of moisture in the coarse sandy soils of the natural habitat by sampling the surface energy of the water. The dispersal unit is flat on one side and convex on the other. Dark germination at 20° C. was 39% and 88% respectively when the moist substrate was in contact with the convex or flat sides, and 91% when it was in contact with both. Decreasing the surface tension by a surfactant (0·01% Tween 80) increased germination (of randomly placed grains) from 57 to 86%. Apparently the dispersal unit is coated with a water-repellent, which helps to localize the pathways for entry of water into the grain, and thereby makes germination dependent on the degree of contact with water (Koller & Roth, 1963).

Moisture regime controls the subsequent capacity for germination also in seeds of the desert water-melon, *Citrullus colocynthis* (Cucurbitaceae). These seeds imbibe readily, yet are almost incapable of germinating unless first subjected to several diurnal cycles of wetting and drying. Seven cycles of 12 hr. wetting at 15° C. alternating with 12 hr. drying (over $CaCl_2$) at 37° C. increased subsequent dark germination from under 10% to 100% (Shur, 1965). This plant inhabits coarse sandy soils in the most arid parts of our deserts. In its natural habitat diurnal cycles of condensation of water vapour may occur in the subsurface layers of the soil as a result of the inversion in the temperature gradient. If such diurnal cycles usually precede conditions favourable for establishment, their perception is of survival value, by enabling the seed to predict these favourable conditions.

The role of heterogeneity

There is a natural tendency to treat physiological heterogeneity within a population as a manifestation of genetical variability. Some of it undoubtedly is. However, quantitative variability in dormancy may be a deliberate countermeasure against the 'uncertainty factor' in survival. We shall examine a few of the more overt examples.

Seeds with water-impermeable coats are common in the Leguminosae. Under field conditions such seeds gradually become permeable, sometimes over a period of several years. This does not take place in a random fashion. On the contrary, within the same pod the seeds show a gradient in responsiveness to such 'weathering', and the gradient may be opposite in direction in different species (Table 4).

Table 4. *Positional differences in susceptibility of water-impermeable seeds to weathering*

	Position of seed in pod†	Imbibition after weathering (%)*		
		0 months	5 months	17 months
Hymenocarpus circinnatus	I	4	8	43
	II	8	69	95
Onobrychis cristagalli ssp.	I	1	2	2
eigii	II	0	2	2
	III	2	3	79
O. crista-galli var. *sub-*	I	0	1	2
inermis	II	0	1	6
	III	0	83	99
Medicago tuberculata	I	20	98	100
	II	0	10	95
	III	0	0	40
	IV	0	0	7
	V	0	0	2
	VI	0	0	0

* At Be'er Sheva, Israel, under 10 mm. loess, protected from rain.
† Starting from base.

Asteriscus pygmaeus (Compositae) is a dwarf desert annual. The seeds are retained on the dead mother plant by infolding of the involucral bracts over the receptacle of the inflorescence, which is situated at the soil surface. These bracts are hygroscopic, opening up and exposing the achenes when moist, closing again when dried. When the achenes are exposed, the peripheral ones are detachable, but only by splashing raindrops. The detachable achenes are non-dormant. The undetachable remainder are dormant, even when excised from the receptacle, but they imbibe freely and their embryos are not dormant. After the peripheral achenes had become detached and the drying bracts had closed up over the receptacle, the newly peripheral achenes become detachable and non-dormant. The sequence appears to be obligatory. The changes which make the achenes detachable and non-dormant are apparently oxidative, since they can be artificially induced by oxidants (sodium hypochlorite, hydrogen peroxide, bromine water). It is likely that dispersal of the peripheral achenes opens the way for such oxidative processes in the newly peripheral achenes. At the functional level, repeated attempts at propagation and establishment are made during consecutive rainfalls, but material for propagation is metered out sparingly.

Heterogeneity may express itself also in quantitative differences in specificity. *Atriplex dimorphostegia* (Chenopodiaceae) is a desert annual. Each individual produces an array of dispersal units, ranging from those with completely flat bracts to those with completely humped ones. Germination in this species exhibits a photoperiodic response, but marked differences in this response exist between the 'flat' and the 'humped' types of fruit (Fig. 2). The 'flat' fruits are much more responsive than the 'humped' fruits to the promotive action of a single short irradiation. This response was markedly enhanced by pretreatment at a supra-optimal temperature, which nearly eliminated the difference between the two types of fruit (Table 5). Obviously, a stronger block to the expression of phytochrome action exists in the 'humped' fruits. When the efficiency of the promotive photoreaction was increased by dark-pretreatment at a supra-optimal temperature, the inhibitory effect of prolonged irradiation was diminished. However, the 'humped' fruits required a much longer pretreatment in order to increase efficiency of the promotive photoreaction to the extent of complete suppression of the inhibitory photoreaction (Fig. 5).

Table 5. *Enhancement of promotive effects of irradiation on germination of Atriplex dimorphostegia seeds by high-temperature pretreatment*

Temperature during pretreatment†	Germination* at 10° C. (%)			
	'Flat' fruits		'Humped' fruits	
	Dark	Irradiated‡	Dark	Irradiated
10° C.	30	58	13	16
25° C.	28	97	15	86

* Dark germination of both types of fruit at 25° C. did not exceed 2%.
† Nineteen days incubation.
‡ Thirty min. at end of pretreatment.

Heterogeneity in dormancy may be linked to other properties which add to the survival potential. *Gymnarrhena micrantha* (Compositae) provides an example of this kind (Koller & Roth, 1964). It is a dwarf desert annual which produces two types of inflorescence: a 'subterranean' type, which is embedded within the underground tissues of the mother plant and carries 1–2 large achenes, and an 'aerial' type at the soil surface, which carries numerous small achenes. The former are never dispersed, but germinate through the dead maternal tissues, forming colonies. The latter become detached by a series of complex hygroscopic movements, which necessitate several cycles of wetting and drying, and are dispersed by wind. When these germinate they form the nucleus for new colonies. Quantitative differences exist between the germination responses of the two types of achenes: the subterranean ones start germinating after a shorter incubation and their germination is less sensitive to inhibition by high temperature. In darkness, the germination falls off almost linearly with increase of temperature from 10° to 25° C., in the aerial achenes from 82% to 4% and in the subterranean achenes from 74% to 30%, corresponding to 5·2% and 2·9% per degree, respectively. In light, germination is reduced from 88% to 38% in the aerial achenes and from 95 % to 87 % in the subterranean ones as temperature is raised from 25° C. to 30° C.; this is 6·3 % and 1·6 % per degree respectively. The subterranean achenes are not only larger than the aerial ones, they also give rise to bigger seedlings. However, their initial relative growth-rate is slower, so that eventually the plants reach the same size (Table 6). Finally, seedlings from the aerial achenes are much more susceptible to adverse

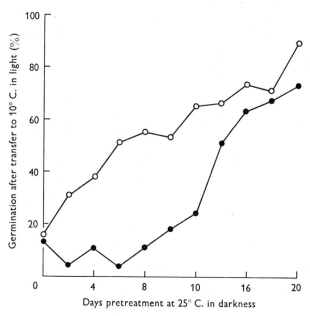

Fig. 5. Germination of 'flat' (O) and 'humped' (●) fruits of *Atriplex dimorphostegia* in continuous irradiation at 10° C., as influenced by duration of dark-pretreatment at 25° C.

Table 6. *Early growth of seedlings from aerial and subterranean achenes of* Gymnarrhena micrantha *(from Koller & Roth, 1964)*

	Aerial achenes	Subterranean achenes
A. Fresh wt. (mg.) unimbibed achene	0·4	6·5
B. Fresh wt. (mg.) seedlings*	4·4	26·3
C. Relative growth: (B−A)/A	11	4

* After 6 days growth (from germination) at 15° C. in darkness.

conditions of soil moisture than those from subterranean achenes, exhibiting a higher mortality at reduced soil moisture content as well as in response to withdrawal of irrigation (Table 7). The aerial achenes are clearly adapted to dispersal of the species, by virtue of their greater abundance, their efficient dispersal mechanism and their more specific requirements for germination. The subterranean achenes, on the other hand, are as clearly adapted to survival of the species, because of their smaller number, their

Table 7. *Seedling mortality in* Gymnarrhena micrantha, *as influenced by soil water content, and by duration of withdrawal of irrigation (from Koller & Roth, 1964)*

	Seedling mortality (%)	
	From aerial achenes	From subterranean achenes
A. Soil moisture content*		
0·15	4	0
0·10	12	0
0·06	42	8
0·05	54	12
B. Days without irrigation†		
1	29	0
3	71	0
5	88	0
7	100	38

* g. water/g. dry loess soil, readjusted daily.
† Irrigated daily to field capacity before and after withdrawal.

less specific requirements for germination, their higher tolerance of unfavourable soil moisture conditions, and the fact that they germinate in the exact spot where their mother plant had successfully completed its own life-cycle. As a matter of fact, the subterranean achenes simulate the vegetative regeneration buds of a perennial.

THE ROLE OF REGENERATION BUDS*

Hordeum bulbosum is a perennial grass inhabiting shallow rocky soil in semi-arid Mediterranean climates and exhibits a typical summer dormancy. Its short growing season coincides with the rainy winter, at the end of which the plant flowers, produces seeds, and its foliage and flowering stalks then dry out completely. Growth is resumed at the onset of the next rainy season, from specialized subterranean axillary buds situated at the base of each flowering stalk, in association with the thick internode which forms the 'bulb'. The association between the spike, the bulb and the regeneration buds is obligatory. The positional relationships between these organs are also rather stable (Fig. 6). There are relatively small environmentally inducible variations in the number of regeneration buds above and below the bulb, and in the number of elongated internodes between it and the spike. Formation of these organs therefore appears to be an integral part of the reproductive sequence of the individual shoots. The initiation of the reproductive sequence takes place more rapidly during long days, and is accelerated by vernalization, as well as by initial exposure to short days (Koller & Highkin, 1960). It is almost completely suppressed by high

* Based on Ofir, Koller & Negbi (1967) and Ofir (unpublished).

temperature (Table 8). Until the shoot becomes reproductive, it continues to produce non-dormant axillary buds, which are rapidly activated and grow into tillers. As soon as the apex becomes reproductive, and under certain conditions even before it becomes visibly so, the bulb is initiated and its associated axillary buds develop into specialized regeneration buds. These remain dormant for a period of 6–9 weeks, after which they start growing into tillers, if soil moisture is still available. As a result, the plant as a whole appears to be non-dormant under irrigation. Under field condi-

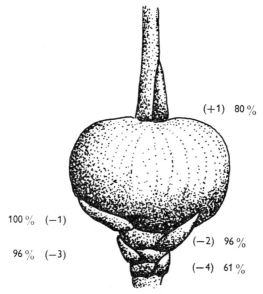

Fig. 6. Bulb (defoliated) and associated regeneration buds on rhizome of *Hordeum bulbosum* (diagrammatic). Percentages represent frequencies of occurrence per bulb in indicated nodal position (numbers in parentheses; in relation to bulb internode). (From Ofir, Koller & Negbi, 1967.)

tions, however, soil water potential eventually becomes too low for the activation of these buds, after which the entire plant becomes dormant.

When dormancy is periodically sampled throughout the summer by supplying water to excised bulbs, the regeneration buds first become capable of activation at lower temperature (Fig. 7). However, while high temperature was most inhibitory at the beginning of summer dormancy, at its termination, when percentages became equally high at low and high temperature, rate of activation was by far the highest at the highest temperature (Table 9). In other words, dormancy imposes the most stringent restrictions specifically on the capacity for most rapid activation, i.e. a high temperature.

Table 8. *Effects of photoperiod and temperature on tillering and on bulb formation in* Hordeum bulbosum *(from unpublished data of Ofir)*

Photoperiod	8 hr.		16 hr.	
Temperature regime*	22°/16° C.	28°/22° C.	22°/16° C.	28°/22° C.
Shoots per plant†	5·8	6·0	6·2	7·7
Bulb-containing shoots†	60%	3%	97%	12%

* Eight-hr. daily at the higher temperature, coinciding with 8 hr. photoperiod, midways at 16 hr. photoperiod.

† 95 ± 20 days after planting.

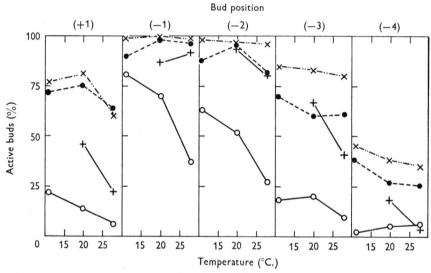

Fig. 7. Seasonal changes in potential of the different regeneration buds of *Hordeum bulbosum* to become active at different temperatures. Sampled on 19 May (○), 16 June (+), 20 July (●) and 14 Sept. (×). (From Ofir, Koller & Negbi, 1967.)

Though the plants have no live organs above the ground throughout the summer, viability of their buds depends absolutely on a continuous moisture supply from deep layers of the soil (possibly situated in rock crevices and fissures), by the seemingly dry root system. Excised bulbs, or bulbs on plants which remain *in situ* but whose root system has been artificially disconnected, lose about 90% of their moisture content and their regeneration buds lose viability entirely. Bulbs on intact plants *in situ* maintain a moisture content of about 100% (dry weight basis) throughout the summer, during which time their buds go through the process of relaxation of dormancy. Throughout this season the non-dormant fraction of the bud population can be induced to sprout by the mere supply of moisture at a suitable temperature. Thus, in addition to the endogenous control of their dormancy, the physiology of the regeneration buds is also governed by a

Table 9. *Seasonal changes in potential rates of sprouting and rooting of regeneration buds of* Hordeum bulbosum *at different temperatures (from Ofir, Koller & Negbi, 1967)*

		Rate of activation* (%/day)		
	Sampling date	11° C.	20° C.	28° C.
Rooting	19 May	3·5	3·7	1·7
	20 June	5·0	9·4	7·9
	14 Sept.	8·7	9·8	19·2
Sprouting	19 May	2·4	2·9	1·3
	20 June	7·3	13·2	13·7
	14 Sept.	12·8	20·0	25·3

* During approximately linear phase.

fine balance of their water potential. At a certain critical level they maintain viability and are able to relax the endogenous control of their dormancy; below that level they become non-viable, while above that level the non-dormant buds sprout and commence active growth.

Under field conditions in the natural habitat the reproductive phase, which includes the formation of regeneration buds, is initiated by the transition from the low temperatures and short days of winter to the longer days of spring. The dormancy which ensues in the regeneration buds is quite flexible, allowing the buds to sprout after a certain delay, provided soil water potential remains sufficiently high and temperatures remain sufficiently low. With the advent of summer, the entire plant becomes dormant. During the latter half of summer the capacity of the regeneration buds to sprout most rapidly at high temperature is unmasked. This capacity can thus be realized in the very first autumn rains, when temperatures may still be high. Growth will remain strictly vegetative until temperatures fall sufficiently low to permit the initiation of the reproductive phase. Such a mechanism of dormancy therefore allows the maximal use to be made of the favourable soil moisture conditions for as long as they prevail, and contribute to the survival of the species in a semi-arid environment.

DORMANCY AND SURVIVAL OF PLANTS IN NON-ARID ENVIRONMENTS

We have speculated at some length on the possible role of some germination-regulating mechanisms for the survival of plants under desert conditions. We must now put these speculations in their proper perspective by examining briefly the regulation of germination and its relation to survival under non-arid conditions. So far the most intensive study has failed to identify any single germination-regulating mechanism which is specific to the desert environment. Regulation by temperature level as well as by temperature alternation, regulation by presence, absence, or periodicity of light, have all been described in species of diverse habitats. Paradoxically, even the various types of regulation by moisture which have been dealt with above with respect to their survival value under desert conditions are not restricted to desert environments. However, we need not be discouraged by this apparent lack of habitat specificity in the ecological distribution of the different germination-regulating mechanisms. This lack of specificity does not disprove their value for survival. Our understanding of the regulation of germination has so far not progressed beyond the recognition and study of some of the components of the complex mechanism by which the seed can obtain as accurate an estimate as possible of its physical environment. The components by which the seed can measure temperature, estimate its own depth in the soil (or water), the intensity, spectral composition or periodicity of light, or the water economy of its micro-environment, or determine the physical properties of the surrounding soil, are all the same everywhere. Under some environmental circumstances it may be of greater importance to identify a certain factor more positively, or with greater accuracy than others. This will express itself in the predominance of a certain component in the regulatory mechanism.

The survival value of seed dormancy may be evaluated by at least the two following criteria. A regulatory mechanism should be one which prevents germination from taking place in an unfavourable environment, without impairing viability of the seed and its capacity to germinate subsequently, under the proper circumstances. Most of the regulatory mechanisms which have been discussed above answer this criterion, though not always in a simple straightforward manner. A more difficult criterion to determine is the actual reduction in survival potential which would be incurred in a species which is well adapted to a certain habitat, by eliminating from its genetical complement the germination-regulating mechanism being studied. Should it be possible to produce such mutants, it would be advisable to remember that they should be tested not under the average environmental conditions of the habitat, but under the most extreme which may be encountered in it, even at intervals of several years.

REFERENCES

AMEN, R. D. (1968). *Bot. Rev.* **34**, 1.
ASCHOFF, J. (1960). *Cold Spring Harb. Symp. quant. Biol.* **25**, 11.
BORTHWICK, H. A. & CATHEY, H. M. (1962). *Bot. Gaz.* **123**, 155.
EVENARI, M., KOLLER, D. & GUTTERMAN, Y. (1966). *Aust. J. biol. Sci.* **19**, 1007.
GALIL, J. (1965). *Israel J. Bot.* **14**, 184.
HARTMANN, K. M. (1966). *Photochem. Photobiol.* **5**, 349.

Heslop-Harrison, J. (1963). Meristems and differentiation. *Brookhaven Symp. Biol.* **16**, 109. Upton, New York: Brookhaven Laboratory (Publications).
Hyde, E. O. C. (1954). *Ann. Bot.* N.S., **18**, 241.
Isikawa, S. & Yokohama, Y. (1962). *Bot. Mag., Tokyo* **75**, 127.
Itai, C., Richmond, A. & Vaadia, Y. (1968). *Israel J. Bot.* **17** (in the Press).
Juhren, M., Went, F. W. & Phillips, E. (1956). *Ecology* **37**, 318.
Kahn, A. (1960). *Pl. Physiol., Lancaster* **35**, 1.
Khan, A. A. (1966). *Physiologia Pl.* **19**, 869.
Koller, D. (1956). *Ecology*, **37**, 430.
Koller, D. & Highkin, H. R. (1960). *Am. J. Bot.* **47**, 843.
Koller, D., Mayer, A. M., Poljakoff-Mayber, A. & Klein, S. (1962). *A. Rev. Pl. Physiol.* **13**, 437.
Koller, D., Poljakoff-Mayber, A., Diskin, T. & Berg, A. (1963). *Am. J. Bot.* **50**, 597.
Koller, D. & Roth, N. (1963). *Israel J. Bot.* **12**, 64.
Koller, D. & Roth, N. (1964). *Am. J. Bot.* **51**, 26.
Koller, D., Sachs, M. & Negbi, M. (1964). *Pl. Cell Physiol., Tokyo* **5**, 85.
Letham, D. S. (1967). *A. Rev. Pl. Physiol.* **18**, 349.
Mohr, H. & Appuhn, U. (1963). *Planta* **60**, 274.
Negbi, M., Black, M. & Bewley, J. D. (1968). *Pl. Physiol., Lancaster* **43**, 35.
Negbi, M. & Koller, D. (1964). *Pl. Physiol., Lancaster* **39**, 247.
Ofir, M., Koller, D. & Negbi, M. (1967). *Bot. Gaz.* **128**, 25.
Rollin, P. & Maignan, G. (1966). *C. r. hebd. Séanc. Acad. Sci., Paris* **263**, 756.
Roth-Bejerano, N. (1967). Ph.D. thesis, Hebrew University, Jerusalem. (Hebrew.)
Sachs, T. & Thimann, K. V. (1967). *Am. J. Bot.* **54**, 136.
Shur, I. (1965). M.Sc. thesis, Hebrew University, Jerusalem (Hebrew).
Smith, H. & Kefford, N. P. (1964). *Am. J. Bot.* **51**, 1002.
Van Overbeek, J. (1968). *Scient. Am.* **219**, 75.
Van Wijk, W. R. & De Vries, D. A. (1963). In *Physics of Plant Environment*, ch. 4. Ed. R. W. Van Wijk. New York: Wiley.
Vegis, A. (1964). *A. Rev. Pl. Physiol.* **15**, 185.
Wareing, P. F. (1965). *Handb. PflPhysiol.* **15** (2), 909.
Wells, P. V. (1959). *Science, N.Y.* **129**, 41.
Went, F. W. (1949). *Ecology* **30**, 1.

THE DORMANCY AND SURVIVAL OF PLANTS IN THE HUMID TROPICS

By K. A. LONGMAN

Department of Botany, University of Ghana, Legon*

INTRODUCTION

It is widely believed that plants in tropical forests grow continuously, producing new leaves and stems without any period of rest. While this is probably true of a small proportion of species, including, for instance, a number of palms, conifers and tree ferns, the great majority of the trees exhibit periodicity of shoot growth, once they have passed the seedling stage (Coster, 1923; Alvim, 1964). Despite the apparently favourable climate throughout the year, extension of new shoots may even be confined to a period of a few weeks only (Njoku, 1963), although careful records of individual specimens are necessary to demonstrate this, since they are often out of phase with each other.

This misconception of continuous growth in the tropical forest is due partly to the lack of precise observations and experiments, but also to a neglect of the early work by German and Dutch authors, in spite of reference books such as those by Schimper (1935) and Richards (1952). It must also be said that the net effect of hundreds of travel articles and books on the tropics is to give an impression of innumerable plants constantly straining upward towards the light in a dark, steamy and impenetrable jungle.

In fact the idea of an unvarying climate is as misleading as that of continuous growth. The microclimate in and around the crowns of the bigger trees, where of course most of the shoot extension growth takes place, shows considerable diurnal and seasonal fluctuations (Cachan & Duval, 1963). On the forest floor, of course, the conditions are usually more equable, but Jeník & Hall (1966) have recently recorded an exceptionally low minimum temperature of 12° C. in riverain forest in Ghana, only $8\frac{1}{2}°$ from the equator and 320 m. above sea-level. In a grassy clearing a diurnal temperature range as great as 26·8° C. was recorded, and it is evident that when the dry, northerly 'Harmattan' wind blows for a few days or weeks, extreme climatic changes may be experienced. Even in rain forests, the evaporative power of the air may occasionally approach or

* Present address: Forestry Commission Research Branch, Department of Forestry and Natural Resources, The University, Edinburgh 9, Scotland.

exceed the values typical of deserts (Jeník & Hall, in the press), while in monsoon forest and savanna areas there are sometimes abrupt climatic changes in addition to the normal dry and rainy seasons.

Against this background, a study of the physiology of dormancy is clearly relevant, as an experimental basis for ecological prediction as well as for its fundamental interest. Attention will be concentrated here on the factors affecting bud dormancy in forest trees, and those influencing seed dormancy in certain widespread, colonizing weeds of open, disturbed ground in the tropics. In the former case, it was already clear that a number of high forest species were very sensitive to changes in day-length (Downs, 1962; Njoku, 1964) and also night temperature (Longman, 1966). A fuller study of the ecology and physiology of the tropical forest and its environment will be published later (Longman & Jeník, 1970), but it may be mentioned here that there is only sparse evidence at the moment for the view that the growing and dormant phases of trees are controlled by endogenous rhythms (Bünning, 1948, 1956).

BUD DORMANCY

One of the problems inherent in growing sizeable tree seedlings under controlled environments is that it is difficult and very expensive to provide adequate light intensities, particularly on the lower leaves. In the experiments to be described, this was overcome by growing the plants in a lightly shaded glasshouse for approximately 7–8 hr. of the day, and then moving them to the appropriate growth room for the remainder of the 24 hr. Accurate control of day-length and night temperature was thus possible, with less precise temperature control in the glasshouse. In one such experiment, 6-month-old seedlings of *Terminalia superba* showed significant effects of photoperiod within 3 days, while after 4 weeks the plants receiving short days (9 hr. 10 min.) had made only one-fifth of the height increment of those receiving long days (17 hr. 10 min.) or short days with a light-break in the middle of the night (Fig. 1).

The greatly reduced growth made under short days in this case was due in part to the internodes being shorter. During the 4th–6th weeks, however, half the plants in this treatment ceased to produce new leaves, and their apices became dormant. A few weeks later, bud-break and renewed leaf expansion occurred, although they were leafy and still receiving short days. It appears that in this case the trees underwent pre-dormancy, to use the definitions suggested by Wareing (see p. 243 in this volume), but that this was only a 'light' or temporary form of pre-dormancy, since growth was resumed without the conditions having been changed.

Seedlings of eight other tropical trees have been found to show considerably reduced shoot-extension growth in response to shortened days and/or cooler night temperatures, both the number of leaves produced and the internode length being reduced in most cases. In the three species illustrated in Fig. 2, pre-dormancy occurred temporarily in a proportion of the seedlings receiving certain treatments. In *Ceiba pentandra*, for

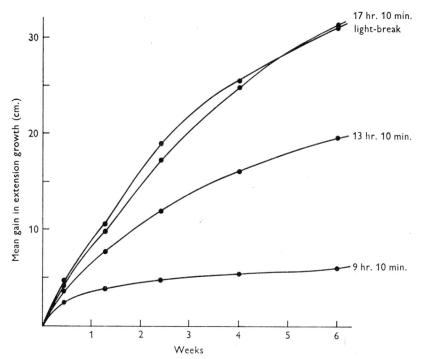

Fig. 1. *Terminalia superba*. Effect of day-length on shoot extension growth. 6-month-old seedlings, approximately 60–80 cm. in height at the start of the experiment. Greenhouse temperatures $31 \pm 3°$ C.; growth-room temperatures $32 \cdot 5 \pm 0 \cdot 5°$ C. Light-break treatment: 9 hr. 10 min. light plus a light break of 2 hr. in the middle of the night. Total light energy received was comparable in all treatments. Effect of day-length highly significant by 4th week.

instance, short days had this effect, while in *Gmelina arborea* night temperatures of 26° C. led to dormancy in some plants, while warmer temperatures did not. In the case of *Bombax buonopozense* there is some evidence of effects, and perhaps an interaction, of both these treatments.

The onset of dormancy in these tropical species thus appears to be favoured by the same factors—short days and cool nights—which reduce shoot growth in general. The photoperiodic responses of such plants are thus broadly similar to those of many temperate trees (Wareing, 1956; Nitsch, 1957), though the dormancy may well not be so deep in many cases.

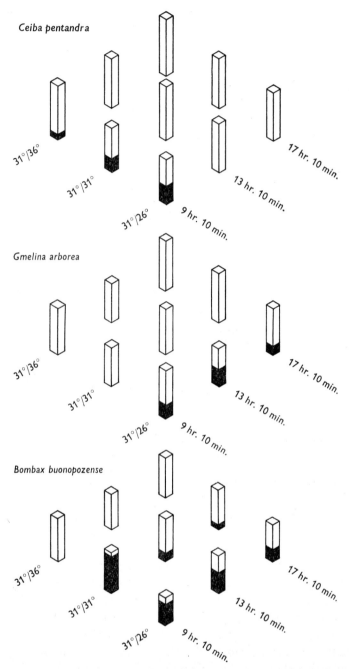

Fig. 2. Effect of day-length and night temperature on dormancy. The height of each histogram represents the number of trees in each treatment (between 8 and 15). The shaded zone indicates the number of trees which became recognizably dormant during the experimental period (3½–4 months). Greenhouse temperatures 31±3° C.; growth-room temperatures 31±0·5° C.

Furthermore, it is quite possible that the control of growth and dormancy in the tropical forest is mediated partly through natural changes in day-length and temperature. Care must be taken, however, in attempting to extrapolate from results with controlled environments. In the first place, the dormancy described in the foregoing experiments was temporary, the leafy plants resuming growth after a short period, while in the forest, as has already been indicated, long periods of dormancy occur. Secondly, although many of the temperatures used in these experiments were in the range commonly experienced in the tropical forest, most of the day-lengths were not, as the primary aim was to test whether there were any photoperiodic effects at all. It is interesting to note, however, that in Fig. 1 the growth under 13 hr. 10 min., equivalent to the longest day in southern Ghana, was intermediate between that attained by plants receiving temperate-region long days and short days, and significantly different from both. Moreover, Njoku (1964) has shown that seedlings of *Hildegardia barteri* are very sensitive to small changes in day-length, becoming dormant rapidly under $11\frac{1}{2}$ hr. days, more slowly under 12 hr., and not at all under $12\frac{1}{2}$ hr. It is interesting to note that rice has even been reported to respond to differences of a few minutes in day-length (Njoku, 1959).

The third difficulty with ecological predictions from such experiments is that they are carried out with seedlings, whose growth characteristics probably differ from those of large, mature trees of the same species. This can partly be overcome by the use of vegetatively propagated material which has been obtained from the upper part of an older tree. An experiment of this kind was carried out with *Cedrela odorata* L. by Miss A. H. Oppong, in which 'mature' plants, obtained by grafting buds from the crown of three 40-year-old trees on to seedling root-stocks were compared in their photoperiodic responses with seedlings pruned and then allowed to grow to the same size as the grafts ('juvenile').

The plants receiving short days became dormant and remained so, unlike the species already mentioned. As soon as most or all the leaves had senesced or fallen off, however, bud-break promptly took place and new shoots appeared while the plants were still under short days. Presumably, although the dormancy was more profound than in the previous cases, it was still pre-dormancy, imposed by the expanded leaves.

Under long days the rate of shoot growth was much greater, and juvenile plants did not stop growing. The mature plants, on the other hand, grew in a series of rapid 'flushes', separated by periods of terminal bud formation. It should be noted that the curves shown in Fig. 3 are the resultant of a number of replicates which were not always in phase with each other, which naturally blurs the picture. Nevertheless, the important conclusion can be drawn that there are distinct differences in the growth habits and responses to photoperiod of the budded plants and seedlings, the mature tissue apparently responding more quickly and to a greater extent than the juvenile to a change in day-length. The growth habit of mature grafts was also studied under natural day-lengths (about 12½–13 hr.) after the completion of the experiment, when it was observed that the cycles of flushing and rest continued, leaving clear evidence on the shoots of inter-

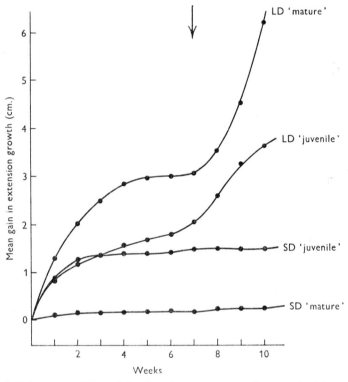

Fig. 3. *Cedrela odorata*. Effect of day-length on shoot extension growth of 'mature' and 'juvenile' plants. Date of budding of 'mature' plants 8 weeks before start of experiment. Long days (LD), 17 hr. 10 min.; short days (SD), 9 hr. 10 min. Greenhouse temperatures $31 \pm 3°$ C.; growth-room temperatures $31 \pm 0.5°$ C. Arrow marks date of addition of extra soil to all pots. Effect of day-length highly significant in 'mature' plants (10th week).

mittent activity (see Plates 1(*a*) and (*b*)). Such a growth habit was never observed in potted seedlings, which did not become dormant.

The foregoing experiments were designed primarily to investigate the onset of bud dormancy. Bud-break, which signals the re-starting of shoot extension growth, was studied only where plants had been rendered dormant by short-day treatment. Nevertheless it seems clear that in mature plants of *Cedrela*, for instance, bud-break will occur under long days and under natural day-lengths of about 13 hr., but does not take place under short days unless the plants have lost most or all of their leaves.

Confirmation of the importance of photoperiod in bud-break, and of the inhibitory effect of the mature leaves, is provided by an experiment carried out by Mr A. A. Enti with *Hildegardia barteri*. Seedlings of this species show dormancy even in the first year, and one series of leafy and one of leafless plants were selected which were dormant at the time. The leafy plants showed bud-break and continued shoot growth in long days, but did not break their dormancy under short days until, as in *Cedrela*, the old leaves were senescent or had fallen off. Bud-break then occurred, followed

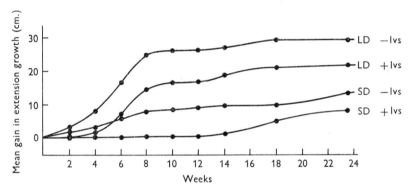

Fig. 4. *Hildegardia barteri*. Effect of day-length on shoot extension growth. All plants dormant at start of experiment, half being leafy (+lvs) and half deciduous or nearly so (−lvs). Experimental conditions as for Fig. 3. Effects of both leaves and day-length highly significant by 10th week. (LD, long day; SD, short day).

by a limited amount of shoot growth. Where leaves were absent at the beginning of the experiment, bud-break took place at the same time under both long and short days, though the growth was much greater in the former case (Fig. 4).

DORMANCY AND SURVIVAL IN THE TROPICAL FOREST

It is difficult to generalize about the survival value of these dormancy mechanisms, since the basic physiological and ecological data is as yet very scanty. In addition, of course, there are many different types of tropical forest, and hundreds of species of woody plant growing in them, some of which may reach the upper part of the canopy, while others never do. As has been seen, the microclimate in the crowns, on the forest floor and in gaps is generally very different, and may vary diurnally and seasonally. Moreover, the response of a mature shoot to a particular environment may not be the same as that of a seedling. Nevertheless, two points of general application emerge from records and observations in the tropics. First, it is clear that the length of the dormant period typically increases as a seedling becomes older. Thus 3-year *Bombax buonopozense* was dormant for 3–4 months, while older trees did not make any shoot extension growth for 9–10 months (Njoku, 1963, 1964). Secondly, there is abundant evidence that many older trees have their rather short period of shoot growth at a particular time of year, which is generally not the wettest period. Where there is a distinct dry season, as for example in eastern Java, some of the trees may show peaks of bud-break at the start of the rains, while others flush well before the rains start (Coster, 1923). A 'pre-rains flush' has also been reported from East Africa by Jeffers & Boaler (1966), and is very marked in West Africa, occurring chiefly in February and March (Taylor, 1960; Njoku, 1963), while a few species show a second growing season about October; see for instance Greenwood & Posnette (1949) for a study of cocoa under shade. Alvim (1964) has drawn attention to the fact that in many parts of the tropics flushing occurs particularly at the equinoxes, though other work, notably that of Holttum (1953, 1968), shows that there are also trees, particularly where climatic fluctuations are smaller, which are out of phase with the 12-month cycle.

A very provisional attempt is made in Table 1 to indicate the chief seasonal changes in a monsoon forest in Ghana, and climatic fluctuations which may or may not be correlated with them. It can be seen from this that the young leaf is most likely to reach full size and maximum photosynthetic capacity in the equinoctial period preceding the rainy season. It is possible that an ecological explanation for this may be found in the fact that both temperatures (Clarke, 1966) and total radiation (Black, 1966) tend to show peaks at the equinoxes. Survival amongst the intense competition of the tropical forest may well depend to a considerable extent on efficiency in utilizing the total radiation, and it may perhaps be that for a sizeable, well-rooted tree radiation is a more critical factor than water.

The physiological factors responsible for 'triggering' bud-break in advance of the rains could possibly be the increasing day-length, or the greater temperatures or diurnal range of temperature (see Hardy, 1964; Murray & Sale, 1967). Similarly, the onset of dormancy might be controlled partly by decreased temperature, though it is likely that internal factors within the shoot are important. Day-length cannot be considered here, except in the case of young plants, since it is still increasing when the shoots stop growing.

Other processes related to bud dormancy may be briefly touched upon. The factors affecting leaf-fall appear to be very diverse, and the timing of the process in nature to be rather variable. However, there seems to be a

Table 1. *Some seasonal changes in monsoon forest in Ghana*

Generalized picture for trees after the seedling stage

	Jan.	Feb.	Mar.	Apr.	May	June	July	Aug.	Sept.	Oct.	Nov.	Dec.
Bud-break			—	-						-		
Onset of dormancy										-	-	
Leaf-fall (deciduous)	-	-	-							-	-	-
Leaf-fall (leaf-exchanging)			-	-					-	-		
Leaf-fall (evergreen)	-	-	-	-	-	-	-	-	-	-	-	-

Climatic fluctuations

	Jan.	Feb.	Mar.	Apr.	May	June	July	Aug.	Sept.	Oct.	Nov.	Dec.
Mean temperature			Max	Max			Min			Max		Min
Total radiation							Min			Max		Min
Cloud and haze	Haze			Clear	-	-	Cloudy	-	-	Clear		Haze
Rainfall				-	-	Max			Decreasing	-		
Day-length	Min		Increasing			Max						Min
Diurnal range of temperature	Max					Min						Max

clear peak of leaf-fall in deciduous species in Ghana in November, in the early part of the dry season (Taylor, 1960), and in the predominantly evergreen trees of the rain forest in March (litter studies by Bray & Gorham, 1964). The timing of leaf-fall in relation to the time of bud-break is obviously important in that jointly they determine whether the tree or branch is deciduous, leaf-exchanging (with the two processes roughly simultaneous) or truly evergreen (Table 1). Broadly speaking, there is an increase in deciduousness with height in the forest canopy and with dryness of climate (Richards, 1952), but the roles of water stress and yearly photosynthesis in the survival of these three types of tree still await careful evaluation. Similarly, the physiological interactions of leaves and buds urgently require study.

Dormancy of the cambium has been demonstrated in normally leafless and experimentally defoliated trees by Coster (1927, 1928), who also prevented the resumption of cambial activity by disbudding and ringing. Very little is known about the relationship of bud dormancy to root growth, nor about the dormancy of flower buds, although this has an important bearing upon flowering times and thus on fruiting, seed dispersal and survival.

SEED DORMANCY

The ecological significance of seed dormancy in the tropics appears to be more closely related to survival during dry periods than is the case with bud dormancy. Where there is a distinct dry season, germination typically occurs when the rains begin in earnest, but the dormancy mechanisms do not necessarily involve a direct effect of the rain-water. For example, in *Hildegardia* the seeds from freshly shed fruits germinate freely, and when stored at 90% relative humidity (R.H.) Enti (1968) has shown that they retain this ability for at least 7 months. However, when they have been stored at 70% and 50% R.H. for 5 weeks, the subsequent germination percentage is much reduced, and the seeds which do germinate take as long as a month to do so. This secondary dormancy has been shown by Miss A. Ako-Addo to be due to the seed-coat becoming impermeable to water under the dry conditions, and it can be largely removed by storing the dormant seeds for a few weeks at 90% R.H.

The most likely interpretation of the ecological role of such a type of dormancy is that, on the steep, dry, rocky slopes occupied by *Hildegardia*, the seeds are soon rendered dormant by drying in the sun and 'Harmattan' wind. They are thus protected from germination in response to occasional storms or heavy dew. The moister conditions prevailing in the rainy season presumably remove the dormancy and allow germination at a time when there is some likelihood of survival. An additional practical conclusion from these results is that somewhat moister storage conditions might well be tried for other tropical forest seeds, many of which are known to be difficult to keep for more than short periods.

Seed dormancy mechanisms of various kinds are to be expected in many other species of forest and savanna which show delayed germination. There are also a number of cosmopolitan weeds which colonize lawns, roadsides and other open, disturbed ground in the tropics, of which two at least have been shown to possess unusual types of dormancy. The stimulus for these investigations was a study by Dr J. Jeník of the effect of watering 1m square plots on a lawn at Legon every day, at the height of the dry season, when only a very scanty cover of perennial plants existed. Amongst the seedlings appearing on the watered plots was the composite *Tridax procumbens*, which is an apparently mesophytic plant which nevertheless survives the extremely drying conditions on these 'lawns', probably because of the presence of absorbing hairs on the leaves and stems which take up considerable quantities of dew at night (Jeník & Longman, 1965). On the other hand the annual grass *Dactyloctenium aegyptium*, which had seeded heavily in the previous rainy season, and indeed appeared in very large numbers on the unwatered parts of the lawn when the next rainy season commenced 2 or 3 months later, occurred very sparsely on the watered plots.

The most likely explanation appeared to be that the breaking of dormancy in *Dactyloctenium* depended on some change in temperature associated with heavy rainfall which did not occur when the ground was merely watered. Preliminary experiments indicated that the 'seeds' would remain dormant for at least 10 weeks when imbibed in Petri dishes under laboratory conditions, but would then germinate freely if exposed to a diurnal temperature fluctuation of 15–$20°$ C. for a few days.

As it was anticipated that there might be changes during storage, freshly harvested seeds were stored at $-20°$, $20°$ and $40°$ C. for 10 weeks, and then sown in de-ionized water under continuous fluorescent light at three constant and three fluctuating temperatures. The histograms in Fig. 5 show the germination percentages finally achieved when the experiment ended after 6 weeks. The seeds which had been stored below freezing point showed virtually no germination at any constant temperature. A fluctuating temperature of $36°/20°$ C, however, led to over 95% germination, but none at all germinated in a series kept in the dark. Daily fluctuations of only $8°$ C. were less effective at lower temperatures and ineffective at higher temperatures.

Assuming no changes in the frozen seeds it appears then that freshly

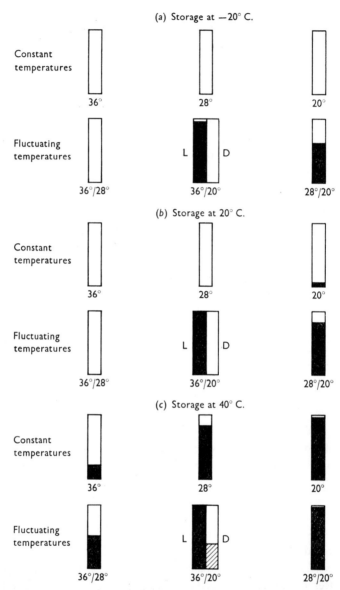

Fig. 5. *Dactyloctenium aegyptium*. Effect of storage and germination temperatures on seed dormancy. The height of the black part of the histograms is proportional to the germination percentage attained by the end of the experiment (6 weeks). Germination under continuous fluorescent illumination except for 36°/20° C., where D = continuous darkness; L = continuous light. Germination temperatures ±1° C.; fluctuation of temperatures: 8 hr. in warmer conditions, 16 hr. in cooler. (*a*) Stored at −20±3° C. for 10 weeks; (*b*) stored at 20±3° C. for 10 weeks; (*c*) stored at 40±2° C. for 10 weeks.

harvested *Dactyloctenium* has an absolute light requirement and a requirement for a substantial daily temperature change. In June and July, when the seeds are shed from the mature plants, the natural temperature fluctuation is at its lowest, owing to the generally cloudy conditions. It is reasonable to suppose that this keeps all but a very small proportion of the seeds dormant at a time when the area is fully occupied with competing vegetation, with the chances of survival to maturity therefore low.

Seeds which had been stored at 20° C. showed substantially the same pattern of germination as those which had been kept frozen, but there was evidence of a slight relaxation of dormancy. Germination at 28°/20° C. was 20 % higher, and a few seeds even germinated at a constant temperature of 20° C.

Storage at 40° C. for the same period, however, had a pronounced effect upon subsequent germination. The values for constant temperatures were nearly as high as those for fluctuating, and the chief influence of temperature is a high-temperature inhibition. In addition, the absolute requirement for light has been relaxed, as 40 % of the dark-sown seeds at 36°/20° C. germinated.

Storage under warm, dry conditions is well known as a method of 'after-ripening' of cereals, including tropical species (Roberts, 1962). In this case, requirements for both light and for temperature fluctuation were reduced by 10 weeks' after-ripening to such an extent that a substantial proportion of seeds were non-dormant over a wide range of conditions. It may be supposed that after-ripening takes place naturally in the dry season, during the day-time at least, and so many seeds are free to germinate when the rains start again.

Another effect of dry storage was to speed up the rate of germination. It can be seen from Fig. 6 that even under the most favourable germination temperatures seed previously stored in the deep-freeze was very slow to start. Most germination took place in the third and fourth week of imbibition and the final percentage was obtained only after 5 weeks. Seeds stored at 20° C. germinated some 3–4 days earlier than this, but those which had been after-ripened reached a value of over 50 % germination on the third day of imbibition. An interesting point is that attainment of the full germination percentage, although much quicker than with seeds not after-ripened, still took another 2 weeks. A possible explanation might be that not all the seeds had completed the after-ripening process, and that these still required the more lengthy period of fluctuating temperatures for germination. Some evidence for this is provided by the curve for seeds stored at 40° C. and germinated at constant 28° C. (dotted line), which is roughly comparable with a fluctuating temperature of 36°/20°. Here, although 80% of the seeds germinated in the first 3 days, very few did so later, indicating that some dormancy still remained.

Extracts from some data, recorded by Mr J. B. Hall, of germination and survival of *Dactyloctenium* seedlings on the lawn at Legon during 1967/8 are shown in Table 2. Seed had been shed naturally in June and July, in the rainy season, and periods of rainfall exceeding about 15 mm. between September and April generally led to the germination of some *Dactyloctenium*. It would appear that during each intervening dry spell, a further

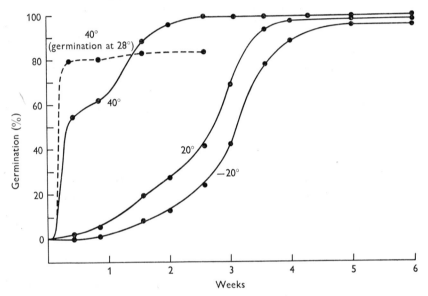

Fig. 6. *Dactyloctenium aegyptium*. Effect of storage temperatures on the rate of germination. All curves represent germination at 36°/20° C. except the dotted line for 28° C. constant. Conditions as in Fig. 5.

small proportion of the seeds became non-dormant, and it was noticeable that there was no further germination when rainy periods were separated by only a few days. Further studies are required to find out which type of release from dormancy is operating, but it seems on the available evidence to be more likely that it is after-ripening rather than fluctuating temperature which is important during the dry season. It was noticeable for instance that the time for first emergence of the seedlings on the lawn decreased from 3 to 4 days in September to 2 days in January and 1 day in late February and April, which is reminiscent of the hastening of germination found with seeds stored at 40° C. Moreover, it may be noted that although considerable diurnal fluctuation of temperature did occur throughout the period of record, the soil surface probably became dry again before the influence of the fluctuating temperatures could be felt.

Table 2. *Germination and survival of* Dactyloctenium *on a 10 cm.^{2} plot*

Date	Rainfall (mm.)	Min. temp.	Max. temp.	Dactyloctenium seedlings
Sept.				
11–13	Heavy	20·0	32·5	—
14	—	—	34·5	25
15	—	—	—	95
16–21	Showery	—	—	
26	9·4	—	—	20
Oct.				
7	20·3	22·5	—	
20	3·0	—	—	(Seedlings from Sept. and Oct. rain did not survive)
22	—	22·0	49·5	
Dec.				
15	3·8	22·8	—	
Jan.				
2–3	31·5	—	—	(New plot)
4	(Ground still moist)	19·0	32·7	(Just emerging)
5	—	18·9	—	123
18	(Harmattan: R.H. 28% at 14.30 hr.)	—	—	3
Feb.				
1–2	4·5	23·5	34·0	3
18–19	9·6	—	37·6	(Did not survive)
22	(Ground dry)	22·3	—	
27–28	19·5	21·4	34·8	(Just emerging 17.00 hr. 28th)
29	(Ground surface dry 16.00 hr.)	20·8	36·0	31
Mar.				
1	—	21·6	40·6	31
7	—	—	—	10
16–17	5·3	—	—	(Did not survive)
Apr.				
18	6·6	—	—	
19–20	35·5	22·2	36·0	(Emerging 10.00 hr. 20th)
22–23	21·9	19·8	36·3	150
24–25	48·2	22·0	37·3	
28–29	(Ground surface dry 15.00 hr.)	22·2	39·6	129 (to 4 cm. height)
May				
11–12	26·0	—	39·7	74 (no more new seedlings) (Many survived to maturity)

From records by Mr J. B. Hall in 1967/8 on a lawn at the University of Ghana, Legon. Temperatures (° C.): minimum thermometer laid on soil surface; maximum thermometer with bulb just buried.

In addition to these two types of dormancy there is the high-temperature inhibition, which presumably results in a higher rate of germination when the weather is generally cooler, and the light requirement, which will prevent germination in seeds which have become completely covered by soil. Altogether, this species is seen to be equipped with a formidable array of dormancy mechanisms which lead to a great deal of variability in the time of germination. As Koller points out elsewhere in this volume, such heterogeneity of seeds may be of considerable survival value, and in the competition between many annual and perennial species for the 'recolonization' of these dry lawns, a genotype which allows a little germination at any rainy period, while conserving the majority of the seeds as a reserve, is likely to survive.

Preliminary experiments with *Tridax procumbens* did not suggest any striking effects of temperature, but it was found that there was a near-absolute light requirement, only about 0·4 % of the 'seeds' germinating in the dark. Further work with seeds pre-soaked for a week in darkness showed that this was a quantitative light effect, since seeds given periods of light of up to 24 hr. and then returned to darkness showed only slight germination. After 2 days' continuous illumination the value was only 16 %, but over half the seeds germinated with 4 days' light, while after 8 days nearly three-quarters of them had done so (Table 3 *b*).

Table 3. *Tridax procumbens*

(*a*) Effect of the number of cycles of light (4 hr.) and darkness and cycle length on germination

Cycle length	Number of cycles		
	2	4	8
12 hr.	(1 day) 7·3	(2 days) 19·2	(4 days) 79·8
24 hr.	(2 days) 21·7	(4 days) 79·5	(8 days) 82·3
48 hr.	(4 days) 42·0	(8 days) 76·6	(16 days) 79·0

(*b*) Effect of continuous illumination

Number of days of light treatment		
2 days	4 days	8 days
15·6	54·0	73·6

The seeds were presoaked in de-ionized water for a week in the dark at 30° C. and any dark germinators removed. After exposure to fluorescent light (warm white) at 30° C. for the appropriate period(s), they were returned to the dark. The data above are the means of 10 replicates of 25 seeds, and represent the final germination percentages on day 17.

Light for a substantial number of days is thus needed to break the dormancy of most seeds, but the situation in *Tridax* is somewhat more complex than so far indicated, involving also a timing mechanism and promotion of germination by a dark period. In Table 3 (*a*) is given the germination resulting from cycles of 4 hr. light and 20 hr. dark given for 2, 4 and 8 days, which was invariably greater, particularly in the case of 4 days' treatment, where the dark-period promotion was highly significant. Seeds whose germination is promoted by repeated cycles of light and dark are termed 'photoperiodic' and *Tridax* appears to be of the short-day type, since it gives maximum germination at day-lengths less than 24 hr., i.e. continuous illumination (Evanari, 1965).

Attention was focused subsequently on the effect of the dark period, with light period kept at a standard 4 hr., which was assumed on the basis of preliminary experiments to be sufficiently long. The length of the dark period was therefore determined by the total cycle length used, and this was studied in relation to the number of such cycles given. For example, it was found that when 4 cycles were given, a dark period of 8 hr. (12 hr. cycle) greatly reduced the germination, compared with dark periods of 20 hr. or 44 hr. When only 2 cycles were given, the germination was of course much reduced, but again the longer dark periods were more favourable. Indeed, it appears that a sub-optimal number of cycles can be partly offset by a very long dark period (44 hr.). Conversely, a dark period which is shorter than optimal can be completely offset by increasing the number of cycles (see 8 cycles of 12 hr. in Table 3).

This rather complex dormancy mechanism is of considerable interest physiologically, since there are rather few examples of photoperiodic seeds. From the point of view of survival on the lawn, with its extremely desiccating conditions during the dry season, the dormancy seems to protect the great majority of seeds from germination until conditions are favourable for seedling growth. At the beginning of the rainy season, and also in watered plots, the surface soil remains moist for a considerable period and the dormancy is then broken, but after dew or occasional storms in the dry season the seeds presumably dry up too quickly for this to happen.

Thus, although the mechanism of seed dormancy is very different in these three species, the end result is rather similar in that they are all found rapidly recolonizing open spaces in the early part of the rainy season (Plate 2). Naturally, survival depends on many factors, but dormancy of various plant organs appears to be of particular importance, conferring the ability to control the timing and extent of active growth according to a variable and only partly predictable environment.

I should like to thank my colleagues and students in Ghana for their stimulation and assistance with these studies, and in particular Mr J. B. Hall for ecological help, Mr J. S. Adomako for his careful work with the numerical data, Mr N. Jones for carrying out the collection and budding of *Cedrela* and Dr I. D. J. Phillips for commenting on the manuscript. Especially my thanks go to Dr J. Jeník, without whose enthusiasm and knowledge of the tropical forest this paper could not have been written.

REFERENCES

ALVIM, P. DE T. (1964). In *Formation of Wood in Forest Trees*, p. 479. Ed. M. H. Zimmermann. New York: Academic Press.
BLACK, J. N. (1966). *Forestry* (Supplement on Physiology in Forestry), p. 98.
BRAY, J. R. & GORHAM, E. (1964). *Adv. Ecol. Res.* **2**, 101.
BÜNNING, E. (1948). *Entwicklungs- und Bewegungsphysiologie der Pflanze*. Berlin: Springer.
BÜNNING, E. (1956). *A. Rev. Pl. Physiol.* **7**, 71.
CACHAN, P. & DUVAL, J. (1963). *Ann. Fac. Sci. Dakar* **8**, 5.
CLARKE, J. I. (1966). (Ed.). *Sierra Leone in Maps*. University of London Press.
COSTER, C. (1923). *Ann. Jard. bot. Buitenz.* **33**, 117.
COSTER, C. (1927). *Ann. Jard. bot. Buitenz.* **37**, 49.
COSTER, C. (1928). *Ann. Jard. bot. Buitenz.* **38**, 1.
DOWNS, R. J. (1962). In *Tree Growth*, p. 133. Ed. T. T. Kozlowski. New York: Ronald Press.
ENTI, A. A. (1968). *Bull. Inst. fr. Afr. noire* A **30**, 881.
EVANARI, M. (1965). *Handb. Pflanzenphysiol.* **15** (2), 804.
GREENWOOD, M. & POSNETTE, A. F. (1949). *J. hort. Sci.* **25**, 164.
HARDY, F. (1964). *Cacao* **9**, 1.
HOLTTUM, R. E. (1953). *Symp. Soc. exp. Biol.* **7**, 159.
HOLLTUM, R. E. (1968). *J. appl. Ecol.* **5**, 5P.
JEFFERS, J. N. R. & BOALER, S. B. (1966). *J. Ecol.* **54**, 447.
JENÍK, J. & HALL, J. B. (1966). *J. Ecol.* **54**, 767.
JENÍK, J. & LONGMAN, K. A. (1965). *Jl W. Afr. Sci. Ass.* **10**, 64 (abstr.).
LONGMAN, K. A. (1966). *Jl W. Afr. Sci. Ass.* **11**, 3.
LONGMAN, K. A. & JENÍK, J. (1970). *Tropical Forest and Its Environment*. Edinburgh: Oliver and Boyd. (In the Press.)
MURRAY, D. B. & SALE, P. J. M. (1967). *Conf. int. rech. agronom. cacaoyères*, Abidjan, 1965. Paris.
NITSCH, J. P. (1957). *Proc. Am. Soc. hort. Sci.* **70**, 526.
NJOKU, E. (1959). *Nature, Lond.* **183**, 1598.
NJOKU, E. (1963). *J. Ecol.* **51**, 617.
NJOKU, E. (1964). *J. Ecol.* **52**, 19.
RICHARDS, P. W. (1952). *The Tropical Rain Forest*. Cambridge University Press.
ROBERTS, E. H. (1962). *J. exp. Bot.* **13**, 75.
SCHIMPER, A. F. W. (1935). *Pflanzengeographie auf physiologischer Grundlage*, revised by F. C. Faber. Jena. (See also English translation by Fisher (1903) of earlier edition.)
TAYLOR, C. J. (1960). *Synecology and Silviculture in Ghana*. (See appendix.) Edinburgh: Thomas Nelson.
WAREING, P. F. (1956). *A Rev. Pl. Physiol.* **7**, 191.

PLATE I

(a) *Cedrela odorata.* A 'mature' graft, showing dormant condition under natural daylengths. (b) As above, showing the scars left after intermittent growth and dormancy (arrows). (c) General view of the lawn at the University of Ghana, Legon, on which watering experiments were carried out. Photograph taken 16 May 1968, showing lawn completely recolonized 1 month after the heavy rainfall of 19/20 April. Whitish areas include flowering plants of *Tridax procumbens.* Uniform patches of *Dactyloctenium aegyptium* in middle distance.

(*Facing p.* 488)

HYPERRESPONSIVENESS IN HIBERNATION*

By C. P. LYMAN AND R. C. O'BRIEN

Department of Anatomy, Harvard Medical School, Boston, Mass.
and the Museum of Comparative Zoology, Harvard
University, Cambridge, Mass., U.S.A.

INTRODUCTION

It now seems reasonably well established that hibernation in mammals, in all its phases, is under remarkably precise physiological control. This is a relatively new concept in the study of hibernation, for it had been assumed for many years that mammals hibernated because they could not avoid it, implying that hibernators have inferior control of body temperature. The consensus at the present time is that temperature control in hibernators is 'set' as it is in other mammals, but that this setting can be changed by certain as yet poorly understood environmental and physiological factors. At the outset this concept seems more acceptable, for every hibernator can rewarm itself from temperatures near 0° C. using only heat from endogenous sources, and this feat would presumably be impossible for an animal that entered hibernation because it was unable to maintain a high body temperature.

There are at least two mental obstacles which make it difficult to accept the idea of physiological control in hibernation. The first is the natural tendency to equate mammalian or avian hibernation with the aestivation or hibernation seen in reptiles, amphibians and fishes, and most biologists have seen the helpless condition of some hibernating poikilotherm. However, if hibernation in homeotherms is a specialized adaptation to carry the species through an unfavourable period, there is no reason to assume that it has much in common with hibernation in poikilotherms with the exception of a cold body temperature.

The second obstacle to accepting the thesis of physiological control is the well-known effect of cold on biochemical reactions. It is common knowledge that cold can block the functions of the organs and tissues of many homeotherms at temperatures well above the usual temperatures of hibernation. However, it has been shown that the tissues and organs of mammals which hibernate are peculiar in their ability to maintain their

* This research was supported by grants GM 05611 and GM 05197 from the U.S. Public Health Service, and U.S. Air Force contract F 41609-67-C-0052.

functions at low temperatures (Chatfield, Battista, Lyman & Garcia, 1948; Lyman & Blinks, 1959; South, 1961) and Marshall & Willis (1962) have given a partial explanation of this specialization. It is well established that the metabolic rate of a mammal in hibernation at a body temperature of 5° C. may be one-fiftieth or less of the rate at 37° C. Most biochemical reactions have a Q_{10} of 2 or higher, and it is difficult to accept the thought that an animal can maintain control of its internal environment if the speed of its reactions is greatly reduced by cold. In this paper we will present evidence to show that the responsiveness of the animal in hibernation can be actually augmented by cold.

The evidence for homeostasis in hibernation has already been presented by many investigators, and will be reviewed only briefly here (Strumwasser, 1960; Lyman, 1963, 1965). During the non-hibernating season mammals which hibernate maintain body temperatures as steady as those of non-hibernators of comparable size. With the approach of the hibernating season many species of hibernators become lethargic and the body temperature becomes more labile. Most hibernators either store food in their burrows or become fat before hibernation, or do both, but neither is an absolute prerequisite for the hibernating state. The physiological mechanisms which permit the animal to abandon its warm-blooded state are not known, though Hammel (1967) has presented a very plausible theory on the changes that may occur in the central nervous system. In all species in which the measurements have been made, it is found that when the animal begins entrance into hibernation, the heart rate, and probably the metabolic rate also, decline before any decline in body temperature occurs. Tucker (1965) has derived evidence that *Perognathus californicus* (the California pocket-mouse) cools simply because the animal ceases to regulate its body temperature. However, in ground squirrels (*Citellus*), which are members of a different family of rodents, skipped heart beats typically occur as the animal enters hibernation, and this is one of the factors which serves to slow the heart. Atropine is a specific blocker of parasympathetic action, and if this is infused into the circulation of the undisturbed animal via an indwelling cannula, the skipped beats are abolished (Lyman & O'Brien, 1963). It is difficult to explain why the heart rate should decrease so abruptly before the drop in body temperature, or why atropine should abolish the skipped beats, without invoking the chronotropic action of the autonomic nervous system.

Strumwasser (1960) has reported that the California ground squirrel (*Citellus beecheyi*) enters hibernation with a series of 'test drops', during which the body temperature sinks to a definite level and rises again. With each bout of hibernation, the temperature drops to a lower level, and

Strumwasser maintains that this sort of conditioning to lower temperatures is necessary in order to achieve deep hibernation (Strumwasser, Gilliam & Smith, 1964). Attractive though this idea is, 'test drops' are not universal in all species of hibernators. Chronic records of body temperature in thirteen-lined ground squirrels (*Citellus tridecemlineatus*) from our laboratory (unpublished) show that many animals in the warm room (22° C.) can maintain a body temperature of 33–37° C. for months at a time. When transferred to an environmental temperature of 5° C. some of these animals exhibit a steady decline in body temperature to 6° C. with no 'test drops'.

At steady ambient temperatures between approximately 3° and 15° C. the body temperature of the hibernator remains about 1° C. above that of the environment and changes passively with any change in environmental temperature. However, if the ambient temperature drops to a potentially lethal level, the animal may increase its metabolism and either arouse from hibernation or maintain a larger gradient between its body temperature and that of the environment (Wyss, 1932; Lyman, 1948). The hibernating rodent responds to an increase in inhaled CO_2 by increasing its respiratory rate, and exhibits about the same sensitivity as the human at a body temperature of 37° C. (Lyman, 1951). When blood *in vitro* at constant pCO_2 is cooled from 37° to 5° C. the pH rises from 7·4 to 7·5. The blood in hibernating animals has been shown repeatedly to have a pH of about 7·4 (Kent & Pierce, 1967), which also implies respiratory control during hibernation.

Although the heart rate in hibernation can be as low as 2–3 beats/min. the blood pressure remains relatively high. There is evidence that vascular tone is maintained and that peripheral resistance actually is increased as the animal enters hibernation (Lyman, 1965). Many investigators have reported that parasympathetic influence is minimal, if not completely lacking, in the hibernating animal or the mammal in hypothermia (Biewald & Raths, 1959; Johansen, Krog & Reite, 1964; Lyman, 1965). The heart cannot be slowed by intravascular infusion of acetylcholine or electrical stimulation of the vagus. Peculiarly, after the animal is disturbed and the heart has accelerated for a few minutes, parasympathetic influence is again effective, though no detectable change in heart temperature has occurred (Lyman & O'Brien, 1963). Unpublished results from this laboratory suggest that the refractoriness to parasympathetic influence is not due to a change in the heart itself. No difference was found in the response to acetylcholine of isolated perfused hearts, removed in a few seconds from hibernating animals, when compared either to the response of hearts which had accelerated for several minutes before removal from the hibernator, or the response of chilled hearts from active animals.

It seems likely that sympathetic influence maintains vascular tone during hibernation and, in the absence of parasympathetic effects, controls the heart rate, which may vary from hour to hour or day to day. If the hibernating animal is infused intra-arterially with the sympatholytic drug β-TM 10 ([2-(2,6-dimethylphenoxy)propyl]-trimethylammonium chloride hydrate) it cannot arouse from hibernation, the heart rate slows, blood pressure drops, and death ensues (Lyman & O'Brien, 1963). Thus the sympathetic system appears to remain on guard during hibernation, ready to be sparked into full action when the arousal begins.

Touch, pressure, heat, cold, or intravascular infusion of a variety of agents, notably acetylcholine, will initiate the coordinated process of arousal. During arousal, the animal mobilizes all its heat-generating mechanisms and warms as rapidly as possible. In the first stages, the blood flow is restricted to the anterior portions of the body, thus warming the essential organs and reducing heat loss. Contrasted to the steady state of hibernation, this is a rapidly changing state which produces a maximum of warming in a minimum time (Lyman, 1965).

In addition to arousal induced by external stimuli, all mammals which hibernate undergo periodic spontaneous arousals during the hibernating season. The reason for these arousals is not known, but the length of the period of hibernation varies with the species and the period tends to be longer during the middle of the hibernating season. Twente & Twente (1965) have shown that there is a linear relationship between the length of the period of uninterrupted hibernation and the temperature of the hibernating animal (*Citellus lateralis*). This suggests that the metabolic rate is influencing the length of the period, and that the animal may arouse after having exhausted some essential store, or after having accumulated a definite amount of some metabolite. This concept again involves physiological control during the hibernating state.

If the animal in deep hibernation is to react to some external or internal stimulus by arousing from hibernation, it is reasonable to expect that some measurable physiological signal would precede any marked physiological change. Our interest in this possibility was heightened by the observation that intra-arterial infusion of very small amounts of acetylcholine (ACh) into the deeply hibernating ground squirrel produced a burst of muscle action potentials (MAP) followed by cardioacceleration. A similar reaction resulted if the animal was poked gently or otherwise physically disturbed (Fig. 1). In either case, if the stimulus was of sufficient magnitude, spontaneous MAP followed the initial burst, cardioacceleration continued, and the animal carried out a normal arousal. With a less vigorous stimulus no further MAP occurred, the heart rate slowed, and the animal returned to its previous state. Long-term records of spontaneous arousals showed that they were always preceded by several bursts of MAP, though in this situation the heart rate sometimes accelerated prior to the MAP. If the hibernating animal was exposed to high concentrations of ambient CO_2 or to subzero ambient temperatures, MAP occurred, followed by cardio-acceleration (Lyman & O'Brien, unpublished). Again, if the stimulus was of sufficient magnitude, MAP and cardioacceleration continued and arousal followed. The MAP can be blocked by curare and this, in turn, blocks the

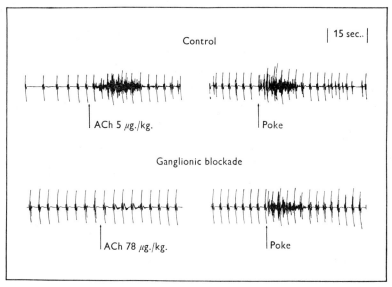

Fig. 1. Cannulated, hibernating thirteen-lined ground squirrel, body temperature 7° C., dermal electrode recording. Upper panels: muscle action potentials in the normal animal caused by infusion of acetylcholine (left), or mechanical stimulation by poking with a stiff wire (right). Lower panels: 18 min. after infusion of 10·8 mg./kg. hexamethonium chloride to produce ganglionic blockade. Effect of acetylcholine is blocked (left), while effect of mechanical stimulation is not (right).

cardioacceleration (Lyman & O'Brien, 1963). We have observed the MAP in four species of rodents from three families and Kulzer (1967) reports it in bats.

Thus the MAP gave every indication of being a signal which presages the onset of arousal or indicates a temporary change in the hibernating state. Our efforts to clarify the origin of this signal are chronicled below.

MATERIALS AND METHODS

Although most of this investigation was carried out using the thirteen-lined ground squirrel (*Citellus tridecemlineatus*), the golden-mantled ground squirrel (*C. lateralis*), the European dormouse (*Glis glis*) and the Turkish hamster (*Mesocricetus auratus brandti*)* were also used for some experiments.

Two important preparations form the backbone of this study: the 'cannulated hibernator' and the 'spinal hibernator'. The cannulated hibernator is fitted with an aortic and sometimes a venous cannula using the method of Still & Whitcomb (1956) or Popovic & Popovic (1960). Three silver electrodes are sewn into the skin of the back for oscillograph recording. One or more thermocouples are embedded in suitable areas, and their leads, along with the cannulae, are made to exit from a common opening at the interscapular region. The animal is prepared during the hibernating season and usually returns to hibernation some days after the operation. Once in hibernation, the animal may be prepared for an experiment if care is taken to avoid arousal. The thermocouple leads are spliced and temperature† monitored with a thermoelectric recorder accurate to 0·2° C.† The electromyogram and electrocardiogram are recorded on a Grass oscillograph‡ with the leads attached to the dermal electrodes by small clips. The short polyethylene tube is joined to a longer tube which leads outside the cold box so that drugs may be introduced with a Krogh–Keyes pipette with the assurance that the animal is not disturbed by mechanical stimuli. Within the range of the volumes used, changes in aortic pressure do not affect the animal, as can be demonstrated by control infusions of isotonic saline. A small air bubble introduced into the distal end of the lengthened tube serves to separate the saline from the drug.

In preparing the spinal hibernator the trachea is cannulated for artificial respiration after a stout thread has been passed dorsal to the trachea and around the neck. This thread is used as a garotte to cut off cerebral circulation and the spinal cord is divided at C_1. With practice, this procedure can be completed within less than 5 min. after picking up the deeply hibernating animal, and the arousal process is essentially stopped at that point. After severing the spinal cord the animal is usually fitted with an aortic cannula and dermal electrodes. Under ideal circumstances, this preparation has lasted and provided useful information for over 3 days at 5° C.

In testing the sensitivity of skeletal muscle to acetylcholine the extensor digiti quarti was removed as quickly as possible from the forefoot. A variety of anaesthetics was used in the active animals, and the hypoxic-hypercapnia technique of Giaja & Andjus (1949) was used to produce hypothermia. No anaesthesia was used on animals in hypothermia or in hibernation. The muscle was suspended in a temperature-controlled bath of a modified

* Identification of this animal to the subspecific level is tentative, and based on geographical location according to Ellerman & Morrison-Scott (1951).
† Speedomax, type G. Leeds and Northrup, Philadelphia, Pa.
‡ Polygraph, model 5D, Grass Instrument Co., Quincy, Mass.

Ringer solution and gassed with 5 % CO_2/95 % O_2. The tendon at each end was tied with a fine thread and the muscle stretched with a tension of 1·5 g. Measured amounts of acetylcholine were introduced with almost immediate mixing, and the tension which was developed was recorded isometrically with a Grass strain-gauge appropriately amplified on the Grass oscillograph. The muscle was washed twice after each test, and many of the muscles were tested with and without eserinization (physostigmine sulphate, 0·6 mg./100 ml.). Electrical stimulation was via stainless steel electrodes with a Grass model SD5 stimulator.

Close arterial infusion of the gastrocnemius muscle was carried out on spinal animals using the close arterial infusion technique of Brown, Dale & Feldberg (1936). The popliteal artery was fitted with a fine glass cannula (0·15 mm. O.D.) and measured amounts of acetylcholine were introduced using a Krogh–Keyes pipette. Tension was measured with the apparatus described above.

The sensitivity of muscle at the region of the motor end-plate was measured using the method of Fatt (1950). With this preparation a long, thin muscle is necessary for proper oxygenation and recording, with no damaged fibres and the end-plates located in discrete bands. The omohyoideus muscle was used, which required long dissection, during which the hibernating animal was kept chilled, anaesthetized with sodium pentobarbital, 80 mg./kg. body weight, and respirated artificially. The muscle is gently held at origin and insertion by two ties which are fastened to the tendon or bone. It is suspended in a bath of Krebs–Ringer–bicarbonate with 0·2 % glucose, and gassed as described for the muscle of the forefoot. One wick electrode is placed at the end of the muscle nearest the surface of the solution and another is placed in the solution. A known amount of ACh is added to the bath, and the muscle is drawn out of the bath at a fixed rate with its long axis perpendicular to the surface of the solution and the wick electrode still resting against it. Normally the muscle surface is equipotential, so no voltage change is recorded as the fluid level scans the surface. When the ACh depolarizes the end-plate regions, an area of negativity is measured as the muscle is withdrawn. With this method both the threshold dose of ACh necessary to produce depolarization and the extent of depolarization along the length of the muscle can be measured.*

The activity of cholinesterase in blood and skeletal muscle was determined using the colorimetric method described by Augustinsson (1957) based on the chemical determination of unreacted ACh.

* We are greatly obliged to the Department of Pharmacology, Harvard Medical School, to whom we constantly turned for advice during the course of this research. To Dr Douglas R. Waud we owe additional thanks for the loan of the recording device used in these experiments.

To test the sensitivity of the hibernating animal to temperature, a circular brass disk 1 cm. in diameter was placed gently on the previously shaved dorsal surface and held in position with adjustable clamps. The brass disk was hollow and liquid was pumped through it from a heated or chilled reservoir. Two fine thermocouples cemented to the surface of the disk which rested on the animal recorded the temperature. With this arrangement the temperature of the disk could be raised from 5° to 55° C. in 15–60 sec. As soon as the animal responded to the stimulus with MAP the disk was removed from the back. To chill the disk a mixture of alcohol and solid CO_2 was used, and the most rapid cooling to $-19°$ from 5° C. was in 1·5 min. Body temperatures in these experiments were obtained by carefully sliding two thermocouples under the ground squirrel, so that they lay between the animal and its nest. Comparisons after the experiment showed that temperatures recorded in this manner were within 0·5° C. of the rectal temperature taken at a depth of 5 cm. Temperatures and electromyograms were recorded as with the cannulated hibernator.

RESULTS

Sensitivity of muscle

Under ideal conditions the amount of intra-arterially infused ACh necessary to produce MAP in the cannulated hibernator is very small. Doses as low as 0·001 mg./kg. will produce a small burst of MAP in a thirteen-lined ground squirrel hibernating at a body temperature of 5–8° C. while ten times that amount is necessary to produce MAP followed by cardioacceleration. The threshold of ACh necessary to produce MAP or cardioacceleration may change from day to day, and sometimes from hour to hour, and infusions of as much as 0·011 mg./kg. may be necessary to elicit only the muscle action potentials. As was explained previously, ACh has no parasympathomimetic effect on the heart in the hibernating animal and bradycardia does not occur. In contrast, as much as 1·5 mg./kg. ACh intra-arterially infused into the unrestrained, active animal causes no MAP. In the eserinized (0·24 mg./kg. physostigmine sulphate) awake animal 0·30 mg./kg. ACh produces bradycardia, but no MAP, and three times this dose is necessary to produce MAP.

The peculiar effect of ACh on the hibernating animal suggested that the drug was acting directly on the skeletal muscle, and producing MAP by its nicotinic action. The hibernating animal is virtually motionless for days at a time, and it seemed possible that skeletal muscle during this period might undergo the same sort of changes that are known to occur in denervated muscle in the active animal. After denervation, skeletal muscle not only undergoes atrophy but also becomes increasingly sensitive to a variety of drugs and chemicals, including ACh. The possibility of increasing sensitivity in muscle during a bout of hibernation had not been previously explored.

For this reason, a series of tests was made in an attempt to contrast the dose–response curves to ACh of the extensor digiti quarti from active animals, active animals which had been made hypothermic, and hibernating animals. Consistent dose–response curves of the amount of tension produced could not be obtained, but the doses necessary to produce a threshold response are shown in Table 1. There was no significant difference between the muscle from active animals tested in a bath at 10° C. and the muscle from hypothermic animals under the same conditions, and both of these groups are combined. In contrast, the threshold to ACh of muscle from hibernators is significantly lower, in spite of the variability of the results. There was no indication that the sensitivity increased as the bout of hibernation continued, for muscles removed from animals in the first day of hibernation were as sensitive as the muscle from animals which had hibernated steadily for 6 days. The muscle from hibernating animals was never as sensitive as the uneserinized muscle from active animals whose foreleg had been denervated 14 or more days before. In some experiments the temperature of the bath was raised to 25° C. The increase in temperature did not lower the threshold of the muscle to ACh, though the threshold to electrical stimulus was always lowered.

Table 1. *Threshold response of isolated muscle to acetylcholine*

	n	μg. ACh/100 ml.
Normal active	31	585 ± 591
Eserinized active	19	155 ± 136
Normal hibernator	16	127 ± 155
Eserinized hibernator	9	$35 \cdot 2 \pm 15 \cdot 8$
Denervated active	6	$36 \cdot 4 \pm 20 \cdot 9$

In an attempt to obtain less variable results, we turned to the close arterial infusion technique using the gastrocnemius muscle. The results with these experiments were even more inconsistent, probably because of the smallness of the artery (0·15 mm. or less) and the lack of uniformity in the blood supply to the muscle. The data, presented in Table 2, are only possibly significant, but the extremely low threshold found in three of the hibernators corroborates the results for isolated muscle.

In a third attempt to obtain more precise results, we turned to the method of Fatt (1950) for testing the isolated muscle. In spite of the elegance of the technique and its potential to answer our questions, we were unable to produce satisfactory results. The dissection of the omohyoideus is extremely difficult, taking as long as 4 hr. During this time, in spite of all precautions, some changes must take place in the muscle, and the possibility of mechanical injury with resultant abnormal depolarization is always present. However, although dose–response curves or thresholds could not be established, we obtained no evidence to indicate that the sensitivity to ACh in the muscles from the hibernating animal moved away from the regions of the motor end-plates, as Elmqvist & Thesleff (1960) and Miledi (1960) have shown to occur in the case of denervated muscle.

Table 2. *Threshold response of gastrocnemius muscle to close-arterial infusion of acetylcholine*

Active spinal chilled (μg. ACh)	Hibernating spinal cold (μg. ACh)
0·22	0·0088
0·55	0·11
1·1	0·11
1·1	0·55
1·1	—
2·2	—
2·2	—
4·4	—
4·4	—
4·4	—
8·8	—
Mean ± S.D. 2·77 ± 2·55	0·1947 ± 0·2416

In the assay of cholinesterase it was found that for the hibernating animal an average of 159 μmoles of ACh were hydrolysed per hr. per ml. of whole blood at 37° C., while the average was 163 μmoles for the active animals. In skeletal muscle the average for the hibernating animals was 6·8 μmoles per 100 mg. of tissue, while the figure was 6·2 for the active animals. The differences in either case between hibernating and active animals are not significant. In determining the Q_{10} it was found that K 7–37° was 1·22 for ground squirrel blood and 1·21 for skeletal muscle.

The spinal reflex

In working with close arterial infusion, it became clear that the spinal hibernator offered a new approach to the study of the sensitivity of the animal in hibernation. To our knowledge, the spinal hibernator has not been studied in other laboratories, and it came as a surprise to find how viable it remained.

The responses of the spinal hibernator are not identical to those of the intact animal in deep hibernation, but are sufficiently similar to provide much useful information, particularly if the experiments can be repeated on the cannulated hibernator. The great advantage in using the spinal hibernator is that a large series of experiments may be run in a single day, without the necessity of waiting between experiments for the animal to return to the deeply hibernating state before another test is undertaken, and without the danger of causing arousal from hibernation.

When the respiratory rate is properly adjusted, the heart rate remains faster than that of the deeply hibernating animal. Also, in contrast to the deep hibernator, a continuous background of muscle action potentials is often recorded. As in the intact hibernating animal, a burst of MAP occurs if the spinal hibernator is touched or infused with ACh. In contrast, no cardioacceleration follows the burst of MAP, presumably because connection with the brain has been severed. Experience has shown that both background MAP and the ability to respond to stimuli with bursts of MAP gradually diminish as the preparation ages, and that the electrocardiogram is a poor indication of the condition of the preparation as a whole. Occasionally we have observed animals with a normal-appearing electrocardiogram, and rigor mortis in the posterior third of the body. Unquestionably the preparation deteriorates because homeostasis is not maintained, but this condition is very difficult to achieve except by empirical means. By trial and error we have established a rate and volume for artificial respiration which keeps the preparation in good condition for a maximum period of time at a body temperature of 5–10° C. At higher temperatures, however, this respiration is not adequate and the correct ventilation has not been determined. Attempts to maintain homeostasis by monitoring the condition of the blood are difficult because of the small size of the animal.

During the studies on the sensitivity of the muscle in the spinal hibernator it became clear that the amount of artificial respiration was critical in maintaining a condition which most resembled the hibernating state. If the animal was either under- or over-ventilated, slow convulsive movements occurred which could be abolished only by changing the total respiration. Similar convulsive movements have been observed in non-hibernating spinal animals at high body temperatures. However, in view of the sensitivity of the animal in hibernation to inspired CO_2, it seemed possible that slight changes in acid-base balance might alter the sensitivity of the spinal hibernator to infusion of ACh or to mechanical stimuli. For this reason experiments were undertaken to change the acid-base balance by infusing dilute hydrochloric acid or sodium bicarbonate at molarities which are customarily used to cause changes in intact, homeothermic animals (Held, Fencl & Pappenheimer, 1964).

The experiments on the respiratory equilibrium and hormonal balance and their relationship to sensitivity in the intact and spinal hibernator are still in progress and cannot be detailed here. However, the immediate effect of infusion of 0·15 M-HCl or 0·3 M-NaHCO$_3$ at rates as low as 0·037 ml./min. was a burst of MAP which continued as long as the infusion lasted. In the intact, cannulated hibernator hydrochloric acid and sodium bicarbonate, as well as acetic acid, lactic acid and ammonium chloride, caused MAP followed by arousal. There was no reaction when 0·15 M-HCl or 0·3 M-NaHCO$_3$ was infused into cannulated active ground squirrels at the same or somewhat faster rates.

Although it was possible that acid or base caused MAP by direct action on sensitized muscle, it seemed more likely that the action was a reflex one initiated by the action of the infused chemical on receptors. Using the spinal hibernator, it was found that MAP which were previously elicited by 1·5 μg./kg. ACh could no longer be induced by doses as high as 118 μg./kg. if the animal was treated with 1 mg./kg. D-tubocurarine chloride. Furthermore, pinching the animal or intra-arterial infusion of 0·5 ml. 0·15 M-HCl or 0·3 M-NaHCO$_3$ also failed to produce MAP after curarization. Although the experiment cannot be clear-cut because of paralysis of respiration, we had previously shown that ACh would not produce MAP in the curarized intact, cannulated hibernator (Lyman & O'Brien, 1963).

In order to trace the reflex to the spinal cord, spinal hibernators were shown to respond with MAP to infusion of ACh, or acid, or to tactile and mechanical stimuli. The spinal cord was then destroyed by passing a roughened wire repeatedly down the spinal canal. After this operation, no MAP could be produced by stimuli of any magnitude. This experiment could not, of course, be repeated on the intact, cannulated hibernator.

Though the evidence was compelling that MAP were caused by a hyperresponsive or hypersensitive spinal reflex, the site where acid, base or ACh initiated the reflex was unknown. It has been known for some time that ACh will stimulate receptors. Douglas & Ritchie (1960), in examining the concept that ACh might be the neurochemical transmitter between the adequate stimulus to the receptor and the initiation of the nervous discharge, were able to stop the effect of ACh by using the ganglionic blocking agent, hexamethonium. They found, however, that the receptors in question still responded to their respective adequate stimuli.

Similar experiments were carried out on spinal and on cannulated intact hibernators. The results, which were the same in both preparations, are illustrated in Fig. 1. It may be seen that 15 times the threshold dose of ACh produces only very slight MAP after the animal is treated with hexamethonium. On the other hand, the effect of gently poking the animal with a steel wire is undiminished. This does not prove that ACh, acid and base

HYPERRESPONSIVENESS IN HIBERNATION 501

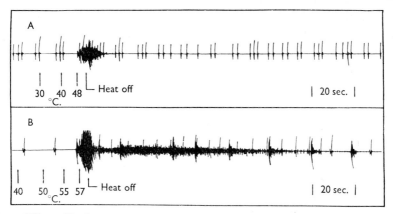

Fig. 2. Effect of body temperature on response to heated disk in hibernating thirteen-lined ground squirrel. Panel A, body temperature 15·6° C. Panel B, body temperature 3·8° C. Figures in lower left corner of each panel are temperature of heated disk which is higher than the temperature of the skin and the receptors.

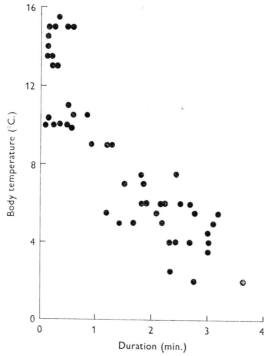

Fig. 3. Effect of body temperature on duration of muscle action potentials in thirteen-lined and golden-mantled ground squirrels. Stimulus of heated disk which induced muscle action potentials was removed within 5 sec. after the beginning of the response.

are acting on peripheral receptors such as those for touch, pain, and temperature, but the parallelism between these results and those of Douglas & Ritchie (1960) and others is persuasive that such is the case.

In order to study further the effect of temperature and hibernation on the spinal reflex, a series of experiments was undertaken in which the body temperature of hibernating animals was slowly changed by altering the ambient temperature. A variety of graded stimuli was attempted, including pressure, intra-arterial infusion of ACh and electric shock, but the most reproducible results were obtained using thermal stimuli with the brass disk described under Materials and Methods above. Similar experiments with spinal animals were not satisfactory because of the change in steady state with changing body temperature (see above).

As can be seen in Figs. 2 and 3 the duration of the MAP increases with the decrease in body temperature. It is also obvious that the voltage of the first burst of MAP is greater when the hibernating animal is cooler (Fig. 2). There was no significant difference in the threshold of the response to the temperature of the disk at different body temperatures. Thresholds varied from 44 to 64° C. and were usually quite reproducible during a given experiment. A weakness in experimental design was the fact that the disk could not be returned to precisely the same place on the back with each experiment, and this will cause some small changes in the threshold. Further work may show that the threshold is lower at high body temperatures, but it is certain that the difference is not very great between 3° and 15° C.

A few experiments were done using the stimulus of a cold disk at a body temperature of 5–6° C. A stimulating temperature of -12 to $-19°$ C. was necessary to produce MAP.

DISCUSSION

The anomalous effect produced by infusion of small doses of ACh into hibernating mammals appeared at the outset to be of far-reaching importance in the study of hibernation. It was clear from previous work (Lyman & O'Brien, 1963) that the muscle action potentials which followed ACh infusion were always associated with cardioacceleration, and, if the dose was large enough, with arousal from hibernation. MAP which were indistinguishable from those induced by ACh occurred if the hibernating animal was physically disturbed or if a spontaneous arousal took place.

One of the more puzzling aspects of the hibernating cycle is that all mammals arouse from time to time during the hibernating season, although it has been calculated that the energy expended in one arousal is equivalent to that spent in 10 days of hibernation (Adolph & Richmond, 1955). It has often been suggested that the hibernator becomes increasingly sensitive as a bout of hibernation continues, and recent reports (Twente & Twente, 1968 a, b) indicate that such may be the case.

The concept that skeletal muscle underwent a sensitization during hibernation which was similar to the sensitization of denervated muscle in the active animal seemed to fit the facts. The reason for sensitization in denervated muscle is unknown, but it is presumably due to the absence of some influence which is present when the nerve is intact and active. If a similar situation existed in the hibernating animal, the skeletal muscles would increase in sensitivity as hibernation continued until some minimal stimulus would produce a burst of MAP which was of sufficient magnitude to start the arousal process. The low doses of ACh necessary to produce MAP in the hibernator could be a reflection of the increasing sensitivity of skeletal muscle to the neuromuscular transmitter. A high Q_{10} of cholinesterase would result in less ACh being destroyed as it was carried in the blood from the site of the infusion to the skeletal muscle, thus augmenting the response.

The data presented here indicate that this concept is not valid. Skeletal muscle from the hibernating animal, when tested *in vitro* at 10° C., is significantly more sensitive to ACh than muscle from hypothermed or from active animals. Although this information was enough to encourage further research, it is not, in itself, evidence that muscle in hibernating animals becomes more sensitized as hibernation progresses. The muscles which were used for comparison had all undergone a violent change. The muscles from the 'active' animals were taken from the warm forefeet of anaesthetized ground squirrels and immediately placed in bathing solution at 10° C. The muscles from the chilled animals were taken from ground squirrels which had undergone enforced hypothermia, which is clearly not the same as the balanced physiological state of hibernation. In either case the enforced transition to cold may have affected their sensitivity.

The increased sensitivity in the muscles from hibernators showed no similarity to the known characteristics of sensitivity in denervated muscle. Sensitivity did not increase as hibernation progressed, nor was there any evidence that sensitivity to ACh had moved from the region of the motor end-plate in comparison with the two types of controls. Furthermore, the threshold sensitivity to ACh of uneserinized muscle from hibernators is higher than the sensitivity of denervated muscles from active animals.

Since there is no difference in the amount of cholinesterase in blood or muscle from the hibernating animal in comparison to the active ground squirrel the marked effect of ACh during hibernation is not due to differences in concentration of this enzyme. Our measurements of the Q_{10} of cholinesterase confirmed the results of others (Clark, Raventós, Stedman & Stedman, 1938; Glick, 1939) that this enzyme is little affected by low temperatures, so that the effect of infused ACh in the intact hibernator could not be due to inactivation of the enzyme.

It is unfortunate that the response to infused ACh over a varied range of body temperatures could not be obtained in either the cannulated or the spinal hibernator. Many experiments to this end convinced us that only very large changes in threshold, such as those found in the active animal compared to those in the deep hibernator, could be repeated reproducibly. Clearly the effect of infused ACh depends on other factors as well as the threshold of the end organ. Perhaps the most important of these factors is the condition of the circulation at the time of infusion, for it is necessary for the blood-borne ACh to reach the critical area as quickly as possible. There is evidence (Strumwasser, 1960; Lyman, 1965) that the vascular bed in the hibernating animal changes periodically so that blood flow is not constant to a given region. If this is the case the threshold to infused ACh as measured by the MAP response would change accordingly.

The importance of the circulation is illustrated by variations in the placement of the cannula. The tubing is inserted in a slit made above the common iliac artery and pushed rostrad until the tip is above the renal arteries and about at the level of the coeliac artery. In this position, MAP are produced with low doses of ACh. Sometimes, due to difficulties in the operation, the tube cannot be pushed this distance, and the tip comes to rest 1 cm. or more caudad to the usual position. In these cases, larger doses of ACh are needed to cause MAP routinely. The same is true if the ACh is introduced via a cannula in the carotid artery or the jugular vein. We know of no way of tracing the exact path of the small infusions of ACh, but it would seem that the standard operation fortuitously placed the tip of the cannula where the circulation to the critical region would be most effective.

In considering the effect of infused ACh on the active and hibernating animal it is likely that the rapid flow and thorough mixing of the drug and blood in the former case might reduce the effect of the drug when it reached the critical area and passed rapidly by it. The heart rate during hibernation is usually below 10 beats/min., while that of the active animal is 200 to 300 beats/min. Circulation time in the active animal can be compared with the animal in the first stages of arousal by measuring the time between infusion of ACh and the resultant bradycardia. With heart rates between 200 and 300 beats/min. the period is about 8 sec., while at a rate of 30 beats/min. this time is doubled. Although the vasoconstriction which occurs during arousal weakens this comparison, still the difference in circulation time between the active and hibernating animal is not large enough to suggest that circulation is the cause of the marked differences in the threshold for producing MAP. This conclusion is fortified by the observation that we find no correlation between heart rate and sensitivity to ACh in the hibernating animal, though the rate may vary threefold during the experimental period. It is reasonable to suppose that rate would reflect cardiac output, and that the higher heart rate would mean a faster circulation.

Finally, one factor in the slow blood flow of hibernation would tend to reduce the effect of ACh rather than increase it. The infused ACh is exposed to cholinesterase as it moves to the critical area. Since the low temperature of hibernation does not appreciably reduce the effect of the enzyme, more ACh would be hydrolysed in the slow blood flow of the hibernating animal compared to the destruction of ACh in the active animal.

All the evidence indicates that the action of ACh and other infused chemicals is not a direct one on the skeletal muscle, but rather an indirect effect via a spinal reflex. It is conceivable that infused acid or base could cause MAP by depolarizing the skeletal muscle directly, but the observation that curare blocks not only the action of ACh but also the action of acid is strong evidence that this is not the case. Even more convincing is the fact that destruction of the spinal cord in the spinal hibernator completely abolishes the effect of ACh and acid. Though this experiment cannot be performed on the intact hibernator, the MAP in both preparations is so similar that there is no reason to doubt that they are physiologically identical.

The afferent arm of the spinal reflex which causes the MAP when the hibernating animal is physically disturbed must originate at the various dermal receptors. MAP can be produced by touch, electric shock, heat and cold, but it is not known whether the latter two are mediated by temperature or by pain.

The receptors which are activated by the infusion of ACh and other chemicals have not been precisely identified. It is clear that these receptors act in the hibernating animal as they do in the active animal, for in both the effect of ACh is blocked by hexamethonium while the effect of a mechanical stimulus is not. The effect of infused ACh is in contrast to the previously reported effect of methacholine, which is devoid of ganglionic effect except in very high doses. Infusion of the latter drug in hibernating animals causes no reaction for 24 sec. or more, when several respiratory-like movements take place, followed by cardioacceleration (Lyman & O'Brien, 1963).

With this drug, the response may be due to the direct action on skeletal muscle.

It is possible that ACh activates several types of receptor, for this is its effect in animals that are not hibernating. Hydrochloric acid at a pH below 3 causes a painful sensation when applied to raw skin in the human (Armstrong, Dry, Keele & Markham, 1953) and this is possibly the action of infused acid or base in the hibernating animal. It is of some interest that local cooling of the skin produces MAP in the hibernating ground squirrel only when the temperature of the cooled plate is below $-12°$ C. Experiments with various species (Wyss, 1932; Lyman & Chatfield, 1955) have shown that animals in hibernation increase their metabolism and protect themselves from freezing when the ambient temperature is only a degree or two below $0°$ C. The present experiments indicate that the peripheral receptors must be well below $0°$ C. before they are activated. Thus the protective reaction probably originates in the central nervous system, as do so many other responses in the regulation of body temperature in the homeothermic animal. Raths & Hensel (1967) report a response in hibernating European hamsters to local stimulation of $0-5°$ C. at a body temperature of $10°$ C., but the method of measuring the temperature of the stimulus is not given.

In the experiment using temperature-induced dermal stimulation during hibernation there was no clear-cut evidence that the threshold was lower at the higher temperatures, though we expected this to be the case. On the other hand, as is shown in Fig. 2, the MAP from a given stimulus not only increases in voltage, but also in duration (Fig. 3) as the body temperature of the animal is decreased. Thus the animal in hibernation becomes hyperresponsive at lower body temperatures.

Koizumi, Brooks & Ushiyama (1959) have reported in detail on the effect of cold on the spinal reflex in the chilled homeotherm. Using cats, they found that excitability decreased but responsiveness increased as the temperature of the cord was lowered from $37°$ to $20°$ C. Below $18-20°$ C. the response ceased completely. They attribute the hyperresponsiveness to: '(1) repetitive discharge, probably contributed to by failure of accommodation and prolongation of excitatory potentials; (2) the greater participation of interneurons in reflex action; and (3) a greater power of recruitment at synaptic junctions because of the changes just mentioned.' They add that this probably causes a greater spread of activation, overflow of excitation into systems not ordinarily invaded, and general failure of confinement of activity to normal pathways.

Nervous tissue from animals that hibernate can function at much lower temperatures than homologous tissue from non-hibernating mammals

(Chatfield *et al.* 1948). If the hyperresponsiveness which Koizumi *et al.* observed as temperatures were lowered to about 20° C. continues to augment in the hibernator down to 0° C., then the animal in hibernation has a built-in alarm system which becomes more responsive as the temperature becomes lower. This increase in responsiveness may well compensate for the loss of sensitivity so that the animal is able to maintain a steady state in hibernation, even if its body temperature approaches the freezing point. Although it has been possible to show that the responsiveness increases with lowered temperature in the intact hibernating animal over the natural range of body temperature of 15–2° C., as yet we have not been able to trace it from 37–15° C. The spinal animal should be used for such an experiment, and we have yet to fix the correct steady-state respiratory rates for the entire temperature range in this preparation.

From the point of view of survival of the animal, it seems logical that the spinal reflex should be involved in guarding the animal during the hibernating period. In the active animal the precise spinal reflex is a first line of defence against traumatic stimuli. In the hibernating animal, precision is sacrificed for hyperresponsiveness and presumably the hyperresponsiveness is sufficient to activate a system rendered sluggish by cold. The change in specificity with changing temperature is illustrated by an experiment with a hibernating *C. lateralis*. At a body temperature of 10° C. a temperature of 53° C. on the stimulating heat disk caused a relatively long MAP, a slight rocking movement of the body, and cardioacceleration. At 17° C. the animal responded to 55° C. local heating with a well-defined scratching movement of the left hind leg which removed the heat disk.

The exact effect of the MAP on the central nervous system has yet to be described. We have previously presented evidence to indicate that the increase in heart rate which follows the MAP is mediated via the cardio-accelerator fibres of the sympathetic nervous system, but the intermediate pathways are unknown (Lyman & O'Brien, 1963). Since MAP never results in cardioacceleration in the spinal animal, one must assume that an intact central nervous system, at least to the level of the mid-brain, is necessary for arousal. Using the duration of cardioacceleration after the burst of MAP as a criterion for the response of the central nervous system, it is clear that the long burst of MAP in an animal with a body temperature of 3° C. produces about the same effect as the short burst which occurs when the body temperature reaches 15° C., but more extended observations cannot be made on the intact hibernator. In due time we hope to clarify these matters using the spinal hibernator with a wide range of body temperatures.

If our thesis is correct, the hyperresponsive spinal reflex is vital to the mammal during its period of hibernation. Although the physiology of the onset of hibernation has interested biologists for many years, from the point of view of the survival of the species the ability to arouse from hibernation is of paramount importance. It is paradoxical that the hibernator by losing, as he chills, part of the specificity which is the hallmark of the mammalian nervous system, actually may be augmenting the ability to arouse from the hibernating state.

The hyperresponsive spinal reflex may act as a protective device and cause evoked arousal if external circumstances become unfavourable for continued hibernation. If the hibernaculum were not well protected both thermally and physically, the hyperresponsiveness of the hibernating rodent to external stimuli would have little survival value, for the animal would die before it attained sufficient coordination to fend for itself. Actually, however, the hibernating burrows of most rodents are below several feet of frozen earth. We have noted that hibernating animals are remarkably sensitive to vibrations such as those made by a digging predator, for in the laboratory hibernaculum MAP may occur each time a well-damped cooling compressor starts its cycle. Thus responsiveness to vibrations or thermal changes in the wild may well be extremely important for survival during hibernation, because the site of the hibernaculum assures that dangerous situations will occur slowly.

In spontaneous arousals it is reasonable to assume that the initial stimulus originates in the brain, probably in the region of the heat-regulating centres of the hypothalamic area. If cold decreases the specificity of neuronal pathways, then the original stimulus could give rise to a 'mass activation of the centres of the hypothalamus which govern heat production and conservation and which give rise to maximal functional activity of the sympathetico-adrenal and somatic motor systems' as postulated 18 years ago (Chatfield & Lyman, 1950).

REFERENCES

ADOLPH, E. F. & RICHMOND, J. (1955). *J. appl. Physiol.* **8**, 48.
ARMSTRONG, D., DRY, R. M. L., KEELE, C. A. & MARKHAM, J. W. (1953). *J. Physiol., Lond.* **120**, 326.
AUGUSTINSSON, K.-B. (1957). *Meth. biochem. Analysis* **5**, 1.
BIEWALD, G.-A. & RATHS, P. (1959). *Pflügers Arch. ges. Physiol.* **268**, 530.
BROWN, G. L., DALE, H. H. & FELDBERG, W. (1936). *J. Physiol., Lond.* **87**, 394.
CHATFIELD, P. O., BATTISTA, A. F., LYMAN, C. P. & GARCIA, J. P. (1948). *Am. J. Physiol.* **155**, 179.
CHATFIELD, P. O. & LYMAN, C. P. (1950). *Am. J. Physiol.* **163**, 566.
CLARK, A. J., RAVENTÓS, J., STEDMAN, E. & STEDMAN, E. (1938). *Q. Jl. exp. Physiol.* **28**, 77.
DOUGLAS, W. W. & RITCHIE, J. M. (1960). *J. Physiol., Lond.* **150**, 501.
ELLERMAN, J. R. & MORRISON-SCOTT, T. C. S. (1951). *Check List of Palaearctic and Indian Mammals 1758 to 1946*, p. 1. British Museum (Natural History).

ELMQVIST, D. & THESLEFF, S. (1960). *Acta pharmac. tox.* **17**, 84.
FATT, P. (1950). *J. Physiol., Lond.* **111**, 408.
GIAJA, J. & ANDJUS, R. (1949). *C. r. hebd. Séanc. Acad. Sci., Paris* **229**, 1170.
GLICK, D. (1939). *Proc. Soc. exp. Biol. Med.* **40**, 140.
HAMMEL, H. T. (1967). In *Mammalian Hibernation*, III, p. 86. Edinburgh: Oliver and Boyd.
HELD, D., FENCL, V. & PAPPENHEIMER, J. R. (1964). *J. Neurophysiol.* **27**, 942.
JOHANSEN, K., KROG, J. & REITE, O. (1964). *Suomal.-ugr. Seur. Aikak.* (Annales Academiae scientarum fennicae), ser. A, IV, **71**, 243.
KENT, K. M. & PIERCE, E. C. II (1967). *J. appl. Physiol.* **23**, 336.
KOIZUMI, K., BROOKS, C. McC. & USHIYAMA, J. (1959). *Ann. N.Y. Acad. Sci.* **80**, 449.
KULZER, E. (1967). *Z. vergl. Physiol.* **56**, 63.
LYMAN, C. P. (1948). *J. exp. Zool.* **109**, 55.
LYMAN, C. P. (1951). *Am. J. Physiol.* **167**, 638.
LYMAN, C. P. (1963). In *Temperature—Its Measurement and Control in Science and Industry*, vol. III, p. 453.
LYMAN, C. P. (1963). In *Handbook of Physiology*, sect. 2, vol. III, p. 1967. Washington, D.C. American Physiological Society.
LYMAN, C. P. & BLINKS, D. C. (1959). *J. cell. comp. Physiol.* **54**, 53.
LYMAN, C. P. & CHATFIELD, P. O. (1955). *Physiol. Rev.* **35**, 403.
LYMAN, C. P. & O'BRIEN, R. C. (1963). *J. Physiol., Lond.* **168**, 477.
MARSHALL, J. M. & WILLIS, J. S. (1962). *J. Physiol., Lond.* **164**, 64.
MILEDI, R. (1960). *J. Physiol., Lond.* **151**, 24.
POPOVIC, V. & POPOVIC, P. (1960). *J. appl. Physiol.* **15**, 727.
RATHS, P. & HENSEL, H. (1967). *Pflügers Arch. ges. Physiol.* **293**, 281.
SOUTH, F. E. (1961). *Am. J. Physiol.* **200**, 565.
STILL, J. W. & WHITCOMB, E. R. (1956). *J. Lab. clin. Med.* **48**, 152.
STRUMWASSER, F. (1960). *Bull. Mus. comp. zool. Harv.* **124**, 285.
STRUMWASSER, F., GILLIAM, J. J. & SMITH, J. L. (1964). *Suomal.-ugr. Seur. Aikak.* (Annales Academiae scientarum fennicae), ser A, IV, **71**, 399.
TUCKER, V. A. (1965). *J. cell. comp. Physiol.* **65**, 405.
TWENTE, J. W. & TWENTE, J. A. (1965). *Proc. natn. Acad. Sci. U.S.A.* **54**, 1058.
TWENTE, J. W. & TWENTE, J. A. (1968a). *Comp. Biochem. Physiol.* **25**, 467.
TWENTE, J. W. & TWENTE, J. A. (1968b). *Comp. Biochem. Physiol.* **25**, 475.
WYSS, O. A. M. (1932). *Pflügers Arch. ges. Physiol.* **229**, 599.

SOME INTERRELATIONS OF REPRODUCTION AND HIBERNATION IN MAMMALS

By W. A. WIMSATT

Division of Biological Sciences, Cornell University

INTRODUCTION

Among mammals two principal seasonal patterns of dormant behaviour have long been recognized, designated 'hibernation' (winter dormancy), and 'aestivation' (summer dormancy) respectively. Hibernation is associated with low ambient temperatures and is characterized by profound torpor that endures with only short periodic arousals throughout the colder months of the year. Aestivation occurs at high ambient temperatures, and is associated with environmental aridity and diminished food availability; torpidity in this case is shallower than in hibernation, although it is not necessarily of shorter duration.

These 'traditional' seasonal patterns of dormancy appear to be connected through an intermediate series of thermoregulatory patterns, observed in some presumed non-hibernators and non-aestivators (as well as in some hibernators). These include so-called 'daily torpor' and 'irregular torpor', variants which may, or may not, show seasonal peaks in their frequency of occurrence in the different species involved. While these various dormancy patterns may differ in physiological detail (Hudson & Bartholomew, 1964), they are clearly not mutually exclusive: many aestivators also hibernate (e.g. certain ground squirrels), some hibernators experience daily torpor during the season of normal activity (e.g. bats), or daily and seasonal torpor may merge together (e.g. pocket mice). As emphasized by Hudson & Bartholomew (1964) the various patterns of dormancy all share the same adaptive significance—energy conservation, and hence may be expected, directly or indirectly, to influence or be influenced by reproductive activities. Hudson & Bartholomew (1964) and Hudson (1967) have reviewed the dormancy patterns of mammals and described their physiological similarities and dissimilarities, and Hoffman (1968) has provided a succinct summary of the major physiological parameters of hibernation.

It is reasonable to suppose that in mammals generally the many physiological adjustments involved in reproduction are facilitated by maintenance of a homeothermic condition near normothermic levels. Torpidity not only suppresses ordinary behaviour patterns such as courtship and mating which depend upon coordinated locomotor functions, but depressed metabolism and lowered body temperatures also have more subtle effects, such as marked retardation, if not frank inhibition, of numerous vital functions at all levels of bodily organization from the subcellular to the organismic. Although some organs, most notably nerves and certain muscles, have become specially adapted in hibernators and retain their functional capabilities at lower body temperatures than do the comparable tissues in non-hibernators (Chatfield, Battista, Lyman & Garcia, 1948; Lyman & Chatfield, 1955), nevertheless, many interrelated functions central to the reproductive process, such as hormone production and the capacity of target organs to respond to humoral stimulation, are markedly depressed or inhibited at the low body temperatures of hibernating mammals (Lyman & Dempsey 1951; Wimsatt, 1960). The extent to which this is true during torpidity at higher body temperatures nearer normothermic levels, as for example in aestivators, remains to be investigated.

Given these depressive effects of deep torpor on regulative reproductive processes, it is perhaps logical to suppose that the periods of reproduction and hibernation in mammals should be temporally dissociated, and that in evolution selective pressure might be exerted to this end. However, among most of the relatively few hibernating species concerning which there is sufficient knowledge of the reproductive sequence to make a valid judgement, this logical expectation is only partially fulfilled. Reproduction is a seasonal phenomenon in all of them, as is hibernation, but the two cycles do, in fact, overlap to a greater or lesser degree, and they can, and do, directly interact. If we accept with Hudson (1967) the notion that hibernation 'represents a phyletically recent but highly specialized modification of the basic mammalian homeothermic machinery', rather than a mere survival with modifications, of a primitive physiological organization (as suggested by Cade, 1964), it would seem more probable that the major evolutionary trend involves an adjustment of reproductive sequences to the cycle of dormancy than the converse.

In what follows I shall review first the chronology of seasonal reproductive and hibernation cycles in selected representatives from the three orders of placental mammals in which hibernating forms are known, in order to illustrate the basic temporal patterns of overlap thus far observed. This will be followed by a characterization of gonadal and secondary sexual characters in preparation for, and during, hibernation, and of endocrine and neurophysiological states of hibernators in so far as they bear upon reproductive functions. A possible reproductive role for brown fat will be discussed, followed by a statement concerning the present status of our knowledge of sperm storage mechanisms in the female reproductive tract of hibernating bats and the focus of current studies. Lastly, the influence of natural torpidity cycles on embryonic development will be briefly considered.

TEMPORAL INTERRELATIONS OF REPRODUCTIVE AND DORMANCY CYCLES

Hibernators

Most mammals known to be deep hibernators belong to one of three Eutherian orders: Insectivora, Chiroptera and Rodentia. Among insectivores only the European hedgehog (*Erinaceus europaeus*) has been extensively investigated and is definitely known to hibernate (e.g. Suomaleinen 1953, 1960, 1964). Hibernating bats occur in both hemispheres, but are restricted to two families, Rhinolophidae and Vespertilionidae. In temperate regions speciation has been greater among Vespertilionids, and most hibernating bats belong to this family (Eisentraut 1934, 1956). Quasi-hibernation ('irregular torpor') may occur irregularly in more northerly dwelling species of the family Molossidae (Herreid 1961, 1963 a,b), but in these cases there is no regular pattern of overlap between reproduction and dormant episodes. Rodents display by far the greatest number and variety of hibernating species. These are distributed among several separate lineages, including Dipodoidea, Gliridae, Cricetidae and Sciuridae (Cade, 1964).

Recently Hickman & Hickman (1960) and Bartholomew & Hudson (1962) have reported the occurrence of natural hibernation among certain Australian phalangers (Metatheria). However, so little is known at present concerning the ecological and temporal aspects of either hibernation or reproduction in these marsupials that it would be unprofitable to attempt to discuss their possible interdependence at this time—despite the fact that the subject might hold unusual interest from an evolutionary standpoint.

As previously stated, the cycles of reproduction and hibernation among Eutherians show a distinct seasonal periodicity, and they overlap temporally, although in different ways and to a varying degree in different species. It is important to note that hibernating mammals are all either monoestrous (e.g. ground squirrels, bats) or seasonally polyoestrous (e.g. the hedgehog and some smaller hibernating rodents). Nevertheless, two fundamental patterns may be discerned characterizing the hedgehog and rodents on the one hand, and bats on the other (Fig. 1). In the first case overt reproductive activity follows hard upon the arousal from hibernation in spring, and the entire reproductive sequence from insemination to the rearing of the young is completed before the ensuing hibernal cycle resumes in autumn. Contrastingly, in bats overt reproductive activity, namely oestrus and copulation, is only initiated in autumn, just prior to hibernation. Strikingly, all those phases of the reproductive sequence which follow insemination—specifically ovulation, fertilization, pregnancy and lactation—are held in abeyance through hibernation and proceed only after the animals have resumed normal activity in the spring (Courrier, 1927; Guthrie, 1933; Wimsatt, 1944a, 1960; Herlant, 1953; and many others).

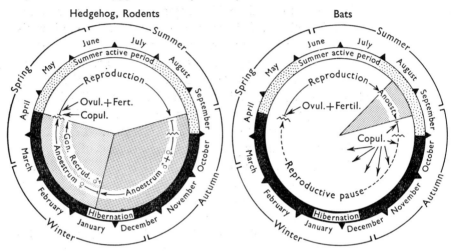

Fig. 1. Schema depicting the temporal relationships between the reproductive and hibernal cycles in representative species of Eutherian mammals. The times of initiation and termination of both cycles are only approximate for they vary somewhat with latitude (cf. also Fig. 2). Schema based on data of many authors (cited in text).

However, an interesting and unique variant of this pattern is observed in hibernating members of the genus *Miniopterus* (Fig. 2). Whereas breeding is again initiated during the prehibernal period, copulation in autumn is followed immediately by ovulation, fertilization and embryogenesis, and the females enter hibernation in a pregnant condition. During hibernation embryonic development is retarded, and does not resume at a normal rate until after the bats emerge in the spring. This condition was first described by Courrier (1927) in the European species *Miniopterus schreibersii* and later confirmed by Planel, Guilhem & Soleilhavoup (1961) and Peyre & Herlant (1963). A similar pattern has more recently been recorded by Dwyer (1963a, b) in *M. schreibersii* and a smaller species, *M. australis*, residing below the equator at 30° S. in New South Wales. The seasons are inverted in the southern hemisphere, but the cycles are nevertheless initiated at times which correspond to late autumn and early winter in Europe, about a month or two later in a seasonal sense than in the European race. On the other hand, nearer the equator—specifically in the New Hebrides at approximately 15° S.—Baker & Bird (1936) found that while both these species retained a clear-cut seasonal periodicity in their breeding cycles, neither hibernated, and gestation was several months shorter than in the European race. Paradoxically, they also noted that the cycle was initiated at a time which corresponds to the northern spring, rather than autumn or early winter as in the European and more southerly races respectively. It is important to note in Fig. 2 that gestation is pro-

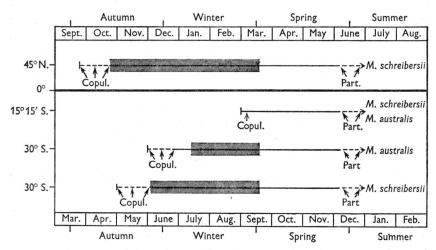

Fig. 2. Schematic representation of reproductive patterns in the genus *Miniopterus* (Vespertilionidae). The arrows indicate the length of the gestation period as a function of the latitude at which the two species depicted (*M. schreibersii* and *M. australis*) reside; the dashed extensions represent the reported spread in the initiation and termination of the pregnancy cycle within the populations studied. The shaded intervals represent a period of retarded embryonic development, and coincide with the hibernal periods. Schema based on data of Courrier (1927), Baker & Bird (1936) and Dwyer (1963 a, b).

longed in *Miniopterus* approximately in proportion to the length of the hibernal period, which in turn coincides with the period of retarded embryogenesis. A cause and effect relationship here seems strikingly apparent.

While it is most likely that these seasonal differences in the initiation of the reproductive cycle in relation to the hibernal cycle represent independently derived genotypic responses to patterns of environmental stimuli among the different lineages of hibernating mammals, it is worth noting that the adaptation nevertheless provides in every case for the birth of the young during a time of year most favourable for their survival.

Aestivators

Some hibernating rodents, such as certain ground squirrels and many pocket mice, also aestivate (Cade, 1964; Hudson & Bartholomew, 1964), but as yet no concurrent patterns of aestivation and reproduction have been discovered. This is not to say they may not exist, for knowledge of natural aestival and reproductive patterns in many supposed aestivators is scanty. It is clear, at least in respect to some aestivating ground-squirrels including *Citellus beldingi, C. columbianus, C. townsendi, C. mohavensis*, and others (Cade, 1964) which bear but a single litter each year—typically in early summer (Asdell, 1964)—that reproduction is completed before the arrival of the aestival period. On the other hand there are some Heteromyids—known aestivators such as *Perognathus baileyi* and *P. penicillatus* (Cade, 1964)—which show two peaks of reproductive activity over the summer, one in May and June, and the second in August (Asdell, 1964). This could reflect an accommodation of reproductive cycles to an intervening aestival period, though it must be emphasized that this is not definitely known to be the case. The subject warrants further investigation by ecologists and reproductive biologists.

The question of the effects of daily torpor on reproduction will be considered later.

Other temporal effects

Further reproductive adaptations of a temporal sort have been observed in various ground-squirrels which live under different environmental conditions and accordingly display different seasonal patterns of behaviour. They involve, first, lengthening or shortening of seasonal reproductive periods correlative with the length of hibernal and/or aestival cycles and, secondly, variations in the actual patterns and rate of post-natal development in relation to environmentally imposed variations in seasonal behaviour.

The first is exemplified by the different patterns of reproductive behaviour described by McKeever (1966) in two sympatric species of California ground-squirrels—*Citellus lateralis* and *C. beldingi*. Both squirrels are hibernators, but a more diversified dietary habit permits *C. lateralis* to remain active for nearly 8 months of the year in contrast to a 4-month active period for *C. beldingi*. Both have a single breeding season shortly after they emerge from hibernation, but the length of this season in both species is proportional to the period of seasonal activity.

The second condition has been described by Pengelley (1966). He studied various post-natal developmental patterns in four species of ground-squirrels: *Citellus lateralis*, which lives above 6000 ft. and hibernates; *C. mohavensis*, a desert species which hibernates and aestivates;

C. tereticaudatus, an aestivating desert species; and *C. leucurus*, a desert-dwelling form which neither hibernates nor aestivates. The parameters measured included gestation length, relative development at birth, postnatal growth-rates, and time of weaning. Pengelley found that the four species presented a graded response in respect to their rates of development, with *C. lateralis* being the fastest, *C. leucurus* the slowest and the others intermediate. He presents a reasonable case for concluding that differences in development rate are adaptive to the animals' particular environments and seasonal behaviour patterns. One could also include in this series the complementary *C. undulatus* (arctic ground-squirrel), a deep hibernator studied by Mayer & Roche (1954) in which a speedy development of the young is equatable with the short summer season of the far north.

GONADAL FUNCTION AND STATUS OF SEX ACCESSORY ORGANS IN HIBERNATORS

The varying seasonal patterns just noted between reproductive and hibernal sequences are accompanied by corresponding differences in the functional cycles of the gonads and sex accessory organs. In most respects the cycles in the hedgehog and rodents are similar in general outline. Minor differences are observed between species in the timing of events, but these are generally attributable to latitudinal and individual variations in the relative length of dormant and active periods, variations in the patterns and frequency of spontaneous arousals during hibernation, depth of torpor, age, and possibly other factors. Contrastingly, the cycles in bats differ markedly from those of rodents and hedgehog—not just qualitatively during hibernation itself, but in over-all temporal sequence resulting from the phase difference in the timing of the reproductive season relative to the initiation of the hibernal cycle. Moreover, the cycles of the gonads and accessory organs in bats differ as between the Miniopterines on the one hand, and other Vespertilionids and the Rhinolophids on the other.

Testis and male accessory glands
Hedgehog and rodents

The cycles of the testis and accessory glands are schematically shown in Fig. 3; the testicular changes may be directly visualized in Fig. 5 (*a, b*). When hibernation begins the two functional components of the testis (seminiferous epithelium and Leydig cells) are virtually quiescent, and the accessory glands (seminal vesicles, prostate, etc.) appear fully involuted. Involution occurs before hibernation begins, but the prehibernal pattern of regression shows temporal variations among the rodents, and a minor qualitative difference between rodents and the hedgehog. In the arctic

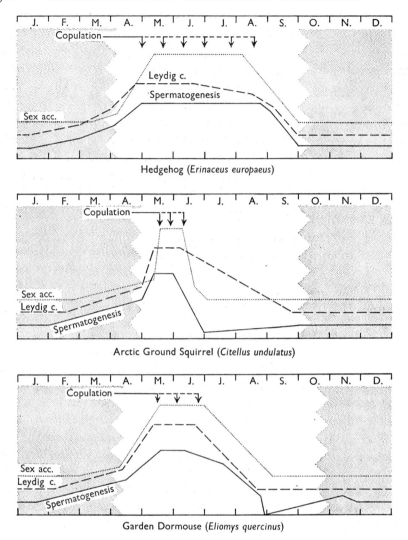

Fig. 3. Annual cycles of testis and male accessory organs in hedgehog and representative rodents. The cycles of the Leydig cells and sex accessories in *Citellus undulatus* are extrapolated from the better-known situation in other species (*C. lateralis*, *C. beldingi*, *C. tridecemlineatus*), in which, however, they are more spread out over a longer period though retaining essentially the same interrelationships. See text for details. ......, accessory glands; – – –, Leydig cells; —, spermatogenesis.

ground-squirrel (*C. undulatus barrowensis*), for example, post-breeding involution of the testis is abrupt, reaching its lowest ebb by late June. However, this is immediately followed, during July, by a partial regeneration of the seminiferous epithelium, involving proliferation of spermatogonia and formation of primary spermatocytes. Regeneration beyond this stage is halted by the onset of hibernation according to Mitchell (1959), but Hock (1960) maintains that prehibernal development goes further, to the formation of secondary spermatocytes. He also postulates that spermatogenesis continues, though slowly, throughout hibernation. A similar but more leisurely pattern of spermatic regression followed by partial regeneration to the primary spermatocyte stage has been noted in the thirteen-lined ground-squirrel, *Citellus tridecemlineatus* (Wells, 1935), two sympatric California ground-squirrels, *C. lateralis* and *C. beldingi* (McKeever, 1966), the woodchuck *Marmota monax* (Rasmussen, 1917), the garden dormouse *Eliomys quercinus* (Gabe *et al.* 1963), the European hamster, *Cricetus cricetus* (Kayser & Aron, 1938), and the golden hamster, *Mesocricetus auratus* (Smit-Vis & Akkerman-Bellaart, 1967). Thus a prehibernal pattern of full spermatogenic regression followed by partial recovery seems to characterize most, if not all, hibernating rodents. The situation in the hedgehog differs only in that testicular involution is relatively delayed (September) and is less profound, being arrested at a stage when primary spermatocytes are still present (Courrier, 1927; Allanson, 1934; Ciani, 1961). The abortive recovery phase characteristic of rodents is not apparent in the hedgehog before hibernation.

However, once hibernation has begun, subsequent developments are basically the same in the hedgehog and rodents. During the earlier months of torpidity there is no substantial change in testicular condition (except perhaps in the arctic ground-squirrel—Hock, 1960). But usually shortly after the mid-hibernal interval (January to early March in various species), spermatogenic recrudescence begins, and continues at a gradually accelerating pace through the remaining weeks of dormancy. As a consequence, peak spermatogenic activity is achieved soon after the end of hibernation. This peak is reached almost immediately, for example, in far northern species with short breeding seasons such as the arctic ground-squirrel, but somewhat later in the hedgehog and in rodents living at more southerly latitudes. In young dormice, *Eliomys* (Gabe *et al.* 1963), and perhaps in other species in which the males mature sexually during the first spring following their birth, initiation of spermatogenesis during hibernation is retarded somewhat relative to the time of its beginning in adults.

Smit-Vis & Akkerman-Bellaart (1967) have recently studied the effects of hibernation on testicular function in the golden hamster in which the seasonal testicular cycle generally resembles that of other hibernating rodents. In hibernating animals they noted the usual autumnal decline in testis weight (relative to body weight) followed by a gradual increase beginning in January and progressing through the remainder of hibernation; in non-hibernated control animals there was only a slightly greater increase in testis weight. However, they noted a marked difference in respect to cell differentiation; specifically, spermiogenesis was much retarded in the hibernating specimens in comparison to the non-hibernating ones. The authors infer that the increasing testicular weight was a consequence of cell proliferation (which cells not specified), and that the discrepancy between proliferation and differentiation in the hibernating specimens possibly reflects an independent control of proliferative and differentiative functions involving a 'different endocrine balance'. The authors do not effectively exclude alternative explanations for the increasing testicular weights during the later months of hibernation, nor in fact do they actully demonstrate cell proliferation in the hibernating specimens. In view of the known depressive effects of the low body temperatures of hibernators on cell proliferation kinetics in intestinal tract, liver, haemopoietic organs, etc. (Thomson, Straube & Smith, 1962; Mayer & Bernick, 1958; Adelstein, Lyman & O'Brien, 1967), it is questionable whether Smit-Vis & Akkerman-Bellaart have demonstrated an actual differential response to hibernation between cell proliferation and differentiation in the testis of the hibernating hamster.

It is probable that the sexual recrudescence observed during the later stages of hibernation in the hedgehog and rodents is greatly facilitated, and perhaps even dependent upon the increasing tempo of temporary spontaneous arousals that occur at this time, for the capacity of reproductive organs at least to respond to hormonal stimulation while subject to the low body temperatures of the deeply hibernating animal is virtually nil (Lyman & Dempsey, 1951; Wimsatt, 1960). Repeated brief returns of body temperature to near normothermic levels, on the other hand, permits hormone synthesis and release and accelerates the responses of the target organs.

The interstitial cell cycle in the testis of the hedgehog and rodents is synchronized more or less with that of the seminiferous tubules, although both species-specific and individual variations are noted. Histological examination seems to indicate that the Leydig cells are most involuted— and presumably non-functional, when hibernation begins in the fall (Rasmussen, 1917; Courrier, 1927; Allanson, 1934; Ciani, 1961; Gabe *et al.* 1963). They remain atrophic in appearance until relatively late in the hibernal period—often later than the beginning of spermatic growth, when proliferation of Leydig cells coupled with gradual hypertrophy and lipid accumulation signal the start of a new growth phase. The cells seem to achieve maximal development in most species shortly after the spermatogenic peak. Typically, however, the post-breeding involution of the interstitial tissue is less abrupt than that of the spermatic tubules, although it is more prolonged in some species than in others. It differs also in lacking an abortive recovery phase before hibernation. The apparent length of the observed 'histological cycle' of the Leydig cells is obviously greater in all cases than the duration of their effective functional phase. Ordinary histological methods, which by and large have been the only ones used, are inadequate for determining precisely when secretion begins and ends, and histochemical methods based on lipid reactions are not much better for they lack real specificity. The responses of the accessory glands are better indicators of effective hormone synthesis and release. It is evident in most species that visible growth of the accessories begins after arousal, later than the first appearance of hypertrophy in Leydig cells during hibernation, and usually quite some time after spermatic recrudescence has begun. It is also true that regression of the accessories is usually terminated before the Leydig cells complete their histological involution in the fall. Gabe et al. (1963) have reported a slight stimulation of male accessory glands late in hibernation in the dormouse, but this has not been noted in most other hibernating rodents. Precise determination of sequential seasonal changes in effective androgen levels by modern histochemical, biochemical or bioassay methods has not to my knowledge been attempted in any mammalian hibernator under natural conditions, and they are sorely needed.

Bats

The annual cycle of the testis and accessory glands in bats and the discrepancies between the Miniopterines and the other hibernating species have been summarized fully elsewhere (Wimsatt, 1960). The essential points are depicted in Fig. 4 and are based on the findings of Courrier (1927), Miller (1939), Pearson, Koford & Pearson (1952), Krutzsch (1956), Peyre & Herlant (1963), Gaisler & Titlbach (1964), and others.

In all genera studied thus far (*Myotis, Eptesicus, Pipistrellus, Plecotus, Rhinolophus, Miniopterus*) the cycle of the germinal epithelium is the same. Throughout hibernation the tubules are fully involuted, containing only Sertoli cells and a few spermatogonia (Fig. 5 c, d). The typical mid-hibernal recrudescence of spermatogenic activity observed in the hedgehog and rodents definitely does not occur. Within a few weeks of emergence from hibernation, however, spermatogenesis begins, and gradually accelerates through the summer months to a peak in late August. Thereafter involution is abrupt, and the atrophic hibernal condition is achieved by late September or October.

Some discrepancy is noted in the accounts of earlier authors in respect to the cycle of the interstitial cells in bats (Wimsatt, 1960). In the genera *Miniopterus, Myotis* and *Plecotus* most authors seem to agree that the Leydig cells are almost fully involuted soon after hibernation begins. This condition is maintained through the period of dormancy and into early summer,

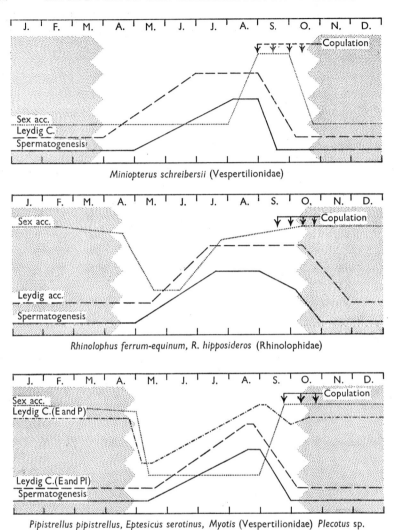

Fig. 4. Annual cycles of testis and sex accessory glands in vespertilionid and rhinolophid bats. See text for details.

at which time the cells begin to hypertrophy in phase with renewed spermatogenic activity. They achieve maximal development in late August, and then enter the involutionary phase which continues into October. According to Courrier (1927), however, the pattern is somewhat different in European members of the genera *Eptesicus* and *Pipistrellus*. Involution of the Leydig cells is slight during September, and the animals enter hibernation with cells still showing substantial evidence of functional activity. This condition is maintained throughout hibernation, but is followed by an

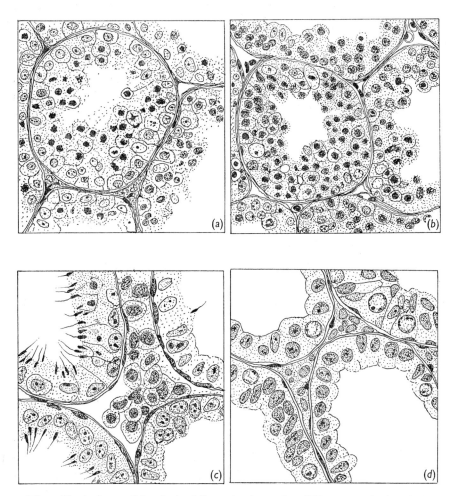

Fig. 5. Testicular condition during hibernation in a rodent (*Eliomys quercinus*) and a bat (*Corynorhinus rafinesquei*). In (*a*) and (*b*) the condition of the seminiferous tubules in the garden dormouse is shown at the beginning of hibernation in late October, and at the time of arousal in March. In the October specimen prehibernal revival has progressed to the formation of numerous spermatogonia and primary spermatocytes. The March specimen is characterized by a great increase in primary spermatocytes, which are highly proliferative. (Drawn from photomicrographs of Gabe et al. 1963.) In (*c*) and (*d*) the condition of the tubules and Leydig cells of the bat are shown as they appear at the beginning of hibernation in early October and near mid-hibernation in late December. Spermatogenesis has ceased and only a few sperm, spermatids, and spermatogonia remain. By December only spermatogonia and Sertoli cells remain, and this condition prevails until some time after arousal in early summer. The Leydig cells are already involuting in the October specimen in comparison with their condition in late summer, and are seen to be further reduced in the December specimen. Nevertheless, both conditions shown coincide with maximum hypertrophy of the sex accessory glands. (Drawn from photomicrographs of Pearson et al. 1952.)

abrupt and complete regression coinciding with arousal from hibernation in the spring. However, their pattern of rejuvenescence during the summer resembles that of other vespertilionids. These discrepant patterns among vespertilionids are separately depicted in Fig. 4.

Except in the genus *Miniopterus*, the cycle of the male accessory glands in bats differs fundamentally from that in the other mammalian hibernators in a way that possibly indicates a uniquely exaggerated temporal disjunction of the spermatogenic and endocrine activities of the testis. During the summer when the Leydig cells are experiencing rapid hypertrophy the accessory glands remain small and non-functional. Beginning in early September—following the collapse of spermatogenesis, and coinciding with the apparent involution of the interstitial tissue (except in *Pipistrellus* and *Eptesicus*, as noted)—the accessories suddenly experience a rapid and luxuriant growth, attaining full size and functional status within 2 weeks. This condition is maintained as the males enter hibernation and in most species visible involution of the accessories is not observed until late in the hibernal period. Their complete collapse, however, is usually postponed until the time of arousal from hibernation in the spring.

A different picture is seen in hibernating members of the genus *Miniopterus*, in which the hypertrophy of the male accessory glands corresponds temporally, and in luxuriance, to that of other vespertilionids, but the regressional cycle differs. Profound involution begins immediately after the autumn breeding period and the accessories are fully involuted when the males enter hibernation. Moreover, the masses of sperm which are retained in the epididymis of other vespertilionids throughout most of hibernation (Plate 1) are greatly reduced or absent in hibernating *Miniopterus*. Thus the reproductive cycle in male *Miniopterus* shows no real overlap with the hibernal period and is not qualitatively different from that of many mon-oestrous non-hibernating mammals. This contrasts with the other vespertilionids in which, except for the spermatogenetic phase, the male cycle carries into, and substantially through, the hibernal period. In bats at least *Miniopterus* may well be a key animal in reflecting the directions of evolutionary adaptation between reproduction and hibernation.

Ovary and female accessory organs

Hedgehog and rodents

Systematic accounts of conditions in the female reproductive organs during natural hibernation and of the effects of hibernation on the reproductive cycle unfortunately are relatively few in so far as hibernating mammals other than bats are concerned; the cycle in male hibernators has attracted

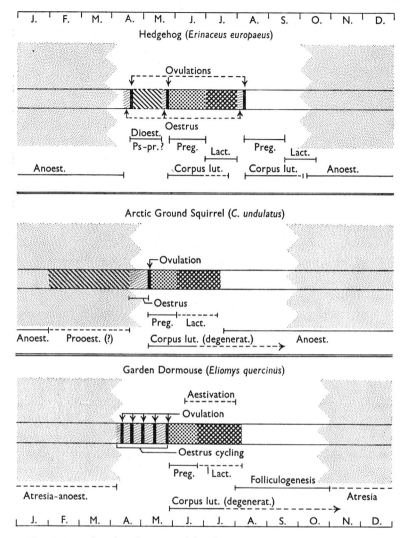

Fig. 6. Annual cycles of ovary and female accessory organs in the hedgehog and representative rodents. See text for details.

far more attention. Some authors have, however, provided bits of information concerning the hedgehog and a few rodents which enable us to discern some overlap and possible interaction between dormancy and female reproductive processes in some species at least. Interspecific variations in the reproductive sequences during the summer active period need not concern us here; only the prehibernal changes and those that have been noted during hibernation itself will be considered. The patterns are depicted in Fig. 6.

The most recent detailed account of the female cycle in the hedgehog is that of Girod, Dubois & Cure (1967) who report folliculogenesis in the ovary following parturition, coinciding with involution of the corpora lutea of pregnancy. After hibernation begins, from October onwards, these elements and the ovarian interstitial ('thecal') cells show progressive atresia, a process which is still under way when hibernation ends. The principal effect of dormancy here seems to be no more than a slowing down of a generalized ovarian involution which began in the autumn. The reproductive tract, as might be expected, is wholly quiescent during hibernation.

Among some rodent hibernators, however, a different picture obtains. Rasmussen (1918) described in the woodchuck (*Marmota monax*) an annual periodicity of ovarian interstitial cells characterized by a gradual hypertrophy during winter, followed by a more rapid enlargement after hibernation. He also noted a marked folliculogenesis in autumn, with many follicles developing to an advanced Graafian stage. Some of these are still apparent near the end of hibernation, and although their subsequent fate was not recorded, they presumably degenerate and are succeeded by the definitive generation of ovulatory follicles following arousal.

A still more intimate and positive relation between hibernation and the female reproductive sequence has been noted in at least two species of ground-squirrel, *Citellus tridecemlineatus* (Foster, 1934; Moore *et al*. 1934) and *C. undulatus* (Hock, 1960). In the thirteen-lined ground squirrel the ovaries and all accessories have become inactive by late summer and autumn. At mid-hibernation, however (January), the ovaries begin to show signs of activity, notably the gradual development of smaller follicles, and the appearance of antra in the larger ones preceding arousal in the spring. The gradual follicular growth is said to be accompanied by a slow development of oviducts, uterus and vagina. All of these changes are accelerated upon arousal, and oestrus occurs soon thereafter. Moore *et al*. (1934) also noted that older females emerge in a more advanced state of reproductive activity than do the young of the preceding summer. The picture in the arctic ground-squirrel is similar, but apparently ovarian development is even more precocious during hibernation. According to Hock (1960) females at emergence already possess ovarian follicles ready to rupture. Hoch also maintains that follicular growth continues slowly through the winter, but provides no evidence in support of this contention.

Finally, Gabe *et al*. (1963) have briefly examined changes in the ovaries of the dormouse (*Eliomys quercinus*) in relation to hibernation. They report a marked prehibernal folliculogenesis following lactation which leads to the formation of a number of Graafian follicles, some of which are observable in the ovaries throughout the hibernal period. They consider that the further development of these persisting follicles is arrested during the dormant period but resumes after the dormice emerge from hibernation. However, in the photomicrographs of Gabe *et al.* all of the vesicular follicles shown—which are presumed by the authors to be normal—display evident signs of atresia. It appears more likely that dormancy has slowed down the atresia of the follicles in question and thereby assured their persistence over a longer period, and that the actual ovulatory follicles are formed late in hibernation or following arousal. A more detailed study of the female reproductive cycle in *Eliomys* is clearly needed.

Bats

In bats hibernation and the reproductive cycle in the female are intimately interwoven—an inevitable consequence of the full seasonal overlap of the two cycles (Fig. 7). Moreover, the effects of dormancy on reproductive processes appear more profound in the female than in the male, and presumably require more complex physiological adjustments.

The simplest condition is that observed in *Miniopterus*, in which females enter hibernation already pregnant, resulting simply in a retardation and prolongation of embryogenesis. It may be that the embryo has evolved special physiological adaptations enabling it to survive long periods of cold stress, but this possibility has not been experimentally investigated. Most of the recent information concerning the reproductive biology of the two known hibernating species of Rhinolophidae (*R. ferrum-equinum, R. hipposideros*) derives from the work of Matthews (1937), Gaisler & Titlbach (1964) and Gaisler (1966). The general chronology of events in both sexes resembles that of hibernating vespertilionids, but an important difference will be noted in respect to the condition of the ovary during hibernation (see below). However, most work to date has concentrated on members of the ubiquitous genus *Myotis*—*M. myotis* in Europe and *M. lucifugus* in America—and the details of reproduction are best documented in respect to this genus (cf. Guthrie, 1933; Wimsatt, 1944a, 1960; Wimsatt & Kallen, 1957; Wimsatt & Parks, 1966; Herlant, 1953, 1956).

The truly unique features of the female reproductive cycle in hibernating bats (except *Miniopterus*, see above) involve (*a*) the prolonged storage of spermatozoa in the reproductive tract of the dormant female without loss of viability or fertilizing capacity, and (*b*) in vespertilionids postponement until spring of the terminal maturation and ovulation stages of the definitive Graafian follicle which is already well developed in the ovary at the beginning of hibernation in the autumn (Fig. 7). A unique chemical adaptation of the overwintering follicle has been described elsewhere (Wimsatt & Kallen, 1957; Wimsatt & Parks, 1966) and will not be elabo-

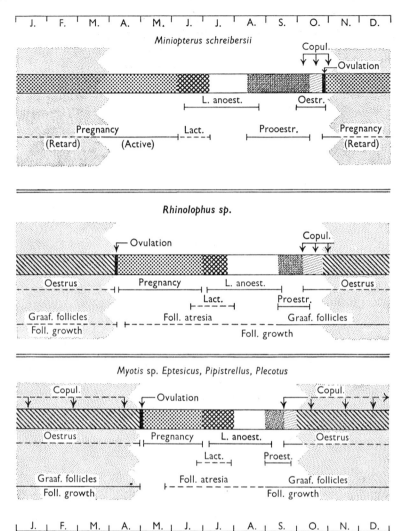

Fig. 7. Annual cycles of ovary and female accessory organs in vespertilionid and rhinolophid bats. See text for details.

rated upon here, except to say that it involves provision within the follicle of a special energy source which is presumably adaptive to follicular longevity under hibernal conditions. Rhinolophids apparently differ from vespertilionids in that the functional ovary contains numerous Graafian follicles during hibernation, a few of which gradually gain ascendancy over the others, which degenerate. But by spring only one of these is said to survive to ovulate (Matthews, 1937). Apparently none of these follicles display the conspicuous chemical marker (glycogen-laden discus cells) so characteristic of the surviving follicles of vespertilionids, and it must be admitted that no one has really demonstrated that the ovulatory follicle has actually persisted in the ovary throughout hibernation, although this is generally assumed to be the case. The facts may actually be otherwise, for Gaisler (1966) writes in reference to *Rhinolophus hipposideros* that '*the formation of a Graafian follicle*, the ovulation and development of (*a*) corpus luteum takes place in April, shortly after the end of hibernation' (italics mine). It is obvious therefore that important differences in detail relative to ovarian structure and function do indeed exist between hibernating rhinolophids and vespertilionids, despite the general similarities of their cycles in most other respects.

Despite its obvious theoretical and practical interest, the phenomenon of sperm survival in the female reproductive tract which has been known for almost 90 years (cf. Hartman, 1933) has attracted relatively little attention until recently, but we shall defer discussion of it until later.

SOME ENDOCRINE FACTORS AND REPRODUCTION IN HIBERNATORS

The involvement of the endocrine organs in the physiology of natural hibernation has long been a focal point of interest and speculation, and continues to be vigorously pursued. The field has yet to be comprehensively reviewed, but limited summations may be found in the writings of various authors (e.g. Kayser & Aron, 1950; Lyman & Chatfield, 1955; Kayser, 1957a; Popovic, 1960; St Girons et al. 1961; Smit-Vis, 1962; Gabe et al. 1963; Hoffman, 1964a, b; Agid, Gabe & Martoja, 1967). A general coverage of this vast subject lies beyond the scope of the present discussion, which will stress only some of the broader aspects of endocrine involvement in reproductive processes that are more or less peculiar to hibernating species.

By way of background we need to recall that in all known mammalian hibernators the prehibernal period of 'preparation' is characterized by a generalized functional involution of the major endocrine glands, especially involving the adenohypophysis, thyroids and, except in female bats, the gonads (Lyman & Chatfield, 1955, and Kayser, 1957a, review the earlier literature). Formerly it was generally believed that the adrenal cortex shared in this involutionary pattern, but more recent findings have necessitated some modification of this view (Deane & Lyman, 1954; Engel, Raths & Schulze, 1957; Kayser, 1957b; Petrovic & Kayser, 1958; Suomalainen, 1960; Smit-Vis, 1962; Gabe et al. 1963). Other endocrine glands not directly dependent on the anterior pituitary, such as the adrenal medulla and the islets of Langerhans, demonstrate during hibernation in most species an accumulation of hormonal secretion antecedents, but there is little evidence that they are regularly discharged except in conjunction with periodic or permanent arousals (Zajaczek & Kaminski, 1947; Skowron & Zajaczek, 1947; Kayser & Aron, 1950; Mosca, 1956; Suomalainen & Uuspää, 1958; St Girons et al. 1961; Smit-Vis, 1962; Gabe et al. 1963; Agid et al. 1967).

It is not to be anticipated of course that the basic hormonal mechanisms regulating reproductive functions are any different in hibernators than in non-hibernators, and it has been well documented in numerous experimental studies that the reproductive organs of hibernators respond comparably to those of non-hibernators to castration, hypophysectomy, and parenteral administration of sex hormones, gonadotrophins, etc. (Herlant, 1932; Moore et al. 1934; Johnson, Gann, Foster & Coco, 1934; Courrier, 1924; Allanson & Deanesley, 1934; Caffier & Kolbow, 1934; Wells, 1935; Zajaczek & Kaminski, 1947). Moreover, in the hedgehog and in a number of hibernating rodents, it has been demonstrated that the endogenous seasonal rhythms of reproduction can be overturned by appropriate hormonal manipulations and breeding behaviour induced at other times of the year (Allanson & Deanesley, 1934; Moore et al. 1934; Wells & Moore, 1936).

Thus the major differences between hibernators and non-hibernators do not involve different hormonal mechanisms *per se*, but rather variations in the patterns of reproductive periodicity, in the relative rates of sexual development, and in the synchronization of specific endocrine functions which directly influence sexual development. Many of these differences would seem to be directly related to the nature and degree of seasonal overlap between the hibernal and reproductive phases, the immediate depressant effects of dormancy and low body temperatures on the synthesis and release of hormones and the responsiveness of target organs as already mentioned, and perhaps most importantly, to presumably different patterns of response of the neuroendocrine centres regulating hypophyseal activities to endogenous and/or environmental stimuli. This last has been the most neglected thus far in hibernators, at least in so far as reproductive parameters are concerned.

Reciprocal functional interrelationships between the reproductive organs and endocrine glands which directly regulate their activities—as, for example, between the adenohypophysis and the gonads, and between the gonads and the sex accessory organs, have of course been established in numerous hibernators (Rasmussen, 1921; Courrier, 1927; Herlant, 1932, 1956; Wells, 1938; Azzali, 1953; Petrovic & Kayser, 1956, 1957; Hoffman & Zarrow, 1958; Wimsatt, 1960; Peyre & Herlant, 1963; Gabe et al. 1963; Girod et al. 1967; and others). They are similar in all respects to the corresponding functions in non-hibernators. However, many more subtle factors relating to the physiological interdependence between seasonal dormancy and reproductive development have been little explored. One such involves the extent to which the varying reproductive patterns observed in hibernators actually depend upon, or are directly regulated by, a hibernal 'preconditioning'.

The substance of this question and some of its implications for future investigations can be illustrated by two examples. Moore et al. (1934) found that female thirteen-lined ground-squirrels maintained for prolonged periods at room temperature without hibernation exhibited little or no tendency to reproductive development, while other females kept in cold and darkness for several months and allowed to hibernate did exhibit normal sexual development—and at any time of year. Contrastingly, they also observed that males displayed normal seasonal sexual development with either treatment, except that breeding condition was achieved somewhat earlier in non-hibernating than in hibernating males. Similar observations were also made by Wells (1935) and Landau & Dawe (1960). It might be concluded that either the neural centres regulating gonadotrophin release from the hypophysis are differentially responsive to endogenous or environmental stimuli in males and females, or they are responsive in males to an endogenous ('circadian') rhythm which is lacking in females. An answer to this question has not yet been provided.

The second example involves the ovulation responses of the vespertilionid bat *Myotis lucifugus*. When females of this species (and most other hibernating vespertilionids, cf. Wimsatt & Kallen, 1957) enter hibernation in the autumn one ovary already contains the definitive Graafian follicle destined to ovulate the following spring. If the animals are brought out to room temperature during the earlier months of hibernation the follicle fails to ovulate, and eventually degenerates (Wimsatt, 1944b). This indicates that increased metabolism and elevated body temperature are not sufficient in themselves to allow ovulation to occur. Injections of thyroxine, ACTH, somatotrophin and other hormones involved in the regulation of a variety of metabolic functions were likewise ineffective in producing follicular rupture in early winter (Wimsatt, 1960), although injections of pituitary gonadotrophins or abdominal infusions of homogenized bat pituitary glands readily induced it overnight provided the animals were maintained at room temperature following the injections. These treatments failed to induce ovulation, however, if the bats were immediately returned to hibernation after the injections. By contrast, bringing females to room temperature any time after mid-hibernation resulted in the spontaneous ovulation of the persistent follicle, and the more promptly the closer to the end of hibernation the bats were awakened. Herlant (1956) concluded, on the basis of a thorough histophysiological study of seasonal changes in the adenohypophysis of a related European species *Myotis myotis*, that ovulation in the bat as in other mammals is precipitated by an abrupt release of luteinizing hormone (LH) from the hypophysis, and Wimsatt (1960) later suggested that the characteristic ovulatory delay in bats results from a seasonally asynchronous release of follicle sensitizing hormone (FSH) and LH. Thus, in the bat, premature arousal from hibernation in spring precipitates LH release, whereas arousal during the autumn does not. This phenomenon again suggests the possibility that a period of hibernal 'preconditioning' is required to provoke a response from the hypothalamic neurosecretory centres controlling LH release from the hypophysis, or alternatively, the existence of an endogenous seasonal rhythm which triggers hypothalamic activity. In either case some needed stimulus is lacking during the autumnal 'refractory period', or conceivably a seasonally conditioned inhibitory influence on the hypothalamic neurosecretory centres is operative early in hibernation, and gradually dissipates during the later months of dormancy. Correlative studies of reproductive and hypothalamic functions have unfortunately not yet been attempted in bats.

It might also be mentioned here that a seasonal asynchrony in the release of pituitary gonadotrophins (FSH and LH) is likewise implicit in the hedgehog and those male rodents in which spermatic recrudescence appears to be dissociated from the hypertrophy of the sex accessory glands by a considerably longer time interval than is characteristic of related non-hibernators. It is thus evident that natural hibernation not only 'dampens' the hormonal responses of target organs, but often tends to exaggerate temporal asynchronies in the synthesis and release of hormones by the endocrine glands.

Further effects of cold exposure (with or without torpidity) on the periodic reproductive behaviour of hibernators have been detected by experimental approaches involving alteration of the normal seasonal periods of cold exposure, and by other means. Wells (1935) and Wells & Zalesky (1940) found, for example, that if male thirteen-lined ground-squirrels were exposed to a temperature of 4° C. during the breeding season they failed to show the testicular regression characteristic of normal males, and that this condition could be maintained indefinitely at low temperatures. They further found that exposure of males to low temperatures during the sexually quiescent period failed to hasten appreciably the normal onset of sexual maturation. Further analysis also indicated that pituitary gonadotrophin levels did not decline in animals exposed to cold during the breeding season and subsequently maintained at low temperature, whereas the converse effect was noted in animals maintained at high temperatures. They could find no evidence that light was an effective stimulus on the male sex cycle, but concluded that low environmental temperature at least partially stimulated sexual maturation in the male ground-squirrel.

While the status of the common American chipmunk, *Tamias striatus*, as a regular deep hibernator has not been fully clarified, an effect of prolonging the normal seasonal period of cold exposure on reproductive development has been reported by Panuska (1965). Chipmunks are polyoestrous, but breeding is first initiated in the northern United States in March, a time which coincides with the peak of spermatogenic activity. Male chipmunks were maintained at 3° C. from early February into June; control animals were maintained at 25° C. The controls became sexually active at the normal time during March and April, but the cold-exposed males displayed a 2-month delay in the onset of spring spermatogenesis.

Lyman & Dempsey (1951) castrated adult male hamsters, and after they had been induced to hibernate, injected them with 5 mg. of testosterone. Some of the testosterone-injected animals resumed hibernation, whereas others did not. In the latter the injected male hormone stimulated a substantial growth of the seminal vesicles, but the accessories of those which re-entered hibernation were virtually unaffected. They drew the obvious and important conclusion that the level of circulating hormone in these cases was a less significant factor than the transient condition of the target organs themselves.

Foster, Foster & Meyer (1939) were unable to induce hibernation in the thirteen-lined ground-squirrel during the breeding season, and noted that the tendency to hibernate in spring diminished in both males and females from the time the testes became scrotal and the vagina became swollen, respectively. They also found that gonadectomy during the active period failed to enhance hibernating ability, whereas hypophysectomy at this time rendered the animals 'poikilothermic'. However, because of the widespread effects of hypophysectomy on other endocrine glands this experiment tells us little concerning the immediate effects of the gonads on the induction or suppression of torpor. Johnson (1927, 1930) observed that castrated male and female ground-squirrels hibernate more (fewer spontaneous arousals) than non-castrates—a difference which was restricted, however, to the spring phase of the hibernal cycle.

Recently Grindeland & Folk (1958, 1962) found that in female hamsters prolonged exposures to cold of 30–90 days suppressed oestrous cycling and induced anoestrum, although by 90 days partial adaptation of reproductive function to cold had occurred. The effect was reversed, and normal cycling resumed, when the animals were returned to room temperature. Barnett & Manly (1959) found that maintenance of laboratory mice for long periods (28 weeks) at 3° C. led to a delay in the onset of breeding, high infant mortality, and greater variation in reproductive performance than among the controls.

These various experimental results nearly all point in the same direction, namely, they indicate that there is a considerable measure of mutual antagonism between hibernation on the one hand and reproduction on the other. Hoffman (1964a) has defined the problem in more specific terms, albeit tentatively. He suggests that 'hibernation does not occur in the presence of elevated levels of LH or ICSH', and that 'a certain level of these gonadotrophins, or indirectly the sex hormones which they regulate, maintains or conditions homeothermic existence'. He correctly reminds us, however, that a cause-and-effect relationship cannot be definitely established here until the annual cycle and physiological changes in long-term castrates have been assessed. It is nevertheless tempting to speculate that in hibernators reproductive recrudescence in spring, or more likely the reciprocal endocrine activities which induce it, may be one of the important causative links in the sequence of physiological changes which culminate in the seasonal termination of the dormant state.

THE NEUROENDOCRINE SYSTEM AND REPRODUCTION IN HIBERNATORS

Unfortunately the involvement of hypothalamic neurosecretory centres in the regulation of hypophyseal gonadotrophic activities has not been directly assessed as yet in hibernating mammals; I could find not a single experimental study directed to this end, despite the enormous progress that has been made with a variety of common non-hibernating mammals and lower vertebrates (cf. van Tienhoven, 1968). To be sure there have been a number of essentially anatomical studies involving the hedgehog, ground-squirrel, dormice and bats, but these have focused mainly on the organization and interrelations of hypothalamic nuclei and nerve tracts and on gross seasonal changes in the over-all distribution and concentration of neurosecretory materials (*hedgehog*: Azzali, 1955; Suomaleinen, 1960; Kiernan, 1964; Girod, Dubois & Cure 1966; Campbell & Holmes, 1966; *dormouse*: Azzali, 1954; Ottaviani & Azzali, 1954; Gabe *et. al.* 1963; Legait, Legait & Burlet, 1966; *ground-squirrel*: Kolaczkowski & Wender, 1958; Benetato, Daneliuc, Nestianu & Gabrielescu, 1965; *bats*: Azzali, 1953, 1954; Barry, 1954, 1956; Barry, Besson & Lamarche, 1958; Troyer, 1960, 1961, 1964, 1965). Those physiological correlations that have been suggested relate to non-reproductive functions such as salt and water balance and species differences in the functional characteristics of metabolism, etc. (cf. Azzali, 1954; Troyer, 1961).

The various cytological methods currently used for visualizing neurosecretory granules (Gomori's chromalum-haematoxylin-phloxine, paraldehyde fuchsin, alcian blue at pH 2·0, sulphhydryl and disulphide reactions, etc.) are relatively non-specific, and do not discriminate between the different neurohumors (or their carriers) such as the posterior lobe hormones on the one hand and gonadotrophin releasing factors on the other. Progress by means of the cytological approach must await the development of more selective staining or histochemical procedures, and/or a more concerted effort than has been attempted up to now in hibernators to relate neurosecretory changes in specific areas of the hypothalamo-hypophyseal complex to experimentally defined reproductive states. Progress will also require the utilization of other experimental approaches that have proved so effective in localizing within particular hypothalamic areas of non-hibernating mammals and birds the sources of specific neurohumoral substances involved in regulation of anterior pituitary functions. Such studies are likely to be very rewarding in hibernators, for the seasonal prolongation of various cyclic phenomena greatly extends the temporal range of pituitary and hypothalamic interaction and neurosecretory functions, perhaps making them more accessible for analysis in respect to some functions at least than in many non-hibernators. In all hibernators thus far examined the dormant state is characterized by a marked accumulation of neurosecretory products in such principal centres as the supra-optic and paraventricular nuclei, their associated fibre tracts, and in the pars nervosa of the hypophysis. From what is already known concerning the generally low gonadotrophic potency of the anterior pituitary during seasonal hibernation, it is reasonable to suggest that some of these stored neurosecretory materials may represent gonadotrophin-releasing factors, but this remains to be demonstrated. Arousal from hibernation, either artificially induced or spontaneous, seems to result in a rapid and generalized decline in the amounts of stainable neurosecretory materials, and presumably signals their release into the circulation. Marked species differences in the localizations and abundance of stored neurosecretions, and in their patterns of discharge upon arousal, have not been adequately defined, although the garden dormouse (*Eliomys quercinus*) is said to show less over-all evidence of neurosecretory activity during hibernation and upon arousal than other hibernators (Azzali, 1954; Gabe *et al.* 1963; Legait *et al.* 1966).

The possible involvement of the pineal gland in the normal reproductive processes of hibernators has not been assessed as yet, although progress is being made in respect to non-hibernators. Since the pineal may have an inhibitory influence on gonadal function, and may be subject to endogenous control through its sympathetic innervation, it could conceivably be an important link in the endogenous, regulative complex controlling reproductive periodicity in hibernators.

A REPRODUCTIVE ROLE FOR BROWN FAT ('HIBERNATING GLAND')?

'Brown fat' is an unusual form of adipose tissue, dark in colour, gland-like in texture and appearance, more restricted in its bodily distribution than white fat, present in some mammals but not in all, and in general better developed in hibernators than in non-hibernators. This enigmatic tissue, whose functional role was so long obscure, has been subjected in the last three decades to searching analysis, and its physiological individuality at least now appears firmly established (cf. reviews of Wells, 1940; Wertheimer & Shapiro, 1948; Johansson, 1959; Joel, 1965; Smalley & Dryer, 1967). The best documented of the many suggested functions for brown fat is its apparent role as a thermogenic source (presumably through the *in situ* oxidation of its own energy supply—Wells, Makita, Wells & Krutzsch, 1965) during cold exposure, the effect being especially noticeable in hibernating mammals during the initial phases of the arousal sequence (Smith & Hock, 1963; Smalley & Dryer, 1963; Smith, 1964; Hayward & Lyman, 1967). There are some indications, however, that brown fat may have other, non-thermogenic functions, and I shall discuss briefly a few evidences from the literature which suggest a possible role for brown fat in the reproductive processes of some hibernators.

The first concerns the now nearly 30-year-old observations of Sweet & Hoskins (1940) on the so-called 'hibernating gland' (interscapular and axillary brown fat) of the woodchuck, *Marmota monax*. These authors demonstrated in bioassay that in summer mixed brown fat of males and females contained 100 μg. of androgenic substance (androsterone reference) per 50 g. of tissue, an amount equivalent to the androgen concentration in bull testis, the richest natural source known. By comparison extraction of an equivalent amount of androgen from axillary fat in men and women required 1 and 3 kg. of fat respectively. Based on the knowledge that brown fat is metabolized at a much slower rate than white fat during hibernation, and that it involutes very rapidly following arousal in the spring coincident with the mating season, they inferred that androgen may be stored in brown fat after its formation in the testes, though to what end they did not speculate. Assuming the correctness of their basic finding, I might do it for them and suggest the possibility that the very rapid sexual development of the woodchuck after emergence from hibernation could be accelerated, at least initially, by release of this extra-gonadal source of 'stored' androgen.

Recently Krutzsch & Wells (1960) prepared non-saponifiable lipoidal extracts from brown fat tissue of adult male hibernating bats, *Myotis lucifugus*. These were injected into weanling rats to determine possible androgenic effects on the growth of the seminal vesicles in comparison with various concentrations of injected testosterone. They record a 'notable androgenic stimulus' by brown-fat extracts in this bioassay procedure, and calculated it to be equivalent to 678 μg. testosterone/g. of tissue, a remarkable amount of androgenic activity vastly exceeding the concentration in bull testis. Control experiments undertaken with similarly prepared extracts of testis, adrenal and liver of male bats revealed only 'insignificant' quantities of androgen in these organs. They also carried out a preliminary analysis of the brown-fat extracts by gas chromatography and separated a compound having the retention time characteristics of C-19 steroids, though the retention time was unlike that of any previously reported steroid, nor did they find it to be identical to that of testosterone or androsterone analysed at the same time as reference compounds. The material does not yet appear to have been further identified.

It is regrettable that Krutzsch and Wells did not assay also the brown fat of immature males and females, for if the androgenic material were to be found only in adult males it might have special significance in relation to a puzzling feature of the male reproductive cycle in *Myotis*. I remarked earlier that enlarged and functional sex accessory glands and continuing libido coexist during hibernation with what appear to be fully involuted testes. At least four theoretical explanations for this suggest themselves: (1) histological appearances of Leydig cells notwithstanding, androgen production might continue in the testis, perhaps at reduced levels consistent with low body temperature, but in still sufficient amounts to maintain the accessories, which are themselves less responsive to stimulation in the cold; (2) the testis might cease androgen synthesis before hibernation, but sufficient 'residual' hormone—protected from being metabolized because of the cold—might persist in the testes or elsewhere to sustain the accessory glands; (3) the accessories may be maintained by androgens from a non-gonadal source such as adrenal cortex or brown fat; or (4) they may not be maintained at all in a real sense, but simply have their normal involution passively retarded by the low body temperature of the torpid animal (see below).

None of these possibilities can be effectively ruled out at present, at least not in *Myotis*. In regard to the first—a continuance of hormone production by the testis at reduced levels, Pearson *et al.* (1952) failed to find evidence of steroids in the Leydig cells of hibernating *Corynorhinus* (= *Plecotus*) by histochemical tests. This in combination with the failure of Krutzsch & Wells (1960) to obtain an androgenic response in bioassay with testicular extracts of *Myotis* suggests that the testis itself in these forms is not a significant source of androgen during hibernation. The question needs re-examination, however, by recently developed and more sensitive histochemical and quantitative techniques of steroid analysis. The second and third alternatives—maintenance of the sex accessories by 'stored' hormone, or by androgens synthesized in non-gonadal sites—simply have not been adequately explored. The adrenal cortex has not been directly assayed for androgens in the hibernating bat, and although Krutzsch and Wells found high concentrations of androgenic material in brown fat, no one has shown as yet that it is released from the tissue under physiological conditions or in a physiologically active form, nor has its effect on the accessory organs of the bat been directly assayed. The final possibility, simple cold-retarded involution, is complicated by the very real possibility of a species difference. Courrier (1927) and Wimsatt (unpublished observations) have both noted in respect to members of the genus *Eptesicus* (*E. serotinus* and *E. fuscus* respectively) that if adult males are returned to hibernation soon after bilateral castration, involution of the sex accessories does not occur even after several months. This appears to provide another striking demonstration of the retardation effects of low body temperature on vital functions, including the essentially 'negative' one of glandular involution. Courrier (1927) reported further that in animals kept at 20° C. subsequent to bilateral castration the accessories did undergo obvious involution over a period of 6–8 weeks. This did not occur in unilaterally castrated or non-castrated males maintained at 20° C. over a corresponding period. He concluded, of course, that the testis of *Eptesicus* continues to produce androgen throughout most of hibernation. I have earlier mentioned Courrier's finding that the Leydig cells in *Eptesicus* seem not to experience as profound atrophy in the autumn as do those of other vespertilionids, and it may well be that the testis continues to produce hormone in *Eptesicus* but fails to do so in other genera, such as *Myotis*. Nevertheless, maintenance of functional integrity of the accessory glands and libido during hibernation is common to all vespertilionids (except *Miniopterus*, see above). The finding in *Myotis lucifugus* of a potent extra-testicular locus of androgen in brown fat may possibly have some significance in relation to the maintenance of sex accessory glands and libido in this species during hibernation when the testis is atrophic.

Lipid analyses have been undertaken of the brown fats of many species, both hibernating and non-hibernating (cf. reviews cited earlier and also Wells et al. 1965), and in general it appears that total cholesterol levels are low. Nevertheless Ratsimamanga, Rahandraha, Nigeon-Dureuil & Rabinowicz (1958) have shown that brown fat of mice can convert a preformed steroid nucleus (progesterone) to a more polar steroid, and Ptak (1965) has demonstrated that this tissue in mice is capable of synthesizing steroid hormones. Admittedly, the evidence obtained thus far is circumstantial, but if brown fat can truly synthesize steroids, alter their configurations, or 'store' them for subsequent release in a biologically active form, then it could very well play a significant role not only in reproduction, but in other endocrine functions as well. Furthermore, when seasonally cyclic phenomena are involved, the periods during the cycle when analyses are carried out could obviously make a great deal of difference in the nature of the results, and accordingly tests must be made through the year. The whole question of the possible steroidal functions of brown fat in relation to other endocrine integrative mechanisms is a most interesting one, and invites further investigation.

SOME SPECIAL REPRODUCTIVE ADAPTATIONS OF HIBERNATING BATS

Survival of spermatozoa in the female reproductive tract

The survival of sperm in the female reproductive tract of hibernating vespertilionid and rhinolophid bats for many months without loss of viability or fertilizing capacity (Hartman, 1933; Wimsatt, 1942, 1944a) is one of the most striking reproductive adaptations to hibernation known. Remarkable as the phenomenon may seem in a mammal, far more dramatic examples are virtually a commonplace among poikilothermous lower vertebrates and invertebrates (cf. Hartman, 1933; L. Hoffman, 1968) and it is perhaps physiologically relevant that the only mammals to have developed this adaptation are themselves essentially poikilothermic during the effective period of sperm retention.

The unusual longevity of bat sperm in the female tract has been appreciated for nearly 100 years. In view of current interest in sperm physiology and in the development of better procedures to potentiate sperm viability *in vitro* for purposes of artificial insemination, it is surprising that the mechanisms have not been more vigorously pursued. Attempts to explain the phenomenon have of course been made, e.g. Hartman's (1933) contention that low body temperature of the hibernating bat reduces sperm motility and thereby fosters their survival, Redenz's (1929) suggestion that high CO_2 levels resulting from sperm metabolism and crowding *in utero* favour survival, and Nakano's (1928) assertion that survival depends upon glycogen deposits in the uterine epithelium which provide nourishment for stored sperm. None of these factors have actually been demonstrated by appropriate experimentation to be immediately involved, although it would be difficult not to believe that low temperatures, at least, are important. Earlier concepts were based on inadequate knowledge of sperm organization and physiology; the complexities of sperm morphology and metabolism revealed in more recent studies (cf. Bishop, 1962; Mann, 1964) were not appreciated, so it is perhaps not surprising that earlier students sought oversimplified explanations for the phenomenon.

It is not my intention to review the subject here in critical detail—for actually little progress has been made over the last 90 years, but simply to offer some definitions of the problem which appear to represent logical avenues of approach, and which when implemented it is hoped will ultimately lead to a better understanding of sperm survival mechanisms in hibernating bats. As I see it, any experimental approach to the question must entertain three logical alternatives: (1) the possibility that bat sperm may have evolved special physiological capabilities which endow them with greater 'survival potential' than the sperm of other mammals, (2) the possibility that prolongation of viability is attributable to the evolution of special physiological adaptations within the female reproductive tract, or (3) the possibility that both have occurred, and that retention of viability involves a co-specialization, and therefore a dynamic interaction between stored sperm and the female tract.

As to the 'reasonableness' of the first possibility—that bat sperm may have evolved special characteristics—one need not look far for precedents; interspecific variations in sperm metabolism and physiology are legion among and within the various vertebrate classes (cf. Mann 1964). Bats possess great antiquity, having evolved independently of other Eutherian stocks possibly since the late Cretaceous. Moreover, heterothermy, as opposed to hibernating capacity *per se*, is widespread among bats, even among tropical species, and there are indications that it, too, is of ancient origin (Twente & Twente, 1964). Evolutionary isolation and the development in some lineages of a propensity for heterothermic existence could well provide the opportunity for development of special adaptive mechanisms in spermatozoa. The same argument, of course, justifies the conceptual validity of the other two alternatives.

In exploring these possibilities one might expect to gain insight from comparative studies of sperm and sperm storage organs in lower vertebrates, especially reptiles (cf. L. Hoffman, 1968), but comparisons with the usual mammalian condition will obviously remain the most pertinent. The existence of special adaptive mechanisms in bat sperm presumably can be sought through direct *in vitro* comparisons with other mammalian sperm involving a full range of structural, chemical, metabolic and physiological properties. Parallel studies of the reproductive tract may likewise be undertaken with particular attention to the possible existence of specialized areas of sperm–uterine interactions, the histochemical and biochemical localization of potential nutrient sources and relevant enzyme systems, and the influence of sex and other hormones on the morphological and biochemical conditioning of the uterine environment preceding and during the sperm storage phase. The tendency in the past has been to seek or postulate nutritive support for stored sperm *in utero* on the presumption that this must constitute a major condition for their survival. It may be an important factor, of course, but consideration needs also to be given to the possibility that specialized repressive or inhibitory influences of the female tract on sperm metabolism could be equally, or even uniquely, responsible for sperm longevity *in utero*.

We have very recently initiated studies of this sort and have made some progress. A first step involved attempts to ascertain whether bat sperm possessed any adaptive morphological peculiarities. Fawcett & Ito (1965) independently examined in detail the ultrastructure of epididymal sperm from hibernating *Myotis* and *Eptesicus*. A concurrent study of our own (Wimsatt, Krutzsch & Napolitano, 1966) confirmed most of their findings and extended the analysis to presumably mature sperm lodged within the reproductive tract of hibernating females of the species *Myotis lucifugus*. Structural differences between sperm from epididymis and uterus were minor, and no special features were noted in comparison with other mammalian sperm which would suggest that bat sperm possess unique characteristics. We concluded that if bat sperm are 'specialized' it is at a physiological level which involves no obvious morphological differentiation detectable by our cytological and ultrastructural techniques. *In vitro* studies, when initiated, may prove more rewarding.

Preliminary studies of the female reproductive tract have produced evidence that in *Myotis* at least there may exist a specialized site for sperm storage, a 'sperm receptable region' which constantly differs from the rest of the tract, (*a*) in the relative numbers of sperm present, (*b*) in the nature of their physical association with the uterine epithelium, and (*c*) in the possession of special histochemical characteristics (Wimsatt, Krutzsch & Napolitano, unpublished observations). It is probable that some sperm destruction also occurs here, but it is less evident than elsewhere in the uterus. Earlier studies had led to the general assumption that the uterus as a whole serves as the primary storage organ (Guthrie, 1933; Nakano, 1928; Redenz, 1929; Hartman, 1933; Wimsatt, 1944a), and indeed this may be so in certain genera such as *Pipistrellus* where the uterus becomes greatly distended by the quantities of sperm introduced (cf. Nakano, 1928; Redenz, 1929), but this no longer appears likely in *Myotis*.

Preliminary biochemical characterization has indicated the probable occurrence of non-seminal fructose in the uterus, and we have had some success in manipulating its levels by parenteral administration of sex hormones (Wimsatt & Krutzsch, unpublished observations). Current studies also focus on the localization of enzymes involved in glycogen synthesis and metabolism in the sperm receptacle area, and the possibility of a 'sorbitol pathway' for derivation of fructose from glucose via sorbitol as intermediate. We have so far demonstrated one of the enzymes known to be involved—DPN-dependent sorbitol dehydrogenase (Mann, 1964)—in the middle piece of stored spermatozoa, probably associated with the mitochondria which are confined to this segment of the flagellum.

Effects of torpor on embryonic development

Among most hibernators the seasonally recurrent periods of dormancy do not coincide with the gestation period and there is presumably no need for developmental processes to become adapted to the exigencies of lowered body temperatures and reduced metabolism. This is not the situation, however, in bats, in which two different patterns of interaction between pregnancy and dormancy have been recorded. The first involves the miniopterines, in which, as already noted, the females typically enter hibernation in a pregnant condition; there results a prolongation of the 'normal' gestation period by approximately the length of the hibernal episode. The second pattern, observable in other vespertilionids, likewise involves a lengthening of gestation, but on a completely fortuitous basis, usually in response to abnormal climatic (temperature) conditions during the normal period of gestation. Its immediate basis is the presumably obligatory heterothermy or 'daily torpor' of bats in summer whereby when at rest their body temperatures decline to within a degree or two of the ambient (Hock, 1951; Stones & Wiebers, 1967). Unseasonal low air temperatures can presumably induce lower-than-normal body temperatures in pregnant females, leading to a slowing of embryonic development and prolongation of gestation.

Such variation in gestation length from one year to the next, or in a given year from one population to another, as a result of varying climatic conditions, has been documented in wild populations of the California lump-nosed bat

(*Corynorhinus rafinesquei*) by Pearson, Koford & Pearson (1952); they estimate that gestation length may vary within the range of 56–100 days. Moreover, Eisentraut (1937) has demonstrated experimentally that temperature can directly influence gestation length. He observed that when pregnant specimens of *Myotis myotis* were kept in cold rooms their body temperatures declined and embryonic development was slowed, roughly in proportion to the degree of temperature depression. Conversely, development was accelerated when the bats were warmed. Smith (1956) and Stones & Wiebers (1967), working with *Myotis lucifugus*, maintain that pregnant and lactating specimens thermoregulate better than non-pregnant females or males, but neither claimed that torpidity could not occur in pregnant animals. On the other hand, Kolb (1950) noted in respect to the lesser horseshoe bat (*Rhinolophus hipposideros*) that pregnant and lactating animals never became torpid and remained active in the cold, while other animals became torpid when the ambient temperature was lowered to 25° C. and below. Interestingly, a similar tendency for body temperatures to be kept at high levels during pregnancy has been reported by Morrison (1945) in the normally poikilothermic three-toed sloth (*Bradypus griseus*).

Whether the apparent differences in thermoregulatory capability of pregnant vespertilionids and rhinolophids are really as definite as the few studies available suggest requires further study. Moreover, the physiological bases underlying an enhanced thermoregulatory capability in pregnant females in contrast to non-pregnant females and males still remains obscure. Kolb (1950) has suggested in respect to the horseshoe bat that the improved thermoregulation of pregnant females may be related to hormones which increase thermogenesis, and help prevent body cooling—a plausible enough explanation, but one still lacking experimental documentation. An additional problem requiring analysis concerns the possibility that females may be more susceptible to cold depression before implantation of the blastocyst than during later pregnancy stages, so that effective lengthening of gestation might involve developmental arrest primarily during the earlier, free stages of embryonic development, rather than later when the more physiologically complex placental relationship has become established.

It is not without interest in this regard that embryonic development in hibernating *Miniopterus schreibersii* appears to become arrested completely, and at the stage when a fully developed blastocyst lies free in the uterus. Courrier (1927) first called attention to a retarded embryonic development in hibernating *Miniopterus*, but he could not postulate a complete developmental arrest for his material was restricted to the months of October and November. It remained for Planel *et al.* (1961) and Peyre & Herlant (1963) to demonstrate in material collected throughout the winter that virtually no development beyond the free blastocyst stage occurs during hibernation, and that implantation and true pregnancy are deferred several months until after the animals arouse from hibernation in the spring. We see here therefore a true case of 'delayed implantation', resembling in over-all pattern that occurring in various non-hibernators such as macropod marsupials, roe deer, armadillo, numerous carnivores, and lactating rodents (cf. Enders, 1963). On the surface, delayed implantation in *Miniopterus* differs from all these, however, in that it appears to be temperature-dependent. There can be little question that delayed implantation in recent mammals has had a polyphyletic origin, and it is becoming increasingly evident that its physiological regulation is both varied and extremely complex. The presumed thermal dependence of delayed implantation in *Miniopterus* appears to be a new variant and therefore merits more critical study than it has received thus far.

I believe, however, that it is premature to assume, as has generally been done, that the condition in *Miniopterus* is merely a direct and passive response to cold and depressed metabolism, for no experimental work whatever has yet been carried out in respect to the mechanisms involved. Such a view is probably an oversimplification, and was more easily acceptable before it became known that development is completely arrested at the preimplantation stage, and while it was still thought that regular seasonal retardation of embryonic development was uniquely restricted to hibernating species of *Miniopterus*. We now know of at least two other species of bats—*Macrotus californicus* and *Eidolon helvum*, representatives of widely separated families (Phyllostomatidae and Pteropidae respectively)—which demonstrate delayed implantation of several months duration. Interestingly, one of these (*Macrotus*) is at best only a 'quasi-hibernator' and the other (*Eidolon*) is not only a non-hibernator, living virtually on the equator, but it belongs to a family in which heterothermic capability is apparently not well developed (Morrison, 1959; Bartholomew, Leitner & Nelson, 1964). The two forms also have widely different feeding habits; *Macrotus* is wholly insectivorous, and *Eidolon* subsists on wild and cultivated fruits.

Bradshaw (1961, 1962) has described the annual reproductive cycle in *Macrotus*. He found that females were inseminated in the autumn and most are gravid by the end of October. Embryonic development is 'slow' during winter, with scarcely any swelling of the uterine horn being apparent until March, when growth becomes rapid. Parturition occurs in May, June or July. The period of 'retarded development'—Bradshaw hesitates to call the phenomenon delayed implantation—corresponds to a time of quasi-hibernation in which rectal temperatures of resting bats were recorded between 26° and 30° C. Since Bradshaw presented no photographs or

PLATE I

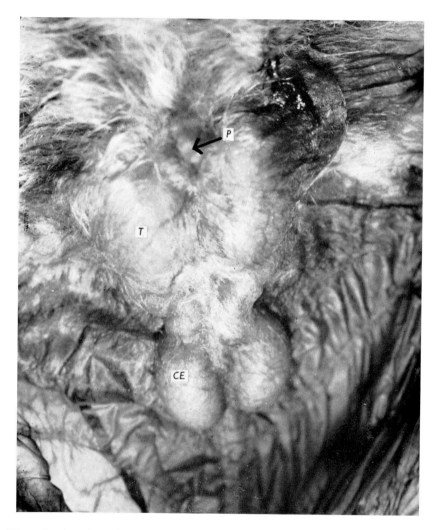

View of enlarged caudae epididymis within uropatagial membrane of hibernating bat (*Miniopterus* sp., November specimen). The epididymis (*CE*) is gorged with stored spermatozoa. The testes (*T*) are reduced in size and inguinal in position. *P*, Penis. (Photo by courtesy of Dr Bernardo Villa-R.)

descriptions of the stages of embryonic development observed, it remains unknown whether the developmental retardation is simply a slow down or a true preimplantation developmental arrest. However, the months-long period of non-engorgement of the uterus, coupled with its sudden rapid expansion in March and in further combination with the relatively high body temperatures recorded, suggest the probability that this may be a true case of delayed implantation. The situation in *Eidolon* is less ambiguous. The annual cycle of this bat, dwelling at 0° 20' N. in Uganda, has been analysed by Mutere (1965, 1967). It exhibits synchronized seasonal breeding, displaying a monoestrous cycle whose rhythm seems to correspond with two annual rainfall peaks. The delay in implantation, which endures approximately 3 months, appears unprecedented in a homoiothermic equatorial species. Mating and fertilization occur April to June, but implanted embryos are not found until October and November; free uterine blastocysts, all at comparable stages of development, were recovered from bats collected in July, August and September. Birth occurs during February and March. Interestingly, the time of implantation coincides with the lower of the two annual rainfall peaks, while birth occurs just prior to the onset of the highest seasonal rainfall peak. This cycle forestalls births during the relatively dry periods of the year, which would presumably be less favourable for the weaning of the young.

These various patterns of delayed implantation in hibernating and non-hibernating bats, which in addition are not very closely related, underscore the probably polyphyletic origin of the phenomenon within the order Chiroptera. The physiological mechanisms of developmental arrest may be, and probably are, different in each case. That maternal dormancy associated with low body temperature may be of primary importance in one of them, however, is of special interest, for it introduces a novel complication into the presumed endocrine control mechanisms usually invoked to explain implantation delay in other, generally homoiothermic, mammals.

Support for the preparation of this review and for current work described herein was provided by a National Science Foundation grant, GB-6435-X. The help of George Batik and Paula DiSanto Bensadoun, who prepared the line figures and testis drawings respectively, is gratefully acknowledged.

REFERENCES

ADELSTEIN, S. J., LYMAN, C. P. & O'BRIEN, R. C. (1967). In *Mammalian Hibernation*, vol. III, p. 398. Edinburgh and London: Oliver and Boyd.
AGID, R., GABE, M. & MARTOJA, M. (1967). *C. r. Séanc. Soc. Biol.* **161**, 459.
ALLANSON, M. (1934). *Phil. Trans. R. Soc.* B **223**, 277.

ALLANSON, M. & DEANESLEY, R. (1934). *Proc. R. Soc.* B **116**, 170.
ASDELL, S. A. (1964). *Patterns of Mammalian Reproduction.* Ithaca, N.Y.: Cornell University Press.
AZZALI, G. (1953). *R. Biol.* **45**, 131.
AZZALI, G. (1954). *Archo. ital. Anat. Embriol.* **59**, 142.
AZZALI, G. (1955). *Z. Zellforsch. mikrosk.* **41**, 391.
BAKER, J. R. & BIRD, T. F. (1936). *J. Linn. Soc.* **40**, 143.
BARNETT, S. A. & MANLY, B. M. (1959). *Proc. R. Soc.* B **151**, 87.
BARTHOLOMEW, G. A. & HUDSON, J. W. (1962). *Physiol. Zoöl.* **35**, 94.
BARTHOLOMEW, G. A., LEITNER, P. & NELSON, J. E. (1964). *Physiol. Zoöl.* **37**, 179.
BARRY, J. (1954). *Bull. Séanc. Soc. sci. Nancy* **13**, 126.
BARRY, J. (1956). *C. r. Ass. Anat.* **42**, 264.
BARRY, J., BESSON, S. & LAMARCHE, M. (1958). *Annls Endocr.* **19**, 1045.
BENETATO, G., DANELIUC, E., NESTIANU, V. & GABRIELESCU, E. (1965). *Rev. Roumaine Physiol.* **2**, 199.
BISHOP, D. W. (1962). *Spermatozoan Motility.* Washington, D.C.: A.A.A.S. Publication, no. 72.
BRADSHAW, G. V. R. (1961). *Mammalia* **25**, 117.
BRADSHAW, G. V. R. (1962). *Science, N.Y.* **136**, 645.
CADE, T. J. (1964). In *Mammalian Hibernation,* vol. II. *Suomal.-ugr. Seur. Aikak.* (Annales Academiae scientarum fennicae), A, IV, **71**, 77.
CAFFIER, P. & KOLBOW, H. (1934). *Z. Geburtsh. Gynäk.* **108**, 185.
CAMPBELL, D. W. & HOLMES, R. L. (1966). *Z. Zellforsch. mikrosk.* **75**, 35.
CHATFIELD, P. O., BATTISTA, A. F., LYMAN, C. P. & GARCIA, J. P. (1948). *Am. J. Physiol.* **155**, 179.
CIANI, P. (1961). *Archo. ital. Anat. Embriol.* **66**, 340.
COURRIER, R. (1924). *C. r. Séanc. Soc. Biol.* **90**, 808.
COURRIER, R. (1927). *Archs Biol., Paris* **37**, 173.
DEANE, H. W. & LYMAN, C. P. (1954). *Endocrinology* **55**, 300.
DWYER, P. D. (1963*a*). *Aust. J. Sci.* **25**, 435.
DWYER, P. D. (1963*b*). *Aust. J. Zool.* **11**, 219.
EISENTRAUT, M. (1934). *Z. Morph. Ökol. Tiere* **29**, 231.
EISENTRAUT, M. (1937). *Biol. Zbl.* **57**, 59.
EISENTRAUT, M. (1956). *Der Winterschlaf mit seinen ökologischen und physiologischen Begleiferscheinungen.* Jena: Gustav Fischer.
ENDERS, A. C. (1963). *Delayed Implantation.* Chicago: University of Chicago Press.
ENGEL, R., RATHS, P. & SCHULZE, W. (1957). *Z. Biol.* **109**, 381.
FAWCETT, D. W. & ITO, S. (1965). *Am. J. Anat.* **116**, 567.
FOSTER, M. A. (1934). *Am. J. Anat.* **54**, 487.
FOSTER, M. A., FOSTER, R. C. & MEYER, R. K. (1939). *Endocrinology* **24**, 603.
GABE, M., AGID, R., MARTOJA, M., ST GIRONS, M. C. & ST GIRONS, H. (1963). *Archs Biol., Paris* **75**, 1.
GAISLER, J. (1966). *Bijdr. Dierk.* **36**, 45.
GAISLER, J. & TITLBACH, M. (1964). *Vĕst. čsl. Spol. zool.* **28**, 268.
GIROD, C., DUBOIS, P. & CURE, M. (1966). *Annls Endocr.* **27**, 452.
GIROD, C., DUBOIS, P. & CURE, M. (1967). *Annls Endocr.* **28**, 581.
GRINDELAND, R. E. & FOLK, G. E. JR. (1958). *Anat. Rec.* **132**, 447.
GRINDELAND, R. E. & FOLK, G. E. JR. (1962). *J. Reprod. Fert.* **4**, 1.
GUTHRIE, M. J. (1933). *J. Mammal.* **14**, 199.
HARTMAN, G. C. (1933). *Q. Rev. Biol.* **8**, 185.
HAYWARD, J. S. & LYMAN, C. P. (1967). In *Mammalian Hibernation,* vol. III, p. 346. Edinburgh and London: Oliver and Boyd.
HERLANT, M. (1932). *Archs Anat. microsc.* **28**, 335.

HERLANT, M. (1953). *Ann. Soc. Roy. Belge*, **84**, 87.
HERLANT, M. (1956). *Archs Biol., Paris* **67**, 89.
HERREID, C. P. (1961). *Diss. Abstr.* **22**, 2520.
HERREID, C. P. (1963a). *J. cell. comp. Physiol.* **61**, 201.
HERREID, C. P. (1963b). *Science, N.Y.* **142**, 1573.
HICKMAN, V. V. & HICKMAN, J. L. (1960). *Proc. zool. Soc. Lond.* **135**, 365.
HOCK, R. J. (1951). *Biol. Bull. mar. biol. Lab., Woods Hole* **101**, 289.
HOCK, R. J. (1960). In *Mammalian Hibernation*, vol. I. *Bull. Mus. comp. Zool. Harv.* **124**, 155.
HOFFMAN, L. H. (1968). Cornell University thesis, Ithaca, N.Y.
HOFFMAN, R. A. (1964a). In *Handbook of Physiology*, sect. 4, p. 379. Washington, D.C. American Physiological Society.
HOFFMAN, R. A. (1964b). In *Mammalian Hibernation*, vol. II. *Suomal.-ugr. Seur. Aikak.* (Annales Academiae scientarum fennicae), A, IV. **71**, 201.
HOFFMAN, R. A. (1968). *Fedn Proc. Fedn Am. Socs exp. Biol.* **27**, 999.
HOFFMAN, R. A. & ZARROW, M. X. (1958). *Anat. Rec.* **131**, 727.
HUDSON, J. W. (1967). In *Mammalian Hibernation*, vol. III, p. 30. Edinburgh and London: Oliver and Boyd.
HUDSON, J. W. & BARTHOLOMEW, G. A. (1964). In *Handbook of Physiology*, Chapt. 34, sect. 5. Washington, D.C. American Physiological Society.
JOEL, C. (1965). In *Handbook of |Physiology*, sect. 5, p. 59. Washington, D.C. American Physiological Society.
JOHANSSON, B. W. (1959). *Metabolism* **8**, 221.
JOHANSSON, B. W. (1960). In *Mammalian Hibernation*, vol. I. *Bull. Mus. comp. Zool. Harv.* **124**, 233.
JOHNSON, G. E. (1927). *Anat. Rec.* **37**, 125.
JOHNSON, G. E. (1930). *Biol. Bull. mar. biol. Lab., Woods Hole* **59**, 114.
JOHNSON, G. E., GANN, E. L., FOSTER, M. A. & COCO, R. M. (1934). *Endocrinology* **18**, 86.
KAYSER, C. (1957a). *C. r. Séanc. Soc. Biol.* **151**, 982.
KAYSER, C. (1957b). *Revue can. Biol.* **16**, 303.
KAYSER, C. & ARON, M. (1938). *C. r. Séanc. Soc. Biol.* **130**, 225.
KAYSER, C. & ARON, M. (1950). *Archs Anat. Histol. Embryol.* **33**, 23.
KIERNAN, J. A. (1964). *Jl R. microsc. Soc.* **83**, 297.
KOLACZKOWSKI, J. & WENDER, M. (1958). *Folia morphol.* **9**, 299.
KOLB, A. (1950). *Zool. Jb.* (Systematik, Ökologie und Geographie der Tiere) **78**, 547.
KRUTZSCH, P. H. & WELLS, W. W. (1960). *Proc. Soc. exp. Biol. Med.* **105**, 578.
KRUTZSCH, P. H. (1956). *Anat. Rec.* **124**, 321.
LANDAU, B. R. & DAWE, A. R. (1960). In *Mammalian Hibernation*, vol. I. *Bull. Mus. comp. Zool. Harv.* **124**, 173.
LEGAIT, E., LEGAIT, H. & BURLET, C. (1966). *C. r. Séanc. Soc. Biol.* **160**, 1657.
LYMAN, C. P. & CHATFIELD, P. O. (1955). *Physiol. Rev.* **25**, 403.
LYMAN, C. P. & DEMPSEY, E. W. (1951). *Endocrinology* **49**, 647.
MANN, T. (1964). *Biochemistry of Semen and of the Male Reproductive Tract*. New York: Wiley.
MATTHEWS, L. H. (1937). *Trans. zool. Soc. Lond.* **23**, 224.
MAYER, W. V. & BERNICK, S. (1958). *Anat. Rec.* **130**, 747.
MAYER, W. V. & ROCHE, E. T. (1954). *Growth* **18**, 53.
MCKEEVER, S. (1966). In *Comparative Biology of Reproduction in Mammals. Symp. Zool. Soc. Lond.* **15**, 365. London: Academic Press.
MILLER, R. E. (1939). *J. Morph.* **64**, 267.
MITCHELL, O. G. (1959). *J. Mammal.* **40**, 45.

Moore, C. R., Simmons, G. F., Wells, L. J., Zalesky, M. & Nelson, W. O. (1934). *Anat. Rec.* **60**, 279.
Morrison, P. R. (1945). *J. Mammal.* **26**, 272.
Morrison, P. (1959). *Biol. Bull. mar. biol. Lab.*, Woods Hole **116**, 484.
Mosca, L. (1956). *Q. Jl exp. Physiol.* **41**, 433.
Mutere, F. A. (1965). *Nature, Lond.* **207**, 780.
Mutere, F. A. (1967). *J. zool. Res.* **153**, 153.
Nakano, O. (1928). *Folia anat. jap.* **6**, 777.
Ottaviani, G. & Azzali, G. (1954). *V. Neoroveg. Symp.*, Vienna (Abst.), p. 29.
Panuska, J. A. (1965). *Am. Zool.* **5**, 739.
Pearson, O. P., Koford, M. R. & Pearson, A. K. (1952). *J. Mammal.* **33**, 273.
Pengelley, E. T. (1966). *Growth* **30**, 137.
Petrovic, A. & Kayser, C. (1956). *C. r. Séanc. Soc. Biol.* **150**, 1990.
Petrovic, A. & Kayser, C. (1957). *C. r. Séanc. Soc. Biol.* **151**, 996.
Petrovic, A. & Kayser, C. (1958). *J. Physiol.*, Paris, **50**, 446.
Peyre, A. & Herlant, M. (1963). *Gen. Comp. Endocrinology* **3**, 726.
Planel, H., Guilhem, A. & Soleilhavoup, J.-P. (1961). *C. r. Ass. Anat.* **47**, 620.
Popovic, V. (1960). In *Mammalian Hibernation*, vol. I. *Bull. Mus. comp. Zool. Harv.* **124**, 105.
Ptak, W. (1965). *Experientia* **21**, 26.
Rasmussen, A. T. (1917). *Am. J. Anat.* **22**, 475.
Rasmussen, A. T. (1918). *Endocrinology* **2**, 253.
Rasmussen, A. T. (1921). *Endocrinology* **5**, 33.
Ratsimananga, A. R., Rahandraha, T., Nigeon-Dureuil, M. & Rabinowicz, M. (1958). *J. Physiol.*, Paris, **50**, 479.
Redenz, E. (1929). *Z. Zellforsch. mikrosk.* **9**, 734.
St Girons, H., St Girons, M., Martoja, M., Agid, R. & Gabe, M. (1961). *C. r. hebd. Séanc. Acad. Sci.*, Paris **253**, 2259.
Skowron, S. & Zajaczek, S. (1947). *C. R. Séanc. Soc. Biol.* **141**, 1105.
Smalley, R. L. & Dryer, R. L. (1963). *Science, N.Y.* **140**, 1333.
Smalley, R. L. & Dryer, R. L. (1967). In *Mammalian Hibernation*, vol. III, p. 325. Edinburgh and London: Oliver and Boyd.
Smith, E. (1956). *Am. J. Physiol.* **185**, 61.
Smith, R. E. (1964). In *Mammalian Hibernation*, vol. II. *Suomal.-ugr. Seur. Aikak.* (Annales Academiae scientarum fennicae), A, IV, **71**, 389.
Smith, R. E. & Hock, R. J. (1963). *Science, N.Y.* **140**, 199.
Smit-Vis, J. H. (1962). *Archs néerl. Zool.* **14**, 513.
Smit-Vis, J. H. & Akkerman-Bellaart, M. A. (1967). *Experientia* **23**, 844.
Stones, R. C. & Wiebers, J. E. (1967). In *Mammalian Hibernation*, vol. III, p. 97. Edinburgh and London: Oliver and Boyd.
Suomalainen, P. (1953). *Proc. Finn. Acad. Sci. Letters* **63**, 131.
Suomaleinen P. (1960). In *Mammalian Hibernation*, vol. I. *Bull. Mus. comp. Zool. Harv.* **124**, 271.
Suomaleinen P. (1964). *Ann. Acad. Sci. Tenn.*, ser. A, IV, **71**, 417.
Suomaleinen, P. & Uuspää, V. J. (1958). *Nature, Lond.* **182**, 1500.
Sweet, J. E. & Hoskins, W. H. (1940). *Proc. Soc. exp. Biol. Med.* **45**, 60.
Thomson, J. F., Straube, R. L. & Smith, D. E. (1962). *Comp. Biochem. Physiol.* **5**, 297.
van Tienhoven, A. (1968). *Reproductive Physiology of Vertebrates*. Philadelphia: W. B. Saunders.
Troyer, J. R. (1960). *Anat. Rec.* **136**, 350.
Troyer, J. R. (1961). *Anat. Rec.* **139**, 281.
Troyer, J. R. (1964). *Anat. Rec.* **148**, 405.

Troyer, J. R. (1965). *Anat. Rec.* **151**, 77.
Twente, J. W. & Twente, J. A. (1964). *Mammalian Hibernation*, 11, *Suomal.-ugr. Seur. Aikak.* (Annales Academiae scientarum fennicae) A, IV, **71**, 433.
Wells, H. J. (1940). *J. Am. med. Ass.* **114**, 2177.
Wells, H. J., Makita, M., Wells, W. W. & Krutzsch, P. H. (1965). *Biochim. biophys. Acta* **98**, 269.
Wells, L. J. (1935). *Anat. Rec.* **62**, 409.
Wells, L. J. (1938). *Endocrinology* **22**, 588.
Wells, L. J. & Moore, C. R. (1936). *Anat. Rec.* **66**, 181.
Wells, L. J. & Zalesky, M. (1940). *Am. J. Anat.* **66**, 429.
Wertheimer, E. & Shapiro, B. (1948). *Physiol. Rev.* **28**, 451.
Wimsatt, W. A. (1942). *Anat. Rec.* **83**, 299.
Wimsatt, W. A. (1944a). *Anat. Rec.* **88**, 193.
Wimsatt, W. A. (1944b). *Am. J. Anat.* **74**, 129.
Wimsatt, W. A. (1960). *Mammalian Hibernation*, vol. 1. *Bull. Mus. comp. Zool. Harv.* **124**, 249.
Wimsatt, W. A. & Kallen, F. C. (1957). *Anat. Rec.* **129**, 115.
Wimsatt, W. A., Krutzsch, P. H. & Napolitano, L. (1966). *Am. J. Anat.* **119**, 25.
Wimsatt, W. A. & Parks, H. F. (1966). *Comparative Biology of Reproduction in Mammals. Symp. zool. Soc. Lond.* **15**, 419. London: Academic Press.
Zajaczek, S. & Kaminski, Z. (1947). *C. r. Séanc. Soc. Biol.* **141**, 1104.

PRINCIPLES AND FURTHER PROBLEMS IN THE STUDY OF DORMANCY AND SURVIVAL*

By LAURENCE IRVING

Institute of Arctic Biology,† University of Alaska College, Alaska

INTRODUCTION

I take it to be a common notion that present forms of life are adapted results of evolution and that they continue to pass tests of natural selection to serve populations in their current environments. In speculating upon how the diversity and fitness of life developed, we are looking downward from existing twigs of organic differentiation for larger stems that we can partially discern toward branches that seem to have disappeared. The record of the circumstances of even recent ancestral forms of life is dim, and yet we presume that survival of populations means that they were adapted to their former environments. Speculation upon the processes in evolutionary survival involves projection backward in negative chronology, for which there are attractive models for help but few empirical quantities.

The idea that ontogeny recapitulates phylogeny is still a pretty fancy. Many biologists have pointed at the persistence of physical and chemical attributes that are related and differentiated in the systematic order of life and in geographical distribution. Some of these attributes are considered to be so ancient that they must have come into operation when lands and seas with their climates and seasons were quite different from conditions in the world at present. As land forms were raised, eroded and perhaps massively translated, animals and plants held to stability in form that has exceeded the firmness of the rocks on which they lived.

I am fascinated by the philosophical speculation of biologists about adaptation, but I am more impressed by the boldness of biologists' imagination than by their logic. My own speculations are uncertain philosophy, but it is clear that of all conditions to which animal and plant populations must adapt for survival, they must always adapt to specific natures that are derived from their past. How do populations so readily modify the constraints imposed by heredity to meet the geographical diversity of environments and their seasonal and secular variations? It is apparent that dormancy of one or many kinds is an inherited imposition, but that its dimensions are in some degree modifiable for local and current

---

* Supported in part by N.I.H. grant GM-10402.
† Institute publication no. 91.

conditions. The experience of populations in dormancy appears to accumulate memories that allow predictive preparations for each new episode of dormancy to run its course in time in accord with specific requirements and adaptively to varying and even novel external conditions.

After many distinguished presentations of aspects of dormancy in this symposium, we can see that it occurs in many plants, animals and micro-organisms in diverse environments. I am not sure about the importance of principles in biology, for I have often seen fashions of experimentation or even prejudices in thinking proposed as principles. There can be no question, however, that the state of dormant individuals presents exciting opportunities for considering the survival of natural populations. It is easy to see that non-motile periods are required for the revolutionary developmental changes through which insects proceed. More obscure are the slow processes in dormant seeds, plants and spores or cysts of micro-organisms. The lapse of time is a common requirement at some stage in life-cycles, at which period the individual is said to be dormant because it is not in motion or growth at the expense of its environment.

We are inclined to see in the diminished metabolic requirements and tolerance of cold, heat and desiccation by dormant organisms, adaptations that enable them to live through periods when external conditions do not permit growth and reproduction. For dormancy to secure survival of a population it must have dimensions of duration conformable with the period of adverse growing conditions. Tolerance and use of dormancy for survival differ so much among forms of life that we still have to describe its adaptive use for survival explicitly for a species, or much more restricted breeding population, in relation to its own particular environment. Knowing the instability of environments, we have to consider and usually puzzle over how established environmental adaptations do not lock the organism in its past.

There is another aspect of dormancy that engages interest. While tolerant of cold or heat and desiccation, the metabolism affords but feeble power for maintaining the organization of a dormant individual. Nevertheless, in conditions unfavourable for incremental activity, each dormant organism retains its individuality and the regulatory information necessary for co-ordinating its programme with that of its neighbours in the local population. Study of the state of enzymes, substrates, solutions and membranes of dormant organisms under experimental manipulation can be quite unrelated to natural survival but may, hopefully, indicate how the organized components operate.

ARCTIC HIBERNATORS

Having made an apology for generalization on dormancy, I will pass to illustrations of its operation in some animals in their natural environment in Alaska, with which I am familiar through studies of my colleagues. Near Fairbanks (latitude 65° N.) winter cold makes food inaccessible during 7 months for two species of bears (*Ursus*), two or three species of marmots (*Marmota*), two species of ground-squirrels (*Citellus*), and two species of jumping mice (*Zapus*). At latitude 68° in winter, food is unavailable for almost 9 months, and this period is passed by the hibernators in prepared shelters in selected dry places. These mammals differ in size from some 20 g. to 500 kg., covering about the size range of the larger array of mammalian species that continue actively to find food during the Arctic winter. Since neither hibernation nor winter feeding ability is peculiar to the Arctic, it is difficult to say that each mode of wintering is employed to suit a variety of environments by adaptive specific capability.

Arctic ground-squirrels (*Citellus*) at Barrow, Alaska (latitude 71° N.), rapidly gained weight in August. They fed actively only in temperatures above freezing. Placed in a metabolism chamber shortly after feeding, respiratory quotients commonly rose above 1·0 at temperatures above 0° C. and were all above 1·0 when warmer than 20° C. (Fig. 1). These high respiratory quotients (R.Q.s), dependent upon temperature, are indicative of synthesis of the fat which elevated the weight of some squirrels by one-third during the last 3 weeks in August. When awakened from hibernation in March, the squirrels were not hungry and warmth did not elevate their R.Q.s (Erickson, 1956a). Erickson (1956b) recorded rectal temperatures around 1° C. during several days in hibernating squirrels. In summer, however, the lively squirrels regulated their warm body temperature homoiothermously in air, between +22° and −19° C.

Elaborate physiological and behavioural preparations are required of mammals in anticipation of hibernation in an Arctic climate. Accurate foresight of the local season is especially required of small hibernators with their limited reserves for energy and feeble metabolic power at low temperature. Since southern populations of ground-squirrels pursue different schedules of hibernation, we can infer that local populations gain by experience as residents, the ability to foresee the coming winter where they live and make appropriate physiological preparations.

In winter hibernation the dormant Arctic ground-squirrel must accurately dole out from its reserves metabolic power at a sufficient but low rate. The hibernating squirrels pass through periods of true torpor alternating with periodic brief reawakened activity. My colleagues Galster

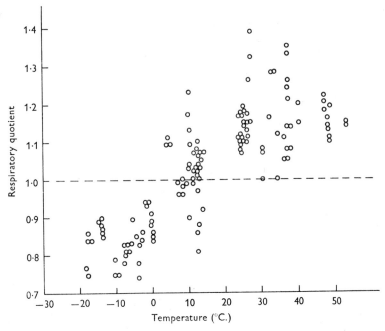

Fig. 1. R.Q. of Arctic Alaskan ground-squirrels (*Citellus*) at different temperatures in August 1948 (Erickson, 1956).

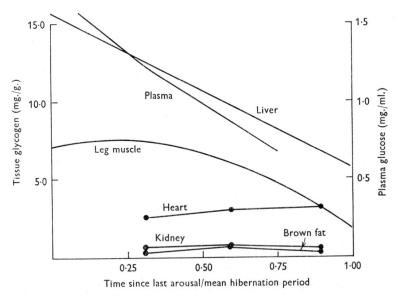

Fig. 2. Plasma and liver glucose in hibernating Arctic Alaskan ground-squirrels declining through time of inactivity in relation to mean duration of inactivity (Galster & Morrison, 1968).

& Morrison (1968) are finding that stores of glucose become depleted during the cold torpid periods and that replenishment of glucose levels is effected during the periodic spells of rewarming as they awaken (Fig. 2). Galster and Morrison regard this as biochemical evidence that the squirrel monitors its metabolic condition and recognizes the necessity for its regulation during the torpor.

The well-prepared mammalian hibernator is still an organized individual while dormant, perceptive of its internal state, reactive to environmental changes and the lapse of time within the capability for which it had earlier chosen a suitably tempered shelter. It had long been remarked that component organs of mammalian hibernators when excised retained reactivity in cold temperatures that block the activity of excised organs from non-hibernators. Considering the activity of excised nerves, which is nicely measurable by action potentials, Kehl & Morrison (1960) showed that conduction continued as the sciatic nerves of thirteen-lined ground-squirrels (*Citellus*) from Wisconsin were cooled to 5° C. At this temperature nerves from animals in hibernation were still conducting and by extrapolation would reach the failure point near 0° C. Early in hibernation, or while squirrels were active in summer, cold block occurred at about 5° C.

The faculty of retaining ability for coordinated reactions while cold that is exhibited by dormant hibernating mammals leads to the contrast of their tissues with those of non-hibernators. General hypothermia below +30° C. is not well tolerated by most mammals. To be sure the recovery of small mammals from experimental cooling to or below their freezing point is amazing, and the preservation of frozen tissues is of enormous practical importance and great scientific interest. But neither is now clearly naturally employed in the survival of populations.

PERIPHERAL COOLING IN MAMMALIAN NERVES

Peripheral nerves from the bare and often cold leg of the herring gull (*Larus argentatus*) in winter were found able to conduct impulses at lower temperature than when the legs were kept warm as in summer. Also, the peripheral nerves from the bare, often cold tarsal region continued to operate at cold temperatures that blocked nerve function in the part from within the feather-covered homoiothermous warm femoral portion of the leg (Chatfield, Lyman & Irving, 1955). Evidently the cold limits of nerve function are modifiable in a fashion adaptive to requirements for conservation of heat.

For some time we have observed the utilization by northern mammals of cooling peripheral tissues in thermoregulation. My colleague L. K. Miller (unpublished) has found that nerves excised from the thinly furred and often cold extremities of several northern mammals continue to show action potentials indicative of excitation and conduction well below their freezing point (Fig. 3). Nerves from extremities well insulated with fur and nerves of the homoiothermous interior naturally exposed to little natural cooling were blocked at warmer temperatures.

A recent survey of nerves from ice-breeding populations of seals during the March 1968 cruise of Scripps R.V. *Alpha Helix* in the ice of the Bering Sea illustrates the comparative operations of cold peripheral and warm internal nerves. Nerves from the often cold phalangeal region, naturally

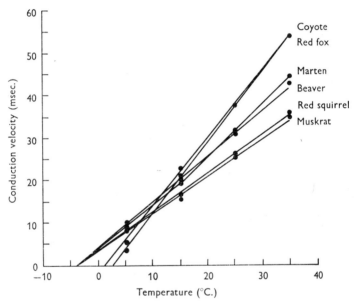

Fig. 3. Change in conduction with temperature in tail nerves of several land mammals (L. K. Miller, unpublished).

exposed to winter sea water at $-1.8°$ C., ceased conduction only when they froze, supercooled at $-5°$ C. (Fig. 4). By comparison, a more central part of the nerve from the warmer pelvic region showed a rapid lengthening of refractory period below $+5°$ C., from which we suspect a large decrement in capacity to transmit information by frequency modulation. Sections from the truly homoiothermous phrenic nerve failed to conduct below $+7°$ C. (Fig. 5) (L. K. Miller, unpublished).

The low-temperature operation of peripheral nerves illustrates the capacity of tissues of some non-hibernating mammals to operate at various and varying temperatures in the interest of heat regulation. At any one time the northern homoiotherm may be heterothermous from surface to

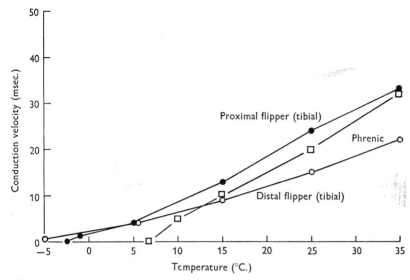

Fig. 4. Change in conduction with temperature in nerves from Bearded Seal (L. K. Miller, unpublished).

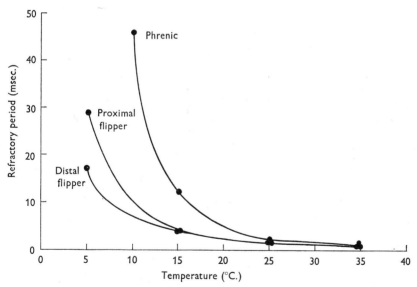

Fig. 5. Change in refractory period with temperature in nerves from Bearded Seal (L. K. Miller, unpublished).

interior and still continue to function as an integrated organism in spite of the changing velocity functions in the system. Hibernation differs in that the entire organism cools. Mammals utilize the great versatility of their tissues for operating at various temperatures in dormancy and in temperature regulation.

Since mammalian nerves can generate action potentials when cooled below their freezing point, the movement of charged particles in membranes persists in supercooled tissue solutions. Upon freezing the action potential can no longer be evoked. But if the nerve does not remain frozen too long or too cold, upon thawing and rewarming its ability for conduction is restored. The tolerance of brief mild freezing is only an example of the demonstrated ability of many mammalian tissues to be preserved after storage at low temperatures. I am inclined to doubt whether ability of mammalian tissues to survive freezing is utilized in the survival of natural populations. I must admit, however, that restoration of function after localized frostbite (a common Arctic human experience) is certainly useful. But measurements of the extent and duration of tolerable freezing are uncertain.

Surfaces of bare-skinned seals and pigs and feet of birds near 0° C. do remain integrated for effective operational control from the warm interior. In the matter of velocity of conduction and capability for frequency transmission we do see signs of a smaller decrement with cooling in some of the nerves frequently exposed to cold. We are withheld from calling this an adaptation because we can only infer its physiological usefulness in survival. (In this company, I suspect we are all unashamed teleologists as we speculate upon ways for better observation.) We can see that below 10° C. there is some compensatory adaptation in these cold nerves toward maintaining their conduction velocity and capacity for frequency modulation when cold, but it is by no means a complete compensation. Without further observational grounds for speculation upon how this system can serve its function, I cannot imagine a model upon which to design observations toward answering the key question of integration of cold with warm parts of the body.

Héroux (1959) found the cell divisions in the margin of rats' ears to be blocked by exposure to cold. In the course of time for acclimation, cellular divisions and restoration of initial damage improved. Monitoring temperature in skin cannot be carried out through prolonged experimental exposure, and it is difficult to estimate whether skin cells are forced by cold into dormancy from which they recover by periodic rewarming or whether they become truly adapted for growth while cold.

In experiments the body skin of seals (*Phoca vitulina*) has been observed to remain near 0° C. in ice water (Irving & Hart, 1957). Seals dwelling on Arctic ice survive in water colder than $-1.5°$ C. during many months. When at rest in ice water during experiments, their skin is near the temperature of ice. We do not know the seals' programme of skin temperature during several winter months when the few Arctic observers do not even sight seals above the sea ice. As far as is known, they remain in the water and do not then emerge from breathing holes. It may be imagined that while vigorously generating heat by swimming their skin warms a few degrees periodically, as we suspect that the cold feet and hooves of Arctic land mammals may warm during exercise and when sheltered during sleep. It may then be that the cold surfaces periodically rewarm to effect metabolic recovery that counteracts the depressing influences of cold.

Feltz & Fay (1966) found that tissue cultures prepared from epidermis of northern seals showed no cell division when colder than 17° C., the lowest temperature at which division also occurred in two cultures from human lung and carcinoma. A culture derived from a harbour seal did resume growth at 37° C. after 26 weeks at 4° C. The human cultures did not survive 3 weeks cold storage. This indicates capability of the seal tissue cultures to survive through a long period of ostensible but artificially imposed cold dormancy. Assumption of periodic dormancy in cold mammalian tissues avoids the question of how their integration can be maintained with central tissues nearly 40° C. warmer than peripheral tissues when the usual Q_{10} leads to rate and velocity functions depressed 20 or 30 times by cold.

In relating conditions in cold mammalian tissues to dormancy we do not much strain that elastic word, but we soon run out of physiological information quantitatively related to survival of populations. I think that crucial questions are: how long do the tissues remain cold; do they require certain periodic rewarming; and, most difficult, how can they maintain integration during cold periods when processes are slowed by cold?

WINTER PROGRAMMES OF ARCTIC INSECTS

We have been much interested in the winter state of Arctic insects and plants which in the vicinity of Fairbanks remain dormant in appearance from mid-September to early May. It has been our special curiosity to discover what changes occurred in the winter months, because, I suppose, of a bias toward thinking that even poikilothermous life is not necessarily dull during Arctic winter. Some physiological happenings seen in natural dormancy we consider to be utilized or even essential for the existence of populations. Adult forms of insects are more conspicuous and attractive to a comparative physiologist than are the less lively larval forms. Bringing wood for heating a winter tent usually revives some insects near the stove to visible activity. We see flies and beetles resume activity as we rewarm a cold unoccupied cabin. In spring moths, bumble bees, and mosquitoes suddenly appear as flying adults before the snow melts. These observations led us to believe that many adult Arctic insects could be aroused from dormancy at any time.

Late in September near Barrow (latitude 71° N.) I watched caddis flies (*Grensia*) (Weber, 1948) walking on the ice of a small stream and deliberately crawling down beneath the snow along grass stems. In air temperature about $-5°$ C. the flies crawled toward me as I circled them at a yard distance. Well-integrated and apparently purposive activity of insects over cold snow in autumn and spring has been frequently noticed, but the thermal capabilities of Arctic insects have seldom been measured.

Our colleague T. Kaufmann (unpublished) has introduced system into our casual observations by identifying some 30 species of adult insects obtained in winter from around stumps and bark of trees near Fairbanks. We have especially sought for populations that would be accessible to us above the snow, where they may be exposed to untempered winter cold as low as $-50°$ C. At ground surface under a good snow cover temperatures may be only a few degrees below freezing, although frost usually penetrates several metres underground.

A population of carabid beetles (*Pterostichus brevicornis*) living in rotting stumps has been obtainable in hundreds. Its small size (7 mg.) and availability permits observation of samples representative of a population in its natural habitat as well as under experimental conditions. T. Kaufmann examined examples from natural habitats throughout the winter and found that the proportion with guts empty increased until all were empty in January. Thereafter they gradually refilled with woody detritus. Eggs visibly began to grow in mid-February and expanded through spring. This is evidence that not all is quiet during the natural dormant period. In this winter of observation (1967-8) temperatures recorded in the stumps were somewhat tempered above the cold spells in air and we are not certain that the beetles remained uninterruptedly colder than their rather low melting or super-cooling levels.

L. K. Miller has observed the rapid recovery of properly dormant beetles after experimental exposure to $-85°$ C. Beetles aroused from dormancy by warming and then exposed to gradual cooling still walked in coordinated fashion at $-5°$ C. and made ineffective movements of appendages at $-12°$ C.

Dormancy and awakening

In autumn Baust (1968a) found a build-up of glycerol in haemolymph to the impressive concentration of about 20%, with subsequent decline in spring (Fig. 6). When taken from deep cold to room temperature the beetles soon disposed of their glycerol. In the midst of dormancy, with large concentrations of glycerol, the haemolymph could be cooled to $-12°$ C. before freezing. This was evidently a supercooling limit, for, upon warming and recooling, freezing points appeared about $-3.5°$ C.

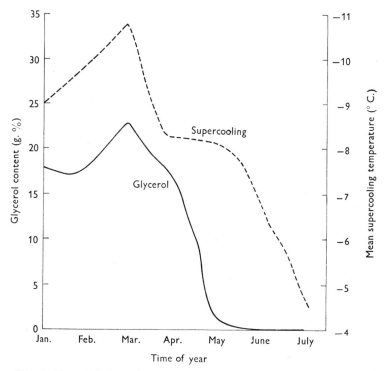

Fig. 6. Glycerol in haemolymph of *Pterostichus brevicornis* (Baust, 1968a).

We can see that description of the circumstances of natural dormancy can become complicated and the natural and tolerable programmes will involve laborious search into rates of cooling, levels attained and duration of extrinsic influences from the environment and intrinsic conditions that can be referred to time or sequence scale in the individual's life-history. The beetle's dormant programme asuredly involves it in lively changes that raise interesting and possibly determinable questions as to what changes occur below the freezing or supercooling points.

Excitation processes

In some fashion the dormant insect regulates its internal state and probably appreciates the lapse of time. My colleague J. Baust has succeeded in recording apparently spontaneous spike potentials from within a proximal process of the awakened beetle's modified trochanter on the last thoracic leg. As the beetle is cooled these spikes show peculiar rises and declines in frequency (Fig. 7). The characteristics of these rather regular discharges suggest that several nerve fibres, which visibly enter the region, are indivi-

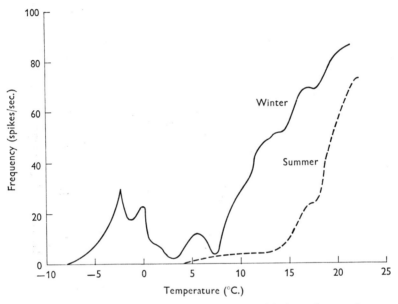

Fig. 7. Frequency of spontaneous potentials in trochanter of *Pterostichus brevicornis* (Baust, 1968b).

dually responsive to separate degrees of temperature. In combination they could represent the influence of temperature over a wider range than could be conveyed by a single fibre, or a group of fibres of like but limited frequency transmission. Pursuit from this speculation to actual demonstration is presently frustrated by the visible presence of muscle fibres and apparent sense organs in combinations difficult to sort out in the small part of an appendage of a 7 mg. beetle.

The potential spikes are seen over a considerable range of temperature, being visible until the haemolymph freezes at −12° C. In reaching this temperature the nerves have passed through the freezing point about −3·5° C. We are especially interested to determine explicitly what processes or tissues are the source of the spike potentials. They appear typical of nerve action potentials. They do indicate that phenomena in solution generating potentials are in operation under supercooled conditions. We have felt safe in attributing the well-characterized action potentials generated upon stimulation of supercooled mammalian nerves to ionic transport in membranes below the freezing point. So we are inclined to believe that spike potentials in supercooled beetles represent generated impulses that might be of use in nervous signalling of intrinsic or extrinsic conditions.

A GLANCE AT QUATERNARY CONDITIONS FOR DORMANCY

As Tertiary climates cooled toward the succession of Pleistocene glacial episodes we can imagine that northern species of plants and animals were confronted with new climatic requirements to which they must adjust if they were to retain their places of residence, or as they retreated southward. The sequence of Pleistocene glacial and inter-glacial episodes, climatic changes and consequent migrations required large and repeated adaptations of annual life-cycles for survival of populations. Evidently, after a prolonged Tertiary evolution without extensive glaciation, the northern species could rather suddenly adapt the physiology of dormancy to new and changing Pleistocene climatic regions.

Throughout the Pleistocene and into Recent time expressions of capabilities for dormancy have had to change patterns at rates that would seem to tax the old-fashioned slow genetic machinery of natural populations. Plant geneticists are facing the problems of physiological adaptability of populations with geographical and experimental indications for amazing physiological versatility within the constitution of populations of some species. Genetics in physiological adaptation processes of animal populations are harder to trace, but they may be sought in relation to measured reactions and conditions that we can reasonably consider essential for survival of populations through dormancy.

CONCLUDING COMMENTS

I have touched upon evidence for activity within the cold tissues of northern animals. Many more examples could be presented to show that dormant life can count the passage of time, is sensitive to change within and without, remembers its specific character or heredity and its progress in life, and, as an individual, anticipates the seasonal changes in its locality. There is no doubt of my bias from experience in pointing out activities in the cold winter tissues of Arctic animals or that often on faith I regard them to be adaptive in the lives of populations. I believe, however, that limits upon adaptability are imposed by the past experience of the individual and its store of evolutionary relics.

I have not dwelt upon the transition from the tolerance of cold in winter and dormancy to the lesser tolerance of cold that is common in summer. I would point out that in a high Arctic climate air-dwelling plants and poikilothermous animals remain frost-tolerant throughout the summer. In the rapidly changing Arctic spring and autumn they are frequently exposed to freezing and thawing temperatures. During short warm periods Arctic poikilotherms must spring to brief incremental activity. As their activity suddenly subsides in a brief cold spell, it cannot relax into the dormancy that involves a slow adaptive change. Each brief incremental period is not reversed by freezing and progress in the life-cycle is maintained as if held by a ratchet. The physiology of this Arctic climatic opportunism deserves study.

In the deeply frozen state of dormancy, energy expenditure is so extremely small that it is still useless to speculate upon its employment for survival of the organization of the individual. Natural northern populations in spring emerge from dormancy to visible incremental activity near the freezing point of water. If they can operate for growth 1° below 0° C. their annual productive programme in the Arctic can add many hours to the short time-span of summer when Arctic water is liquid. So we have looked for active processes in the survival of cold as well as dormant life. There appear to be several that are measurable and that could reasonably contribute to the successful maintenance of populations. The persistence of definable vital reactions in cold and supercooled tissues also presents an extended range of characteristics related to solutions and membranes.

REFERENCES

BAUST, J. (1968a). Abstract for Alaska Science Conference on glycerol on haemolymph of *Pterostichus brevicornis*.
BAUST, J. (1968b). Abstract for Alaska Science Conference on frequency of spontaneous potentials in trochanter of *Pterostichus brevicornis*.
CHATFIELD, P. O., LYMAN, C. P. & IRVING, L. (1955). *Am. J. Physiol.* **172** (3), 639.
ERICKSON, H. (1956a). *Acta physiol. scand.* **36**, 66.
ERICKSON, H. (1956b). *Acta physiol. scand.* **36**, 79.
FELTZ, E. T. & FAY, F. H. (1966). *Cryobiology* **3**, 261.
GALSTER, W. A. & MORRISON, P. R. (1968). Abstract for Alaska Science Conference on changes in glycerol content in hibernating ground squirrels (*Citellus*).
HÉROUX, O. (1959). *Can. J. Biochem. Physiol.* **37**, 1247.
IRVING, L. & HART, J. S. (1957). *Can. J. Zool.* **35**, 497.
KEHL, T. H. & MORRISON, P. R. (1960). *Bull. Mus. comp. Zool. Harv.* **124**, 387.
WEBER, N. A. (1948). *Ent. News* **59**, 253.

AUTHOR INDEX

Figures in bold type indicate pages in which references are listed

Aaronson, S., 129, **141**
Abdalla, A. A., 219, 232, **237**
Addicott, F. T., 222, **239**, 253, 258, **261**
Adelstein, S. J., 378, **393**, 520, **545**
Adkisson, P. L., 263, 268, **283**, 288, 289, 290, 292, 293, 294, **300**, 301, 310, 312, 315, 326, **328**, **329**
Adolph, E. F., 354, 356, 366, **393**, 503, **508**
Agid, R., 519, 520, 521, 526, 527, 529, 530, 534, 535, **545**, **546**, **548**
Aigner, P., 373, **394**
Akazawa, T., 409, 423, **447**, **448**
Åkerman, A., 419, **445**
Akkerman-Bellaart, M. A., 519, 520, 548
Ako-Addo, A., 480
Albin, F. E., 84, **97**
Alcorn, S. M., 205, **216**
Alexandrov, V. Y., 403, 436, 441, 442, **445**
Allanson, M., 519, 520, 530, **545**, **546**
Allard, H. A., 245, **261**
Allen, M. B., 128, **140**
Allen, R. M., 221, **237**
Allende, J. E., 146, **160**
Allfrey, V. G., 79, **80**
Alpers, D. H., 41, **47**
Alvim, P. de T., 471, 478, **488**
Amen, R. D., 219, 236, **237**, 450, **468**
Anagnostidis, K., 138, 140, **140**
Anchori, S., 201, **216**
Anderson, C., 33
Anderson, J. W., 426, **445**
Andjus, R. K., 352, 353, 354, 356, 358, 360, 361, 366, 367, 370, 372, 377, 379, 383, 385, 387, 388, 390, 391, **393**, **394**, 494, **509**
Andreeva, E. V., 390, **394**
Andrewartha, H. G., 266, **283**
Andrews, S., 421, 424, **445**
Anfinsen, C. B., 415, 420, **445**
Appleman, C. O., 220, 228, 229, **237**
Appleman, W., 175, **191**
Appuhn, U., 193, 197, 207, 214, **217**, 456, **469**
ap Rees, T., 180, 181, **190**
Arai, N., 409, 410, 426, **447**
Armstrong, D., 506, **508**
Armstrong, R. L., 17, 41, **47**
Aron, M., 519, 529, 530, **547**
Arrandondo, J., 72, **80**
Artschwager, E. F., 231, **237**
Asahina, E., 331, **350**

Aschoff, J., 450, **468**
Asdell, S. A., 516, **546**
Ashhurst, D. E., 346, **350**
Aubertot, T., 95, **97**
Augustinsson, K.-B., 495, **508**
Augusti-Tocco, G., 78, **81**
Awad, E. S., 440, **445**
Azzali, G., 530, 534, 535, **546**, **548**

Baard, G., 279, **284**
Babcock, G. E., 427, **448**
Bachmetiev, P. E., 352, **394**
Bacon, J. S. D., 187, **191**
Baker, H., 129, 132, 138, **140**, **141**
Baker, J. R., 515, **546**
Baker, R. F., 42, 43, **47**
Balassa, G., 17, **47**
Balfour, W. M., 370, **394**
Ballard, L. A. T., 175, **190**
Bamberg, S., 432, **445**
Band, R. N., 58, 59, **80**
Barker, J., 182, **190**
Barker, R. J., 310, 326, 326, **328**
Barner, H. D., 72, 78, **80**
Barnett, S. A., 534, **546**
Barry, J., 534, **546**
Barskaya, E. N., 225, **237**
Bartholomew, G. A., 511, 513, 516, 544, **546**
Barton, L. V., 161, 176, **190**, 223, **237**
Bather, R., 96, 97, **97**
Battista, A. F., 498, 507, **508**, 512, **546**
Baust, J., 561, 562, **564**
Beament, J. W. L., 87, **97**
Beck, S. D., 293, **299**, 310, 311, 326, **328**
Beermann, W., 79, **80**
Beevers, H., 143, **160**, 180, 181, 182, **190**
Beevers, L., 258, **261**
Belcher, J. H., 126, 127, **141**
Belderok, B., 162, 164, 169, 179, 182, **190**
Belehradek, J., 433, 434, **445**
Bell, E., 11, **47**
Bell, R. A., 263, **283**
Bello, J., 402, **445**
Benetato, G., 534, **546**
Bennet-Clark, T. A., 221, 222, **237**
Bennett, J. P., 424, **446**
Benoit, J., 263, **283**
Benton, W. F., 52, 54, 59, 60, 61, 67, **80**, **81**
Berberich, M. A., 40, **48**
Berendes, H. D., 233, **237**

Berg, A., 455, **469**
Berg, P., 45, **48**
Bergmeyer, H. U., 114, **120**
Bergquist, 130
Berns, D. S., 440, **445**
Berwick, S., 520, **547**
Berrie, A. M. M., 166, **190**
Besson, S., 534, **546**
Bewley, J. D., 203, 204, 205, 207, 208, 209, 210, 212, **216, 217**, 456, **469**
Bidault, Y., 206, **217**
Biebl, R., 125, 129, 130, 134, 136, 137, **140**, 396, 397, **446**
Biewald, G.-A., 491, **508**
Bigelow, C. C., 410, 436, 437, **446**
Billen, D., 34, **48**
Bird, T. F., 515, **546**
Bishop, D. W., 540, **546**
Black, M., 163, 164, 165, 182, 186, **190**, 194, 196, 197, 198, 202, 203, 204, 207, 208, 209, 210, 212, **216, 217**, 245, **262**, 456, **469**
Black, J. N., 478, **488**
Black, S. H., 12, **48**
Blank, G. B., 133, 140, **140**
Blattler, D. P., 421, **446**
Blinks, D. C., 490, **509**
Blommaert, K. L. J., 221, **237**
Bloom, G., 181, 185, **190**
Blum, P. H., 170, **190**
Blumenthal-Goldschmidt, S., 222, 223, 224, 225, 226, 228, 235, **237, 238, 239**
Boaler, S. B., 478, **488**
Bock, R. M., 40, **48**
Boczkowska, M., 265, **283**
Boer, J. A. de, 275, 276, 277, 278, 280, 281, **284**
Bonga, H., 266, **284**
Bonner, J., 189, **192**, 225, 232, 236, **237**, 249, 260, **261, 262**
Boo, L., 223, 224, **237**
Borriss, H., 175, **190**
Borthwick, H. A., 167, 176, **192**, 193, 194, 195, 197, 206, 207, 211, **216, 217**, 243, 244, 247, **261**, 455, **468**
Bosher, J. E., 88, **97**
Boulter, D., 173, **191**
Boulter, D. W. K., 198, **217**
Böventer-Heidenhain, B. von, 99, **121**
Bowers, B., 55, 57, 58, 59, 79, **80**
Bowler, K., 354, **394**
Bradbeer, J. W., 182, **190**
Bradshaw, G. V. R., 544, **546**
Brain, M. C., 417, **446**
Brandts, J. F., 398, 399, 400, 401, 403, 404, 408, 412, 414, 421, 422, 431, 435, 436, 437, **446**
Braun, R., 68, **80**
Bravo, M., 146, **160**

Bray, J. R., 480, **488**
Bredow, K. von, 205, **216**
Bremer, H., 44, **48**
Brendel, W., 373, **394**
Bretz, C. F., 231, **239**
Brian, P. W., 202, **216**, 223, **237**
Briggs, D. R., 418, **448**
Bristol, B. M., 134, **140**
Brittinger, G., 55, **81**
Britton, S. W., 351, 352, **394**
Brock, T. D., 128, 132, **140**, 396, 434, **446**
Brooks, C.McC., 505, **509**
Brown, D. S., 223, **237**
Brown, G. L., 495, **508**
Brown, R., 162, **190**
Bruce, A. L., 417, **446**
Bruce, V. G., 312, 314, **328**
Bruener, R., 44, **48**
Bruin, W. J., 165, **190**
Bruinsma, J., 183, **190**
Buch, M. L., 222, **237**
Buchwald, H. D., 297, **299**
Butenandt, A., 285, **299**
Budd, K., 106, 110, **120**
Bula, R. J., 428, **446**
Bullough, W. S., 52, **80**
Bünning, E., 131, **140**, 309, 310, 311, 314, 327, **328, 488**
Bünsow, R., 205, **216**
Bünsow, R. C., 310, **328**
Bunt, J. S., 126, **140**
Burlet, C., 534, 535, **547**
Burlington, R. F., 366, 378, 379, **394**
Burma, D. P., 181, **190**
Burr, H. K., 107, **120**
Burris, R. H., 69, **81**
Burton, W. G., 220, 222, 233, 234, 236, **237**, 249, **261**
Büsgen, M., 242, **261**
Busnel, R. G., 270, **283**
Butler, W. L., 195, **216**
Butt, V. S., 180, **190**

Cachan, P., 471, **488**
Cade, T. J., 512, 513, 516, **546**
Caffier, P., 530, **546**
Caldwell, F., 162, 172, **190**
Cameron, R. E., 133, 140, **140**
Campbell, D. W., 534, **546**
Campbell, L. L., 101, **121**, 433, 434, **446**
Carr, D. J., 162, **190**, 202, **217**
Carrell, R. W., 417, **446**
Carroll, W. R., 402, **446**
Caspary, R., 193, **216**
Castelfranco, P. A., 427, **448**
Castenholtz, R. W., 129, 138, **140, 142**
Cathey, H. M., 207, **217**, 455, **468**
Chakravorty, M., 181, **190**
Chalmers, B., 344, **350**

AUTHOR INDEX

Chamberlin, M., 45, **48**
Chance, B., 117, **120**
Chandler, C., 223, **237**
Chapman, V. J., 136, **141**
Chater, A. P. J., 428, **448**
Chatfield, P. O., 489, 506, 507, 508, **508**, **509**, 512, 529, **546**, 555, **564**
Cheatum, G. G., 424, **448**
Chen, S. S. C., 165, 189, **190**
Cherry, J. H., 143, 146, **160**
Chetram, R. S., 169, **190**
Ching, T. M., 164, **190**
Chrispeels, M. J., 224, 235, **237**, 258, **261**
Christophersen, J., 395, **446**
Church, B. D., 12, **48**
Ciani, P., 519, 520, **546**
Ćirković, T., 383, **394**
Clark, A. J., 504, **508**
Clarke, J. I., 478, **488**
Clarkson, D. T., 215, **216**
Clegg, J. S., 84, 94, 97, **97**
Clegg, M. D., 225, 226, 228, 230, 231, **237**, **239**
Clements, J. B., 202, **217**
Clive, A. L., 40, **48**
Cochrane, J. C., 104, **120**
Cochrane, V. W., 104, **120**
Coco, R. M., 530, **547**
Cohen, C. F., 326, 327, **328**
Cohen, D., 8, **10**
Cohen, S. S., 72, 78, **80**
Cole, L., 1, **10**
Coleman, E. A., 428, **446**
Collyer, D. M., 135, **141**
Colman, B., 182, **190**
Colvill, A. J. E., 42, **48**
Conn, E. E., 188, **190**
Connelly, C. M., 117, **120**
Contaxis, C. C., 421, **446**
Cornforth, J. W., 222, **237**, 252, 253, **261**
Coster, C., 471, 478, 480, **488**
Cotrufo, C., 220, **237**, 402, **446**
Courrier, R., 514, 515, 519, 520, 521, 522, 530, 538, 543, **546**
Cox, R. P., 40, **48**
Crane, J. C., 224, **238**
Crocker, W., 161, 176, **190**, 250, **261**
Cumming, B. G., 198, **216**
Cummins, J. E., 77, **80**
Cunningham, T. A., 96, 97, **97**
Cure, M., 526, 530, 534, **546**
Curran, H. R., 12, **48**

Dacil, J. V., 417, **446**
Dahl, N. A., 370, **394**
Daily, W. A., 124, **141**
Dale, H. H., 495, **508**
Daneliuc, E., 534, **546**
Daniels, F., 339, **350**, 396, **446**

Danilevskii, A. S., 309, 318, 327, **328**
Danylievsky, A. S., 264, **283**
Davidkova, E., 426, **447**
Davidson, B., 432, 436, **446**
Davidson, T. M. W., 232, 233, **237**
Davis, R. L., 428, **446**
Davis, W. E., 250, **261**
Dawe, A. R., 531, **547**
Deane, H. W., 529, **546**
Deanesley, R., 530, **546**
Demain, A. L., 13, **48**
Dempsey, E. W., 512, 520, 533, **547**
Denny, F. E., 219, **237**
Deranleau, D. A., 440, **445**
Desrosier, N. W., 12, **48**
Deutschar, M. P., 20, **49**
De Vries, D. A., 453, **469**
Dickinson, E., 21
Digby, J., 245, 256, **261**
Dische, Z., 63, 65, **80**
Diskin, T., 455, **469**
Distler, J. R., 103, 104, 119, **121**
Dodge, B. O., 100, **121**
Doi, R. H., 16, 19, **48**
Donnellan, J. E., Jun., 17, 41, 45, **48**, **49**
Donoho, C. W., 223, **237**
Doorenbos, J., 223, **237**, 247, **261**
Dörffling, K., 252, **261**
Dorner, R. W., 423, **446**
Douglas, W. W., 500, 502, **508**
Douthit, H. A., 24, 25, **48**
Downs, R. J., 195, **216**, 243, 244, 247, **261**, 472, **488**
Drennan, D. S. H., 166, **190**
Dry, R. M. L., 506, 508
Dryer, R. L., 536, **548**
Dubnau, D., 41, 43, **48**, 49
Dubois, J., 424, **448**
Dubois, P., 526, 530, 534, **546**
Dugger, W. M., Jun., 426, **447**
Duintjer, J. C. S., 267, 268, **284**
Duncan, C. J., 354, **394**
Dure, L., 146, **160**
Durham, V. M., 164, **190**
Durkee, A. B., 235, **239**
Duthie, H. C., 125, **141**
Duval, J., 471, **488**
Dwyer, P. D., 514, 515, **546**

Eagle, D. J., 126, **141**
Eagles, C. F., 221, 223, 224, 236, **237**, 240, 253, 255, **261**
Eaks, I. L., 410, **446**
Edelman, J., 224, 231, **238**
Edsall, P. C., 425, **447**
Edwards, R. L., 307, **328**
Effer, W. R., 181, 182, **190**
Ege, R., 274, **283**

Eilers, F. I., 106, 110, 111, 112, 114, 115, 116, 118, **120**
Eisentraut, M., 513, 543, **546**
El Antably, H. M. M., 222, 224, **238**, **240**, 248, 253, 256, 258, **261**
Eldjarn, L., 427, **446**
El Hilali, M., 358, 360, **394**
El Ibrashy, M. T., 272, 278, 282, **283**
Ellenby, C., 83, 84, 85, 86, 88, 89, 90, 91, 92, 93, **97**
Ellerman, J. R., 494n, **508**
Elliot, B. B., 164, **190**
Ellis, R. J., 187, **191**
Elmquist, D., 498, **509**
Emerson, M. E., 100, 102, **120**
Emilsson, B., 220, 229, 232, **238**
Enders, A. C., 544, **546**
Endo, B. Y., 85, **97**
Engel, R., 529, **546**
Enser, M., 427, **448**
Enteman, C., 65, **80**
Enti, A. A., 477, 480, **488**
Epstein, R., 14, 15, 17, 18, 21, 40, 41, **48**, **49**
Erickson, H., 553, **564**
Esashi, T., 205, **217**, 224, **238**
Eskin, A., 327, **328**
Essery, R. E., 169, 172, **190**, **192**
Evans, F. R., 12, **48**
Evans, J. H., 134, 135, 136, 140, **141**
Evans, L. T., 235, **238**
Evenari, M., 181, 190, 192, 193, 196, 198, 201, **216**, 451, **468**, 486, **488**
Everson, E. H., 163, **191**

Falaschi, A., 46, **48**
Falk, M., 96, **97**
Farmer, J. L., 34, **49**
Farr, A. L., 60, 61, 65, **80**
Fasman, G. D., 432, 436, **446**
Fatt, P., 495, 497, **509**
Fawcett, D. W., 541, **546**
Fay, F. H., 559, **564**
Fay, P., 130, 140, **141**
Feeley, J., 11, **49**, 143, 146, 150, 153, 155, 156, **160**, 236, **238**
Feinbrun, N., 201, **216**, 232, **238**
Feldberg, W., 495, **508**
Feldman, N. L., 436, 438, **446**
Feltz, E. T., 559, **564**
Fencl, V., 499, **509**
Fennema, O., 411, **446**
Ferenczy, L., 221, 222, **239**
Ferket, P., 263, 267, 281, **284**
Feughelmann, M., 413, 417, **446**
Fielding, M. J., 84, 88, **97**
Filov, V. A., 390, **394**
Fink, O. F., 265, **283**
Finlayson, L. H., 298, **299**
Fischwich, O., 172, **191**

Fisher, H. G., 297, **299**
Fiske, C. H., 60, **80**
Fitz-James, P. C., 17, 34, **48**
Flaks, J. G., 72, 78, **80**
Flescher, D., 180, **190**
Flint, L. H., 193, **216**
Foda, H. A., 162, **192**
Fogg, G. E., 123, 126, 127, 128, 131, 135, 138, 140, **141**
Folk, G. E., Jun., 533, **546**
Fondeville, J. C., 206, **216**
Foote, W. H., 164, **190**
Forro, J. R., 46, **49**
Foster, M. W., 526, 530, 533, **546**, **547**
Foster, R. C., 533, **546**
Fowden, L., 139, **141**
Frankland, B., 188, **191**, 206, **216**
Franklin, M. T., 86, **97**
Fredericq, H., 206, **216**
Frémy, P., 136, **141**
French, M. E., 109, **121**
Fried, J. H., 285, **299**
Friedman, S. M., 436, 437, **446**
Fries, N., 130, **141**
Fritsch, F. E., 134, 135, **141**
Frowein, R. A., 373, **394**
Fugii, R., 143, **160**
Fuhrman, F. A., 297, **299**
Fujisawa, H., 157, **160**
Fukuda, I., 131, **141**
Fukuda, S., 305, **328**
Fukumoto, S., 205, **217**
Furlenmier, A., 285, **299**
Fürst, A., 285, **299**

Gabe, M., 519, 520, 521, 526, 527, 529, 530, 534, 535, **545**, **546**, **548**
Gaber, S. D., 170, 171, 172, 176, **191**, **192**
Gabrielescu, E., 534, **546**
Gadd, L., 178, **191**
Gaevskaya, M. S., 382, **394**
Gaisler, J., 521, 527, 529, **546**
Galil, J., 452, **468**
Galster, W. A., 553, 554, **564**
Galston, A. W., 206, **217**
Gann, E. L., 530, **546**
Gänshirt, H., 379, **394**
Garcia, J. P., 489, 507, **508**, 512, **546**
Garner, W. W., 245, **261**
Gassner, G., 193, **216**
Gehenio, P. M., 347, **350**
Geiduschek, E. P., 42, **48**
Genci, R., 376
Gerhardt, P., 12, **48**
Gerloff, E. D., 428, **446**
Gessner, F., 128, **141**
Getman, F. H., 339, **350**, 398, **446**
Gfeller, F., 402, 429, 431, **448**
Gibbs, M., 180, **191**

AUTHOR INDEX

Giaja, J., 494, **509**
Gifford, R. H., 157, **160**
Gilbert, L. I., 278, **283**, 286, 287, **299**
Gilbert, S. G., 170, **190**
Gilliam, J. J., 491, **509**
Ginsburg, A., 402, **446**
Girault, A. A., 265, **283**
Girod, C., 526, 530, 534, **546**
Glade, R., 130, 140, **141**
Glick, D., 504, **509**
Goddard, D. R., 100, 102, 104, **120**, 162, **191**
Godnev, T. N., 410, **446**
Godward, M. B. E., 123, **141**
Goldberger, R. F., 40, **48**
Good, J., 257, **262**
Good, N., 224, **240**
Goodwin, P. B., 221, **238**
Gorham, E., 480, **488**
Gorman, J., 40, **48**
Goryshin, N. I., 266, 267, **283**, 312, 326, **328**
Goss, J. A., 196, **217**
Grahl, A., 172, **191**
Grant, J., 60, 66, **81**
Grant, N. H., 412, **446**
Greenwood, M., 478, **488**
Greiff, P., 423, **446**
Grif, V. G., 402, **446**
Griffin, M. J., 40, **48**
Griggs, W. H., 223, **237**
Grigsby, B., 176, **192**
Grindeland, R. B., 533, **546**
Grison, P., 265, 268, 281, **283**
Grobstein, C., 40, **48**
Gross, P. R., 11, **49**
Grutzmann, K., 424, **447**
Guilhem, A., 514, 543, **568**
Guillard, R. R. L., 137, **141**
Gumma, M. R., 383, **394**
Guthrie, M. J., 514, 527, 542, **546**
Gutman, A. B., 69, **80**
Gutman, E. B., 69, **80**
Guttenberg, H. von, 252, **261**
Gutterman, Y., 451, **468**
Guyon, D., 133, **142**

Haber, A. H., 201, 202, **216**, 232, **238**
Hackett, D. P., 182, **192**
Hageman, R. H., 180, **190**
Hagen, I., 424, **447**
Haines, F. M., 134, 135, **141**
Hale, 60, 61
Hall, C. E., 11, **47**
Hall, J. B., 471, 472, 484, 485, **488**
Hall, M. A., 224, **238**
Halmann, M., 13, **48**
Halvorson, H., 11, 12, 13, 14, 15, 17, 18, 19, 20, 21, 22, 24, 25, 26, 27, 30, 31, 32, 33, 36, 37, 39, 40, 41, 45, **48**, **49**, 68, **80**, 99, 104, 109, **121**, 184, **192**
Hamilton, J. B., 352, **394**
Hammel, H. T., 490, **509**
Hamner, K. C., 310, 313, **328**
Hamner, W. M., 326, **328**
Hanawatt, P., 34, **49**, 77, **80**
Hansen, V. K., 126, **142**
Hanson, J. B., 187, **191**
Hanson, R. S., 21, **49**
Harder, W., 402, **446**
Hardy, F., 478, **488**
Harel, E., 181, **191**
Harper, J. L., 161, 177, **191**, **192**
Harper, J. P., 428, **446**
Harrell, W. K., 12, **48**
Harrison, I. T., 285, **299**
Hart, J. S., 360, **394**, 559, **564**
Hartman, G. C., 529, 539, 542, **546**
Hartman, K. A., Jun., 96, **97**
Hartmann, K. M., 214, **216**, 456, **468**
Hartmann, K.-V., 38, **48**
Hartsema, A., 219, **238**
Hartwell, L. H., 30, **48**
Hashimoto, T., 202, **217**, 226, **238**
Hasman, M., 427, **447**
Hasselberger, F. X., 365, 370, 373, 377, 378, 380, 381, 382, **394**
Hatch, F. T., 417, **446**
Haupt, W., 206, **216**
Hay, J. R., 165, 178, 182, **191**
Hayashi, F., 222, 223, 226, **238**, **239**
Hayashi, J., 166, 173, 182, **192**
Hayashi, M., 143, **160**
Hayward, J. S., 536, **546**
Heber, U., 419, 423, 426, **446**, **448**
Heidelberger, C., 38, **48**, 72, 74, **80**
Heiligman, F., 12, **48**
Heitmann, P., 415, **446**
Held, D., 499, **509**
Hemberg, T., 220, 221, 222, 235, **238**, 252, **261**
Hemming, H. G., 223, **237**
Henckel, P. A., 138, **141**, 220, **238**
Hendricks, S. B., 167, 176, **192**, 183, 194, 195, 197, 206, **216**, **217**, 310, 312, **328**
Henneberg, W., 95, **97**
Henney, H. R., 11, **48**, 106, **120**
Hensell, H., 506, **509**
Herdtle, H., 131, **140**
Herlant, M., 513, 514, 521, 527, 530, 532, 543, **546**, **547**, **548**
Herman, W. S., 286, 286, **299**
Héroux, O., 558, **564**
Herreid, C. P., 513, **547**
Herring, P. T., 352, **394**
Heslop-Harrison, J., 449, **469**
Heslop-Harrison, Y., 249, **261**
Hess, U., 134, 135, 140, 141

Hewitt, R., 34, **48**
Heydecker, W., 169, **190**, **192**
Heymans, J. F., 352, **394**
Hickman, J. L., 513, **547**
Hickman, V. V., 513, **547**
Hidaka, H., 427, **447**
Higa, A., 11, 18, 21, 22, 30, 41, 45, **48**, **49**
Higgins, J. J., 117, **120**
Higham, B., 189
Highkin, H. R., 464, **469**
Hill, E. P., 107, 108, **120**
Hillman, J., 222, 224, **238**, 253, 258, **261**
Hillman, W. S., 206, 214, 215, **216**, **217**
Hiltner, L., 197, **191**
Hirano, M., 124, **141**
Hirschhorn, R., 55, **81**
Hoad, G. V., 254, **261**
Hoch, G., 126, **140**
Hock, R. J., 519, 526, 536, 542, **547**, **548**
Hocks, P., 285, **299**
Hodek, I., 268, **283**
Hof, T., 136, **141**
Hoffman, L. H., 539, 540, **547**
Hoffman, R. A., 511, 529, 530, 534, **547**
Hofmann, T., 415, **447**
Hoffmeister, H., 284, **299**
Holdgate, M., 127, **141**
Holmes, R., L., 534, **546**
Holmes, W., 117, **120**
Holm-Hansen, O., 124, 125, 130, **141**
Holmsen, T. W., 408, **446**
Holton, R., 99, 100, 104, 120, **120**, **121**
Holttum, R. E., 478, **488**
Hopkins, C. R., 307, **328**
Horecker, B. L., 427, **448**
Horikoshi, K., 11, **48**
Horne, A. J., 123, 126, 127, **141**
Horneland, W., 124, 125, **141**
Horwitz, J., 307, **329**
Hoskins, W. H., 536, **548**
Hotchkiss, R. D., 59, **80**
House, W., 383, **394**
Hoyem, T., 24, 25, **48**
Hozic, N., 383, 387, 390, **393**, **394**
Hryniewiecka, L., 69, **80**
Hsu, A., 226, **239**
Hudson, J. W., 511, 512, 513, 516, **546**, **547**
Huffaker, R., 157
Huggins, C., 69, **80**
Hummel, H., 285, **299**
Humphreys, T., 11, **47**
Hurwitz, J., 42, **49**
Hutner, S. H., 129, 132, 138, **140**, **141**
Huystee, R. B. van, 146, **160**
Hwang, S.-W., 124, 125, **141**
Hyatt, M. T., 13, **49**

Ibanez, M. L., 410, **446**

Ichikawa, T., 59, **81**
Idle, D. B., 345, **350**
Idriss, J., 20, 21
Igarashi, R. T., 16, 19, **48**
Ikeda, Y., 11, **48**
Ikuma, H., 114, 196, 201, 202, 203, 207, **216**, **217**
Iljin, W. S., 413, **446**
Imamoto, F., 41, 42, 43, 44, **48**
Ingraham, J. L., 124, 126, 138, **141**
Inoue, T., 13, **48**
Irving, L., 555, 559, **564**
Irving, R. M., 425, **446**
Isaac, 130
Isemura, T., 419, **448**
Ishii, T., 202, **217**
Isikawa, S., 194, 197, **217**, 455, **469**
Itai, C., 450, **469**
Ito, S., 541, **546**
Ivins, J. D., 219, **238**
Iwakiri, B. T., 223, **237**
Iyer, V. N., 37, **49**, 71, 78, **81**

Jachymczyk, W., 146, **160**
Jackson, K. A., 344, **350**
Jacob, H. S., 417, **446**
Jacobi, G., 423, **446**
Jacobsen, L., 424, **446**
Jacobson, K. B., 438, **446**
Jaffe, M. J., 206, **217**
James, T. E., 70, 78, **80**
James, W. O., 173, 179, **191**
Jann, G. J., 14, **49**
Jansson, G., 170, **191**
Jaswal, A. S., 182, 183, **191**
Jayme, G., 61, **80**
Jeffers, J. N. P., 478, **488**
Jeník, J., 134, 136, **141**, 471, 472, 481, **488**
Jenner, C. E., 310, **328**
Jensen, T., 69, **80**
Jensen, W. A., 59, **80**
Jermy, T., 266, **283**
Joel, C., 536, **547**
Joerrens, G., 310, 327, **328**
Johansen, K., 491, **509**
Johansson, B. W., 536, **547**
Johnson, G. E., 530, 533, **547**
Johnston, F. B., 146, **160**, 235, **239**
Johri, M. M., 234, **238**
Jollès, P., 439, **446**
Joly, P., 275, **284**
Jones, E., 60, **81**
Jones, J. W., 167, **191**
Jones, R. L., 256, **261**
Jørgensen, E. G., 126, **141**, **142**, 404, 405, **446**, **448**
Juhren, M., 453, **469**
Jung, G. A., 428, 433, **446**

AUTHOR INDEX

Kahn, A., 196, 203, 207, **217**, 423, **446**, 456, **469**
Kaku, S., 341, 345, 346, **350**
Kalabukhov, N. I., 387, 388, **394**
Kallen, F. C., 527, 531, **549**
Kaminski, Z., 530, **549**
Kämpfe, L., 86, **97**
Kao, C. Y., 297, **299**
Kaplan, N. O., 109, **121**
Karlson, P., 285, **299**
Karssen, C. M., 211, **217**
Kates, M., 109, **120**
Kaufmann, T., 560
Kawase, M., 221, **238**
Kawata, T., 13, **48**
Kay, C. M., 146, **160**
Kayser, C., 366, 379, **394**, 519, 529, 530, **547, 548**
Keele, C. A., 506, **508**
Kefford, N. P., 221, 222, **237**, **239**, 450, **469**
Kehl, T. H., 555, **564**
Keilin, D., 84, 96, **97**
Kent, K. M., 491, **509**
Kerb, U., 285, **299**
Kerjean, P., 42, 45, **48**
Kessmanly, N. W., 390, **394**
Keynan, A., 13, 14, 18, 26, 39, 41, 45, **48, 49**
Khan, A. A., 165, **191**, 202, **217**, 450, 451, **469**
Khan, A. W., 426, **447**
Kiernan, J. A., 534, **547**
Kim, S., 427, **447**
Kinbacher, E. J., 436, 439, **447, 448**
King, P. E., 307, **328**
Kinzel, W., 193, **217**
Kiovsky, T. E., 412, **448**
Kirschner, M., 45
Kirsop, B. H., 169, 170, 172, 173, 179, **190, 191, 192**
Kirtley, M. E., 402, 414, 430, **447**
Kjeldgaard, N. O., 77, **80**
Klain, G. J., 378, **394**
Klebs, G., 247, **261**
Klein, A., 176, **191**
Klein, R. L., 55, 62, 64, **80**
Klein, R. M., 425, **447**
Klein, S., 201, **216**, 232, **238**, 453, 456, **469**
Kleinsmith, L. J., 79, **80**
Klotz, I. M., 96, **97**
Knull, H. R., 436, **447**
Kobayashi, Y., 11, 18, 19, 20, 21, 22, 30, 41, 45, **48**
Koblitz, H., 424, **447**
Koch, A. L., 428, **446**
Kodis, F. K., 352, **394**
Koffler, H., 403, 436, 440, **447**
Koford, M. R., 521, 523, 538, 543, **548**
Kogure, M., 305, **328**

Köhler, D., 202, 203, **217**
Kohn, H., 425, **447, 448**
Koizumi, K., 506, **509**
Kolaczkowski, J., 534, **547**
Kolb, A., 543, **547**
Kolbow, H., 530, **546**
Koller, D., 6, 161, 176, **191**, 194, 195, 196, 213, **217**, 451, 453, 455, 456, 460, 463, 464, 465, 466, 467, **468, 469**, 485
Kondrashova, M. N., 410, **448**
Konrad, M., 44, **48**
Korn, E. D., 55, 57, 58, 59, 79, **80**
Kornberg, A., 20, 46, **48, 49**
Kornberg, H. L., 143, **160**
Kort, C. A. D. de 269, 270, 274, 279, 280, **284**
Kosar, W. F., 196, **217**
Koshland, D. E., 402, 414, 430, **447**
Koukkari, W. L., 206, **216, 217**
Koukol, J., 426, **447**
Kovach, J. S., 40, **48**
Krakow, J. S., 103, 104, 119, **121**
Kramer, P. J., 248, **261**
Krauss, R. W., 131, **142**
Krishna-Murty, G. G., 12, **49**
Krog, J., 491, **509**
Krueger, W. A., 187, **191**
Krull, E., 427, **447**
Krutzsch, P. H., 521, 536, 537, 538, 539, 541, 542, **547, 549**
Kubín, S., 133, **141**
Kudo, S., 170, **191**
Kuempel, P. L., 40, **49**
Kulzer, E., 493, **509**
Kumagai, T., 205, **217**
Kunetsova, J. A., 264, **283**
Kupila-Ahvenniemi, S., 225, **238**
Kuraishi, S., 409, 410, 426, **447**
Kurtz, E. B., 205, **216**
Kutham, T., 72, **80**
Kuzuya, H., 427, **447**

LaBerge, M., 40, **48**
LaCroix, L. J., 182, 183, **191**
Lamarche, M., 534, **546**
Landau, B. R., 531, **547**
Lane, W. C., 195, **216**
Lang, A., 196, 202, 213, **217**, 219, 231, **238, 239**
Lang, F., 61, **80**
Langemann, A., 285, **299**
Lanphear, F. O., 423, 425, **446, 448**
Lapides, J., 69, **80**
Larsen, H., 136, 137, 138, **141**
Laties, G. G., 234, **238**
Laufman, H., 352, **394**
Lawson, G. W., 134, 136, **141**
LeBerre, J., 268, 271, 272, 274, 283, **284**
Lees, A., D., 263, 282, **284**, 293, **299**, 302, 305, 312, 319, 320, 324, 325, 326, **328**

Lees, E., 95, **97**
Legait, E., 534, 535, **547**
Legait, H., 534, 535, **547**
Lehmann, H., 417, **446**
Leike, H., 252, **261**
Leitner, P., 544, **546**
Leopold, A. C., 164, **190**
Letham, D. S., 450, **469**
Levins, R., 3, **10**
Levinson, H. S., 13, 17, 41, **48**, **49**
Levinthal, C., 11, 18, 21, 22, 25, 30, 41, 44, 45, **48**, **49**
Levitt, J., 8, 402, 405, 407, 410, 413, 415, 416, 419, 421, 422, 424, 425, 428, 429, 432, 433, 435, 436, 439, 440, 441, 442, **445**, **446**, **447**, **448**
Levy, H. M., 442, **447**
Lewin, R. A., 128, **141**
Lewis, D. A., 219, **238**, 249, **261**
Lewis, J. C., 107, **120**
Lewontin, R. G., 1, **10**
Li, P. H., 246, **262**, 428, 429, **447**
Libbert, E., 256, **261**
Lichtenstein, J., 72, 78, **80**
Lievesley, P. M., 424, **448**
Light, A., 430, **447**
Lindbloom, H., 220, 232, **238**
Lindner, A., 72, 78, **80**
Lingappa, B. T., 104, 105, **120**
Lingappa, Y., 102, 104, **120**
Lipe, W. N., 224, **238**
Lipman, C. B., 133, **141**
Lipp, A. E. G., 175, **190**
Lippert, L., 223, **238**, **239**
Lison, L., 59, **80**
Livingston, C. H., 223, **239**
Lockwood, S., 129, **141**
Loeb, M. R., 72, 78, **80**
Loeffler, J. E., 224, 235, **239**, 258, **261**
Loggins, M., 63, 67
Longman, K. A., 472, 481, **488**
Loof, A. de, 274, 279, **284**
Lord, R. C., 96, **97**
Lovelock, J. E., 354, 361, 372, 387, **394**
Löwenstein, A., 129, **142**
Lowry, O. H., 60, 61, 65, **80**, 365, 368, 370, 373, 377, 378, 380, 381, 382, **394**
Lowry, R. J., 99, 100, 101, 102, 107, **120**
Lozina-Lozinskaya, F. L., 390, **394**
Luginbill, B., 153, 155, **160**
Luippold, H. J., 201, **216**, 232, **238**
Lundegårdh, H., 394, **447**
Luyet, B. J., 374, **350**
Lyman, C. P., 378, **393**, 489, 490, 491, 492, 493, 504, 505, 506, 507, 508, **508**, **509**, 512, 520, 529, 533, 536, **545**, **546**, **547**, 555, **564**

Maaløe, O., 77, **80**

McAlister, E. D., 193, **216**
MacDonald, I. R., 182, 187, **191**, 234, **238**
McDonough, W. T., 197, 203, **217**, 427, **447**
McGugan, W. A., 235, **239**
McGuire, J. M., 124, **141**
MacKechnie, I., 21, **49**
McKeever, S., 516, 519, **547**
McLean, R. J., 129, 132, 134, 136, 138, **142**
MacLeod, A. M., 166, **191**
McManus, J. F. A., 59, **80**
McNaughton, S. J., 435, **447**
Madden, D., 189
Madison, M., 220, 222, 224, 226, 233, **238**
Magasanik, B., 30, 48
Maggio, R., 146, **160**
Mahler, I., 44, **49**
Maignan, G., 456, **469**
Maisel, H., 439, **447**
Maitra, U., 42, **49**
Major, W., 162, 166, 172, 173, 175, 176, 179, 184, 185, 186, 187, **191**
Makita, M., 536, 539, **549**
Malan, A., 379, **394**
Malcolm, N. L., 434, **447**
Mancinelli, A. L., 196, 197, 207, 211, **217**
Manly, B. M., 534, **546**
Mann, J. E., 128, **142**
Mann, L. K., 219, 232, **237**, **238**, 249, **261**
Mann, T., 540, 542, **547**
Mantini, E., 12, **48**
Manuel, J., 257, **262**
Mapson, L. W., 182, **190**, **191**
Marchetti, J., 42, 45, **48**
Marcus, A., 11, **49**, 143, 146, 150, 153, 155, 156, **160**, 236, **238**
Markham, J. W., 506, **508**
Marmur, J., 41, 43, **48**, **49**
Marré, E., 132, 133, 137, **142**
Marshall, J. M., 490, **509**
Martoja, M., 510, 520, 521, 526, 527, 529, 530, 534, 535, **545**, **546**, **548**
Marzusch, K., 271, **284**
Mason, B. J., 336, **350**
Mason, M. I. R., 224, 235, **239**
Massey, V., 415, **447**
Masters, M., 40, **49**
Matić, O., 352, 356, 361, 366, 370, **393**
Matsko, S. N., 388, **394**
Matsubara, H., 438, 440, **447**
Matthews, L. H., 428, **547**
Mavrides, C. A., 109, **121**
Maximov, N. A., 419, **447**
Mayer, A., 326, 327, **328**
Mayer, A. M., 161, 162, 175, 176, 181, **191**, **192**, 453, 456, **469**
Mayer, W. V., 517, 520, **547**
Mazur, P., 345, **350**

Mefford, R. D., Jun., 101, **121**
Mehta, P. D., 439, **447**
Melchers, G., 310, **328**
Menaker, M., 327, **328**
Merry, J., 162, **191**
Messmer, K., 373, **394**
Meyer, R. K., 533, **546**
Meyers, D., 69, **80**
Michener, H. D., 220, **238**
Mikkelsen, D. S., 166, **191**
Milborrow, B. V., 222, **237**, 252, 253, **261**
Milborrow, V. T., 235, **238**
Miledi, R., 498, **509**
Miller, C. O., 195, 196, **217**
Miller, L. K., 555, 556, 557, 560
Miller, R. E., 521, **547**
Millett, M. A., 61, **81**
Milthorpe, F. L., 219, **238**
Minder, I. P., 265, 283, **284**
Minimakawa, T., 409, **447**
Minis, D. H., 310, 311, 314, 315, 325, 326, 327, **328**
Minks, A. K., 280, **284**
Mirsky, A. E., 79, **80**
Mishiro, Y., 442, **447**
Mitchel, E. D., 165, **191**
Mitchell, O. G., 519, **548**
Mitsui, E., 224, **238**
Mittermayer, C., 68, **80**
Miyamoto, T., 162, **191**
Mohr, H., 193, 197, 202, 207, 214, **217**, 456, **469**
Monroy, A., 146, **160**
Monroy, G. C., 422, **448**
Mook, J., 267, 268, **284**
Moore, C. R., 526, 530, 531, **548**, **549**
Moore, J. W., 297, **299**
Moore, S., 61, **80**
Moore, W. E., 61, **81**
Morel, G., 223, **238**
Morell, P., 41, 43, **49**
Morris, L., 410, **446**
Morrison, P. R., 543, 544, **548**, 554, 555, **564**
Morrison-Scott, T. C. S., 494n, **508**
Morton, W., 415, 416, 418, 428, 431, **447**
Mosca, L., 530, **548**
Mosher, H. S., 297, **299**
Moshkov, B. S., 246, **261**
Moyse, A., 133, **142**
Mullah, G. S., 2, **10**
Müller, J., 223, **238**
Müller-Thurgau, H., 220, **238**
Münch, E., 242, **261**
Muntwyler, E., 60, 65, 66, **81**
Murray, D. B., 478, **488**
Mutere, F. E., 545, **548**
Myers, J., 127, 129, 131, **142**

Nachmony-Bascombe, S., 249, **261**

Nagao, M., 205, **217**, 224, **238**
Nagatsu, T., 427, **447**
Nags, E. H., 17, 41, **48**
Nakano, O., 540, 542, **548**
Nakata, H. M., 14, **49**
Nakayama, N., 423, **448**
Napolitano, L., 541, **549**
Narahashi, T., 297, **299**
Nasr, T. A. A., 245, **261**
Nathan, H. A., 129, **141**
Naylor, J. M., 164, 165, **191**, **192**
Neff, R. H., 51, 52, 54, 60, 61, 67, 78, **80**, **81**
Neff, R. J., 51, 52, 53, 54, 55, 57, 59, 60, 61, 64, 67, 69, 70, 78, **80**, **81**
Negbi, M., 195, 196, 203, 204, 207, 208, 209, 210, 212, 213, **216**, **217**; 455, 456, 466, 467, **469**
Neish, A. C., 188, **191**
Nelson, D. L., 20, **49**
Nelson, J. E., 530, 544, **546**
Nelson, W. O., 526, 531, **548**
Nestianu, V., 534, **546**
Neumann, G., 196, **216**
Newell, R. C., 130, 131, **142**
Newkirk, J. F., 12, **48**
Nezgovorov, L. A., 410, **447**
Nickerson, K., 14, 15, 17, 21, 40, 41, **49**
Nigeon-Dureuil, M., 539, **548**
Nikolaeva, L. V., 410, **448**
Nitsch, J. P., 223, 224, 236, **238**, 243, 244, 245, **261**, 318, **328**, 473, **488**
Njoku, E., 471, 472, 475, 478, **488**
Nossova, E. A., 382, **394**
Novic, B., 60, 65, 66, **81**
Nüesch, H., 298, **299**
Nutile, G. E., 162, **191**, 196, **217**
Nyman, B., 162, **191**, 201, **217**

O'Brien, R. C., 378, **393**, 490, 491, 492, 493, 505, 507, **509**, 520, **545**
Ochi, M., 442, **447**
Ofir, M., 465, 466, 467, **469**
Ogowara, K., 166, 173, 176, 182, **192**
Ogur, M., 63, **81**
Ogura, Y., 436, 440, **448**
Ohkuma, K., 222, **239**, 253, 258, **261**
Ohta, Y., 436, 440, **448**
Ohtaka, Y., 11, **48**
Ohtaki, T., 286, **299**
Oishi, M., 41, 43, 45, **49**
Okagami, N., 224, **238**
Okazawa, Y., 223, **238**
Okinina, E. Z., 225, **237**
Olien, C. R., 138, 139, **142**
Olney, H. O., 183, **192**, 224, **239**
Ono, K., 176, **192**
Oostra, H. G. M., 278, **284**
Oota, Y., 143, **160**
Opitz, E., 379, **394**

Oppenheim, A., 427, **448**
Oppong, A. H., 475
Orphanes, P. I., 169, **192**
Orton, N. L., 427, **448**
Osawa, S., 143, **160**
Osborne, D. J., 258, **261**
Oserkowsky, J., 424, **446**
Oshima, N., 223, **239**
O'Sullivan, A., 21, 31, 44, **49**
Ottaviani, G., 534, **548**
Owen, E. B., 161, **192**
Owens, O. van H., 126, **140**
Ozrina, R. D., 410, **448**

Pace, B., 433, 434, **446**
Paik, W. K., 427, **447**
Paleg, L. G., 224, **239**
Palmer, G., 415, **447**
Palmer, G. H., 166, **191**
Palmer, R. L., 426, **447**
Panuska, J. A., 533, **548**
Papenfuss, G. F., 124, **142**
Pappenheimer, J. R., 499, **509**
Pardee, A. B., 40, **49**
Paris, O. H., 310, **328**
Park, H. Z., 57, 67, 68, **81**
Parker, M. W., 193, **216**
Parkes, A. S., 356, 387, **394**
Parks, H. F., 527, **549**
Passonneau, J. V., 365, 370, 373, 377, 378, 380, 381, 382, **394**
Paulsen, G. M., 428, **446**
Pavlović, M., 393, **393**
Pavlovic-Hournac, M., 391, 393, **393**
Pearson, J. A., 259, **261**
Pearson, A. K., 521, 523, 538, 543, **548**
Pearson, O. P., 521, 523, 538, 543, **548**
Peary, J. A., 129, **142**
Pengelley, E. T., 516, 517, **548**
Petersen, R. A., 129, **141**
Peterson, D. M., 326, **328**
Petrova, D. W., 265, 283, **284**
Petrovic, A., 529, 530, **548**
Petrović, V., 352, 356, 361, 366, 370, **339**
Pettijohn, D., 34, **49**
Peyre, A., 514, 521, 530, 543, **548**
Pfleiderer, G., 373, **394**
Phillips, E., 453, **469**
Phillips, I. D. J., 221, **239**, 252, 256, **261**
Pierce, E. C., II, 491, 509
Pilet, P. E., 424, **448**
Pincock, R. E., 412, **448**
Piper, R. J., 14, 15, 17, 21, 40, 41, **49**
Pittendrigh, C. S., 310, 311, 313, 314, 315, 325, 326, 327, **328**
Plaisted, P. H., 231, **239**
Planel, H., 514, 543, **548**
Poapst, P. A., 235, **239**
Pogo, A. O., 79, **80**

Pogo, B. G. T., 79, **80**
Poljakoff-Mayber, A., 161, 162, 175, 176, 181, **191**, **192**, 453, 455, 456, **469**
Pollock, B. M., 157, **160**, 183, **192**, 220, 224, 236, **239**, 250, **261**
Pollock, J. R. A., 162, 169, 170, 172, 173, 179, **190**, **191**, **192**
Pontremoli, S. S., 427, **448**
Popay, A. I., 177, **192**
Popovic, P., 387, 391, **394**, 494, **509**
Popovic, V., 387, 391, **394**, 494, **509**, 529, **548**
Posnette, A. F., 478, **488**
Powell, J. F., 12, **49**
Precht, H., 271, **284**, 395, **448**
Pregl, F., 60, 65, **81**
Ptak, W., 539, **548**
Pullman, M. E., 422, **448**
Pye, V. I., 130, 131, **142**

Rabinowicz, M., 539, **548**
Rabinowitch, E. I., 126, **142**
Radley, M., 223, **237**, 258, **261**
Rahandraha, T., 539, **548**
Rajevski, V., 352, 356, 361, 366, 370, 391, **393**
Randall, R. J., 60, 61, 65, **90**
Ranson, S. L., 181, 182, **190**
Rappaport, L., 220, 222, 223, 224, 225, 226, 227, 228, 229, 231, 232, 233, 235, **237**, **238**, **239**
Rasmussen, A. T., 519, 520, 526, 530, **548**
Rasmussen, L., 77, **81**
Raths, P., 491, 506, 408, **509**, 529, **546**
Ratsimananga, A. R., 539, **548**
Raunkiaer, C., 241, **261**
Raventós, J., 504, **508**
Ray, S. A., 53, 69, **81**
Redenz, E., 539, 542, **548**
Reich, E., 78, **81**
Reid, D. M., 202, **217**
Reinitzer, F., 221, **239**
Reite, O., 491, **509**
Reithel, F. J., 421, **446**
Reulen, H. J., 373, **394**
Rhéaume, B., 402, 428, 429, 431, **448**
Richards, P. W., 471, 480, **488**
Richardson, M., 186, **190**, 196, 202, **216**
Richmond, A., 450, **469**
Richmond, J., 356, **393**, 503, **508**
Rightsel, W. A., 423, **446**
Rimon, D., 201, **217**
Rinaldi, A. M., 146, **160**
Ring, R. A., 305, **328**
Rissland, I., 202, **217**
Ritchie, J. M., 500, 502, **508**
Robbins, W. E., 285, **300**
Roberts, 250, **261**
Roberts, E. H., 162, 166, 167, 168, 169, 170, 171, 172, 173, 175, 176, 177, 179, 184, 185, 189, **191**, **192**, 483, **488**

AUTHOR INDEX

Robinson, P. M., 223, 236, **240**, 252, 258, **261**
Roche, E. T., 517, **547**
Rodenberg, S., 14, 15, 17, 21, 24, 25, 40, 41, **48**, **49**
Rodriguez, E., 129, **141**
Rogers, W. P., 83, 91, 92, **97**
Rollin, P., 193, 197, 206, 213, 214, **217**, 456, **469**
Rosa, J. T., 232, **239**
Rose, A. H., 402, **448**
Rosebrough, N. J., 60, 61, 65, **80**
Rosen, G., 63, **81**
Ross, I., 373, **394**
Roth, N., 460, 463, 464, **469**
Roth-Bejerano, N., 196, 213, **217**, 457, **469**
Rothman, F., 34, **49**
Roubaud, E., 307, **328**
Rowan, K. S., 426, **445**
Rucker, R., 72, **80**
Rusch, H. P., 68, 77, **80**
Russell, E. W., 177, **192**
Russell-Wells, B., 136, **142**
Ryan, E. M., 442, **447**
Ryan, M. T., 109, **121**
Ryan, R. B., 305, **328**
Ryback, G., 222, **237**, 252, 253, **261**

Saathoff, J., 379, **394**
Sacher, G., 3, **10**
Sachs, J., 401, **448**
Sachs, M., 195, **217**, 223, 231, 233, **239**, 455, 456, **469**
Sachs, R. M., 231, **239**
Sachs, T., 450, **469**
Saeman, J. F., 61, **81**
Sahar, R., 428, **448**
St Girons, H., 519, 520, 521, 526, 527, 529, 530, 534, 535, **546**, **548**
St Girons, M. C., 519, 520, 521, 526, 527, 529, 530, 534, 535, **546**, **548**
Sale, P. J. M., 478, **488**
Salle, A. J., 14, **49**
Salt, R. W., 331, 333, 335, 336, 340, 341, 342, 344, 345, 346, **350**
Samish, R. M., 200, 221, 236, **239**
Samson, F. E., 370, **394**
Sanders, M., 129, **141**
Sankaranarayanan, K., 72, **80**
Santarius, K. A., 419, 423, 426, **446**, **448**
Saringer, Gy., 266, **283**
Sarnat, M., 42, **48**
Satarova, N. A., 225, **239**
Saunders, D. S., 301, 302, 305, 307, 318, 322, 326, **328**
Sayer, D., 347, **350**
Scarano, E., 78, **81**
Scheibe, J., 196, 202, 213, **217**
Schimpfer, A. F. W., 471, **488**

Schippers, A. A., 232, **239**
Schlichting, H. E., Jun., 128, **142**
Schmalhausen, I. I., 4, 5, **10**
Schmidt, P. Yu., 388, **394**
Schmidt, R. R., 78, **81**
Schneider, W. C., 63, 65, **81**
Schneidermann, H. A., 271, 278, **283**, **284**, 307, **329**
Schultz, D. W., 365, 370, 373, 377, 378, 380, 381, 382, **394**
Schultz, G., 285, **299**
Schulze, W., 529, **546**
Schuster, 223
Schwabe, W. W., 249, **261**, 310, **329**
Schwarz, W., 432, **445**
Schwemmle, B., 319, **329**
Schwendiman, A., 177, **192**
Scott, E., 440, **445**
Scrutton, M. C., 117, **121**
Seible, D., 408, 409, **448**
Seifter, S., 60, 65, 66, **81**
Selivanova, V. M., 388, **394**
Servettaz, 132
Severiss, G., 379, **394**
Sexton, Q. D., 1
Seymour, S., 60, 65, 66, **81**
Shabelskaya, E. F., 410, **446**
Shakhbazov, V. G., 293, **299**
Shands, H. L., 177, **192**
Shapiro, B., 536, **549**
Shapiro, S., 427, **448**
Shear, C. L., 100, **121**
Shelton, D. C., 428, 433, **446**
Shih, S. C., 428, 433, **446**
Shihira, I., 131, **142**
Shur, I., 460, **469**
Shuster, L., 157, **160**
Siddall, J. B., 285, **299**
Siegelman, H. W., 195, **216**
Siew, T. C., 283, **284**
Siminovitch, D., 402, 418, 428, 429, 431, **448**
Simmons, G. F., 526, 530, 531, **548**
Simmonds, F. J., 305, **329**
Simon, E. W., 176, **192**
Simon, F. G., 104, **120**
Simpson, G. M., 164, 165, **191**, **192**
Simpson, S., 352, **394**
Sinah, M. N., 166, **191**
Singh, S. P., 231, **238**
Sinha, N. K., 430, **447**
Sisken, J. E., 75, **81**
Sitnikova, O. A., 220, **238**
Skowron, S., 530, **548**
Slama, K., 272, **284**
Slater, J. W., 232, **239**
Slayter, H. S., 11, **47**
Slepecky, R., 21, **49**
Sloviter, A. 379, **394**
Smalley, R. L., 536, **548**

Smith, A. U., 354, 361, 387, 388, 391, 392, **394**
Smith, D., 428, **446**
Smith, D. E., 196, **217**, 520, **548**
Smith, E., 543, **548**
Smith, H., 188, **191**, 221, **239**, 450, **469**
Smith, I., 41, 43, **48**, **49**
Smith, J. L., 491, **509**
Smith, L., 85, **97**
Smith, O., 222, 223, 225, 226, 228, **237**, **239**, 253, 258, **261**
Smith, P., 211, **217**
Smith, R. D., 185, 186
Smith, R. E., 536, **548**
Smit-Vis, H., 519, 520, 529, 530, **548**
Snell, N. S., 107, **120**
Snow, A. G., 206, **217**
Snyder, L., 42, **48**
Sobotka, H., 129, 132, **140**
Söderström, I., 130, **141**
Sokolov, A. K., 410, **447**
Soleilhavoup, J.-P., 514, 543, **548**
Sommerville, R. I., 83, 91, 92, **97**
Sorokin, C., 127, 129, 131, **142**
South, F. E., 383, **394**, 490, 491, **509**
Spaeth, J., 105, **120**
Spiegelman, S., 143, **160**
Spitteller, G., 285, **299**
Spragg, S. P., 424, **448**
Spudich, J. A., 20, 46, **48**, **49**
Staal, G. B., 279, **284**
Stafford, R. S., 45, **49**
Stahl, N., 223, 225, **239**
Stahmann, M. A., 428, **446**
Stanier, R. Y., 99, **121**
Stauffer, J. F., 69, **81**
Stedman, E., 504, **508**
Stedman, E., 504, **508**
Steemann-Nielsen, E., 126, **142**, 404, **448**
Stegwee, D., 271, 272, 273, 274, 277, 280, **284**
Stein, W. H., 61, **80**
Steinbauer, G. P., 176, **192**
Steinberg, W., 11, 13, 14, 15, 17, 18, 19, 21, 22, 24, 27, 30, 31, 32, 33, 36, 37, 38, 39, 40, 41, 45, **48**, **49**
Steiner, G., 84, **97**
Steponkus, P. L., 423, **448**
Stern, H., 146, **160**
Stetten, D., 181, 185, **190**
Stewart, W. D. P., 127, 132, **141**, **142**
Stiles, W., 162, **192**
Still, J. W., 494, **509**
Stones, R. C., 542, 543, **548**
Storck, R., 11, 106, **48**, 106, **120**
Strange, R. E., 12, **49**
Straube, R. L., 520, **548**
Strumwasser, F., 490, 491, 504, **509**
Stumpf, P. K., 188, **190**
Stuy, J., 14, **49**

Subbarrow, Y., 60, **80**
Sueoka, N., 17, 21, 31, 41, 43, 44, 45, **47**, **49**
Sugiyama, T., 423, **448**
Suhara, K., 379, **394**
Suit, J. C., 34, **48**
Sullivan, C. Y., 436, 439, **447**, **448**
Sun, C. Y., 102, **121**
Suomaleinen, P., 513, 529, 530, 534, **548**
Sussman, A. S., 11, 45, **49**, 99, 100, 101, 102, 103, 104, 105, 106, 107, 108, 109, 110, 111, 113, 118, 119, **120**, **121**, 184, **192**
Sussman, M., 52, **81**
Swanson, M. A., 69, **81**
Swartz, M. N., 109, **121**
Sweet, J. E., 536, **548**
Szalai, I., 221, **239**, 424, **448**
Szulmajster, J., 42, 45, **48**
Szybalski, W., 37, **49**, 71, 78, **81**

Takagi, A., 13, **48**
Talma, E. G. C., 398, **448**
Tanada, T., 206, **217**
Tanaka, T., 205, **217**
Tanaka, Y., 293, 294, 312, **329**
Tantaway, D. C., 2, **10**
Tata, J. R., 234, **239**
Tauro, P., 40, **48**
Taylor, C. J., 478, 480, **488**
Taylor, N. W., 427, **448**
Taylor, R. E., 51, **81**
Tazaki, T., 409, 410, 426, **447**
Terman, S. A., 11, **49**
Tester, C. F., 146, **160**
Thesleff, S., 498, **509**
Thielbein, M., 172, **191**
Thiessen, W. E., 222, **239**, 253, 258, **261**
Thimann, K. V., 182, **192**, 196, 201, 202, 203, 207, **216**, **217**, 450, **469**
Thomas, T. H., 223, **239**
Thomas, T. H., 258, **261**
Thompson, R. C., 196, **217**
Thompson, R. H., 226, **239**
Thomson, J. F., 520, **548**
Thomson, R. Y., 59, **81**
Thorn, W., 373, **394**
Thornton, N. C., 177, 178, **192**, 220, **239**, 250, **261**
Tietz, N., 201, **217**
Timm, H., 223, **238**, **239**
Titlbach, M., 521, 527, **546**
Tobolsky, A. V., 443, **448**
Tolbert, N. E., 163, 165, **190**, **191**, 202, **216**
Tolkowsky, A., 197, 211, **217**
Tomkins, G. M., 41, **47**
Tomlinson, G., 60, 61, 69, **81**
Tompkins, D. R., 223, **239**

AUTHOR INDEX

Toole, E. H., 167, 176, **192**, 193, 194, 206, **216, 217**
Toole, V. K., 157, **160**, 167, 176, **192**, 193, 194, 206, 207, **216, 217**
Tower, W. L., 265, **284**
Trager, W., 52, **81**
Traniello, S., 427, **448**
Tranquillini, W., 432, **445**
Treadwell, P. E., 14, **49**
Troll, W., 55, **81**
Trouvelot, B., 268, **284**
Troyer, J. R., 534, 535, **548, 549**
Trunova, T. I., 423, **448**
Tsukamoto, Y., 219, **239**
Tuan, D. Y. H., 189, **192**, 225, 232, 236, **239**, 260, **262**
Tucker, V. A., 490, **509**
Tumanov, I. I., 421, **446**
Twente, J. A., 492, 503, **509**, 540, **549**
Twente, J. W., 492, 503, **509**, 540, **549**
Tyrrell, E., 101, **121**
Tyshchenko, V. P., 312, 326, **328**

Umbreit, W. W., 69, **81**
Uritani, I., 409, **447**
Ushakov, D. P., 404, **448**
Ushatinskaya, R., 265, 283, **284**
Ushiyama, J., 506, **509**
Ushyima, T., 409, 410, 426, **447**
Utter, M. F., 117, **121**
Uuspää, V. J., 530, **548**

Vaadia, Y., 450, **469**
Vaartaja, O., 246, **262**
Van den Berg, L., 426, **447**
Van Huystee, R. B., 246, **262**
Van Overbeek, J., 224, 235, **239**, 258, **261**, 406, **448**, 451, **469**
Van Tienhoven, A., 534, **548**
Van Wagtendonk, W. J., 62, **81**
Van Wijk, W. R., 453, **469**
Varga, M. B., 221, 222, **239**
Varner, J. E., 165, 189, **190**, 224, 234, 235, **237, 238**, 258, **261**
Vary, J. C., 13, 14, 15, 17, 18, 19, 21, 40, 41, **48, 49**
Vaughan, D., 187, **191**
Vegis, A., 161, 178, **192**, 219, 233, 236, **240**, 243, 249, 250, 251, 254, 262, 406, 407, **448**, 449, **469**
Velasco, J., 143, **160**
Veldkamp, H., 402, **446**
Vezina, P. E., 198, **217**
Vickermann, K., 55, 57, 58, 59, **81**
Villiers, T. A., 221, 236, **240**
Vinogradov, A. D., 410, **448**
Vinter, V., 21, **49**
Volcani, T., 156, **160**

Volkonsky, M., 55, 57, 59, **81**
Vose, P. B., 177, 178, **192**

Wada, A., 436, 440, **448**
Waisel, Y., 425, **448**
Wake, R. G., 46, **49**
Walcott, C., 292, **300**
Waldvogel, G., 285, **299**
Walker, D. R., 223, **237**
Walker, M. G., 221, **240**
Wallace, H. R., 88, **97**
Walter, H., 397, **448**
Ward, E. W. B., 402, **448**
Wareing, P. F., 162, 163, 182, **190, 192**, 194, 195, 197, 198, 206, **216, 217**, 221, 222, 223, 224, 236, **237, 238, 239, 240**, 243, 244, 245, 247, 248, 250, 252, 253, 255, 256, 257, 258, **261, 262**, 451, **469**, 472, 473, **488**
Warner, D. T., 97, **97**
Warren, J. C., 423, **448**
Watanabe, A., 124, **142**
Waters, L., 146, **160**
Waxman, S., 236, **240**, 247, **262**
Weaver, R. J., 224, **240**
Webb, S. J., 96, 97, **97**
Weber, N. A., 560, **564**
Webster, J. M., 88, 91, **97**
Weeks, D. P., 151, 165, **192**
Weinberg, E., 13, 26, 39, 41, 45, **49**
Weiser, C. J., 246, **262**, 428, 429, **447**
Weissman, G., 55, **81**
Wellington, P. S., 164, **190**
Wells, H. J., 536, 539, **549**
Wells, L. J., 519, 526, 530, 531, 532, **548, 549**
Wells, P. V., 454, **469**
Wells, W. W., 536, 537, 538, 539, **547, 549**
Wellso, S. G., 243, **283**
Wender, M., 534, **547**
Went, F. A. F. C., 102, **121**
Went, F. W., 319, **328**, 453, 458, **469**
Wertheimer, E., 536, **549**
Wesson, G., 198, **217**
Whitcomb, E. R., 494, **509**
Whitton, B. A., 124, 140, **142**
Wiebers, J. E., 366, 378, 379, **394**, 542, 543, **548**
Wiechert, R., 285, **299**
Wigglesworth, V. B., 275, **284**, 285, **299**
Wilborn, M., 53, **81**
Wilde, J. de, 263, 264, 265, 266, 267, 268, 269, 270, 272, 275, 276, 277, 278, 279, 280, 281, **283, 284**, 301, 309, **329**
Wildman, S. G., 423, **446**
Wilhelm, A. F., 408, **448**
Williams, C. M., 263, 268, 271, **284**, 285, 286, 287, 288, 290, 292, 293, 294, 295, 298, **299, 300**, 301, **329**, 349, **350**
Williams, J. T., 177, **192**

Willis, J. S., 373, **394,** 490, **509**
Willmer, E. N., 52, **81**
Wilson, K. M., 424, **448**
Wilson, M. C., 34, **49**
Wiman, L. G., 425, **448**
Wimsatt, W. A., 512, 513, 520, 521, 527, 530, 531, 532, 538, 541, 542, **549**
Woese, C. R., 13, 17, 46, **49**
Wolf, N., 225, 226, 227, 228, 229, 232, **239**
Wolfe, F. H., 146, **160**
Wolfensone, N. W., 390, **394**
Woodstock, L. W., 162, **191**
Wright, B. E., 52, **81**
Wyss, O. A. M., 491, 506, **509**

Yagi, M., 219, **239**
Yamada, M., 202, **217**
Yamaki, T., 202, **217**
Yang, S. F., 226, **239**
Yaniv, Z., 211, **217**

Yannas, I. V., 443, **448**
Yanofsky, C., 42, 43, **47**
Yarwood, C. E., 436, **448**
Yashida, T., 170, **191**
Yasuma, A., 59, **81**
Yocum, C. S., 182, **192**
Yokohama, Y., 194, **217,** 455, **469**
Yoshikawa, H., 21, 31, 44, 45, 46, **49**
Yutani, A., 419, **448**
Yutani, K., 419, **448**

Zahner, R., 244, **262**
Zajaczek, S., 530, **548, 549**
Zalesky, M., 526, 530, 531, 532, 548
Zarrow, M. X., 530, **547**
Zeeuw, D. de, 319, **329**
Zeuthen, E., 77, **81**
Zhmeido, A. T., 388, **394**
Zylka, W., 379, **394**

SUBJECT INDEX

abscisic acid, 222
  and cold-hardiness, 425
  counteracted by auxins, 253; by cytokinins, 253, 451; and by gibberellic acid, 253, 254, 255–6
  in dormancy of buds and seeds, 224, 226, 235, 253–60, 406
Acanthamoeba sp., encystment of, 51–81
Acer spp.
  abscisic acid in, 252–3
  bud scales of, 250
  day-length and growth of, 243, 244, 245, 254
acetaldehyde
  in activated *Neurospora* ascospores, 103, 104
  and seed germination, 173, 178
acetaldehyde dehydrogenase, effects of SH reagents on, 427
acetylcholine, response of hibernators to, 491, 492, 493, 496, 497–8
acetyl-CoA, theory of seed dormancy involving, 178
acid phosphatase
  in *Acanthamoeba* during encystment, 68
  in heat-hardened plants, 436, 438
acriflavine, and respiration of potato tubers, 183–4
Acronycta rumicis, chilling and photoperiodism in, 318
actinomycin D
  and enzyme synthesis in bacterial spores, 25, 28, 29–30
  and germination and growth of bacterial spores, 18, 19, 20, 46
  and RNA synthesis, 68, 78, 260
  and seed germination, 202
  and sprouting of potato buds, 226
activation
  of bacterial spores, 12, 101
  of *Neurospora* ascospores, 100–20
actomyosin, heat inactivation of, 442
adaptability, limits on, 564
adaptation, to extremes of temperature, 395
adenohypophysis, in hibernation, 529
ADP
  in brain during cooling and asphyxia, 362, 364, 372, 373, 375, 379–82
  and survival times in cooling and asphyxia, 368–9
adrenal glands, in hibernation, 529–30
adrenalectomy, recovery from cold after, 356
Aesculus hippocastanum, day-length and growth of, 243

aestivation, 511
  in Colorado beetle, 265
  in rodents, 516
age of tree, and period of dormancy, 478
aggregation of proteins
  in cold-shock, 444
  in cold-hardening, 441
  in freezing injury, 413–15, 444
  in heat injury, 435, 442, 445
Ailanthus glandulosa
  abscisic acid in, 253
  day-length and growth of, 245
alanine dehydrogenase, in outgrowing bacterial spores, 27, 47
Albizzia julibrissin, phytochrome and leaf movements of, 206
alcohol dehydrogenase, from heat-hardened *Drosophila*, 438
aldolase
  heat resistance of, 435, 436
  SH groups in, 423
alfalfa, protein and nucleic acids in, during autumn, 428
algae, survival of, in adverse conditions, 123–42
Allium ampeloprasum, soil depth and sprouting of bulbils of, 453
Amaranthus caudatus, seed dormancy in, 175
amino acids
  and breaking of seed dormancy, 167, 176
  of cyst wall of *Acanthamoeba*, 61
  hydrophilic, in thermophiles, 437
  in lipoproteins of cell membranes, 417
  released by *Acanthamoeba* during encystment, 67; by bacterial spores on activation, 12; and by plants with chilling injury or in the dark, 408, 444
ammonium salts, and breaking of seed dormancy, 167, 172, 176, 177
AMO 1618, and seed germination, 205
amoebae, biochemistry of encystment in, 51–81
AMP (adenylic acid)
  in brain during cooling and asphyxia, 362, 363, 364, 365, 366, 379–83
  and survival times in cooling and asphyxia, 368–70
N-amyl alcohol, and sprouting of potato buds, 234
amylase
  abscisic acid and synthesis of, 258
  renaturation of, 419
  in seed germination, 165

amylase (*cont.*)
  SH groups in, 424
  from thermophiles, 436, 439
Anabaena sp.
  enzymes from, 133
  freezing of, 124
  mechanical disruption of cells of, 138, 139
  spores of, 130, 140
Anacystis nidulans, freezing of, 124
anaerobiosis
  and bud dormancy, 251
  deactivation of activated *Neurospora* ascospores by, 102–3
  and diapause, 274
  and germination inhibitors, 163, 164
  and phytochrome, 201
  and seed dormancy, 177, 178
androgen, in brown fat of hibernating bats, 536–9
Anguina tritici, survival of, in dry state, 84
anoxia of tissues, caused by cooling, 353–4, 361–6
Antheraea mylitta, 296
Antheraea pernyi
  control of diapause in, 288–95
  dark measurer in, 312
  facial cuticle of pupa of, 290, 293, 296
Antheraea polyphemus, 296
antibiotics
  and germination and growth of bacterial spores, 14
  and water sensitivity of seeds, 170–1
  see also individual antibiotics
antimycin A, and respiration of potato tubers, 183
Aphanocapsa thermalis, 132, 133
Aphanotheca halophytica, 136
aphids, abscisic acid in 'honey-dew' of, 254
Arctic, problems of dormancy in, 553–64
arginase, heat stability of, 439
Arrhenius plots, for enzymically controlled reactions, 398, 399
arsenate, and seed germination, 175, 176
Artemia salina, survival of, in dry state, 84, 94, 97
Artemisia monosperma, light and seed germination in, 195, 455, 456
ascorbic acid/glutathione redox system, in oxygen poisoning of potato tuber, 182
ascorbic acid oxidase, 173
ascospores of *Neurospora*, dormancy and germination of, 99–121
asphyxia
  brain metabolism in cooling and, 361–83
  survival times in, 366–73
  total number of respiratory movements in, 370–1, 379

Asteriscus pygmaeus, dispersal of achenes of, 461
ATP
  in brain during cooling and asphyxia, 363, 372, 373, 375, 379–82
  and polysome formation in seed homogenates, 148, 150, 153, 155, 156, 159
  and survival times in cooling and asphyxia, 368–70
ATPase
  in heat-hardened plants, 436, 438
  in mitochondria, 422
  protection of SH groups of, from freezing injury, 423, 426
Atriplex dimorphostegia, day-length and seed germination in, 456, 457, 461–2
aureomycin, and germination of bacterial spores, 14
auxins
  antagonized by abscisic acid, 253
  and bud rest in potatoes, 221
  and water sensitivity in seeds, 170
Avena fatua (wild oats)
  RNA synthesis by dormant and non-dormant seeds of, 189
  secondary dormancy in, 178
  seed dormancy in, 162, 164–5, 182
Avena sativa (oats), seed dormancy in, 164, 175
6-azauracil, and seed germination, 202
azide
  and pentose-phosphate pathway, 180, 182
  and respiration of seeds, 173
  and seed dormancy, 168, 173, 175, 176
  and water sensitivity of seeds, 170

Bacillariophyceae, 123, 134
Bacillus, activation, germination, and outgrowth of spores of, 11
Bacillus cereus
  activation of enzymes in spores of, 13
  DNA synthesis in spores of, 32–9
  oxidation pathways in spores and vegetative cells of, 184–5
  RNA and protein synthesis in spores of, 14–20, 21, 41, 46
Bacillus megaterium, ribosomes from spores of, 20–1
Bacillus stearothermophilus, sRNAs of, 434
Bacillus subtilis
  dissociation of germination and outgrowth in spores of, 13
  DNA synthesis in spores of, 21, 31, 45, 46
  RNA synthesis in spores of, 41
bacteria
  activation of spores of, 12, 101
  breaking of dormancy in, 11–49

SUBJECT INDEX

bacteria (*cont.*)
  psychrophilic, 126
  resistance of, to adverse conditions, 137
  thermophilic, 128, 129, 132, 396
  and water sensitivity of seeds, 170-1
  see also individual spp.
Bangia fuscopurpurea, desiccation and heat resistance of, 130
bats
  hibernation and reproduction of, 514
  ovarian cycle in, 527-9
  storage of spermatozoa in female, 527, 529, 539-42
  supercooling and survival of, 352, 387, 388
  testicular cycle in, 521-4
Begonia evansiana
  light and seed germination of, 205
  rest period of aerial tubers of, 223-4
benzyladenine (cytokinin), 224
Beta vulgaris, water sensitivity in seeds of, 169
Betula
  abscisic acid and gibberellins in, 224, 253, 255
  bud dormancy in, 243, 244, 245, 247, 250
  seed dormancy in, 163, 182, 198, 224
blood
  pH of, in hibernation, 491
  proteins of, in Colorado beetle (after allatectomy) 277; (in diapause) 274-5, 282
body weight
  of Colorado beetles in diapause, 270-1
  of repeatedly cooled rats, 384
Bombax buenopozense, day-length and dormancy in, 473
Bombyx mori, maternal influence in, 305
Bracteococcus cinnabarinus, resistance to freezing of, 124
Bradypus griseus, body temperature of, in pregnancy, 543
brain
  and endocrine system in insects, 285
  in insect metamorphosis, 298-9
  location of insect photoperiodic mechanism in, 263, 268, 289, 290-2, 296
  metabolism of, in rodents, during cooling and asphyxia, 361-83
brain hormone in insects
  acting on prothoracic gland, 285, 286, 287
  day-length and, 289, 291-2
bud dormancy
  in potato tubers, etc., 219-40
  in seed plants, 241-62
  in tropical forest trees, 472-7
bud scales, and dormancy, 242, 250, 251, 255
buds, foliar and stipular, 242, 253

cabbage leaves
  cold-hardening of, 429
  SH groups in, after freezing injury, 416-17
cacao seed, cold-shock and respiration of, 410
calcium
  released by bacterial spores, 12, 13
  role of, in freezing injury, 392-3
Calligonum comosum, light and seed germination in, 198, 455
Calothrix
  desiccation of, 135
  freezing of, 124
canavanine, and encystment of *Acanthamoeba*, 75-7
Capsella bursa-pastoris, seed dormancy in, 177
carbohydrates
  in *Acanthamoeba* during encystment, 60, 61, 66
  in active and diapausing Colorado beetles, 270
  metabolism of, in *Neurospora* ascospores, 104-6
carbon dioxide
  and exsheathment of *Haemonchus* larvae 91, 92
  in hot springs, 132
  sensitivity of hibernators to, 493, 499
  see also respiration, respiratory rate
carbon monoxide
  and pentose-phosphate pathway, 181, 182
  and seed dormancy, 168, 173
Carpocapsis pomonella, 'night interruption' experiments on, 326
carrot roots, SH in callus of, 424
castration, hibernation after, 533, 538
cat, recovery from cold in, 351
catalase
  SH groups in, 422
  from thermophilic algae, 133
cation transport, in hypothermia, 373
Cedrela odorata, day-length and dormancy in, 475-7
Ceiba pentandra, day-length and dormancy in, 473, 474
cell membranes
  freezing and permeability of, 339-40, 345, 417
  ice growth and, 346
  lipoproteins of, 417
  minimum temperature for active transport across, 402
  phytochrome and permeability of, 206
cell walls
  of encysted *Acanthamoeba*, 55, 59
  of *Neurospora* ascospores, 105, 106, 118
  tension on, in freezing, 412-13, 414, 419

cells
  of *Acanthamoeba*, 64, 66, 70
  freezing of, 341, 345, 346
  of seeds during germination, 200–1
  of sprouting potato buds, 226
  temperature and division of, 520, 558, 559
cellulose, in cyst wall of *Acanthamoeba*, 58, 60, 63, 77
cellulose synthetase (UDPG-cellulose transglucosylase), in *Acanthamoeba* during encystment, 69, 70
central nervous system, in arousal of hibernators, 507
Cephus cinctus larvae
  freezing injury to, 349–50
  inoculative freezing of, 344
  nucleative freezing in, 332–5
chemical reactions (non-enzymic), rates of, at freezing temperatures, 411, 412
Chenopodium sp.
  light and seed germination in, 198, 211
  seed dormancy in, 177
chilling
  and chill-hardening, in insects, 347–50
  and egg production in *Nasonia*, 307
  and emergence of buds from dormancy, 246, 248, 249, 254–5
  and insect diapause, 263, 266, 294–5, 296, 301
  and percentage of diapause larvae in *Nasonia*, 307, 318–19, 321–4
  and seed germination, 167, 172, 182, 198, 206
  and water sensitivity of seeds, 169
chilling injury, 444
  to tropical and semi-tropical plants, 407–9
Chlamydomonas nivalis, photosynthesis by, 140
Chlamydomonas pseudococcum, freezing of, 124
chloramphenicol
  and enzyme synthesis in bacterial spores, 25, 26, 28
  and germination and growth of bacterial spores, 14, 20
  and seed germination, 196, 202
  L-threo and D-threo forms of, as dormancy-breaking agents, 186–8
Chlorella (cold-water planktonic), temperature and growth rate of, 126
Chlorella pyrenoidosa
  freezing of, 125
  at high temperatures, 129, 130
  temperature and growth rate of, 127
  temperature and photosynthesis by, 131–2, 404–5
Chlorella sorokiniana (thermophilic), temperature and growth rate of, 127, 131

chlorocholine chloride, and seed germination, 205
Chlorococcaceae, in adverse conditions, 129–30, 133
Chlorophyceae, 137
  in adverse conditions, 123, 124, 128, 129, 134
chloroplasts
  fraction I protein of, 423, 439
  phytochrome and orientation of, 206
  SH groups of, in dehydration, 419
Chlorosphaera antarctica, photosynthesis by, 40
cholesterol, in brown fat of hibernators, 539
cholinesterase, in hibernating and active ground squirrels, 503, 504, 505
Chondrus, temperature and respiration rate of, 131
chromatin, in encysting *Acanthamoeba*, 55, 56, 79
Chrysophyceae, in adverse conditions, 123, 124, 128, 134
chymotrypsin, chymotrypsinogen, SS bonds in, 430–1
Circadian rhythm, endogenous, theories for photoperiodic clock involving, 309, 310, 314, 324, 325–6
Citellus beldingi, breeding season in, 516, 519
Citellus beecheyi (Californian ground squirrel), hibernation in, 490–1
Citellus citellus (ground squirrel)
  brain metabolism in, during cooling and asphyxia, 368, 374, 375, 377–80
  limit of super-cooling and survival in, 387, 389
  recovery from cold in, 356, 357–9, 360
  survival times of, in cooling and asphyxia, 368, 369, 371, 372, 373
Citellus lateralis (golden-mantled ground squirrel)
  hibernation of, 492, 519
  rate of development of, 516, 517
Citellus leucurus, rate of development of, 517
Citellus mohavensis, rate of development of, 516–17
Citellus tereticaudatus, rate of development of, 517
Citellus tridecemlineatus (thirteen-lined ground squirrel)
  breeding season of, 533
  hibernation of, 491, 493
  ovarian cycle in, 526, 531
  temperature and conduction by nerves of, 555
  testicular cycle in, 532
Citellus undulatus (Arctic ground squirrel), hibernation of, 553–5

Citellus (cont.)
  ovarian cycle in, 525, 526
  rate of development of, 517
  testicular cycle in, 518, 519
Citrullus colocynthis, seed germination in, 455, 460
cocoons of insect pupae, effects of removal of, 288, 289, 295–6
Coeloides brunneri, maternal influence in, 305
cold
  growth of algae in, 125–8
  hardening of plants to, *see* hardening
  mechanisms of mammalian tolerance to, 351–94
  response of hibernators to, 493
  survival of algae in, 124–5
  survival of insects in, 331–50, 560
  survival of plants in, 3, 395–433, 441–3
cold-injury, *see* chilling injury, freezing injury
cold-shock, 409–10, 444
coleoptiles of oats, assay of sprouting inhibitor by, 221, 222
Colorado beetle
  diapause and seasonal synchronization in, 263–84
  freezing of, 341
conformation of proteins
  in cold-hardening, 430
  flexibility of, 403
  temperature, and entropy of conversion of, 399, 400, 401, 435
Convallaria, breaking of dormancy of rhizomes of, 249
Cornus stolonifera, day-length and frost resistance in, 246
corpora allata
  cerebral control of, 278, 279, 280
  and diapause, 275, 276–7, 283
Corylus avellana, pentose-phosphate pathway in seeds of, 182
Corynorhinus
  gestation period in, 543
  testicular cycle in, 523, 538
  see also Plecotus
Cosmarium, freezing of, 125
p-coumaric acid, and seed germination, 166
coumarin, and seed germination, 196
creatine phosphate
  in brain during cooling and asphyxia, 362, 363, 366, 372, 373, 375
  and survival times in cooling and asphyxia, 368, 369
Cricetus cricetus (European hamster), involution of testis in, 519
critical temperature of warming, for mammals, 356
'cryophylactic' agents, 354, 391
Cryptophyceae, 128, 134

Cryptus inornatus, maternal influence in, 305
cucumber
  chilling injury to, 410
  phytochrome in seeds of, 211
cuticle, of insects, and freezing of interior, 342–3
cyanide
  and germination of bacterial spores, 14
  and pentose-phosphate pathway, 180, 182, 187
  and respiration of seeds, 173, 184
  and seed dormancy, 168, 173, 175, 179, 181, 187
  and water sensitivity of seeds, 170
Cyanidium caldarium
  in hot springs, 128, 137
  temperature and photosynthesis by, 131–2
Cyanophyceae
  in adverse conditions, 123, 124, 128, 129, 133–4, 136, 137
  maximum temperature for growth of, 304
  spores of, 140
cycloheximide
  inhibition of enzyme formation by, 158
  inhibition of polyribosome formation by, 149, 153
  and seed germination, 202
Cylindospermum majus, at high temperatures, 130
cysteine, in proteins of Cyanophyceae, 139
cysts, egg-containing, of *Heterodera*, 84–5
  see also encystment
cytochrome *c*
  in dormant *Neurospora* ascospores, 120
  heat stability of, 439
cytochrome oxidase
  in chilled barley shoots, 410
  in seed coverings, 179
  and seed dormancy, 168, 169, 181
cytochromes, in muscles of Colorado beetle during diapause, 273
cytokinins
  and bud dormancy, 224, 450
  counteracted by abscisic acid, 253, 451
cytoplasm, of encysted *Acanthamoeba*, 56, 57, 62

Dactyloctenium aegyptium, germination and survival of seeds of, 481–3, 484, 485
dark
  mechanism for measuring duration of, 312, 319
  photochemical clock in abnormal cycles of light and, 313–14
day-length
  and bud dormancy, 224, 243–9, 464–7

day-length (cont.)
  and growth of tropical forest trees, 472, 473–7
  and insect diapause, 263, 266–8, 285–300, 301–29
  and levels of gibberellins and growth inhibitor in *Ribes*, 256
deactivation, reversible, of heat-activated *Neurospora* ascospores, 102–3, 111
defoliation, effect of, on pre-dormant buds, 242, 244
dehydration of proteins
  by freezing, 412–13, 414
  by heat, 435, 443
  protection against freezing injury by, 419
dehydrogenases
  in chilled barley shoots, 410
  of pentose-phosphate pathway, 180
  see also individual dehydrogenases
denaturation of proteins, 398–401, 443
  in chilling injury, 407–9, 444
  at low temperatures, 398, 401, 406, 413, 415
  heat-hardening as resistance to, 441
  at high temperatures, 398, 401, 433, 434–5
deoxyadenosine, and sprouting of potato buds, 226
deoxyribonucleotides
  in encystment of *Acanthamoeba*, 78–9
  enzymes for synthesis of, in dormant bacterial spores, 46
  inhibitors of synthesis of, and encystment of *Acanthamoeba*, 70, 72–5
deoxyuridine monophosphate, reversal of FUdR inhibition by, 73, 74
desert plants
  delayed seed germination in, 8–9
  dormancy and survival of, 459–69
  light sensitivity of seeds of, 198
desiccation
  of algae, 128, 130, 133–6, 138
  and onset of seed dormancy in Papilionaceae, 457
  see also dehydration, drying
development
  autonomous regulative, 5; dependent, 4
diapause, 8–9, 10
  day-length and endocrine aspects of, 285–300
  day-length and, in *Nasonia*, 301–29
  seasonal synchronization and, in Colorado beetle, 263–84
diatoms, in adverse conditions, 126, 128, 134
2,6-dichlorophenol-indophenol, and seed dormancy, 172
dicoumarol, and seed dormancy, 176
DIECA, and seed dormancy, 173, 175

Digitaria exilis, seed dormancy in, 175, 176
dimethyl sulphoxide, protection from freezing injury by, 393, 421
2,4-dinitrophenol
  and germination of bacterial spores, 14
  and seed germination 175, 176
  and seed respiration, 183
Dinophyceae, 134
dipicolinic acid
  released by bacterial spores on activation, 12, 13
  and SH groups, 442
disulphide (SS) bonds in proteins
  confer stability at low temperature, 408
  conformation of proteins and reactivity of, 430–1
  formed from SH groups (in freezing), 139, 412, 413, 415, 416, 420, 442; (in heat injury) 442, 443; (under stress) 413; (in thawing) 444
  in heat-stable and heat-labile enzymes, 438–9
  in proteolytic enzymes, 408
disulphides, activation of enzymes by, 427
Ditylenchus dipsaci larvae
  rates of water uptake and loss by, 89–91
  survival of, in dry state, 84, 88, 94–5
DNA
  conformation of, in bacterial spores and vegetative cells, 45
  content of, in *Acanthamoeba* at different stages, 63–5, 70, 79
  from dormant and non-dormant potato buds, 260
  of thermophiles, 434
DNA synthesis
  blocked by FUdR, 72, 75
  inhibitors of, and encystment of *Acanthamoeba*, 70, 71–2, 78–9
  in outgrowing bacterial spores, 31–9, 42, 44–7
  plant growth hormones and, 224
  possible initiator proteins for, 46, 77
  in sprouting potato buds, 225, 232, 233
dogs, brain metabolism in, during cooling and asphyxia, 382
dormancy
  as an adaptive strategy, 1–10
  breaking of, in bacteria, 11–49
  of buds in seed plants, 241–62
  in desert plants, 449–69
  of fungus spores, 99–121
  general problems of, 551–64
  in nematodes, 83–97
  post-, 243
  pre- (or summer), 243, 244–5, 248, 472, 473, 475
  relation between cold-hardiness and, 425

dormancy (cont.)
  of seeds, light and, 189; oxidation processes and, 161–92
  temperature extremes for growth at entrance to and emergence from, 406–7
  in tropical plants, 471–88
  in tubers, 219–40
  winter (deep or true), 243, 245
  see also hibernation
Drosophila spp.
  heat-hardened, 438
  'night interruption' experiments on eclosion rhythm of, 310, 311, 326, 327
  selection of, for egg production, 1
  temperature/viability-of-eggs relation in, 2
drought resistance, correlation between freezing resistance and, 442
drying, rate of, and survival of nematode larvae, 88, 94, 95–6
  see also dehydration, desiccation
Dunaliella spp., halophiles, 136, 137

ecdysone, 233, 285
  and analogues, 285–7
eggs
  production of, in Nasonia, 304–9 passim
  temperature and viability of, in Drosophila, 2
Eidolon helvum, period of gestation in, 544, 545
Eliomys quercinus
  neurosecretory activity during hibernation of, 535
  ovarian cycle in, 525, 526–7
  testicular cycle in, 519
embryos, delayed development of, in some bats, 514–15, 542–5
encystment in amoebae, biochemistry of, 51–81
endocrine organs and reproduction, in hibernators, 529–34
endoplasmic reticulum
  in Acanthamoeba, 57
  in dormant fungus spores, 100
  in germinating Neurospora ascospores, 102
energy, conservation of, in dormancy, 511
Enteromorpha
  at high temperatures, 129
  temperature and respiration rate of, 131
environments, 3, 5, 6–9
  control of dormancy by, 449
  reduced dependence on, in dormancy, 449, 450
enzymes
  activation of, in seeds, by gibberellic acid, 164

  activity of, at extremes of temperature, 397–402
  activity as, retained by denatured proteins, 401–2, 415
  in encysting Acanthamoeba, 69, 78
  in heat injury, 433–5
  in heat resistance, 435–41
  in hypothermia, 354, 379
  oxidative, breaking of seed dormancy by inhibitors of, 168, 169, 173, 175
  properties of heat-stable, 439, and heat-labile, 440
  stability of, and resistance to adverse conditions, 133, 138
  sulphhydryl, 410, 415, 422
  synthesis of, in outgrowing bacterial spores, 25–7, 35–9, 40–1
  theories of adaptation of, to extremes of temperature, 403–7
  'unmasking' of, in activation of bacterial spores, 13
Epilobium, light and seed germination in, 194
Eptesicus
  castration of males of, 538
  ovarian cycle in, 528
  testicular cycle in, 522, 524
Eragrostis, light and seed germination of, 195
Erinaceus europaeus
  hibernation and reproduction of, 513–14
  ovarian cycle in, 524–7
  testicular cycle in, 517–21
erythrocytes, membrane of, 417
Escherichia coli
  RNA synthesis in, 41, 42, 43
  unbalanced growth in, 78
esterase, in encysting Acanthamoeba, 69
ethanol
  in activated Neurospora ascospores, 104, 115, 116
  and seed germination, 173, 178
ethylene, and sprouting of potato buds, 234, 235
ethylene chlorohydrin, and bud dormancy, 220, 221, 224, 225
Euglena spp., in adverse conditions, 124, 129, 132, 134, 138
Eurosta solidaginis, freezing of large cells in, 345
eyes, not photoperiodic receptor organs in Colorado beetle, 268

facial cuticle, transparent, of Antheraea pupa, 290, 293, 296
Fagopyrus esculentum, pentose-phosphate pathway in germinating seeds of, 181, 182
Fagus, bud dormancy in, 247, 250

far-red light
  and phytochrome, 195, 200, 203, 206
  and seed germination, 194, 197, 207–15
  under leaf canopies, 198, 199, 215
fat, brown, as 'hibernating gland', 536–9
fat cells, freezing of, 345
fats, see lipids
ferrulic acid, and seed germination, 166
fitness, maximization of, in different environments, 3–4, 6–8
flight muscles of Colorado beetle
  after allatectomy, 277
  in diapause, 269, 272–4
  juvenile hormone and, 280
fluoride
  and oxygen uptake of seeds, 173
  and pentose-phosphate pathway, 180
  and seed dormancy, 173
5-fluoro-2'-deoxyuridine (FUdR)
  and encystment in *Acanthamoeba*, 72–5
  and histidase induction in outgrowing bacterial spores, 38–9
follicle-sensitizing hormone, in hibernating bats, 532
forests, tropical, dormancy and survival of plants in, 471–88
Fouquieria splendens (ocotillo), rainfall and sprouting of, 451
Fragilaris sublinearis, oxygen exchange of, at low temperature, 126–7
Fraxinus excelsior
  abscisic acid in, 253
  bud dormancy in, 252
  seed dormancy in, 222
freezing
  conditions for, 331
  inoculative, 342–6
  nucleative, 331–40
  possibility of revival from, 354–5, 390–1, 558
  resistance of plants to, 395, 397, 417–33, 442
freezing injury, 347–50, 411–17, 443, 444
  role of calcium in, 392–3
fructose, in uterus of bats, 542
fructose diphosphatase, activation of, by disulphides, 427
Fucus spp.
  freezing of, 125
  temperature and respiration rate of, 131
fungi
  dormancy and germination of spores of, 12, 99–121
  and water sensitivity of seeds, 170–1
furans, activation of *Neurospora* ascospores by, 100–1
furfural
  activation of *Neurospora* ascospores by, 100, 101–2, 118

differences between spores activated by heat and by, 111–12, 116, 118, 119
effect of, on aged spores, 112, 113, 114

β-galactosidase, heat-lability of, 440
Galeruca tanaceti, diapause in, 283
ganglia (isolated), recovery of, after freezing, 393
ganglionic blockade, in hibernating mammals, 493, 500
gelatin
  dehydration of, 442–3
  thiolated, see thiogel
generation time
  for Colorado beetle in different locations, 265, 267
  and rate of increase of population, 1, 2
genes
  control of potato bud rest at level of, 233, 236, 249
  for diapause, 282
  for synthesis of enzymes allowing growth at extremes of temperature, 406, 407
germarium, of Colorado beetle in diapause, 275
germination of seeds
  of desert plants, 451–64
  inhibitors of, 162–6, 177, 178, 189
  light and, 193–217, 454–8, 468, 483, 486
  protein synthesis during, 156–60
germination of spores
  of bacteria, 12, 13, 14
  of fungi, 12
  of *Neurospora*, 104, 105
gestation, periods of, and hibernation, in *Miniopterus* spp. 514–15, 542–5
gibberellic acid, gibberellin
  antagonized by abscisic acid, 253, 254, 255–8
  and bud dormancy, 223–4, 251, 406
  chilling and level of, 206, 248
  and cold-hardiness, 425
  in light-controlled germination of seeds, 196, 202–15
  and pentose-phosphate pathway, 187
  and RNA synthesis in excised nuclei, 234
  and seed dormancy, 164–5, 171, 175, 186
  and sprouting of potatoes, 223, 226, 229–30, 233–4
  and water sensitivity of seeds, 170
Gladiolus, breaking of dormancy of corms of, 249
Glis glis (European dormouse), 493
glucose
  in brain, of hibernators, 379; of infant and mature animals, 378
  in encysting *Acanthamoeba*, 66–7
  in hibernating *Citellus*, 554, 555

SUBJECT INDEX

glucose (cont.)
  utilization of, by dormant, activated, and germinating *Neurospora* ascospores, 105–6
glucose oxidase, heat-lability of, 440
glucose-1-phosphatase, in encysting *Acanthamoeba*, 69
glucose-6-phosphate, oxidation of, in germinating seeds, 180, 181
glucose-6-phosphate dehydrogenase, in barley seeds, 185
α-glucosidase, synthesis of, in outgrowing bacterial spores, 25–7, 28–30, 37, 47
glutamate dehydrogenase
  freezing and SH groups of, 422
  heat-lability of, 440
glutamate oxaloacetate transaminase, heat resistance of, 435
glutathione
  and bud rest, 224
  conversion of, to reduced form, in growing plants and germinating seeds, 424
  as protector of SH groups in freezing, 422
  and seed dormancy, 425
glyceraldehyde phosphate dehydrogenase
  in diapausing Colorado beetles, 274
  heat-lability of, 440
glycerol
  in haemolymph of *Pterostichus* during hibernation and awakening, 501
  protection from desiccation by, 84, 97
  protection against freezing injury by, 124, 343, 348, 421
α-glycerophosphate dehydrogenase, in diapausing Colorado beetles, 274
glycogen
  in active and diapausing Colorado beetles, 270
  in brain, during cooling and asphyxia, 377
  in encysting *Acanthamoeba*, 65–6
glycol, as protectant against freezing injury, 421
glycollate oxidase, in potato tubers, 183, 184
glycolysis
  in brain tissue from infant and mature hypoxic animals, 378
  during germination of lettuce seeds, 181
  inhibition of enzymes of, and breaking of seed dormancy, 173, 175
Gmelina arborea, day-length and dormancy in, 473
Golgi system, in encysting *Acanthamoeba*, 55, 57
gonads, and accessory organs, in hibernation, 517–29
Graafian follicle, overwintering of, in bats, 514, 527–9, 531

Grensia (caddis-flies), in Arctic, 560
Griffithsia, temperature and respiration rate of, 131
growth
  of algae at low temperatures, 125–8; at high temperatures, 128–33
  associated with high SH content, 424
  inverse relation between hardiness and, 419, 424
  more affected than metabolism by low temperature, 402, 405
  of plants at extremes of temperature, 396–407, 443–4
  sympodial, 242, 253
  of tropical forest trees, 471, 472–3
  unbalanced, 70, 72, 78
  uncoupling of metabolism from, 402, 443
gut, of insects, nucleative freezing from, 335, 346
Gymnarrhena micrantha, germination of aerial and subterranean achenes of, 462, 463, 464

haemolymph of insects, freezing of, 341, 346
Haemonchus contortus, water uptake and loss by larvae of, 91–4, 47
halophilism, 136–7
hamsters
  limit of supercooling and survival in, 387, 388
  see also *Cricetus*, *Mesocricetus*, etc.
Haptophyceae, 123, 134
hardening of plants
  to cold, 395, 407, 409, 417–33, 444
  to heat, 395, 407, 435–41, 445
  to heat and cold compared, 441–3, 445
heart rate
  in hibernating mammals, 490, 504–5; after acetylcholine, 492, 493; after stimulus, 493, 507
  in 'spinal' hibernator preparation, 499
hearts (isolated), recovery of, after freezing, 352, 393
heat
  activation of spores by, 12, 100
  difference between spores activated by furfural and by, 111–14, 116, 118, 119
  hardening of plants to, see hardening
  production of, in mammals during re-warming, 358–61
  survival of algae in, 128–33
  survival of plants in, 433–43
heat injury, 433–5, 445
heat shock, and subsequent resistance of algae to high temperatures, 130
Hedera helix, cold-hardiness of leaves of, 423

SUBJECT INDEX

Helianthus tuberosus
  breaking of dormancy of tubers of, 249
  invertase in tubers of, 224
Heterodera spp.
  humidity and survival of, 84–6
  water uptake and loss by larvae of, 86–8, 94, 95, 96
hexosemonophosphate shunt, in spores of *B. cereus*, 184–5
'hibernaculae' (resting buds), 249
'hibernating gland' (brown fat), 536–9
hibernation, 1, 10
  interrelations of reproduction and, 511–49
hibernators
  Arctic, 553–5
  hyper-responsiveness in, 489–509
  recovery from cold in, 351, 354, 356, 358–9
  tolerance by, of anoxia, 375, 379; and of cold, 366, 367
  in wet environment, 342, 343
Hildegardia bartoni
  day-length and dormancy of, 475, 477
  seed germination in, 480
histidase, synthesis of, in outgrowing bacterial spores, 26, 27, 30, 31, 38–9, 47
Hordeum bulbosum, regeneration buds of, 464–7
Hordeum distichon (barley)
  abscisic acid and protein synthesis in, 235, 258
  cytochrome oxidase in seeds of, 179
  germination inhibitor in seeds of, 165
  gibberellin and enzyme synthesis in seeds of, 224
  pentose-phosphate pathway in seeds of, 185–8
  RNA synthesis by dormant and non-dormant seeds of, 189
  seed dormancy in, 162, 171–5
  water sensitivity of seeds of, 169–71
Hordeum spontaneum, seed dormancy in, 166, 173, 182
hormones
  and bud rest in plants, 220–4, 249, 251–60
  and diapause in Colorado beetle, 275–80
  in hibernation, 529–34
'hour-glass' measurement, theories for photoperiodic clock involving, 309, 312, 325–6
humidity, control of gemination by, 458–60, 468, 480, 481, 484–5
Hyalophora
  brain of pupa of, 286–7, 296
  diapause in, 263, 271, 295–6
Hydrocharis, winter buds of, 249, 251
hydrogen bonds in proteins, 399, 400, 401
  proportion of hydrophobic bonds to, and extremes of temperature for growth, 403–4, 407, 418, 432–3, 436–8, 441, 443–4
hydrogen peroxide
  and seed dormancy, 166, 169, 172
  and water sensitivity of seeds, 170, 171
hydrogen sulphide, and seed dormancy, 168, 173, 175, 179
hydrophobic bonds in proteins, 399, 400, 401
  in cold-shock, 410
  proportion of hydrogen bonds to, and extremes of temperature for growth, 403–4, 407, 418, 432–3, 436–6, 441, 443–4
hydroquinone, and seed dormancy, 179
p-hydroxybenzoic acid, and seed dormancy, 166
hydroxylamine, and seed dormancy, 168, 173, 175, 176, 177
hygroscopic value, in seeds of Papilionaceae, 451, 458
Hyla, supercooling and rewarming of, 389
Hymenocarpus circinnatus, weathering of seeds of, 461
Hyoscyamus desertorium, seed germination in, 456, 457
Hyoscyamus niger, chilling and photoperiodic effects in, 319
hypothermia
  brain metabolism and recovery from, 361–83
  distinction between hibernation and, 352
  induced tolerance to, 356, 383–7
  protection from anoxia by, 353–4
hypoxia
  increased resistance of cold-adapted rats to, 385
  see also asphyxia

ice, in biological systems, 337–40, 346–7, 443
imbibition,
  respiration during, in dormant and non-dormant seeds, 162, 184
  ribosomes during, in wheat embryos, 144
indole-acetic acid, and seed germination, 166
induction
  of encystment in *Acanthamoeba*, 55
  of enzymes in outgrowing bacterial spores, 26, 28–31, 40, 47
β-inhibitor of growth, in potato tubers, 221–3, 252
  see also abscisic acid
inhibitors
  of bud growth, 248, 252, 254
  of respiratory reactions, 174
  of seed germination, 162–6, 177, 178, 189
  of sprouting in potatoes, 221
inoculative freezing, 342–4
  intracellular, 344–6

SUBJECT INDEX

inositol, and survival in dry state, 97
insects
  Arctic, 559–63
  cold-hardiness in, 331–50
  day-length and endocrine aspects of diapause in, 285–300
  recovery from cold in, 352
  see also Colorado beetle, *Nasonia*, etc.
invertase
  in *Helianthus* tubers, 224
  in *Neurospora* conidia, mycelium, and ascospores, 108
iodoacetate
  and germination of bacterial spores, 14
  and pentose-phosphate pathway, 180
  and seed dormancy, 173
iododeoxyuridine, and FUdR block in *Acanthamoeba*, 74
islets of Langerhans, in hibernation, 529
isocitratase, in encysting *Acanthamoeba*, 69

joints, of *Melanoplus* legs, nucleative freezing from, 336–7
Juncus maritimus, light and seed germination of, 198
juvenile hormone in Colorado beetle, and diapause, 278–80, 283

Kalanchoë
  light and seed germination in, 205
  phytochrome and flowering of, 206
kinetin
  and abscisic acid, 256
  and bud dormancy, 251, 254
  and seed germination, 195, 196
Krebs cycle
  in germinating lettuce seeds, 181
  inhibition of enzymes of, and breaking of seed dormancy, 173, 175
  oxygen poisoning of, in potato tubers, 182

lactate
  in brain during cooling and asphyxia, 362, 363, 365, 366, 374, 375–8
  and seed germination, 173, 178
  and survival times in cooling and asphyxia, 368, 369, 370
lactate dehydrogenase, freezing and SH groups of, 422, 423
Larix spp., day-length and growth of, 243, 244, 247
Larus argentatus, temperature and conduction by nerves of, 555
latent heat of crystallization, and ice growth in biological systems, 332, 337–9, 342, 344, 387
Lathyrus, humidity, and imbibition by seeds of, 458, 459, 460

leaves
  abscisic acid and senescence of, 258
  aged, and corpora allata of beetles consuming them, 281–2
  effects of, on bud dormancy, 247, 251–2, 477
  heat-hardening of, 438
  transmission of far-red radiation by, 198–9, 215
lecithin, and oviposition by Colorado beetle, 268, 281, 282
Leguminosae, water-impermeable seed-coats in, 460–1
Lemna
  bud dormancy in, 224
  effects of abscisic acid in, 235, 258
Leptinotarsa decemlineata, *see* Colorado beetle
lettuce
  pentose-phosphate pathway in seeds of, 181
  phytochrome system in, 193, 194
  seed germination in, 175, 197, 202–6, 256
Leydig cells, during hibernation, 520–1, 522, 523, 524, 537, 538
life-span, temperature and, in *Nasonia*, 305–6
light
  and dormancy in tropical forest trees, 478
  far-red, *see* far-red light
  integration of, within cocoon of *Antheraea*, 292–3
  photoperiodic clock in abnormal cycles of dark and, 313–14
  and seed germination, 193–217, 454–8, 468, 482, 486
  wavelengths of, affecting photoperiodic mechanism of insect brain, 292
lipids
  in active and diapausing Colorado beetles, 270–1
  in cyst wall of *Acanthamoeba*, 58, 60
  in encysting *Acanthamoeba*, 63, 65
  in establishment of seed dormancy, 178
  in hibernators, 536–9
  in micro-organisms on desiccation, 135, 136
  metabolism of by *Neurospora* ascospores, 104, 105; in seeds, 201, 251
lipoproteins, of cell membranes, and freezing injury, 417
lipoyl dehydrogenase, 422
  inactivation of, in SH and SS forms, 415–16
Liquidamber styraciflua, day-length and growth of, 243
Liriodendron tulipifera, day-length and leaf fall in, 245

Lolium perenne, seed dormancy in, 175
Lucilia caesar, maternal influence in, 305
Lunularia cruciata, day-length and dormancy in, 249
luteinizing hormone, in hibernators, 532, 534
lysosomes
    in encysting *Acanthamoeba*, 55, 57, 63, 79
    heat-stability of, 438–9
Lythrum salicaria, light and seed germination of, 206

Macrotus californicus, period of gestation in, 544–5
maize
    abscisic acid and gibberellins in shoots of, 257
    chilling resistance of, 410
    seed dormancy in, 175
    temperature/growth curve for, 396
malate, in activated *Neurospora* ascospores, 115, 116, 117, 118
malate dehydrogenase, in heat-resistant plants, 435, 436
malate synthetase, in encysting *Acanthamoeba*, 69
malonate
    and seed respiration, 173
    and seed dormancy, 173, 175
maltose, utilization of, by dormant and non-dormant seeds, 165
Malus, day-length and growth of, 243
mammals, mechanisms of tolerance of low body temperature in, 351–94
man, recovery from cold in, 352
Marmota monax
    brown fat in, 536–7
    ovarian cycle in, 526
    testicular cycle in, 519
Mastigocladus laminosus, at high temperatures, 129, 132
Medicago laciniata, humidity, and imbibition by seeds of, 458, 459
Medicago tuberculata, weathering of seeds of, 461
Megoura viciae (Aphidae)
    dark measurer in, 312, 325–6
    maternal influence in, 305
Melandrium noctiflorum, seed dormancy in, 175
Melanoplus bivittatus, freezing injury to eggs of, 349
Melanoplus sanguinipes, freezing of excised legs of, 336–7
mercaptans, breaking of seed dormancy by, 179
mercaptoethanol, and freezing injury, 422, 427
Mesocricetus auratus, 493–4
    testicular cycle in, 519–20

metabolic rate, in hibernators, 489–90
metabolism
    less affected than growth by low temperatures, 402, 405
    uncoupling of growth from, 402, 443
metabolites, accumulation of, in cold-hardening, 429; at low temperatures, 405–6; and in non-growing plants, 419
metamorphosis of insects, 298–9
Metatheria (phalangers), hibernation in, 513
6-methyl purine, and seed germination, 202
methylene blue
    and pentose-phosphate pathway, 180
    and seed dormancy, 168, 169, 172, 177
Micrococcus chthonoplastes, halotolerant, 136
Micrococcus cryophilus, psychrophile, 434
micro-organisms
    lipids in, on desiccation, 135, 136
    stimulus for hatching of *Heterodera* larvae from, 85
    and water sensitivity of seeds, 170–1
    see also bacteria, fungi, *etc.*
microsomes, enzyme catalysing naturation of RNA-ase in, 420
Mimosa pudica, phytochrome and leaf movements of, 206
Miniopterus spp.
    embryonic development in, 543–4
    hibernation and reproduction in, 514–15
    ovarian cycle in, 527, 528
    testicular cycle in, 521, 524
mitochondria
    of brain of hibernating and other animals, 378
    cold-labile ATPase in, 422
    in dormant fungus spores, 100, 120
    enzymes of, in diapausing Colorado beetle, 272–3
    in encysting *Acanthamoeba*, 55, 57, 63–4, 69–70
    in germinating *Neurospora* ascospores, 102
    inhibited by cyanide, 181
    SH groups and NAD reduction in, 410
mitomycin C
    and DNA synthesis in outgrowing bacterial spores, 32, 43, 44
    and encystment of *Acanthamoeba*, 71–2, 79
    and histidine synthesis in outgrowing bacterial spores, 38, 43
mitoses
    in germinating lettuce seeds, 201, 232
    minimum temperature for, 402
    in potato buds, 226, 231–2

SUBJECT INDEX

monofluoroacetate
  and respiration of seeds, 173
  and seed dormancy, 173
mouse
  brain metabolism of, during cooling and asphyxia, 368, 372, 373, 374, 375
  brown fat of, 539
  survival times of, in cooling and asphyxia, 368, 369, 370
  temperature and reproduction of, 534
muramidase (lysozyme), heat-stability of, 439
muscle action potentials in hibernators
  acetylcholine and, 492, 496, 499
  body temperature and, 501, 502, 507
  stimulation and, 492–3
muscles
  formation of, in insect metamorphosis, 298
  from hibernators, 494–5, 497–8
  see also flight muscles
myofibrillar proteins, effects of freezing on, 426
myokinase, heat stability of, 439
myosin, SH groups in, 422
Myotis
  androgen in brown fat of, 537, 538
  ovarian cycle in, 527, 528, 531
  temperature and gestation period in, 543
  testicular cycle in, 521, 522

NAD-malic dehydrogenase, NADH-cytochrome *c* oxidoreductase, NADP-isocitric dehydrogenase, in encysting *Acanthamoeba*, 69
NADPH
  ascorbic acid/glutathione redox system and, 182
  nitrite and methylene blue and, 180
  ratio of NADP to, in cold-hardening of plants, 409–10, 426, 444
NADPH-cytochrome *c* reductase, from thermophilic algae, 133
naphthalene-acetic acid, methyl ester of, and bud rest in potatoes, 221, 226
Nasonia (*Mormionella*) *vitripennis*
  diapause and photoperiodism in, 301–29
  effect of shortage of hosts on egg production and nature of larvae of, 307–9
Nasturtium, effects of abscisic acid in, 258
Navicula minima, freezing of, 124
Navicula spp. in hot springs, 128
nematodes, dormancy and survival in, 83–97
Nemophila, light and seed germination in, 197
Neomarica gracilis, seed dormancy in, 175, 176

neonates (rodents)
  brain metabolism of, in cooling and asphyxia, 368, 372, 373, 374, 375, 376, 378
  limit of supercooling and survival in, 387, 389, 391
  survival times of, in cooling and asphyxia, 366, 367, 368, 370, 373
  tolerance of, to cold and anoxia, 366, 375
nerves, temperature and conduction by, 555–8
nervous system, primordial function of, 299
neuroendocrine system, and reproduction, in hibernators, 534–6
neuroscretory cells
  in *Antheraea* pupa, 298
  in Colorado beetle, 276, 280, 281, 283
  in *Hyalophora* pupa, 287, 296
  in *Ostrinia nubilalis*, 311
  stimulated by chilling, 351
Neurospora tetrasperma, dormancy and germination of ascospores of, 99–121
Nicotiana, temperature and light-sensitivity of seeds of, 194
Nigella, light and seed germination of, 197
'night interruption' experiments on insects, 310, 314–17, 325–7
nitrate
  and seed dormancy, 167, 169, 172, 176, 177
  and water sensitivity of seeds, 170, 171
nitrate reductase, in cold-hardened wheat, 428
nitrite
  and pentose-phosphate pathway, 180, 187
  and seed dormancy, 167, 169, 172, 176, 177, 187
nitrogen
  in encysting *Acanthamoeba*, 66, 67
  not fixed by algal spores, 140
  temperature and fixation of, by algae, 127, 132
Nitzschia spp., in hot springs, 128
Nostoc spp., in adverse conditions, 125, 130, 133, 135
nucleative freezing, 331–2
  in insects, 332–40
  single and multiple, 340–2
nuclei
  in encysting *Acanthamoeba*, 52, 56, 57, 58, 63
  in germinating *Neurospora* ascospores, 102
  (isolated) gibberellin and synthesis of RNA in, 234
nucleic acid synthesis
  in dormant and germinating *Neurospora* ascospores, 106

nucleic acid synthesis (*cont.*)
  in dormant seeds, 189
  during cold-hardening, 428
  in germinating and outgrowing bacterial spores, 14
  in sprouting potato buds, 227–31, 233
  at temperatures too low for mitosis, 402
  see also DNA synthesis, RNA synthesis
nucleoli, in encysting *Acanthamoeba*, 56
nucleoside phosphorylase, FUdR degraded by, 74
3′-nucleotidase, during imbibition by wheat embryo, 157–9
nutrition
  and algal resistance to adverse conditions, 129, 130, 138
  and diapause of Colorado beetle, 266, 267, 281
  and encystment of *Acanthamoeba*, 53
nyctinastic movements of legume leaflets, phytochrome and, 206
nystatin, control of fungal infection of germinating seeds by, 170, 172

Ochromonas malhamensis, at high temperatures, 129, 138
Oleaceae, day-length and growth of, 243
onion, gibberellin and dormancy of bulbs of, 223, 224, 249
Onobrychis cristagalli, weathering of seeds of, 461
Oryzopsis milicea, temperature and seed germination in, 453
Oscillatoria spp., catalase from, 133
Oscillatoriaceae, desiccation of, 134
Ostrinia nubilalis, 'night interruption' experiments on, 310, 311, 326
ova, of Colorado beetle in diapause, 275
ovalbumin, SH groups in, 422–3
ovary and female accessory organs, in hibernation, 524–9
oxidases (terminal)
  breaking of seed dormancy by inhibitors of, 168, 169, 173
  pentose-phosphate pathway and, 180
oxidation
  of oxidation inhibitors in seeds, 162–3
  and seed dormancy, 161–92, 406, 461
oxidative phosphorylation
  in diapause, 274
  juvenile hormone and, 280
oxygen
  and encystment of *Acanthamoeba*, 53
  exchanges of, by diatoms at low temperatures, 126–7
  and Krebs cycle in potato tubers, 182
  and water sensitivity of seeds, 170, 171
  see also respiration, respiratory rate

Paeonia suffruticosa, gibberellin and physiological dwarfism in, 223
Panagrellus redivivus, water loss of larvae of, 89, 90, 95
Panagrellus silusiae, desiccation of, 95
Pandorina morum, desiccation of, 134
Panicum turgidum, contact with water, and seed germination in, 460
Panonychus ulmi, dark-measurer in, 312
Papilionaceae, humidity, and seed dormancy and germination in, 451, 458
parasites, dormancy in life-history of, 83
parasympathetic system, in hibernators, 491
Paulownia, light and seed germination in, 194
peanut, ribosomes in cotyledons of, during germination, 143–4, 146
Pectinophora gossypiella
  diapause inhibition in, 310, 314, 315, 324, 326, 327
  oviposition rhythm in, 310, 311
penicillin, and germination of bacterial spores, 14
pentose-phosphate pathway of respiratory metabolism
  in breaking of seed dormancy, 180–90
  in spores of *B. cereus*, 184–5
  in tubers, 181–2
pepsin, heat-stability of, 439
Perilla ocymoides, chilling and photoperiodic effects in, 319
permeability
  of bacterial spores during activation, 12
  of cell membranes, freezing and, 339–40, 345, 417; phytochrome and, 206
  of egg-shell of *Heterodera* during drying, 87–8, 91, 94, 95
  of *Neurospora* ascospores, before and after germination, 107
  of sheath of *Haemonchus* during drying, 92–4
Perognathus baileyi, aestivation and respiration in, 516
Perognathus californicus, body temperature of, 490
Perognathus penicillatus, aestivation and reproduction in, 516
peroxidases, in cold-hardened alfalfa, 428
pH
  of blood of hibernators, 491
  and dormancy and water sensitivity of seeds, 170
  and encystment of *Acanthamoeba*, 53
  of hot springs, and species of algae, 128
Phacelia, light and seed germination of, 197
Phaeophyceae, 134
phage T4, RNA polymerase in, 44
Phalaris arundinacea, theory of seed dormancy in, 177–8

SUBJECT INDEX 593

Phaseolus radiatus, pentose-phosphate pathway in seeds of, 181
Phaseolus vulgaris, water sensitivity in seeds of, 169
phenol oxidases
  during breaking of dormancy in potato tubers, 183
  in seed coats, and dormancy, 179
phloridzin, and seed dormancy, 176
Phormidium tenue, halophile, 136
phosphoenolpyruvate, in activated *Neurospora* ascospores, 116
phosphoglucomutase, SH groups in, 422
6-phosphogluconate dehydrogenase, in seeds, 181, 185
Phosphon D, inhibition of seed germination by, 203, 205
phosphoprotein, in cyst wall of *Acanthamoeba*, 62, 63, 77
phosphorus, in outer cyst wall of *Acanthamoeba*, 58, 60, 61, 62
photoperiod, *see* day-length
photoperiodic clock
  accuracy of, 302
  and diapause, 290
  measures from lights on rather than lights off, 324, 325
  situated in mother for larvae of *Nasonia* and other insects, 302–5, 309
  temperature and, 306–7
  theories and models for, 309–24
photoperiodic seeds, 486–7
'photoperiodically inducible phase' (ϕ_i) for diapause inhibition, 311, 313, 315, 319, 320, 324, 327
photosynthesis
  in algae, at different temperatures, 131–2, 402–3; during desiccation, 135–6; snow, 128, 140; thermophilic, 131–2, 133
  in cold-hardened plants, 432
  per unit area of sea surface, 126
phycocyanins, of thermophiles, 133, 434
Physarum, RNA synthesis in, 68
phytochrome, 193, 194
  in germination of seeds, 200–6, 211, 213, 455
  possible second type of, 214–15
phytochrome system, 310
  'hour-glass' model for, 312
phytoecdysones, 285–6
Picea abies, ecotypes of, 246
Pieris spp., 'night interruption' experiments on, 310, 326, 327
Pinguicula grandiflora, bud dormancy in, 249
Pinus spp.
  chilling and growth of, 248
  day-length and growth of, 245

  ecotypes of, 246
  light and seeds of, 201, 206
Pipistrellus
  ovarian cycle in, 528
  testicular cycle in, 522, 524
Pisum sativum, gibberellin and synthesis of RNA by isolated nuclei of, 234
pituitary gonadotrophins, in hibernators, 531, 532, 535
plankton, rate of photosynthesis in, 126
Plantago, light and seed germination in, 194
plants
  comparison of resistance of, to heat and freezing, 441–3
  desert, dormancy and survival of, 449–69
  growth of, at extremes of temperature, 396–407
  survival of, at high temperatures, 433–41; at low temperatures, 407–33
  tropical, dormancy and survival of, 471–88
Platysamia cecropia, freezing injury to pupae of, 349–50
Plecotus
  ovarian cycle in, 528
  testicular cycle in, 521, 523
  see also Corynorhinus
Pleurococcus, desiccation of, 134
poikilotherms
  in Arctic, 564
  hibernation of, 489
polylysine, hydrophobic and hydrogen-bonded forms of, 432, 436
polymerization of proteins, prevented in cold-hardening, 431, 444
polymorphism, 4
polyribosomes
  abscisic acid and, 260
  in dry and imbibed seeds, 146
  in fertilized sea-urchin eggs, 146
  in germinating *Neurospora* ascospores, 106
  in seed homogenates incubated *in vitro*, 146–50, 158
Porphyra, temperature and respiration rate of, 131
potato
  abscisic acid and gibberellins in, 258
  breaking of dormancy in, 182, 183, 249, 250
  dormancy in tubers of, 219–40, 252
  stimulus from roots of, for hatching of *Heterodera* larvae, 85
Prasiola, desiccation of, 134
pregnancy
  body temperature in, 543
  varying length of, in bats, 514–15, 542–5
proctodone, hormone, in *Ostrinia nubilalis*, 311

protein synthesis
  abscisic acid and, 258
  in bacterial spores, 11, 14–20, 21–5, 39–45
  in encysting *Acanthamoeba*, 68, 70, 75–7
  inhibitors of, and seed germination, 188, 202
  by ribosomal preparations from dry and imbibed seeds, 143–6
  in sprouting potato buds, 231
  at temperatures too low for mitosis, 402
proteins
  of *Acanthamoeba* in different phases, 63, 65, 67, 70
  of *Acanthamoeba* cyst wall, 60, 61, 77
  of blood in Colorado beetle, 274–5, 277, 282
  per cell of *Chlorella* at different temperatures, 405
  of chloroplast, 423, 439
  in cold-hardening of plants, 417–33, 441–45
  in growing and non-growing plants, 420, 422
  in heat-hardening of plants, 435–45
  stability of, and resistance to adverse conditions, 138–9
  tension on, in freezing, 413, 414
  of thermophiles, 132–3, 403, 436–8, 440
  see also aggregation of proteins, dehydration of proteins, denaturation of proteins, disulphide bonds in proteins
proteolysis, of denatured proteins, 408–9
prothoracic glands of insects, brain hormone and secretion of ecdysone by, 285, 286, 287
Protosiphon cinnamomeus, desiccation of, 133
Prunus sp., day-length and growth of, 243
Prunus cerasus, pentose-phosphate pathway in seeds of, 182, 183
Prunus persicus
  bud dormancy in, 223, 251
  seed germination in, 224
Pseudomonas sp. psychrophile, 402
psychrophiles, proteins of, 403, 434
psychrophilic bacteria, 126
Pterostichus brevicornis
  in Arctic, 560
  glycerol in haemolymph of, 501
  temperature and nerve action potentials in, 562–3
puromycin
  and protein synthesis, 28, 68
  and seed germination, 202
  and sprouting of potato buds, 226
pyridine nucleotides, reduced, in activated *Neurospora* ascospores, 115
pyrophosphatase, heat-stability of, 439

Pyrrhocoris, ovaries and respiratory rate in, 272
Pyrus, day-length and growth of, 243

Quercus robor, day-length and bud growth in, 247–8

rabbit, brain metabolism of, in cooling and asphyxia, 372, 373, 374
radioautographs, of nucleic acids in sprouting potato buds, 227
radish, abscisic acid and protein and RNA synthesis in, 258
rainfall
  and ovarian cycle in *Eidolon*, 545
  and sprouting of *Fouquieria*, 451
Raphanus, germination of embryos of, 156–7
Raphidonema nivale, photosynthesis by, 140
rat
  brain metabolism of, during cooling and asphyxia, 362–83 *passim*
  limit of supercooling and survival in, 386–7
  recovery from cold in, 351, 356, 358–60
  survival times of, in cooling and asphyxia, 369, 370, 371, 372–3
recrystallization of ice, in frozen biological systems, 347
reducing sugars, in sprouting potato buds, 229–31
refractility, changes of, in germinating bacterial spores, 13, 14
reproduction
  delay of, as response to environmental uncertainty, 1, 9
  interrelations of hibernation and, 511–49
reproductive potential, and dormancy, 7–8
resistance adaptation, 393
resources
  principle of allocation of, 2–3
  principle of optimization of, 3–4
respiration
  and bud dormancy, 249, 250–1
  inhibitors of, as breakers of seed dormancy, 168–9, 173–5, 189
  pentose-phosphate pathway of, *see* pentose-phosphate pathway
  of seeds, 201
respiratory quotients
  of *Citellus* at different temperatures, 553, 554
  of germinating *Neurospora* ascospores, 104
respiratory rate
  of *Acanthamoeba*, 51, 55, 69
  of activated *Neurospora* ascospores, 103, 104, 111–14, 117–18, 119–20
  of algae, 130, 131

respiratory rate (*cont.*)
  of algal spores and vegetative cells, 140
  of Colorado beetles in diapause, 271–2, 277, 280
  of plants after cold-shock, 410
  postulated decrease in, as enforcing seed dormancy, 162
respiratory uncouplers, and seed dormancy, 175, 176
rewarming
  from hibernation, 489, 492
  in induced hypothermia in mammals, 351–2, 356–61
rhesus monkey, recovery from cold in, 351–2
Rhinolophus spp.
  ovarian cycle in, 526
  pregnancy and reaction to cold in, 543
  testicular cycle in, 521, 522, 538
Rhodnius, allatectomized, 275
Rhododendron, day-length and bud growth of, 247
Rhodophyceae, 134
rhubarb, gibberellin and growth of, 223
Ribes nigrum
  abscisic acid and gibberellins in, 248, 253, 254–5, 256, 258
  bud dormancy in, 254
  day-length and frost resistance in, 246; and growth of, 245
ribonuclease
  in bacterial spores, 21
  heat stability of, 439
ribosomes
  in bacterial spores, 20–1, 39, 46–7
  from dry and imbibed seeds, 144–6
  of thermophiles, 434, 437
  in wheat embryos during imbibition, 157, 158
  see also polyribosomes
rice (*Oryza sativa*)
  cytochrome oxidase in seeds of, 179
  effect of cyanide on dormant and non-dormant seeds of, 184
  seed dormancy in, 162, 166–9
RNA
  in cotyledons of germinating peanuts, 143
  in encysting *Acanthamoeba*, 63, 65
RNA synthesis
  abscisic acid and, 224, 258–60
  during cold-hardening, 431, 432
  cytokinin and, 450
  in dormant seeds, 189
  in encysting *Acanthamoeba*, 67–8, 70, 77
  in outgrowing bacteria spores, 14–18
  in *Pseudomonas* sp., maximum temperature for, 402
  in sprouting potato buds, 225, 232, 233

mRNA
  in bacterial spores, after germination, 18, 26, 39, 46; during outgrowth, 28–30, 40
  in encysting *Acanthamoeba*, 68
  in *E. coli*, 41
  in *Neurospora* ascospores, after germination, 106
  in psychrophiles, 434
rRNA
  abscisic acid and, 258–9
  in cotyledons of germinating peanuts, 143
  in dormant and germinating *Neurospora* ascospores, 106
  in thermophiles, 434
sRNA
  abscisic acid and, 258–9
  in dormant and germinating *Neurospora* ascospores, 106
  in outgrowing bacterial spores, 17–18
  of thermophiles, 434
tRNA
  in cotyledons of germinating peanuts, 143
  cytokinins and, 450
  and polysome formation in seed homogenates, 154, 155, 156, 160
  temperature-sensitive syntheases for, 434
RNA-ase
  denaturation of, 401–2
  inhibition of polyribosome formation in seeds by, 148
  naturation of, 420
RNA polymerases, in bacterial spores, 43, 44, 45
Robinia pseudacacia, day-length and frost resistance in, 246; and growth of, 243, 244, 245
rodents
  hibernation and reproduction of, 513–14
  ovarian cycle in, 524–7
  testicular cycle in, 517–21
  see also individual species
root tips, phytochrome and 'stickiness' of, 206
rye, cold-hardiness in, 423

salicylate, and seed dormancy, 176
Salix viminalis, abscisic acid in, 253, 254
Salsola, light and seed germination in, 198
Samia cynthia
  pigment in ganglia of central nervous system of, 296
  temperature and termination of diapause in, 295
Sclerotinia borealis, growth and respiration in, 402
Scytonema, desiccation of, 135

seals
  temperature and conduction by nerves of, 556, 557
  temperature of skin of, 558–9
seed-coats
  and dormancy, 161–2, 167, 169, 172, 178–9, 189, 250
  germination inhibitors in, 163
  in Leguminosae, 460–1
  permeability of, and dormancy, 406, 480
  and water sensitivity, 169, 171
seeds
  conditions of storage of, and germination, 482, 483, 484
  dormancy of, and oxidation processes, 161–92
  dormancy and survival of, in desert plants, 449–69; in tropical plants, 480–7
  germination of, and capacity for protein synthesis, 143–60
  light-controlled germination of, 193–217
  glutathione in, during germination, 424
  photoperiodic, 486
serum albumin, protection of, against heat injury, by picolinate, 442
sex organs
  of Colorado beetles after allatectomy, 277; and in diapause, 269–70
  in hibernation, 517–29
  removal of, and diapause, 277
shivering, rewarming of mammals with and without, 357, 358–60
Silene noctiflora, seed dormancy in, 175
Skeletonema costatum, growth and photosynthesis in, at different temperatures, 126
skin of seals, temperature of, 558–9
snow algae, 123, 128, 395; spores of, 140
sodium, in halophilic algae, 137
sodium/potassium ratio, in vegetative and encysted *Acanthamoeba*, 62
sodium hypochlorite, and dormancy and water sensitivity of seeds, 170
soil
  depth of, and germination of seeds, 456
  and sprouting of bulbils, 453–4
sorbitol dehydrogenase, in spermatozoa, 542
Sorghum vulgare, seed dormancy in, 162
soybean, effect of long dark periods on flowering of, 313
Spalangia drosophilae, maternal influence in, 305
spermatogenesis, after hibernation, 519, 521
spermatozoa, storage of, in female bats, 527, 529, 539–42

Spinacia oleracea
  abscisic acid and gibberellins in, 258
  water sensitivity in seeds of, 169
spinal reflex, in hibernators, 498–502, 505, 507–8
Spirulina subsalsa, halophile, 136
Spongiochloris typica, in adverse conditions, 130, 132, 135, 136, 138
spores of algae, resistance of
  to desiccation, 135, 140
  to high temperatures, 130, 140
spores of bacteria
  changes on germination of, 11–49
  pathways of oxidation in, 184–5
spores of fungi, dormancy and germination of, 99–121
steroids
  in brown fat of hibernating bats, 537–9
  cortical, and recovery from cold, 356
sterols, ecdysone and plant analogues as, 285
Stratiotes aloides, turions of, 249, 250
streptomycin
  control of bacterial infection of germinating seeds by, 170, 172
  and germination of bacterial spores, 14
stress, conditions of, and seed germination, 196, 197
subtilin, and germination of bacterial spores, 14
succinate-cytochrome-*c*-, oxidoreductase, in encysting *Acanthamoeba*, 69
succinate dehydrogenase
  heat-lability of, 440
  in insects in diapause, 274
  SH groups in, 422
succinate oxidase, of mitochondria in diapause, 272
sucrose, as protectant against freezing injury, 421, 423
sugars, as protectants against freezing injury, 419, 420, 421, 427
sulphide, and seed dormancy, 195
sulphhydryl (SH) groups of proteins
  in cold-hardening, 425–6
  conformation of proteins, and reactivity of, 430–1
  during growth of plants, 3, 424–5
  in heat injury, 442
  see also disulphide bonds
sulphhydryl reagents, effects of, on enzymes, 427, 433
supercooling, 331
  in biological systems, 337–40, 338, 344, 354, 561
survival
  of algae in adverse conditions, 123–42
  dormancy and, in desert plants, 449–69; in nematodes, 83–97; in tropical plants, 471–88

SUBJECT INDEX

survival (cont.)
  of mammals at low temperatures, 351–94
  of plants at extremes of temperature, 393–447
sympathetic system, in hibernators, 491–2
Synechococcus, thermophile, 129, 440
Tamias striatus, breeding season of, 533
temperature
  and bud dormancy, 224, 249, 250, 465
  and cell division, 520, 558, 559
  and diapause, 263, 264–6, 267
  and egg production and life-span in *Nasonia*, 305–6, 307, 309
  and encystment in *Acanthamoeba*, 53
  and endocrine factors in hibernators, 531–2, 532–3, 533–4, 538
  and gestation period in bats, 543, 544
  and growth and survival of algae, 124–33
  and growth and survival of plants, 395–447
  of hibernators, 489, 490–1, 492
  lethal minimum for Colorado beetle, 266, 271
  and muscle action potentials in hibernators, 501–2
  and nerve action potentials in *Pterostichus*, 562–3
  and nerve conduction, 555
  nocturnal, and growth of tropical forest trees, 472–7
  and nucleic acid synthesis in sprouting potato buds, 227–9
  and photoperiodic clock, 306–7, 309, 326–7
  in pregnancy, 543
  and rate of photosynthesis in *Chlorella*, 131–2, 404–5
  and respiratory quotient of *Citellus*, 553, 554
  and seed germination, 166–7, 169, 172, 177, 194, 198, 452–4, 456, 468, 481, 485
  sensitivity of hibernators to, 496
  and termination of insect diapause, 294–5
  and water sensitivity of seeds, 169, 171
temperature rebound, in freezing, 337, 340
Tenebrio molitor, allatectomy in, 282
Terminalis superba, day-length and growth of, 472, 473
terramycin, and germination of bacterial spores, 14
testis and male accessory organs, in hibernation, 517–24
tetramethylthiuram disulphide, and resistance of maize to chilling, 410
tetrodotoxin (puffer-fish toxin)
  and insect metamorphosis, 298
  and response of *Antheraea* pupa to day-length, 296–8

thawing injury, 415, 444
thermogenesis in mammals during re-warming, 356–60
thermolysin, heat-resistant protein, 438
thermophilic bacteria, 128, 129, 132, 396
thermophiles
  proteins of, 132–3, 403, 436–8, 440
  theories on heat resistance of, 432
thiogel (thiolated gelatin), formation of SS bonds in
  accelerated by freezing, 412, 413
  favoured by high molecular weight, 432
  prevented by solutes, 421, 423, 424
thioglycollate, and proteins of Cyanophyceae, 138
thiourea
  and bud dormancy, 224
  and seed dormancy, 179
  and seed germination, 196, 208
threonine, in cyst wall of *Acanthamoeba*, 62
thymidine, reversal of FUdR block in *Acanthamoeba* by, 72–3, 74–5
thymidylate synthetase, inhibited by FUdR, 38, 73, 74
thyroid gland, in hibernation, 529
thyroidectomy, recovery from cold after, 356, 360
thyroxine, and recovery from cold, 359, 360
Tilia europea, transmission spectrum of leaves of, 198–9
tissue cultures, unbalanced growth caused by FUdR in, 72, 78
TMV (tobacco mosaic virus), RNA of, as source of mRNA for polysome formation in seed homogenates, 152–6, 159
Tolypothrix distorta, desiccation of, 134
Tolypothrix tenuis, freezing of, 124
tomato
  light and seed germination in, 197
  protein breakdown in, at low temperatures, 409
torpor, daily and irregular, 511, 542
trees
  soluble protein in, during cold-hardening, 428
  in tropical forests, 471–80
trehalase, in *Neurospora* conidia, mycelium, and ascospores, 107–8, 109
trehalose
  metabolism of, in *Neurospora* ascospores, 104–5
  in *Neurospora* conidia and ascospores, 110
Tridax procumbens, seed germination in, 481, 486
Trifolium subterraneum, seed dormancy in, 176
triosephosphate dehydrogenase, freezing and SH groups of, 422

trypsin, trysinogen
  heat stability of, 439
  SS bonds in, 430
tubers
  breaking of dormancy in, 181–2, 183
  dormancy in, 219–40
Turbatrix aceti, drying of larvae of, 95
turions (winter resting buds of water plants), 249, 250
Tylenchus sp., survival of, in dry state, 84

Ulva spp.
  freezing of, 125
  temperature and respiration rate of, 131
UDPG-cellulose transglucosylase (cellulose synthetase), in encysting *Acanthamoeba*, 69, 70
UDPG pyrophosphorylase, in encysting *Acanthamoeba*, 69
uncertainty
  delayed reproduction as response to, 9
  dormancy as response to, 6
urea
  and proteins of Cyanophyceae, 138
  and seed dormancy, 167, 176
  and water sensitivity of seeds, 170, 171
urease
  from heat-hardened plants, 436, 438
  protectants of, against aggregation, 421
uridine, and effect of FUdR on *Acanthamoeba*, 74
uronic acid residues, in outermost coat of *Neurospora* ascospores, 99
uterus, of bats, storage of sperm in, 541–2
Utricularia, turions of, 249, 250

vanillic acid, and seed germination, 166
variability
  in critical temperature of warming for mammals, 356
  in dormancy of seeds, 460–4
Verbascum, seed germination in, 203, 205
vernalization, 464
vibrations, sensitivity of hibernators to, 508
Vigna sesquipedalis, RNA in germination of, 143
Vitis vinifera, hormones and rest period in, 224

water
  determination of content of, in nematodes, 83
  and dormancy of *Neurospora* ascospores, 107
  and dormancy of regenerating buds, 465–7
  in dormant and germinating bacterial spores, 45
  in encysted *Acanthamoeba*, 57, 62
  forming clathrate structure round hydrophobic groups of proteins, 414, 431
  freezing point of, in capillary spaces, 344
  rates of uptake and loss of, by nematode larvae, 86, 90, 92–4
  regulation of dormancy by supply of, 458–60, 468
  sensitivity to, in seeds, 169–71
  transfer of, through cell membranes, in freezing, 339–40, 345
  uptake of, by wheat embryos, 156–9
Weigela, day-length and growth of, 247
wheat
  catechol oxidase in seeds of, 179
  cold-hardiness in, 421
  germination of embryos of, 156–9
  germination inhibitors in seeds of, 163, 164
  mechanism of polysome formation in embryos of, 146–56
  ribosomes in dry and imbibed embryos of, 144–6
  water sensitivity in seeds of, 169
  wound stimulus to breaking of seed dormancy in, 182
Wittrockia superba, light and seed germination in, 195
wound stimulus
  and bud dormancy, 224, 227, 250
  and seed dormancy, 178, 182

xanthine oxidase, heat-lability of, 440
Xanthium
  abscisic acid and protein and RNA synthesis in, 258
  seed dormancy in, 162
Xanthium pensylvanicum, chilling and photoperiodic control of flowering in, 319
Xanthophyceae, 134

yeast, sexual agglutination reaction of, 427

Zygogonium, desiccation of, 134